ADVANCES IN INVERTEBRATE (NEURO)ENDOCRINOLOGY

A Collection of Reviews in the Post-Genomic Era

VOLUME 1: Phyla Other Than Arthropoda

ADVANCES IN INVERTEBRATE (NEURO)ENDOCRINOLOGY

A Collection of Reviews in the Post-Genomic Era

VOLUME 1: Phyla Other Than Arthropoda

Edited by
Saber Saleuddin, PhD
Angela B. Lange, PhD
Ian Orchard, DSc, PhD

APPLE ACADEMIC PRESS

Apple Academic Press Inc.
4164 Lakeshore Road
Burlington ON L7L 1A4
Canada

Apple Academic Press Inc.
1265 Goldenrod Circle NE
Palm Bay, Florida 32905
USA

© 2020 by Apple Academic Press, Inc.

First issued in paperback 2021

Exclusive worldwide distribution by CRC Press, a member of Taylor & Francis Group

No claim to original U.S. Government works

Advances in Invertebrate (Neuro)Endocrinology: A Collection of Reviews in the Post-Genomic Era
International Standard Book Number-13: 978-1-77188-809-7 (Hardcover)
International Standard Book Number-13: 978-0-42926-445-0 (eBook)

Volume 1: Phyla Other Than Arthropoda
ISBN 13: 978-1-77463-177-5 (pbk)
ISBN 13: 978-1-77188-892-9 (hbk)

Library and Archives Canada Cataloguing in Publication

Title: Advances in invertebrate (neuro)endocrinology, two volumes : a collection of reviews in the post-genomic era / edited by Saber Saleuddin, PhD, Angela Lange, PhD, Ian Orchard, DSc, PhD.

Names: Saleuddin, Saber, editor. | Lange, Angela, 1957- editor. | Orchard, Ian, 1951- editor.

Description: Includes bibliographical references and indexes. | Contents: Volume 1. Phyla other than arthropoda.

Identifiers: Canadiana (print) 2019018860X | Canadiana (ebook) 20190188677 | ISBN 9781771888097 (set ; hardcover) | ISBN 9781771888929 (v. 1 ; hardcover) | ISBN 9780429264450 (set ; ebook)

Subjects: LCSH: Invertebrates—Endocrinology. | LCSH: Neuroendocrinology.

Classification: LCC QP356.4 .A38 2020 | DDC 573.412—dc23

Library of Congress Cataloging-in-Publication Data

Names: Saleuddin, Saber, editor. | Lange, Angela, 1957- editor. | Orchard, Ian, 1951- editor.

Title: Advances in invertebrate (neuro)endocrinology : a collection of reviews in the post-genomic era / edited by Saber Saleuddin, Angela Lange, Ian Orchard.

Description: Oakville, ON ; Palm Bay, Florida : Apple Academic Press, [2020] | Includes bibliographical references and index. | Contents: v. 1. Phyla other than Arthropoda -- v. 2. Arthropoda. | Summary: "Advances in Invertebrate (Neuro)Endocrinology: A Collection of Reviews in the Post-Genomic Era (2-volume set) provides an informative series of reviews from expert scientists who are at the forefront of their research into the endocrinology of invertebrates. These two volumes are timely and appropriate in this post-genomic era because of the rapid pace of change brought about by genome projects, functional genomics, and genetics (omics technologies). The volumes show the rich history and strong tradition of cutting-edge research using invertebrates that has opened up our broader understanding of comparative endocrinology and the evolution of regulatory pathways and systems. These reviews set the scene and context for this exciting new era of understanding that has come from this post-genomic revolution. This book undertakes the daunting task of covering most of the diverse endocrine systems that exist among invertebrates. The papers in this book will advance our knowledge of invertebrate endocrinology but also of endocrinology in general, making the book valuable to researchers and students. Key features: Looks at the enormous diversity of species involved and the variety of hormonal pathways covers the diverse endocrine systems that exist among invertebrates makes relevant comparisons of molecular, cellular, and behavioral aspects of invertebrate endocrinology Explores the molecular genetics techniques and new allowing exploitation of these genomes through specific interference with genes, and thereby interference with their phenotypic expression"-- Provided by publisher.

Identifiers: LCCN 2019042135 | ISBN 9781771888097 (set ; hardcover) | ISBN 9781771888929 (v. 1 ; hardcover) | ISBN 9781771888936 (v. 2 ; hardcover) | ISBN 9780429264450 (ebook)

Subjects: MESH: Neurosecretory Systems--physiology | Invertebrates--physiology | Invertebrates--genetics | Hormones--physiology | Neuropeptides--physiology | Neuroendocrinology

Classification: LCC QP356.4 | NLM WL 102 | DDC 612.8--dc23

LC record available at https://lccn.loc.gov/2019042135

Apple Academic Press also publishes its books in a variety of electronic formats. Some content that appears in print may not be available in electronic format. For information about Apple Academic Press products, visit our website at **www.appleacademicpress.com** and the CRC Press website at **www.crcpress.com**

DEDICATION

"Knowledge advances progress, whereas ignorance impedes it."

This book is dedicated to Dr. Berta Scharrer for her pioneering work in (neuro)endocrinology and to our parents for giving us the encouragements and opportunities to become biologists.

About the Editors

Saber Saleuddin, PhD

Saber Saleuddin, PhD, is a University Professor Emeritus in the Department of Biology, York University in Toronto, Ontario, Canada. He received his early education in Bangladesh and his doctorate from the University of Reading in the UK. His studies on biomineralization in mollusks started at the University of Alberta in Edmonton, Canada, and continued at Duke University, Durham, North Carolina, in the laboratory of Karl Wilbur. Though offered a teaching position at Duke University, he accepted a faculty position at York University, where he still teaches. His outstanding contributions in teaching, research, and administration were recognized by York University. He has published extensively in international journals and has co-edited five books on molluscan physiology. He served as co-editor of the *Canadian Journal of Zoology* for 18 years and was president of the Canadian Society of Zoologists, from whom he received a Distinguished Service Medal.

Angela B. Lange, PhD

Angela B. Lange, PhD, is a world leader in the field of insect neuroendocrinology, with over 175 research publications and numerous invited research talks at international and national conferences. Professor Lange also demonstrates leadership in the research community, being a Council Member of the North American Society of Comparative Endocrinology, the European Society of Comparative Endocrinologists, and the International Federation of Comparative Endocrinology Societies. In addition to her international stature in research, Professor Lange was Chair of the Department of Biology, Acting Vice-Principal and Dean, and is currently Vice-Dean, Faculty, at the University of Toronto Mississauga, Ontario, Canada. She obtained her BSc and PhD degrees from York University, Canada.

Ian Orchard, DSc, PhD

Ian Orchard, DSc, PhD, is a Professor Emeritus, Biology, of the University of Toronto, Ontario, Canada. An expert in insect neuroendocrinology with over 200 research publications, Professor Orchard has been funded by the Natural Sciences and Engineering Council of Canada since 1980, has chaired the

NSERC Animal Physiology Grant Selection Committee, and has presented numerous invited keynote and plenary research talks at international conferences. Professor Orchard served as a Vice President of the University of Toronto and Principal of the University of Toronto Mississauga (2002–2010) and Vice President Academic and Provost, University of Waterloo (2014–2017), Canada. He earned a BSc (1972), PhD (1975), and DSc (1988), all from the University of Birmingham, UK.

Contents

Contributors

William G. Bendena
Department of Biology, Queen's University, Kingston, ON, K7L 3N6, Canada

Paul R. Benjamin
Sussex Neuroscience, School of Life Sciences, University of Sussex, Brighton BN1 9QG, UK

Ian D. Chin-Sang
Department of Biology, Queen's University, Kingston, ON, K7L 3N6, Canada

Anna Di Cosmo
Department of Biology, University of Naples Federico II, Italy

Maurice R. Elphick
Professor, Maurice Elphick, School of Biological & Chemical Sciences, Queen Mary University of London, Mile End Road, London E1 4NS, UK, Tel.: +44(0) 20 7882 6664, Fax: +44(0) 20 7882 7732, E-mail: m.r.elphick@qmul.ac.uk

Toshihiro Horiguchi
Ecosystem Impact Research Section, Center for Health and Environmental Risk Research, National Institute for Environmental Studies, Tsukuba, Ibaraki 305-8506, Japan, E-mail: thorigu@nies.go.jp

Ildikó Kemenes
Sussex Neuroscience, School of Life Sciences, University of Sussex, Brighton BN1 9QG, UK

Kazuya Kobayashi
Department of Biology, Faculty of Agriculture and Life Science, Hirosaki University, 3 Bunkyo-Cho, Hirosaki, Aomori 036-8561, Japan, E-mail: kobkyram@hirosaki-u.ac.jp

Valeria Maselli
Department of Biology, University of Naples Federico II, Italy

Toshie Matsumoto
National Research Institute of Aquaculture, 422-1 Nakatsuhamaura, Minami-ise, Mie 516-0193, Japan

Fumihiro Morishita
Department of Biological Science, Graduate School of Science, Hiroshima University, 1-3-1 Kagamiyama, Higashi-Hiroshima 739-8526, Hiroshima, Japan. E-mail: fumi425@hiroshima-u.ac.jp

Spencer T. Mukai
Department of Multidisciplinary Studies, Glendon College, 2275 Bayview Avenue, Toronto, ON, Canada M4N 3M6. E-mail: smukai@yorku.ca

Yasuhiko Ohta
Laboratory of Experimental Animals, Department of Veterinary Medicine, Faculty of Agriculture, Tottori University, Tottori, Tottori 680-8553, Japan, E-mail: yohta1022@gmail.com

Makoto Osada
Graduate School of Agricultural Science, Tohoku University, 1-1Tsutsumidori-Amamiyamachi, Sendai 981-8555, Japan, Tel.: +81227178725, Fax: +81227178727. E-mail: makoto.osada.a8@tohoku.ac.jp

Marina Paolucci
Department of Science and Technology, University of Sannio, Italy

Michel Salzet
INSERM U1192–Laboratoire Protéomique, Réponse Inflammatoire et Spectrométrie de Masse (PRISM), Université de Lille 1, F-59000 Lille , France

Honoo Satake
Suntory Foundation for Life Sciences, Bioorganic Research Institute, 8-1-1 Seikadai, Seika-Cho, Soraku-gun, Kyoto 619-0284, Japan, E-mail: satake@sunbor.or.jp

Kiyono Sekii
Department of Biology, Faculty of Agriculture and Life Science, Hirosaki University, 3 Bunkyo-Cho, Hirosaki, Aomori 036-8561, Japan

Toshio Takahashi
Suntory Foundation for Life Sciences, Bioorganic Research Institute, 8-1-1 Seikadai, Seika-Cho, Soraku-gun, Kyoto 619-0284, Japan

Abbreviations

ABFs	anti-bacterial factors
ACE	angiotensin-converting like enzyme
ACh	acetylcholine
AII	angiotensin II
AKH	adipokinetic hormone
AL	anterior lobe
AM	adrenomedullin
AMY	amylin
AR	androgen receptor
ATRA	all-*trans* retinoic acid
BWM	body wall muscles
CaFl	calfluxin
CBs	chromatoid bodies
CC	canopy cell
CCAP	crustacean cardioactive peptide
CCK	cholecystokinin
CDCH	caudodorsal cell hormone
CDCs	caudodorsal cells
CGRP	CT gene-related peptide
CHO	Chinese hamster ovary
CKR-2	cholecystokinin receptor 2
CNCs	caenacins
CNS	central nervous system
CO_2	carbon dioxide
CPEB	cytoplasmic polyadenylation element-binding protein
CPG	cerebral and pedal ganglia
CREB	cAMP response element-binding protein
CRF	corticotrophin-releasing factor
CRH	corticotropin-releasing hormones
CRSP	CT receptor-stimulating peptide
CT	calcitonin
DAG	diacylglycerol
DAO	D-amino acid oxidase
DBD	DNA-binding domain

DBL-1	transforming growth factor-β homolog
DBs	dorsal bodies
DEG	degenerin
Deg/ENaC	degenerin/epithelial sodium channel
DMSR	dromyosuppressin-like receptor
DTS	developmentally timed sleep
EEA1	early endosome antigen-1
EGF	epidermal growth factor
EH	eclosion hormone
ELFamide	thyrotropin-releasing hormone-like peptides
ELH	egg-laying hormone
ELISAs	enzyme-linked immunosorbent assays
ENaC	epithelial Na^+ channel
ERR	estrogen-related receptor
ERs	estrogen receptors
FARPS	FMRFamide-related peptides
FLP	FMRFamide-like peptide
GABA	γ-aminobutyric acid
GALP	galanin and galanin-like peptide
GC-MS	gas chromatography with mass spectrometry
GFP	green fluorescent protein
GI	gonad index
GPCRs	G-protein coupled receptors
GSS	gamete shedding substance
GSS	gonad stimulating substance
GTH	gonadotropin
GVBD	germinal vesicle breakdown
GVs	germinal vesicles
H3K5	histone H3 lysine 5
HA	heart accessory
HK-1/EKs	hemokinin-1/endokinins
HPG	hypothalamus–pituitary-gonadal axis
HSNs	hermaphrodite specific neurons
INS	insulin
INS-1	insulin-1
INS-Ls	insulin-like peptides
IP3	inositol triphosphate
LBD	ligand-binding domain
LGC-55	tyramine gated channel
LH	luteinizing hormone

LIPs	*lymnaea* inhibitory peptides
LL	lateral lobes
LNPY	*lymnaea* neuropeptide Y
LPS	lipopolysaccharides
LRR	leucine-rich repeat
LSCPR2	second Lys-conopressin receptor
LUQ	left upper quadrant
LYC	light yellow cell
LYCP	light yellow cell protein
MAPK	mitogen-activated protein kinase
MCH	melanin-concentrating hormone
MCT	mutable collagenous tissue
ME	methionine-enkephalin
MERF	methionine-enkephalin arginine phenylalanine
MIP	molluscan insulin-related peptide
MIPR	molluscan insulin-related receptor
MIPs	molluscan insulin-related peptides
MIS	meiosis-inducing substance
MLN	median lip nerves
MMA	metamorphosin A
MPF	maturation promoting factor
MS	mass spectrometry
NKA	neurokinin A
NKB	neurokinin B
NLP	neuropeptide-like peptide
NO	nitric oxide
NPF	neuropeptide F
NPR	neuropeptide receptor
NPS	neuropeptide-S
NPY	neuropeptide Y
NR	nuclear receptor
NT	neurotensin
NTLPs	neurotensin-like peptides
NTR	nematocin receptor
ODS	octadecylsilane
OKs	orcokinins
OL	olfactory lobe
OMP	olfactory marker protein
ORNs	olfactory receptor neurons
OT	oxytocin

OTR	OT receptor
PCs	prohormone convertases
PDF	pigment dispersing factor
PDFR	pigment dispersing factor receptor
PEA	proenkephalin A
PER	period protein
PGCs	primordial germ cells
PGE_2	prostaglandin E_2
PGF2α	prostaglandin F2α
PIP2	phosphatidylinositol-4,5 bisphosphate
PN	penis nerve
PP/OK	pedal peptide/orcokinin
PPAR	peroxisome proliferator-activated receptor
PPs	pedal peptides
PRR	pattern recognition receptors
PTH	parathyroid hormone
QMUL	Queen Mary University of London
RAMPs	receptor activity-modifying proteins
RAR	retinoic acid receptor
RAS	renin-angiotensin system
RGP	relaxin-like gonadotropic peptide
RIA	radio immune assay
RLM	radial longitudinal muscle
RN	ring neuron
RNAi	RNA interference
RNP	ribonucleoprotein
RXR	retinoid X receptor
SCP_A	SGYLAFPRMamide
SG	segmental ganglia
SIS	stress-induced sleep
SMP	starfish myorelaxant peptide
SP	substance P
SRs	steroid receptors
TBT	tributyltin
TcHT	tricyclohexyltin
TeBT	tetrabutyltin
TIR	toll/IL-1 receptor
TKRPs	tachykinin-related peptides
TLRs	toll-like receptors
TPhT	triphenyltin

TPrT	tripropyltin
TRH	thyrotropin-releasing hormone
TRHR	thyrotropin-releasing hormone receptor
VDS	vas deferens sequence
VP	vasopressin
VPF	vitellogenesis promoting factor
YCs	yellow cells

Preface

Neuroendocrine control of cellular activities first evolved in invertebrates, and the presence of endocrine cells was first reported in a mollusk. Both peptidergic and lipid-derived hormones have been found in invertebrates, and a few hormones are unique to invertebrates. Berta Scharrer, along with her husband, Ernst, were the first to coin the term "neuroendocrinology." Later, Berta, based on her work on *Aplysia* (a mollusk) and *Nereis* (an annelid) in the early 1930s, introduced the term "neurosecretion." As a result of her subsequent work on arthropods, the concept (neuroendocrinology/neurosecretion) became accepted by the scientific community.

This series of reviews bring together expert scientists who are at the forefront of their research into the endocrinology of invertebrates. These two volumes are timely and appropriate in this so-called post-genomic era. Timely, because of the pace of change brought about by genome projects, functional genomics, and genetics ('omics technologies), including transcriptomics, peptidomics, proteomics, metabolomics, gene microarrays, and sophisticated mass spectrometry techniques. Appropriate, because of the rich history and strong tradition of cutting edge research using invertebrates that have opened up our broader understanding of comparative endocrinology and the evolution of regulatory pathways and systems. These reviews set the scene and context for this exciting new era we find ourselves in, and the depth of understanding that has come from this post-genomic revolution.

Studies broadly defined as invertebrate endocrinology have been transformed over the last two decades since the original sequencing of the *Drosophila* genome. Sequenced genomes are now available for many invertebrates, and a bold initiative is underway to sequence genomes from 5,000 arthropod species (the i5k project; http://i5k.github.io/). This project, along with other invertebrate projects, is transformative and is consolidating the discipline in this 21st century. These projects have global implications, since invertebrates are so important in; for example, agriculture (pollination, crop destruction), aquaculture and other food sources, medicine (toxins, analgesics), and industry (silk). Researchers are now able to approach questions concerned with evolution and phylogeny, and make relevant comparisons of molecular, cellular, and behavioral aspects of invertebrate endocrinology. These comparisons can be made between invertebrates, but also between invertebrates and vertebrates, since another bold initiative, the Genome 10K project, aims to assemble the genomes of 10,000 vertebrate species (https://genome10k.soe.ucsc.edu/).

Developments in molecular genetics techniques are now allowing exploitation of these genomes through specific interference with genes, and thereby interference with their phenotypic expression. Functional genomics and genetics are, therefore, helping to unravel complex regulatory processes. The genes and proteins at all levels of the endocrine signaling pathway are being defined. Gene and peptide expression can be determined for specific tissues and individual cells, and experimental manipulation in their expression can aid in an understanding of endocrine regulation and physiology.

Editing a review series on invertebrate endocrinology is a difficult task because of the enormous diversity of species involved and the variety of hormonal pathways. Though we are aware it is a daunting task, this book covers the diverse endocrine systems that exist among invertebrates. In spite of our sincere desire, we were unable to find contributions in certain groups because of the unavailability of contributors. Thus, the list of papers in this book is based on expert colleagues willing and able to write within the allotted time frame. In order to compensate for the absence of certain topics/ fields, we have reproduced two already published papers.

We are indeed indebted to those colleagues who contributed to this book and gratefully acknowledged the role of reviewers who, with their thoughtful comments, made our work easier. We hope that the papers in this book will advance our knowledge of invertebrate endocrinology but also of endocrinology in general and that the book will be valuable to researchers and students.

Finally, we are grateful to Sandra Jones Sickels, Ashish Kumar and Rakesh Kumar of Apple Academic Press for their support in every step involving planning, editing, printing, and marketing.

CHAPTER 1

Cnidarian Peptide Signaling Molecules

TOSHIO TAKAHASHI

Suntory Foundation for Life Sciences, Bioorganic Research Institute, 8-1-1 Seikadai, Seika-Cho, Soraku-gun, Kyoto 619-0284, Japan

1.1 INTRODUCTION

The Cnidaria are a group of animals that branched early in the eukaryotic tree of life, and they can be divided into five classes: Scyphozoa (true jellyfish), Cubozoa (box jellyfish), Hydrozoa (the species *Hydra* and *Hydractinia*), Anthozoa (sea anemones, corals, and sea pens), and Staurozoa (stalked jellyfish). Cnidarians have a diffuse nervous system that includes a nerve net in which the sensory and ganglionic neurons, and their processes are interspersed among the epithelial cells of the ectodermal and endodermal layers. Although the cnidarian nervous system is strongly peptidergic, it is not solely peptidergic [1]. Classical molecules such as the biogenic amines that have long been studied as major neurotransmitters in higher eukaryotes are also involved in cnidarian neurotransmission [2]. The prevailing view of the nervous system in the freshwater polyp *Hydra* (Hydrozoa) is that the neuronal network is simple and diffused throughout the animal's body. As to the hydrozoan medusae, marginal nerve rings and ganglion-like structures associated with sensory organs are observed.

Cnidaria such as *Hydra* is composed of multiple cell types that represent the fundamental architecture of multicellular organisms. *Hydra* exhibits a simple body plan with a single body axis. A head composed of tentacles and hypostome is located at one end of the body axis, with a foot composed of a peduncle and a basal disk at the other end. The gastric region is located between the head and foot. The epithelium of *Hydra* consists of two layers, the ectodermal and endodermal epithelial cell layers, which are separated by a basement membrane known as the mesoglea. The cells of both epithelial layers also function as muscles cells. *Hydra* also has multipotent interstitial stem cells from which the nerve cells [3], nematocytes [3], gland cells [4], and germ cells [5] are derived. The capability of *Hydra* for regeneration and asexual reproduction by budding are well known. Indeed, a good deal of our understanding of axial pattern formation into head-and foot-specific tissues has been gained through the exploitation of these features.

Peptides play important roles in signaling in developmental and physiologic processes. For example, a wide variety of peptides function in intercellular communication, neurotransmission, and signaling pathways that spatially and temporally control axis formation and cell differentiation. Recent discoveries in the cnidarian *Hydra* [6]—one of the most basic metazoans and a key model system for studying the peptides involved in these processes—have shown that the *Hydra* peptides act directly on muscle cells to induce contraction and relaxation and also are involved in cell differentiation and morphogenesis. Moreover, epitheliopeptides produced by epithelial cells in *Hydra* exhibit morphogen-like activities and play roles in regulating neuron differentiation through neuron-epithelial cell interactions. The primary aim of the present overview is to describe the structure and functions of peptide signaling molecules such as neuropeptides and epitheliopeptides in cnidarians. The importance of these peptide-signaling molecules in the development and physiology of cnidarians is also discussed.

1.2   CNIDARIAN NEUROPEPTIDES

1.2.1   FLPS (FMR-FAMIDE-LIKE PEPTIDES)

The peptide FMRFamide, which is composed of four amino acid residues with C-terminal amidation, was first isolated from the cerebral ganglion of the clam *Macrocallista nimbosa* [7, 8]. To date, peptides sharing a similar sequence have been isolated from other mollusks and from members of most other phyla. These peptides are now divided into two groups according to the level of structural similarity compared with FMRFamide. One group is known as FMRFamide-related peptides (FaRPs), which contain N-terminal extensions of the C-terminal FMRFamide or FLRFamide core sequence [9]. The other group is the FLPs, which include all peptides with only the RFamide sequence [10]. Thus, FLPs include FaRPs and all other RFamide peptides. An excellent review has examined FaRPs from invertebrates [11]. This overview primarily focuses on cnidarian FLPs.

The evolutionarily ancient nervous system of cnidarians expresses a variety of FLPs (Table 1.1). Peptides with GRFamide at the C-terminus have been found in a scyphozoan [12], three hydrozoans [13–18], and an anthozoan [19], whereas peptides with TRFamide and/or RRFamide at the C-terminus have been described in another anthozoan [20].

TABLE 1.1 FLPs in Cnidaria

Name	Peptide Sequence	Species	References
Antho-RFamide	pQGRFamide	*Anthopleura elegantissima*	[19]
Cyanea-RFamide I	pQWLRGRFamide		
Cyanea-RFamide II	pQPLWSGRFamide	*Cyanea lamarckii*	[12]
Cyanea-RFamide III	GRFamide		
Pol-RFamide I	pQLLGGRFamide	*Polyorchis penicillatus*	[14]
Pol-RFamide II	pQWLKGRFamide		[15]
Hydra-RFamide I	pQWLGGRFamide	*Hydra magnipapillata*	[13]
Hydra-RFamide II	pQWFNGRFamide		
Hydra-RFamide III	KPHLRGRFamide		
Hydra-RFamide IV	HLRGRFamide		
Hydra-RFamide V	pQLMSGRFamide	*Hydra magnipapillata*	[16]
Hydra-RFamide VI	pQLMRGRFamide		
Hydra-RFamide VII	pQLLRGRFamide		
Hydra-RFamide VIII	KPHYRGRFamide		
Hydra-RFamide IX	HYRGRFamide		
Hydra-RFamide X	KPHLIGRFamide	*Hydra magnipapillata*	[17]
Hydra-RFamide XI	pQLMTGRFamide		
He-RFamide	pQWLKGRFamide	*Hydractinia echinata*	[18]
Nv-RFamide I	pQITRFamide	*Nematostella vectensis*	[20]
Nv-RFamide II	VVPRRFamide		

Abbreviations: pQ – pyroglutamate.

All neuropeptides are produced and secreted by highly regulated secretion pathways. In general, the precursors of neuropeptides are incorporated as preprohormones in the endoplasmic reticulum, where they are converted into prohormones. The prohormones are then transported to the Golgi apparatus, where they undergo post-translational modification such as endoproteolysis and C-terminal amidation before assuming their final active peptide forms. Cnidarian FLP cDNAs have been identified in several different animals. A cDNA from *Calliactis parasitica* contains 19 copies of Antho-RFamide (Table 1.1), two copies of FQGRFamide, and one copy of YVPGRYamide [21]. In *Anthopleura elegantissima*, two cDNAs have been isolated, one of which has 13 copies of Antho-RFamide (Table 1.1) and nine other FLPs, whereas the other has 14 copies of Antho-RFamide and eight other FLPs [22].

In *Renilla koellikeri*, 36 copies of Antho-RFamide are present [23]. A cDNA from *Polyorchis penicillatus* has one copy of Pol-RFamide I (Table 1.1) and 11 copies of Pol-RFamide II (Table 1.1), along with another predicted FLP [24]. The Hydra-RFamides are derived from three different preprohormones. Preprohormone-A contains all four Hydra-RFamides (Hydra-RFamide I-IV) (Table 1.1) [16]. Preprohormone-B contains one copy of Hydra-RFamide I, one copy of Hydra-RFamide II, and two putative Hydra-RFamides [16]. Preprohormone-C contains one copy of Hydra-RFamide I and seven copies of putative neuropeptide sequences [16]. In *Hydractinia echinata*, one copy of He-RFamide is present (Table 1.1) [18]. In *Nematostella vectensis*, two FLPs (Nv-RFamide I and II) are present (Table 1.1) [20]. Collectively, precursor-encoding cnidarian FLP cDNAs yield many neuropeptides of great structural diversity, indicating that they have great functional diversity as well.

Cnidarian FLPs mediate a variety of functions, including control of muscle contraction, feeding, sensory activity, reproduction, metamorphosis, and larval movement. In the sea anemone *Calliactis parasitica*, application of Antho-RFamide causes increases in the tone, contraction amplitude, and frequency of slow muscle contraction [25]. The same peptide induces tonic contractions in the rachis and peduncle of the colony in individual autozooid polyps of *Renilla koellikeri*, with a threshold of 5 nM in summer colonies and 1 µM in winter colonies [26]. In *Hydra*, Hydra-RFamide III exhibits a dose-dependent effect on the pumping activity of the peduncle [27]. Because the gastrovascular cavity not only digests food but also delivers nutrients throughout the body, it has been suggested that the contractility of the peduncle is akin to cardiac activity in higher organisms.

A peptide-gated ion channel in snails is regulated by FMRFamide [28, 29]. Three ion channel subunits of the degenerin (DEG)/epithelial Na^+ channel (ENaC) gene family were cloned from *Hydra* and designated *Hydra* Na^+ channel (HyNaC) 2–4 [30]. Subsequently, a new subunit, designatedHyNaC5, was cloned, and expression of the gene was shown to be co-localized with HyNaC2 and HyNaC3 at the base of the tentacles [31]. Co-expression of HyNaC5 with NyNaC2 and HyNaC3 in *Xenopus* oocytes greatly increases current amplitude after peptide stimulation and increases the affinity of the channel for Hydra-RFamide I and II [31]. A combination of HyNaC2/3/5 forms a peptide-gated ion channel of the DEG/ENaC gene family that contributes to fast neurotransmission in cnidarians. Analysis of a HyNaC and ENaC chimera revealed intriguing insights regarding the evolutionary aspects of the ion channel. From analyses of HyNaCs, researchers have speculated that release of Hydra-RFamide I and/or II leads to tentacle contraction, possibly when the animals are feeding [30, 31]. Assmann and co-workers reported the molecular cloning of seven more HyNaC subunits, HyNaC6-HyNaC12, all of which are members of the DEG/ENaC gene family [32]. In *Xenopus* oocytes, these subunits assemble together with the four previously described subunits into 13 different ion channels that are directly gated by Hydra-RFamide I and II with high affinity. An inhibitor of HyNaCs, diminazene, delays tentacle movement in live *Hydra*. It has been shown that *Hydra* has a large variety of peptide-gated ion channels that are activated by a restricted number of FLPs [32]. Thus, *Hydra* may select FLPs for fast neuromuscular transmission. However, the possible function of Hydra-RFamide IV in *Hydra* remains unclear.

In addition to nerve cells, cnidarians have differentiated and highly specialized mechanoreceptor cells that play a pivotal role in prey capture and defense [33]. These are phylum-specific stinging cells known as nema-tocytes. Ultrastructural studies demonstrated the presence of two- and three-cell synaptic pathways in the tentacle epidermis of the sea anemone, including synaptic connections between nematocytes surrounding nerve cells [34, 35]. FLPs likely play a role in the function of cnidarian sensory structures. The presence of immunoreactivity for FMRFamide and RFamide in the tentacles of our classes suggests that FLPs are involved in the chemosensory regula-tion of cnidocyte discharge [36]. FMRFamide immunoreactivity has also been observed in the epidermal sensory cells of the spot ocellus in *Aurelia* [37]. This neuronal control likely decreases the spontaneous firing activity of nematocytes.

FLPs are also involved in cnidarian reproduction, larval movement, and metamorphosis. Colonial octocorals such as *Renilla koellikeri* repro-duce via a two-step spawning and exfoliation process. During spawning, intact gamete follicles are released into the environment, and during exfo-liation, the follicles rupture, freeing the gametes. Antho-RFamide, which is expressed in ciliated neurons within the follicle epithelium of *Renilla koel-likeri*, induces exfoliation of the follicle epithelium, releasing the gametes into the surrounding medium [38]. Furthermore, the potency of the peptide is enhanced by light [38].

Hydractinia echinata is a colonial marine hydroid closely related to the freshwater hydra. Fertilized eggs of this species undergo rapid cleavage divi-sions for about 1 day and develop into spindle-shaped planula larvae in about 3 days [39]. The planula larvae are capable of migrating toward light [40], and they metamorphose into adult polyps when they receive appropriate environmental stimuli [41, 42]. Hydra-RFamide I inhibits the migration of planula larvae, thus modulating phototaxis by inhibiting myomodula-tion [40]. Metamorphosis is also inhibited by the peptide, leading to the suggestion that the endogenous FLPs function in stabilizing the larval stage

[43]. Thus, FLPs may play a role in regulating the movement of the planula larvae prior to metamorphosis, possibly linking movement to chemotactic or phototactic processes [44]. The presence of FLP-expressing sensory neurons in the planula larvae indicate that planula migration and metamorphosis may be regulated by the release of endogenous neuropeptides in response to environmental cues.

1.2.2 GLWAMIDES

GLWamides have characteristic structural features in their N- and C-terminal regions. For example, most of the peptides share a GLWamide motif at their C-terminus (Table 1.2). In *Hydra*, seven GLWamide peptides were isolated and found to have a proline residue at the second position (X-Pro) or at the second and third positions (X-Pro-Pro) of their N-terminal region (Table 1.2) [6, 45]. Metamorphosin A (MMA), which was isolated from the anthozoan *Anthopleura elegantissima*, has a pyroglutamyl residue at the N-terminus (Table 1.2) [46]. Both of these N-terminal structures confer resistance to aminopeptidase digestion [47].

Cnidarian GLWamide cDNAs have also been identified in several different animals. Leviev et al., [48] cloned a cDNA encoding the preprohormone from *Hydra magnipapillata* containing 11 (eight different) immature peptide sequences. The cDNA encodes one copy each of Hym-53 (NPYPGLWamide), Hym-54 (GPMTGLWamide), and Hym-249 (KPIPGL-Wamide), two copies of Hym-248 (EPLPIGLWamide), and three copies of Hym-331 (GPPPGLWamide) along with three other predicted GLWamides (Table 1.2). One of the predicted peptides, designated Hydra-LWamide VIII, likely includes GMWamide at the C-terminus [48]. In *H. echinata*, one cDNA encoding GLWamides was cloned [49]. The cDNA encodes one copy of He-LWamide I and 17 copies of He-LWamide II (Table 1.2). Compared with the preprohormones of GLWamides, two distinct cDNAs were cloned from the anthozoans *Actinia equine* and *Anemonia sulcata* [49]. The cDNA from the *Actinia* species encodes one copy each of MMA, Ae-LWamide IV, Ae-LWamide V, Ae-LWamide VI, and Ae-MWamide, two copies each of Ae-LWamide I and Ae-LWamide III, and four copies of Ae-LWamide II (Table 1.2). By contrast, the cDNA from the *Anemonia* species encodes one copy each of MMA, Ae-LWamide II, and As-IWamide, two copies of As-LWamide II, and four copies of As-LWamide I (Table 1.2) [49]. The original MMA is contained in anthozoan preprohormones but not hydrozoan preprohormones. Thus, MMA is a species-specific peptide.

In addition, the peptide may be a prototype of the family with the protection of the N-terminus by pyroglutamate [50]. Two other peptides that are possibly encoded in the preprohormones of *Actinia* and *Anemonia* are likely processed into -GMWamide (Ae-MWamide) and -GIWamide (As-IWamide) at their C-terminus (Table 1.2). Whether these two peptides and Hydra-LWamide VIII belong to the GLWamide family or not is uncertain, as it has been reported that substitution of the Leu residue in GLWamide with Met or Ile results in complete or almost-complete loss of contractile activity in the retractor muscle of the anthozoan *Anthopleura fuscoviridis* [51]. It is probable that other novel neuropeptide families exist.

Species of the genus *Hydractinia* are colonial and usually live on snail shells inhabited by hermit crabs. In the *Hydractinia* life cycle, only a planula larval stage exists, with no medusa stage. Upon settling on snail shells, the planula larvae undergo metamorphosis and develop into polyps after approximately 1 week [52]. MMA induces this metamorphosis [46]. This finding demonstrates that cnidarian neuropeptides function as neurohormones and control developmental processes in addition to playing roles as neurotransmitters and neuromodulators. All *Hydra* GLWamide peptides also induce the metamorphosis of *Hydractinia serrata* planula larvae into polyps [6, 45]. Studies involving an N-terminal deletion series revealed that a common GLWamide sequence is necessary for the induction of metamorphosis in *Hydractiona*, indicating that induction of metamorphosis is very specific for the GLWamide terminus and that amidation is essential [53]. Furthermore, displacement of the Gly residue of GLWamide with another common amino acid (with the exception of Cys) resulted in either a decrease or complete loss of potency, and displacement of the Leu or Trp residues of GLWamide with another common amino acid except Cys resulted in complete or almost-complete loss of potency in muscle contraction of *Anthopleura fuscoviridis* [51]. However, the precise mechanism of the actions of GLWamide peptides in the induction of metamorphosis is not yet clearly understood. Interestingly, larvae can be induced to undergo metamorphosis in response to a chemical signal secreted by environmental bacteria [46]. This chemical signal is most likely received by the sensory neurons of the planula larvae, which then release endogenous GLWamide peptides that act on the surrounding epithelial cells, resulting in a change in phenotype. Because hydra develops directly from embryos into adult polyps and has no intermediate larval stage, the precise function of the GLWamide peptides in early embryogenesis in *Hydra* remains an open question.

TABLE 1.2   GLWAmide-Family Peptides in Cnidaria

Name	Peptide Sequence	Species	References
MMA	pQQPGLWamide	*Anthopleura elegantissima*	[48]
Hym-53	NPYPGLWamide	*Hydra magnipapillata*	[6]
Hym-54	GPMTGLWamide		[45]
Hym-248	EPLPIGLWamide		
Hym-249	KPIPGLWamide		
Hym-331	GPPPGLWamide		
Hym-338	GPPhPGLWamide		
Hym-370	KPNAYKGKLPIGLWamide		
He-LWamide I	pQRPPGLWamide	*Hydractinia echinata*	[49]
He-LWamide II	KPPGLWamide		
Ae-LWamide I	pQQHGLWamide	*Actinia equine*	[49]
Ae-LWamide II	pQNPGLWamide		
Ae-LWamide III	pQPGLWamide		
Ae-LWamide IV	pQKAGLWamide		
Ae-LWamide V	pQLGLWamide		
Ae-LWamide VI	RSRIGLWamide		
Ae-MWamide	pQDLDIGMWamide		
MMA	pQQPGLWamide		
As-LWamide I	pQQAGLWamide	*Anemonia sulcata*	[49]
As-LWamide II	pQHPGLWamide		
As-IWamide	pQERIGIWamide		
Ae-LWamide II	pQNPGLWamide		
MMA	pQQPGLWamide		

Abbreviations: pQ – pyroglutamate; hP – hydroxyproline.

Sexual reproduction in reef-building corals also involves motile planula larvae, which undergo complex metamorphosis after location of an appropriate substrate, resulting in the founding of a juvenile coral colony. In the coral genus *Acropora*, Iwao, and co-workers found that Hym-248 induces metamorphosis of *Acropora* planula larvae into polyps at high rates (approximately 100%) and that *Acropora* planula larvae respond to the peptide in a dose-dependent manner [54]. Interestingly, however, Hym-248 cannot induce metamorphosis in other coral genera [54, 55]. Therefore, the specificity of ligand recognition by receptors appears to depend on the extent to which peptide(s) with particular structures are recognized by corals. In *Hydractinia*, the specificity is less stringent, and receptors can recognize any peptide belonging to the GLWamide family. Because Hym-248 is a surrogate ligand in *Acropora*, natural ligand(s) that are similar in structure to Hym-248 should be identified.

In *Hydra*, we found that all GLWamide peptides induce bud detachment from the parental polyp due to contraction of the sphincter muscle in the basal disk [6]. Tests of myoactivity typically employ epithelial *Hydra*, which are hydra with no nerve cells or any other cells derived from interstitial stem cells, except for gland cells [56.57]. A similar effect was also observed in normal *Hydra* treated with the peptides. Unexpectedly, one of the *Hydra* GLWamide peptides, Hym-248, not only induces bud detachment but also elongation of the body column [45]. *Hydra* muscle processes extending from the ectodermal and endodermal epithelial cells run perpendicular to each other. Hym-248 may have two types of receptors, one that is common to all GLWamide-family peptides and another that is specific to Hym-248.

In *Anthopleura*, we also found that all GLWamide-family peptides induce contraction of the retractor muscle [45]. Immunohistochemical staining with an antibody specific for the GLWamide motif revealed intensely stained nerve cells in the retractor muscle of the sea anemone and the nervous system of *Hydra* [45].

In *Hydractinia echinata*, migration of planula larvae is regulated by GLWamide and RFamide neuropeptides. One of the GLWamide-family peptides, He-LWamide II, stimulates migration, primarily by lengthening the active period [40]. As mentioned above, Hydra-RFamide I inhibits the migration of planula larvae. Thus, GLWamides and FLPs work antagonistically to regulate migration in planula larvae of *Hydractinia echinata*.

In nature, oocyte maturation and subsequent spawning in hydrozoan jellyfish are generally triggered by light-dark cycles. In sexually mature female medusas of the hydrozoan jellyfish *Cytaeis uchidae*, a light period of 1 sec in duration is sufficient to trigger oocyte maturation and spawning in intact medusas or medusas without umbrellas, but the oocytes cannot resume meiosis unless they remain inside the ovary for at least 4 min after light stimulation [58]. We revealed that the Hym-53–dependent period required for oocyte maturation and spawning is <2 min and that the onset time of spawning after neuropeptide treatment is comparable to that after light stimulation [59]. These observations suggest that neuropeptide(s) work as hormones to mediate the initial step that determines whether oocytes undergo irreversible induction of meiotic maturation after light reception.

1.2.3   HYM-176

A novel myoactive neuropeptide, Hym-176 (APFIFPGPKVamide), was identified in the *Hydra* peptide project (Table 1.3) [6, 60]. The Hym-176–encoding gene is expressed abundantly in peduncle neurons and at lesser levels in neurons scattered throughout the gastric region [61]. This peptide specifically and reversibly induces contraction of the ectodermal muscle of the body column, particularly in the peduncle region of epithelial *Hydra* [60]. This expression pattern correlates with the myoactive function of the peptide, suggesting a localized function.

TABLE 1.3   Hym-176 and Hym-357 in Cnidaria

Name	Peptide Sequence	Species	References
Hym-176	APFIFPGPKVamide	*Hydra magnipapillata*	[60]
Hym-357	KPAFLFKGYKPamide	*Hydra magnipapillata*	[17]

The Hym-176–encoding gene also encodes another peptide, Hym-357 (KPAFLFKGYKPamide) (Table 1.3). This peptide strongly induces both tentacle and body contraction in normal polyps but has no effect on epithelial *Hydra* [17]. These observations indicate that Hym-357 does not act directly on muscle but presumably activates other neurons, which in turn, release neurotransmitters that directly induce muscle contraction.

1.2.4   HYM-355

The primary structure of Hym-355 is FPQSFLPRGamide (Table 1.4) [62]. Muneoka et al., [63] proposed classifying peptides with a PRXamide sequence at their C-terminus as PRXamide peptides, which can be further divided into three sub-groups: (a) pheromone biosynthesis–activating neuropeptides [64] and related peptides, (b) small cardioactive peptides [65, 66], and (c) antho-RPamide [47] and related peptides. Thus, PRXamide peptides are widely distributed in invertebrates. Hym-355 shares some homology with members of sub-group (c): LPPGPLPRPamide (Table 1.4), AAPLPRLamide from the echiuran *Urechis unicinctus* [67], and QPPLPRYamide and pQPPLPRYamide from the snail *Helix pomatia* [68].

TABLE 1.4   PRXAmide Peptides in Cnidaria

Name	Peptide Sequence	Species	References
Hym-355	FPQSFLPRGamide	*Hydra magnipapillata*	[63]
Antho-RPamide	LPPGPLPRPamide	*Anthopleura elegantissma*	[47]

A persistent question regarding the comparative physiology of nervous systems is whether cnidarians express peptides of the oxytocin-vasopressin superfamily, which function as neurohypophysial hormones in vertebrates. In *Hydra*, vasopressin-, and oxytocin-like immunoreactivity in the nervous system has been regarded as evidence for the presence of oxytocin-vasopressin superfamily peptides [69, 70]. Morishita et al., [71] purified two peptides, FPQSFLPRGamide (Hym-355) and SFLPRGamide, from *Hydra magni-papillata* using high-performance liquid chromatography fractionation and immunologic assays. They concluded that Hym-355 (FPQSFLPRGamide) and SFLPRGamide accounted for the vasopressin-like immunoreactivity in the hydra nervous system. As Hym-355 (FPQSFLPRGamide) and vasopressin share the same sequence at the C-terminus (PRGamide) and neither respective antibody discriminates with the other, Koizumi et al.[72] carried out immunohistochemical staining using an anti-Hym-355 antibody and revealed that the antibody labeled the nerve rings in *Cladonema radiatum* and *Turritopsis nutricula* (order Anthomedusae). However, whether Hym-355 (FPQSFL-PRGamide) functions as a neurohypophysial hormone remains unclear.

Hydra tissue is in a dynamic state and constantly undergoes renewal due to continuous growth and differentiation of epithelial cells and interstitial stem cells. Nevertheless, neuronal density is maintained at a constant level. This neuronal homeostasis appears to be positively regulated by the neuropeptide Hym-355 (FPQSFLPRGamide) and negatively regulated by peptides of the Pro-Trp (PW) family [6, 62, 73]. As PW peptides have a common Pro-Trp sequence at their C-terminus, we designated them the PW peptide family. The members are epitheliopeptides produced by epithelial cells [74].

Hym-355 (FPQSFLPRGamide) enhances neuronal differentiation at an early stage, whereas PW peptides, such as Hym-33H (AALPW), inhibit neuronal differentiation [62, 73]. Furthermore, Hym-355 (FPQSFLPRGamide) and Hym-33H (AALPW) show antagonistic properties, as treatment with one peptide nullifies the effect of treatment with the other peptide. Considered collectively, these results suggest that a feedback model can explain the mechanism that regulates the homeostasis of neuronal differentiation in *Hydra* [62]. According to this model, Hym-355 (FPQSFLPRGamide) produced by neurons increases the rate of neuronal differentiation at an early stage in the pathway. To keep this effect in check, epithelial cells produce PW peptides that block neuronal differentiation. We propose the presence of a third factor that controls the production and/or release of PW peptides from epithelial cells and/or neurons. This tripartite mechanism presumably maintains a constant neuronal density in *Hydra*. Hym-355 (FPQSFLPRGamide) is a neuropeptide that specifically enhances neuronal differentiation from interstitial stem cells in *Hydra* and also weakly promotes muscle contraction of the retractor muscle in the sea anemone *Anthopleura fuscoviridis* [62].

As mentioned above, Hym-53 (NPYPGLWamide) (Table 1.2) induces oocyte maturation and spawning. Hym-355 (FPQSFLPRGamide) also triggers these events, but the stimulatory effect is weaker than that of Hym-53 (NPYPGLWamide). An antibody against Hym-355 recognizes neurons located in the ovarian ectodermal epithelium [59]. Considering the effects of Hym-53 (NPYPGLWamide) and Hym-355 (FPQSFLPRGamide) on oocyte maturation and spawning in *Cytaeis*, we speculate that neurons that express neuropeptides function downstream of light receptors in the cnidarians. Future studies will be needed to isolate the endogenous neuropeptides that are responsible for the pathway and clarify the types of cells that both release and respond to the molecules.

1.2.5 FRAMIDE FAMILY

During research aimed at systematic identification of peptide signaling molecules in *Hydra* [6], two novel neuropeptides, FRamide-1 (IPTGTLIFRamide) and FRamide-2 (APGSLLFRamide), were identified (Table 1.5) [75]. Now that Hydra EST and genome databases [76] are available, more rapid identification of signaling peptides and their genes has become possible via searching for amino acid sequences of purified peptides. The two abovementioned peptides and the single gene encoding both peptides were identified using this exact approach.

TABLE 1.5 FRAmide-Family Peptides in Cnidaria

Name	Peptide Sequence	Species	References
FRamide-1	IPTGTLIFRamide	*Hydra magnipapillata*	[76]
FRamide-2	APGSLLFRamide		

FRamide-1 (IPTGTLIFRamide) and FRamide-2 (APGSLLFRamide) exhibit opposing effects even though they are encoded by the same gene.

The former peptide evokes contraction of endodermal muscles, thereby elongating the body column, whereas the latter peptide evokes contraction of ectodermal muscles, thereby contracting the body column [75]. There are at least two possible explanations for these seemingly contradictory observations. One possibility is that the release of each peptide is differentially regulated [77, 78], and the other possibility is that each peptide is processed in a different neuron [79].

1.3 CNIDARIAN EPITHELIOPEPTIDES

1.3.1 PW PEPTIDE FAMILY

Neurons in *Hydra* are differentiating from multipotent interstitial stem cells [3]. Although *Hydra* tissue is in a dynamic state, constantly undergoing renewal as a result of continuous growth and differentiation, neuron density is maintained at a constant level. As mentioned above, this homeostasis appears to be positively regulated by the neuropeptide Hym-355 (FPGSFLPRGamide) (Table 1.4) and negatively regulated by members of the PW peptide family (Table 1.6), which are epitheliopeptides localized in ectodermal epithelial cells [6, 73].

TABLE 1.6 PW-family Peptides in Cnidaria

Name	Peptide Sequence	Species	References
Hym-33H	AALPW	*Hydra magnipapillata*	[6]
Hym-35	EPSAAIPW		[74]
Hym-37	SPGLPW		
Hym-310	DPSALPW		
Hym-1397	TPALHW		

Through combined efforts in the *Hydra* peptide and EST projects, we identified the gene encoding PW peptides [73]. The PW gene encodes three copies of Hym-37 (SPGLPW) as well as other PW peptides identified previously (Table 1.6). There are also three putative peptide sequences within the precursor protein: EPSAALPW, TRTLPW, and IPALPW. The precursor also contains a new peptide with the sequence TPALHW (Hym-1397). Neuropeptides are generated from preprohormones via proteolytic cleavage, most frequently occurring at basic residues. *Hydra* PW epitheliopeptides are also flanked by dibasic residues at the immediate C-terminus. An N-terminal Lys residue precedes all peptides except for the AALPW peptide (Hym-33H), which is preceded by an Ile residue, and the IPALPW, EPSAAIPW, and one of the three copies of the Hym-37 peptide, which are preceded by an Asn residue. This Asn residue has been suggested to function as a processing signal in *Hydra* [48]. However, no enzyme that specifically cleaves at the Asn residue has yet been identified. The AALPW peptide could possibly be natural or a derivative of the IAALPW peptide.

1.3.2 HYM-301

The *Hydra* head is composed of the hypostome and tentacles. The novel epitheliopeptide Hym-301 plays a specific role in the head formation, namely, in determining the number of tentacles formed. Hym-301 is composed of 14 amino acid residues, with one intramolecular disulfide bond and a C-terminal amidation (Table 1.7) [80]. The preprohormone gene encoding Hym-301 contains one copy of Hym-301 [80]. It is also a typical preprohormone as defined for neuropeptide preprohormones.

TABLE 1.7   Hym-301 in Cnidaria

Name	Peptide Sequence	Species	References
Hym-301	KPPRRCYLNGYCSPamide	*Hydra magnipapillata*	[81]

The gene encoding Hym-301 is expressed in epithelial cells, especially in ectodermal epithelial cells in the head region [80]. The expression pattern of the *Hym-301* gene suggests that it plays a role in the head formation. By evaluating the effect of the peptide on tentacle formation during head regeneration and the effect of knockdown of *Hym-301* gene expression using RNAi, we revealed that the Hym-301 peptide and its gene are clearly involved in tentacle formation, especially in determining the number of tentacles formed [80].

It is generally accepted that neuropeptides are localized in the granular cores of dense-cored vesicles. Khalturin et al., [81] showed that a fusion protein composed of the Hym-301 peptide precursor and green fluorescent protein (GFP) is localized in secretary vesicle-like structures beneath the cell membrane. They also revealed that *Hym-301* overexpression affects both the speed of tentacle regeneration and the pattern in which tentacles arise [81]. We attempted to directly determine the intracellular localization of Hym-301 using an immunogold precipitation method; we found that the peptide is stored in vesicles that are located adjacent to the external surface of the ectodermal epithelial cells [82]. Thus, these observations suggest that Hym-301 is stored in such vesicles and released in a manner similar to that described for neuropeptides when tentacle formation occurs in the *Hydra* head.

1.3.3   HYM-323

The *Hydra* foot, including the peduncle and the basal disk, is believed to be composed primarily of terminally differentiated cells. Hym-323, which was identified during the course of the *Hydra* peptide project [6], is composed of 16 amino acid residues, with a free C-terminus (Table 1.8). The Hym-323 preprohormone contains one copy of Hym-323 at its C-terminus. The Hym-323-encoding sequence is preceded by a Thr residue at the N-terminus. It has been shown that a Thr residue is the cleavage site in cnidarians [16], with cleavage at this residue producing mature Hym-323.

TABLE 1.8   Hym-323 in Cnidaria

Name	Peptide Sequence	Species	References
Hym-323	KWVQGKPTGEVKQIKF	*Hydra magnipapillata*	[84]

Both *in situ* hybridization and immunohistochemical analyses with an antibody specific to Hym-323 revealed the presence of Hym-323 in both ectodermal and endodermal epithelial cells throughout the *Hydra* body, with the exception of the basal disk and the head region, indicating that Hym-323 is an epitheliopeptide [83]. During foot regeneration , the peptide disappears when the basal disk cells are formed [83]. As epithelial cells continuously undergo mitosis, and because the body column can always regenerate afoot, the presence of Hym-323 in the epithelial cells of the body column would be maintained at an appropriate level.

Hym-323 treatment enhances the capacity of *Hydra* for ectopic foot formation from tissues of the body column by grafting [83]. For example, epithelial tissue from the body column of a donor hydra that was treated with Hym-323 peptide was grafted onto the body column of a host hydra, after which ectopic foot formation was assayed. The potential for activation of foot formation was enhanced in the peptide-treated tissue as compared with untreated control tissue. Hym-323 significantly enhances foot regeneration of epithelial *Hydra* that lacks nerve cells, strongly suggesting that

Hym-323 directly acts on epithelial cells. Collectively, these data indicate that upon initiation of foot formation, stored Hym-323 peptide is released from epithelial cells, and when the concentration of Hym-323 exceeds a certain threshold, Hym-323 binds to a receptor on these same epithelial cells to increase the foot-forming potential of the tissue, eventually enhancing foot formation.

The target gene of Hym-323 is an astacin matrix metalloprotease known as foot activator-responsive matrix metalloprotease (*Farm1*) [84]. *Farm1* is normally expressed in epithelial cells in the gastric region but not in apical and basal tissues. Treatment of *Hydra* with Hym-323 induces an immediate down-regulation of the*Farm1*gene. Thus, *Farm1* is a transcriptional target of positional signals that specify foot differentiation and appears to exert a potent effect on basal patterning processes.

1.3.4   PEDIN/HYM-330 AND PEDIBIN/HYM-346

Two peptides isolated from *Hydra vulgaris*, pedin, and pedibin, have been shown to increase the rate of foot regeneration [85]. In addition, pedin increases the rate of interstitial cell proliferation and neuron differentiation [85]. Pedin was also found to enhance bud outgrowth [86]. We identified pedin and pedibin peptides that lack the C-terminal Glu residue found in the corresponding molecules in *Hydra magnipapillata*, and these peptides were designated Hym-330 and Hym-346, respectively (Table 1.9) [6]. The preprohormones of pedin/Hym-330 and pedibin/ Hym-346 contain no typical signal peptide sequence in the N-terminal regions [86, 87]. This is also the case for the Hym-323 preprohormone [83]. The pedibin/Hym-346 preprohormone contains one copy of the peptide at its C-terminus. A basic amino acid (Lys) processing site is located at the 26th amino acid residue from the N-terminus. Thus, it is plausible that the two subsequent amino acid residues, Leu, and Ala, are cleaved by a dipeptidyl aminopeptidase to produce the mature peptide [87]. Treatment with either Hym-323 or pedibin/Hym-346 was shown to enhance the capacity of *Hydra* for ectopic foot formation from tissues of the body column by grafting [83, 88]. The target gene of Hym-323 and pedibin/Hym-346 is *Farm1*, and treatment with the peptides induces an immediate down-regulation of this gene [84].

TABLE 1.9   Pedin/Hym-330 and Pedibin/Hym-346 in Cnidaria

Name	Peptide Sequence	Species	References
Pedin	EELRPEVLPDVSE	*Hydra vulgaris*	[86]
Hym-330	EELRPEVLPDVS	*Hydra magnipapillata*	[6]
Pedibin	AGEDVSHELEEKEKALANHSE	*Hydra vulgaris*	[86]
Hym-346	AGEDVSHELEEKEKALANHS	*Hydra magnipapillata*	[6]

A fraction of the secretory pathway of the peptides found in the brain and gut are known to function as signaling molecules. The preprohormones containing Hym-323, pedin/Hym-330, and pedibin/Hym-346 have unusual structures that lack a signal sequence. Recent mass spectrometry-based peptidomics analysis of mouse brain peptides demonstrated that half of the identified peptides are derived from non-secretory pathway proteins [89]. Many of these peptides represent the N- or C-terminus of the protein, rather than an internal fragment [89]. This observation suggests that these peptides are formed by selective processing rather than by protein degradation. Indeed, epitheliopeptides such as Hym-323 and pedibin/Hym-346 are derived from the C-termini of their respective origin proteins. It is plausible that these peptides, including pedin/Hym-330, are secreted from epithelial cells, function in cell-cell signaling, and then induce developmental events in *Hydra*. Fricker [90] proposed that intracellular peptides in neurons that are secreted at levels sufficient to produce a biological effect would be more appropriately called "non-classical neuropeptides." Though further studies are needed to confirm that epitheliopeptides such as Hym-323, pedin/Hym-330, and pedibin/Hym-346 meet the criterion to be considered non-classical epitheliopeptides in the epithelium, an appropriate term to distinguish these peptides from other biologically active epitheliopeptides, such as Hym-301, would be "non-classical epitheliopeptides."

1.4   CONCLUSION

Neuropeptides released from nerve cells and epitheliopeptides released from epithelial cells in response to a variety of stimuli are mandatory for fine-tuned regulation of behavior, reproduction, metamorphosis, and tissue maintenance (Figure 1.1). Here, I described 47 species of neuropeptides and 11 species of epitheliopeptides so far identified in cnidarians. However, the study of neuropeptides and epitheliopeptides are still at their infancy. Further identification of novel peptide signaling molecules is likely to provide a new and effective means to explore the mechanisms that underlie physiological and developmental processes in cnidarians and most likely will inform our understanding of peptide function in other animals as well.

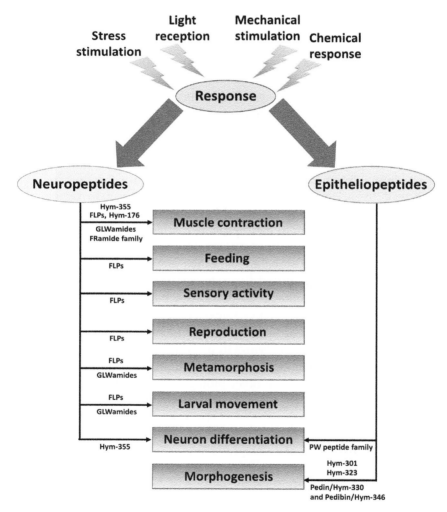

FIGURE 1.1   Summary for roles of neuropeptides and epitheliopeptides in the control of behavior, reproduction, metamorphosis, and tissue maintenance. Cnidarian peptide signaling molecules function together and/or separately to keep their lifestyle in response to stress stimulation, light reception, mechanical stimulation, and chemical stimulation.

ACKNOWLEDGMENT

The author acknowledges a grant from the JSPS KAKENHI to T.T. (grant number 17K07495).

CONFLICTS OF INTEREST

The author declares no conflict of interest.

KEYWORDS

- **degenerin**
- **epithelial cells**
- **epithelial Na⁺ channel**
- **epitheliopeptides**
- **green fluorescent protein**
- **metamorphosin A**
- **neuropeptides**

REFERENCES

1. Grimmelikhuijzen, C. J. P., Leviev, I., & Carstensen, K., (1996). Peptides in the nervous system of cnidrians: Structure, function and biosynthesis. *Int. Rev. Cytol., 167*, 37–89.
2. Kass-Simon, G., & Pierobon, P., (2007). Cnidarian chemical neurotransmission, an updated overview. *Comp. Biochem. Physiol. A Mol. Integr. Physiol., 146*, 9–25.
3. David, C. N., & Gierer, A., (1974). A cell cycle kinetics and development of *Hydra attenuate*. III, Nerve and nematocyte differentiation. *J. Cell Sci., 16*, 359–375.
4. Schmidt, T., & David, C. N., (1986). Gland cells in *Hydra*: Cell cycle kinetics and development. *J. Cell Sci., 85*, 197–215.
5. Bosch, T. C. G., & David, C. N., (1987). Stem cells of *Hydra magnipapillata* can differentiate intosomatic cells and germ line cells. *Dev. Biol., 121*, 182–191.
6. Takahashi, T., Muneoka, Y., Lohmann, Y., De Haro, M. S. L., Solleder, G., Bosch, T. C. G., et al., (1997). Systematic isolation of peptide signaling molecues regulating development in hydra: LWamide and PW families. *Proc. Natl. Acad. Sci. USA., 94*, 1241–1246.
7. Price, D. A., & Greenberg, M. J., (1977). Structure of a molluscan cardioexcitatory neuropeptide. *Science, 97*, 670–671.

8. Price, D. A., & Greenberg, M. J., (1977). Purification and characterization of a cadioexcitatory neuropeptide from the central ganglia of a bivalve mollusk. *Prep. Biochem., 7*, 261–281.

9. Price, D. A., & Greenberg, M. J., (1989). The hunting of the FaRPs: The distribution of FMRFamide-related peptides. *Biol. Bull., 177*, 198–205.

10. Espinoza, E., Carrigan, M., Thomas, S. G., Shaw, G., & Edison, A. S., (2000). A statistical view of FMRFamide neuropeptide diversity. *Mol. Neurobiol., 21*, 35–56.

11. Krajniak, K. G., (2013). Invertebrate FMRFamide related peptides *Protein Pept. Lett., 20*, 647–670.

12. Moosler, A., Rhinehart, K. L., & Grimmelikhuijzen, C. J. P., (1997). Isolation of three novel peptides, the Cyanea-RFamides I-III, from scyphomedusae. *Biochem. Biophys. Res. Commun., 236*, 743–749.

13. Moosler, A., Rhinehart, K. L., & Grimmelikhuijzen, C. J. P., (1996). Isolation of four novel neuropeptides, the Hydra-RFamides I-IV, from *Hydra magnipapillata. Biochem. Biophys. Res. Commun., 229*, 596–602.

14. Grimmelikhuijzen, C. J. P., Hahn, M., Rhinehart, K. L., & Spencer, A. N., (1988). Isolation of pyroGlu-Leu-Leu-Gly-Gly-Arg-Phe-NH$_2$ (Pol-RFamide), a novel neuropeptide from hydromedusae. *Brain Res., 475*, 198–203.

15. Grimmelikhuijzen, C. J. P., Rhinehart, K. L., & Spencer, A. N., (1992). Isolation of the neuropeptide less than Glu-Trp-Leu-Lys-Gly-Arg-Phe-NH$_2$ (Pol-RFamide II) from the hydromedusa *Polyorchis penicillatus. Biochem. Biophys. Res. Commun., 183*, 375–382.

16. Darmer, D., Hauser, F., Nothacker, H. P., Bosch, T. C. G., Williamson, M., & Grimmelikhuijzen, C. J. P., (1998). Three different prohormones yield a variety of Hydra-RFamide (Arg-Phe-NH$_2$) neuropeptide in *Hydra magnipapillata. Biochem. J., 332*, 403–412.

17. Fujisawa, T., (2008). *Hydra* Peptide Project, 1993–2007. *Dev. Growth Differ., 50*, S257–S268.

18. Gajewski, M., Schmutzler, C., & Plickert, G., (1998). Structure of neuropeptide precursors in cnidarian. *Ann. N. Y. Acad. Sci., 839*, 311–315.

19. Grimmelikhuijzen, C. J. P., & Graff, D., (1986). Isolation of Glu-Gly-Arg-Phe-NH$_2$ (Antho-RFamide), a neuropeptide from sea anemones. *Proc. Natl. Acad. Sci. USA., 83*, 9817–9821.

20. Anctil, M., (2009). Chemical transmission in the sea anemone *Nematostella vectensis*: A genomic perspective. *Comp. Biochem. Physiol. D., 4*, 268–289.

21. Darmer, D., Schmutzler, C., Diekhoff, D., & Grimmelikhuijzen, C. J. P., (1991). Primary structure of the precursor for the sea anemone neuropeptide Antho-RFamide (<Glu-Gly-Arg-Phe-NH$_2$). *Proc. Natl. Acad. Sci. USA., 88*, 2555–2559.

22. Schmutzler, C., Darmer, D., Diekhoff, D., & Grimmelikhuijzen, C. J. P., (1992). Identification of a novel type of processing sites in the precursor for the sea anemone neuropeptide Antho-RFamide (<Glu-Gly-Arg-Phe-NH$_2$) from *Anthopleura elegantissima. J. Biol. Chem., 267*, 22534–22541.

23. Reinscheid, R. K., & Grimmelikhuijzen, C. J. P., (1994). Primary structure of the precursor for the anthozoan neuropeptide Antho-RFamide from *Renilla kollikeri*: Evidence for unusual processing enzymes. *J. Neurochem., 62*, 1214–1222.

24. Schmutzler, C., Diekhoff, D., & Grimmelikhuijzen, C. J. P., (1994). The primary structure of the Pol-RFamide neuropeptide precursor protein from the hydromedusa *Polyorchis penicillatus* indicates a novel processing proteinase activity. *Biochem. J., 299*, 431–436.

25. McFarlane, I. D., Graff, D., & Grimmelikhuijzen, C. J. P., (1987). Excitatory actions of Antho-RFamide, an anthozoan neuropeptide, on muscles and conducting systems in the sea anemone *Calliactis parasitica. J. Exp. Biol., 133*, 157–168.

26. Anctil, M., & Grimmelikhuijzen, C. J. P., (1989). Excitatory action of the native neuropeptide Antho-RFamide in muscles in the pennatulid *Renilla kollikeri. Gen. Pharmacol., 20*, 381–384.

27. Shimizu, H., & Fujisawa, T., (2003). Peduncle of *Hydra* and the heart of higher organisms share a common ancestral origin. *Genesis, 36*, 182–186.

28. Cottrell, G. A., Green, K. A., & Davis, N. W., (1990). The neuropeptide Phe-Met-Arg-Phe-NH$_2$ (FMRFamide) can activate a ligand-gated ion channel in *Helix* neurons. *Pflugers Arch., 416*, 612–614.

29. Lingueglia, E., Champigny, G., Lazdunski, M., & Barbry, P., (1995). Cloning of the amiloride-sensitive FMRFamide peptide-gated sodium channel. *Nature., 378*, 730–733.

30. Golubovic, A., Kuhn, A, Williamson, M., Kalbacher, H., Holstein, T. W., Grimmelikhuijzen, C. J. P., & Gründer, S., (2007). A peptide-gated ion channel from the freshwater polyp *Hydra. J. Biol. Chem., 282*, 35098–35103.

31. Dürrnagel, S., Kuhn, A, Tsiairis, C. D., Williamson, M., Kalbacher, H., Grimmelikhuijzen, C. J. P., Holstein, T. W., & Gründer, S., (2010). Three homologous subunits form a high affinity peptide-gated ion channel in *Hydra. J. Biol. Chem., 285*, 11958–11965.

32. Assmann, M., Kuhn, A, Dürrnagel, S., Holstein, T. W., & Gründer, S., (2014). The comprehensive analysis of DEG/ENaC subunits in *Hydra* reveals a large variety of peptide-gated channels, potentially involved in neuromuscular transmission. *BMC Biol., 12*, 84.

33. Tardent, P., (1995). The cnidarian cnidocyte, high tech cellular weaponry. *BioEssays., 17*, 351–362.

34. Holtmann, M., & Thurm, U., (2001). Mono- and oligo-vesicular synapses and their connectivity in a Cnidarian sensory epithelium (*Coryne tubulosa*). *J. Comp. Neurol., 432*, 537–549.

35. Westfall, J. A., Elliott, C. F., & Carlin, R. W., (2002). Ultrastructural evidence for two-cell and three-cell neural pathways in the tentacle epidermis of the sea anemone *Aiptasia pallida. J. Morphol., 251*, 83–92.

36. Anderson, P. A., Thompson, L. F., & Moneypenny, C. G., (2004). Evidence for a common pattern of peptidergic innavations of cnidocytes. *Biol. Bull., 207*, 141–146.

37. Nakanishi, N., Hartenstein, V., & Jacobs, D. K., (2009). Development of the rhopalial nervous system in *Aurelia* sp. 1 (Cnidaria, Scyphozoa). *Dev. Genes Evol., 219*, 301–317.

38. Tremblay, M. E., Henry, J., & Anctil, M., (2004). Spawning and gamete follicle rupture in the cnidarian *Renilla Koellikeri*: Effects of putative neurohormones. *Gen. Comp. Endocrinol., 137*, 9–18.

39. Plickert, G., Kroiher, M., & Munck, A., (1988). Cell proliferation and early differentiation during embryonic development and metamorphosis of *Hydractinia echinata. Development., 103*, 795–803.

40. Katsukura, Y., Ando, H., David, C. N., Grimmelikhuijzen, C. J. P., & Sugiyama, T., (2004). Control of planula migration by LWamide and RFamide neuropeptides in *Hydractinia echinata. J. Exp. Biol., 207*, 1803–1810.

41. Leitz, T., (1998). Metamorphosin A and related compounds: A novel family of neuropeptides with morphogenetic activity. *Ann. N. Y. Acad. Sci., 839*, 105–110.

42. Leitz, T., (1998). Induction of metamorphosis of the marine hydrozoan *Hydractinia echnata* Fleming, 1828. *Biofouling., 12*, 173–187.

43. Katsukura, Y., David, C. N., Grimmelikhuijzen, C. J. P., & Sugiyama, T., (2003). Inhibition of metamorphosis by RFamide neuropeptides in planula larvae of *Hydractinia echinata*. *Dev. Genes Evol., 213*, 579–586.
44. Seipp, S., Schmich, J., Will, B., Schetter, E., Plickert, G., & Leitz, T., (2010). Neuronal cell death during metamorphosis of *Hydractinia echinata* (Cnidaria, Hydrozoa). *Invertebr. Neurosci., 10*, 77–91.
45. Takahashi, T., Kobayakawa, Y., Muneoka, Y., Fujisawa, Y., Mohri, S., Hatta, M., Shimizu, H., Fujisawa, T., Sugiyama, T., Takahara, M., Yanagi, K., & Koizumi, O., (2003). Identification of a new member of the GLWamide peptide family: Physiological activity and cellular localization in cnidarian polyps. *Comp. Biochem. Physiol. Part, B., 135*, 309–324.
46. Leitz, T., Morand, K., & Mann, M., (1994). Metamorphosin A: A novel peptide controlling development of the lower metazoan *Hydractinia echinata* (Coelenterata, Hydrozoa). *Dev. Biol., 163*, 440–446.
47. Carstensen, K., Rinehart, K. L., McFarlane, I. D., Graff, D., & Grimmelikhuijzen, C. J. P., (1992). Isolation of Leu-Pro-Pro-Gly-Pro-Leu-Pro-Arg-Pro-NH$_2$ (Antho-RPamide), and N-terminally protected, biologically active neuropeptide from sea anemones. *Peptides, 13*, 851–857.
48. Leviev, I, Williamson, M., & Grimmelikhuijzen, C. J. P., (1997). Molecular cloning of a preprohormone from *Hydra magnipapillata* containing multiple copies of Hydra-LWamide (Leu-Trp-NH$_2$) neuropeptides: Evidence for processing at Ser and Asn residues. *J. Neurochem., 68*, 1319–1325.
49. Gajewski, M., Leitz, T., Schlosherr, J., & Plickert, G. L., (1996). Wamides from cnidaria constitute a novel family of neuropeptides with morphogenetic activity. *Roux's Arch. Dev. Biol., 205*, 232–242.
50. Leitz, T., & Lay, M., (1995). Metamorphosin A is a neuropeptide. *Roux's Arch. Dev. Biol., 204*, 276–279.
51. Takahashi, T., Ohtani, M., Muneoka, Y., Aimoto, S., Hatta, M., Shimizu, H., Fujisawa, T., Sugiyama, T., & Koizumi, O., (1997). Structure-activity relation of LWamide peptides synthesized with a multipeptide synthesizer. In: Kitada, C., (ed.), *Peptide Chemistry* (pp. 193–196). Protein Research Foundation: Osaka.
52. Takahashi, T., & Hatta, M., (2011). The importance of GLWamide neuropeptides in Cnidarian development and physiology. *J. Amino Acids*, doi: 10.4061/2011/424501.
53. Schmich, J., Trepel, S., & Leitz, T., (1998). The role of GLWamides in metamorphosis of *Hydractinia echinata*. *Dev. Genes Evol., 208*, 267–273.
54. Iwao, K., Fujisawa, T., & Hatta, M., (2002). A cnidarian neuropeptide of the GLWamide family indices metamorphosis of reef-building corals in the genus *Acropora*. *Coral Reefs., 21*, 127–129.
55. Erwin, P. M., & Szmant, A. M., (2010). Settlement induction of Acropora palmate planulae by a GLW-amide neuropeptide. *Coral Reefs., 29*, 929–939.
56. Marcum, B. A., & Campbell, R. D., (1978). Development of *hydra* lacking nerve and intestinal cells. *J. Cell Sci., 29*, 17–33.
57. Campbell, R. D., (1976). Elimination of *Hydra* intestinal and nerve cells by means of colchicines. *J. Cell Sci., 21*, 1–13.
58. Takeda, N., Kyozuka, K., & Deguchi, R., (2006). Increase in intracellular cAMP is a prerequisite signal for initiation of physiological oocyte meiosis maturation in the hydrozoan *Cytaeis uchidae*. *Dev. Biol., 298*, 248–258.

59. Takeda, N., Nakajima, Y., Koizumi, O., Fujisawa, T., Takahashi, T., Matsumoto, M., & Deguchi, R., (2013). Neuropeptides trigger oocyte maturation and subsequent spawning in the hydrozoan jellyfish *Cytaeis uhcidae. Mol. Reprod. Dev., 80*, 223–232.

60. Yum, S., Takahashi, T., Koizumi, O., Ariura, Y., Kobayakawa, Y., Mohri, S., & Fujisawa, T., (1998). A novel neuropeptide, Hym-176, induces contraction of the ectodermal muscle in *Hydra magnipapillata. Biochem. Biophys. Res. Commun., 248*, 584–590.

61. Yum, S., Takahashi, T., Hatta, M., & Fujisawa, T., (1998). The structure and expression of a preprohormone of a neuropeptide, Hym-176 in *Hydra magnipapillata. FEBS Lett., 439*, 31–34.

62. Takahashi, T., Koizumi, O., Ariura, Y., Romanovitch, A., Bosch, T. C. G., Kobayakawa, Y., et al., (2000). A novel neuropeptide, Hym-355, positively regulates neuron differentiation in *Hydra. Development., 127*, 997–1005.

63. Muneoka, Y., Takahashi, T., Kobayashi, M., Ikeda, T., Minakata, H., & Nomoto, K., (1994). Phylogenetic aspects of structure and action of molluscan neuropeptides. In: Davey, K. G., Peter, R. E., & Tobe, S. S., (eds.), *Perspectives in Comparative Endocrinology* (pp. 109–118). National Research Council of Canada: Toronto, ON, Canada.

64. Raina, A. K., Jaffe, H., Kempe, T. G., Keim, P., Blacher, R. W., Fales, H. M., Riley, C. T., Klum, J. A., Ridgway, R. L., & Haves, D. K., (1989). Identification of a neuropeptide hormone that regulates sex pheromone production in female moths. *Science, 244*, 796–798.

65. Morris, H. R., Panico, M., Karplus, A., Lloyd, P. E., & Piniker, B., (1982). Identification by FAB-MS of the structure of a new cardioactive peptide from *Aplysia. Nature, 300*, 643–645.

66. Lloyd, P. E., Kupfermann, I., & Weiss, K. R., (1987). Sequence of small cardioactive peptide A: A second member of a class of neuropeptides in *Aplysia. Peptides, 8*, 179–183.

67. Ikeda, T., Kubota, I., Miki, W., Nose, T., Takao, T., Shimonishi, Y., & Muneoka, Y., (1993). Structures and actions of 20 novel neuropeptides isolated from the ventral nerve cords of an echiuroid worm, Urechis unicinctus. In: Yanaihara, N., (ed.), *Peptide Chemistry* (pp. 583–585). Leiden, the Netherlands.

68. Minakata, H., Ikeda, T., Fujita, T., Kiss, T., Hiripi, L., Muneoka, Y., & Nomoto, K., (1993). Neuropeptides isolated from Helix pomatia. Part 2. FMRFamide-related peptides, S-Iamide peptides, FR peptides and others. In: Yanaihara, N., (ed.), *Peptide Chemistry* (pp. 579–582). Leiden, the Netherlands.

69. Grimmelikhuijzen, C. J. P., Dierickx, K., & Boer, G. J., (1982). Oxytocin/vasopressin-like immunoreactivity is present in the nervous system of hydra. *Neuroscience, 7*, 3191–3199.

70. Koizumi, O., & Bode, H. R., (1991). Plasticity in the nervous system of adult hydra. I. I., I., Conversion of neurons to expression of a vasopressin-like immunoreactivity depends on axial location. *J. Neurosci., 11*, 2011–2020.

71. Morishita, F., Nitagai, Y., Furukawa, Y., Matsushima, O., Takahashi, T., Hatta, M., Fujisawa, T., Tunamoto, S., & Koizumi, O., (2003). Identification of a vasopressin-like immunoreactive substance in hydra. *Peptides., 24*, 17–26.

72. Koizumi, O., Hamada, S., Minobe, S., Hamaguchi-Hamada, K., Kurumata-Shigeto, M., & Nakamura, M., (2015). The nerve ring in cnidarians: Its presence and structure in hydrozoan medusa. *Zoology, 118*, 79–88.

73. Takahashi, T., Koizumi, O., Hayakawa, E., Minobe, S., Suetsugu, R., Kobayakawa, Y., Bosch, T. C. G., David, C. N., & Fujisawa, T., (2009). Further characterization of the PW peptide family that inhibits neuron differentiation in *Hydra. Dev. Genes Evol., 219,* 119–129.

74. Takahashi, T., (2013). Neuropeptides and epitheliopeptides: Structural and functional diversity in an ancestral metazoan *Hydra. Protein Pept. Lett., 20,* 671–680.

75. Hayakawa, E., Takahashi, T., Nishimiya-Fujisawa, C., & Fujisawa, T., (2007). A novel neuropeptide (FRamide) family identified by a peptidomic approach in *Hydra magnipapillata. FEBS, J., 274,* 5438–5448.

76. Chapman, J. A., Kirkness, E. F., Simakov, O., Hampson, S. E., Mitros, T., Weinmaier, T., et al., (2010). The dynamic genome of *Hydra. Nature, 464,* 592–596.

77. Eipper, B., & Mains, R. E., (1980). Structure and biosynthesis of proadrenocorticotropin/endorphin and released peptides. *Endocr. Rev., 1,* 1–27.

78. Rosa, P. A., Policastro, P., & Herbert, E., (1980). A cellular basis for the differences in regulation of synthesis and secretion of ACTH/endorphin peptides in anterior and intermediate lobes of the pituitary. *J. Exp. Biol., 89,* 215–237.

79. Klumperman, J., Spijker, S., Van Minnen, J., Sharp-Baker, H., Smit, A. B., & Geraerts, W. P. M., (1996). Cell type-specific sorting of neuropeptides: A mechanism to modulate peptide composition of large-dense-core vesicles. *J. Neurosci., 16,* 7930–7940.

80. Takahashi, T., Hatta, M., Yum, S., Gee, L., Ohtani, M., Fujisawa, T., & Bode, H. R., (2005). Hym-301, a novel peptide, regulates the number of tentacles formed in hydra. *Development, 132,* 2225–2234.

81. Khalturin, K., Anton-Erxleben, F., Sassmann, S., Wittlieb, J., Hemmrich, G., & Bosch, T. C. G., (2008). A novel gene family controls species-specific morphological traits in *Hydra. PLoS Biol., 6,* 2436–2449.

82. Takaku, Y., Shimizu, H., Takahashi, T., & Fujisawa, T., (2013). Subcellular localization of the epitheliopeptide, Hym-301, in *Hydra. Cell Tissue Res., 351,* 419–424.

83. Harafuji, N., Takahashi, T., Hatta, M., Tezuka, H., Morishita, F., Matsushima, O., & Fujisawa, T., (2001). Enhancement of foot formation in *Hydra* by a novel epitheliopeptide, Hym-323. *Development, 128,* 437–446.

84. Kumpfmeuller, G., Rybakine, V., Takahashi, T., Fujisawa, T., & Bosch, T. C. G., (1999). Identification of an astacin matrix metalloprotease as target gene for *Hydra* foot activator peptides. *Dev Genes Evol., 209,* 601–607.

85. Hoffmeister, S. A. H., (1996). Isolation and characterization of two new morphologically active peptides from *Hydra vulgaris. Development, 122,* 1941–1948.

86. Herrmann, D., Hatta, M., & Hoffmeister-Ullerich, S. A. H., (2005). Thypedin, the multi copy precursor for the hydra peptide pedin, is a β-thymosin repeat-like domain containing protein. *Mech. Dev., 122,* 1183–1193.

87. Hoffmeister, S. A. H., (2001). The foot formation stimulating peptide pedibin is also involved in patterning of the head in hydra. *Mech. Dev., 106,* 37–45.

88. Grens, A., Shimizu, H., Hoffmeister, S. A. H., Bode, H. R., & Fujisawa, T., (1999). The novel signal peptide, pedibin and Hym-346, lower positional value thereby enhancing foot formation in hydra. *Development, 126,* 517–524.

89. Fricker, L. D., (2010). Analysis of mouse brain peptides using mass spectrometry-based peptidomics: Implication for novel functions ranging from non-classical neuropeptides to microproteins. *Mol. Biosyst., 6,* 1355–1365.

CHAPTER 2

Sex-Inducing Substances Terminate Dormancy in Planarian Postembryonic Reproductive Development

KIYONO SEKII and KAZUYA KOBAYASHI

Department of Biology, Faculty of Agriculture and Life Science, Hirosaki University, 3 Bunkyo-Cho, Hirosaki, Aomori 036-8561, Japan, E-mail: kobkyram@hirosaki-u.ac.jp

2.1 INTRODUCTION

Some metazoans occasionally switch between asexual and sexual reproduction based on environmental changes, life cycle phases, or both. In general, asexual animals possess pluripotent stem cells and can regenerate lost body parts by asexual reproduction. When they switch from an asexual to a sexual state, they postembryonically differentiate reproductive organs from pluripotent stem cells and dormant primordial germ cells (PGCs). To study this mechanism, an experimental system was established in an asexual strain of a freshwater planarian, *Dugesia ryukyuensis*. In this assay system, asexual worms are able to switch to a sexual state (sexual induction) when fed minced conspecific sexual worms or sexually mature worms of xenogeneic planarian. This indicates that the dormancy in postembryonic reproductive development from the planarian pluripotent stem cells, neoblasts are broken down by "sex-inducing substances" contained in sexual planarians. Recently, we found that the yolk gland, which is a planarian-specific reproductive organ filled with nurse cells, possibly contains the crucial sex-inducing substances [hydrophilic, heat-stable, and putative low-MW compounds (<500), but not peptides] that can fully sexualize asexual worms. The crucial sex-inducing substances we have targeted may activate all necessary endocrine systems to induce postembryonic reproductive development. This review will introduce insights obtained from our studies on the sex-inducing substances to help elucidate the mechanisms for postembryonic reproductive development during reproductive switching in metazoans.

2.2 ENDOCRINOLOGY IN PLANARIANS

The phylum Platyhelminthes consists of three parasitic classes (Cestoda, Monogenea, and Trematoda) and one free-living class (Turbellaria) [1, 2]. The Platyhelminthes are one of the primitive metazoans. Although worms belonging to this phylum have three tissue layers (triploblasts), bilateral symmetry, and cephalization (central nervous system), they have no coelom and blood circulatory system [3, 4].

Turbellarian species generally have pluripotent stem cells called neoblasts (i.e., Catenulidae [5], Macrostomida [5–7], and Polycladida [8]). They exhibit a remarkable capacity for regeneration owing to the neoblasts. The mechanisms for regeneration have been most extensively examined in freshwater planarians of the order Tricladida [9–23].

Owing to this strong regenerative capacity, which depends on neoblasts, some freshwater planarians are able to reproduce asexually. Asexual reproduction is achieved by successive biological processes from transverse fission to regeneration [24–26]. In general, the term "asexual planarians" (or asexual worms) refers to asexually reproductive worms lacking hermaphroditic reproductive organs. When they undergo this asexual reproduction, it is very important to reconstruct the anteroposterior polarity, because this asexual reproduction cannot be accomplished unless both head and tail fragments correctly reconstruct the anteroposterior polarity after transverse fission [20, 21, 27–39].

Because Platyhelminthes have no blood circulatory system, the nervous system should have a most important role in terms of endocrinology, namely, a neuroendocrine system. Some substances involved in the endocrine system in the study of fission (asexual reproduction) and regeneration have been reported (reviewed by Reuter & Kreshchenko, in 2004 [40]). In planarians, 5-HT (serotonin), its receptor, its precursor, 5-HTP, and its derivative, melatonin have been identified in the body [41–45]. The process of regeneration may be regulated through a signaling pathway via 5-HT receptors, since regeneration is inhibited by antagonists of 5-HT receptors, dihydroergocriptine, and methiothepin [46]. Additionally, melatonin inhibits transverse fission [47], and regeneration [48]. Neuropeptides also should act as either inhibitory or stimulatory signals in the mechanisms of fission and regeneration. It was reported that FMRFamide-related peptides and neuropeptide F (NPF) regulate planarian regeneration [49]. The neuropeptides substance P and K may stimulate cell proliferation during regeneration [50].

Substances involved in the neuroendocrine system for planarian reproductive development have also been reported, i.e., neuropeptide Y (NPY) [51, 52], as described in detail in section 2.4.2. It was reported that LSD (lysergic acid diethylamide), an antagonist (or agonist) of 5-HT receptors, caused reproductive organs to completely regress in sexual planarians [53]. An insulin-like peptide is known to regulate the proliferation of not only neoblasts, but also male germline cells in planarians [54].

On the other hand, the involvement of mammalian sex steroid-like compounds in planarians has also been proposed [55, 56]. Recently, the existence of a nuclear hormone receptor involved in planarian reproductive development was reported [57], as described in section 2.5.3. These findings imply that lipophilic substances regulate planarian reproductive development, in addition to substances in the neuroendocrine system.

2.3 REPRODUCTIVE SWITCHING FROM AN ASEXUAL TO A SEXUAL STATE: POSTEMBRYONIC REPRODUCTIVE DEVELOPMENT

In metazoans, the separation of somatic and germ cells, referred to as the determination of primordial germ cells (PGCs), occurs via three mechanisms: preformation, epigenesis, and postembryonic germ cell development [58, 59]. Many metazoans determine PGCs during embryogenesis by either preformation or epigenesis. They possess immature gonads until birth/hatching, and then become sexually mature under the control of sex hormones at the appropriate age. In contrast, some metazoans can undergo germ cell development containing the determination of PGCs from pluripotent stem cells at postembryonic stages (after hatching), depending upon environmental conditions and the stage of their life cycle. Although the mechanisms associated with preformation and epigenesis have been well-studied in the fly and mouse, respectively, the mechanisms underlying postembryonic germ cell development (postembryonic reproductive development) are largely unknown.

Because planarians have pluripotent stem cells termed neoblasts [12, 15, 16, 60–64], they undergo homeostatic regulation of their body size by "cell turnover" [65]. An oviparous planarian species, *Bdellocephala brunnea,* undergoes the full development of complex hermaphroditic reproductive organs in colder seasons, with subsequent degeneration occurring in warmer seasons [66]. This observation suggests that homeostatic regulation of the reproductive organs derived from neoblasts (mature or immature) is controlled, depending upon environmental conditions such as temperature.

Depending on environmental conditions, some worms are able to switch between an asexual state without reproductive organs and a sexual state [67–71]. The mechanism by which this reproductive switch is achieved is not yet clear, even though the phenomenon was first described more than a century ago [72, 73]. The switch from an asexual state without reproductive organs to a sexual state is termed sexual induction. Understanding the process by which neoblasts differentiate into complicated hermaphroditic reproductive organs is important to the fields of not only reproductive biology, but also developmental biology. Sexual induction by environmental stimulation is unstable, and does not seem feasible as a relatively quick and reliable assay system to study the underlying mechanisms. In *Girardia tigrina*, when the posterior two-thirds of the body of an asexual worm were grafted to the anterior third portion of the body of a sexual worm, differentiation of reproductive organs occurred in the asexual body [74]. This result suggested that hormone-like chemicals derived from the sexual body induce the development of reproductive organs in the asexual body. Many scientists have identified that asexual worms are sexualized if fed with minced sexual worms of conspecific or xenogeneic worms [53, 75–78]. The sexual induction by feeding treatment clearly supports the existence of putative hormone-like chemicals (referred to as sex-inducing substances), as presumed by Kenk [74], and is a powerful bioassay system to elucidate the mechanism underlying the switch between asexual and sexual states in planarians. We consider that the most important points to understand include how planarians prevent, initiate, and maintain the development of reproductive organs from neoblasts and dormant PGCs. These factors are all involved in the mechanism by which the production of sex-inducing substances is regulated. Here, we emphasize that the sex-inducing substances can break down the dormancy in postembryonic reproductive development in the asexual state.

A natural question to ask is whether cannibalization among conspecific and/or xenogeneic worms occur in nature or not. However, it is usually difficult to visually identify if planarians undergo cannibalization. Recently, we found that three planarian species, *B. brunnea*, *Polycelis sapporo*, and *Seidlia auriculata* share the same habitat at Mt. Hakkoda in Aomori Prefecture, Japan (Figure 2.1A). Because the body color of *P. sapporo* is extremely white, it is easy to judge whether *B. brunnea* and *S. auriculata* eat *P. sapporo*. We could collect many worms of *B. brunnea* and *S. auriculata* that ate *P. sapporo* (Figure 2.1B). This suggests that planarians undergo cannibalization involving xenogeneic worms. We have already confirmed that *P. sapporo* has a strong sex-inducing activity in asexual worms of *D. ryukyuensis* (the OH strain) (data not shown). There is a possibility that doping of the sex-inducing substances affects switching from an asexual to a sexual state, and induces maturation in oviparous species such as *B. brunnea* in natural habitats.

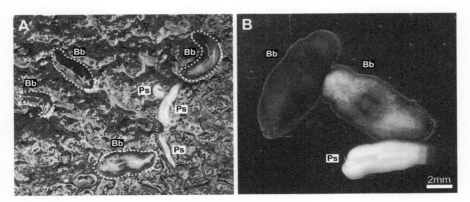

FIGURE 2.1   Planarian cannibalization in nature. (A) *Bdellocephala brunnea* and *Polycelis sapporo* share the same habitat at Mt. Hakkoda in Aomori Prefecture, Japan. (B) After a collection of worms, external observations were carried out under a binocular microscope. We could readily identify *B. brunnea* worms that had recently eaten *P. sapporo* (a worm in the center of the photograph), because the body colors of *B. brunnea* and *P. sapporo* are extremely black and white, respectively. Bb, *B. brunnea*; Ps, *P. sapporo*.

2.4   A BIOLOGICAL ASSAY SYSTEM FOR SEXUAL INDUCTION IN *D. RYUKYUENSIS*

2.4.1   *ESTABLISHMENT OF A BIOLOGICAL ASSAY SYSTEM*

The OH strain of *D. ryukyuensis* was derived from one asexual worm collected in 1984. Since then, asexual worms of the OH strain have only reproduced asexually and have never shown any evidence of sexuality under laboratory conditions. To isolate and identify sex-inducing substances, we first chose the OH strain as the recipient because spontaneous sexual induction has never been observed in this strain, and used sexually mature worms of *B. brunnea* as the source of the sex-inducing substances according to Sakurai [76], who reported sexual induction in asexual worms of *D. japonica*, a closely related species of *D. ryukyuensis*, by being fed with *B. brunnea* [79].

In the established bioassay system for sexual induction using the OH strain, all asexual worms are sexualized within approximately one month, if fed daily with *B. brunnea* [79]. The OH worms consistently develop a pair of ovaries, testes, yolk glands, and a copulatory apparatus, in that order. Morphological changes during sexual induction allowed us to divide the process into five stages (Figure 2.2A). At stage 1, a pair of ovaries becomes sufficiently large to be externally apparent behind the head, although no oocytes are evident. At stage 2, oocytes appear within the ovaries; however, other reproductive organs remain undetectable. At stage 3, the primordial testes and yolk gland primordia emerge, and a copulatory apparatus becomes visible in the post-pharyngeal region. In stage 4, the genital pore in the copulatory apparatus opens on the ventral side of the worm and spermatocytes appear in the testes. By stage 5, mature yolk glands are formed, and many mature spermatozoa are detectable within the testes. By external observation of the test worms, we can identify those with only a pair of ovaries (stage 1–2), those with a copulatory apparatus (stage 3), and those with a genital pore (stages 4–5) (Figure 2.2B).

2.4.2   PLANARIAN GONADS

Because planarians have no blood circulatory system, as described in section 2.2, the nervous system should be the most important to initiate and maintain the development of reproductive organs in response to various environmental stimuli which planarians receive through the central nervous system [5, 81–83]. In the sexual worms, an ovary is adjacent to a ventral nerve cord (Figure 2.3A). The neuropeptide NPY-8, which expresses specifically in the nerves surrounding the testes in *Schmidtea mediterranea,* was reported [51, 52].

In general, a double-transgenic model to pulse label predicted gamete stem cells with fluorescent proteins such as GFP is needed to demonstrate the existence of gamete stem cells [84–86]. Although transgenic techniques and genomic editing are not available in planarians at present, it is presumed that some PGCs which express a homolog of *nanos,* which is well known as a conserved germline determinant in metazoans [87–92], are putative gamete stem cells [93–96]. We considered the most peripheral oogonium-like cell mass in which a *nanos* homolog, *Dr-nanos,* is strongly expressed and is directly adjacent to a ventral nerve cord, as being ovarian stem cells (Os in Figure 2.3A) in *D. ryukyuenisis* [96]. Interestingly, *Dr-nanos* was strongly expressed in testes as well as ovaries in sexual worms (Figure 2.4), whereas it was expressed in only small cell masses of ovarian PGCs (probably ovarian stem cells) in asexual worms (Figure 2.4). Thus, in the asexual state of *D. ryukyuensis*, only ovarian

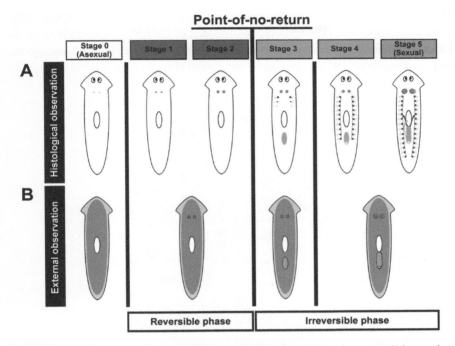

FIGURE 2.2 Five stages of sexual induction and estimation of stages by external observation. (A) Morphological changes during sexual induction allowed us to divide the process into five stages. The colored regions correspond to the reproductive organs: red, ovary; aqua blue, seminal receptacle; blue, testis; yellow, yolk gland; green, copulatory apparatus. Cell masses of female primordial germ cells (primordial ovaries) were histologically recognized in asexual worms (stage 0), although they were hardly visible externally. The ovaries became sufficiently large to be externally apparent behind the head, although no oocytes or other reproductive organs were detectable at stage 1. Oocytes appeared in the ovaries, but other reproductive organs remained undetectable at stage 2. The primordial testes and yolk gland primordia emerged, and a copulatory apparatus became visible in the post-pharyngeal region at stage 3. The genital pore in the copulatory apparatus opened on the ventral side of the worm and spermatocytes appeared in the testes at stage 4. Mature yolk glands formed, and spermatids and spermatozoa were detectable in the testes at stage 5. (B) By external observation of the test worms, we could recognize a pair of ovaries and a copulatory apparatus (with or without a genital pore) on the ventral side of test worms. The colored regions correspond to these reproductive organs: red, ovary; green, copulatory apparatus. Three stages were identified by external observation: worms with only a pair of ovaries (stages 1–2), worms with a copulatory apparatus without a genital pore (stage 3), and worms with a copulatory apparatus with a genital pore (stages 4–5). This sexual induction has a point-of-no-return between stages 2 and 3. In the reversible phase, worms degenerate a pair of developing ovaries to return to asexual status if feeding with sexual worms is stopped. In the irreversible phase, worms continue developing all reproductive organs, even if feeding with sexual worms is stopped. (Reproduced with permission from Nakagawa et al., [80]. http://creativecommons. org/licenses/by/4.0/).

PGCs, but not testicular PGCs, differentiate from neoblasts and immediately enter a dormant state. The testicular expression of *Dr-nanos* appears in stage 3. Later, *Dr-elav1*, a homolog of *Smed-elav1*, a marker gene for spermatogonia and spermatocytes [97], begins to be expressed (Figure 2.4).

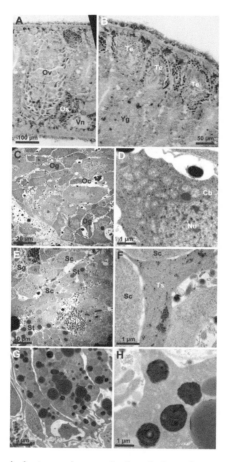

FIGURE 2.3 Light and electron microscopic description of *D. ryukyuensis* reproductive organs. (A, B) Hematoxylin-eosin staining of transverse sections of glutaraldehyde-fixed paraffin-embedded sexual worms. (C–H) Heavy-metal staining of ultrathin sections of osmium tetroxide-postfixed sexual worms. (D, F) High magnification of the area squared by the yellow line in (C, E), respectively. (A, C, D) Ovarian morphology. Os, putative ovarian stem cells; Ov, ovary; Vn, ventral nerve cord in (A). Oc, oocyte; Og, oogonium in (C). Cb, chromatoid body; Nu, nucleus in (D). Yellow asterisk indicates mitochondrion in (D). (B) Morphology of testis and yolk gland. Te, testis; Yg, yolk gland. (E, F) Morphology of testis. Sc, spermatocyte; Sg, spermatogonium, Sp, sperm; St, spermatid; Ts, Testicular supporting cell. (G, H) Morphology of yolk gland.

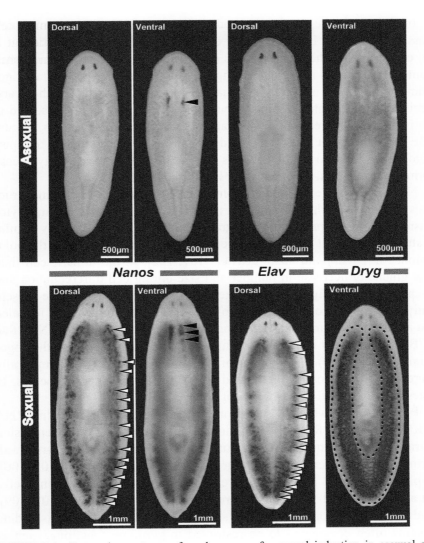

FIGURE 2.4 Expression patterns of marker genes for sexual induction in asexual and sexual worms. Whole-mount *in situ* hybridization analysis was performed using three marker genes for sexual induction in asexual and sexual worms of *Dugesia ryukyuensis*. *Nanos*: a gene expressed in PGCs or putative gamete stem cells in both ovaries and testes, *Dr-nanos* [96], a homologous genes of *Nanos* in *Drosophila melanogaster* [87]; *Elav*: a gene expressed in testes (spermatogonia and spermatocytes), *Dr-elav1*, a homologous gene of *Smed-elav1* in *Schmidtea mediterranea* [97]; *Dryg*: a gene expressed in the yolk glands [98]. The regions indicated by black and white arrowheads are ovaries and testes, respectively. In particular, the signal of *Dr-nanos* in asexual worms (the ventral view) indicates a cell mass of ovarian PGCs. The domains bound by the dashed line in sexual worms (the signal of *Dryg*) are mature yolk glands.

In planarians, the structure of both gonads appears to be a germline cell mass (Figure 2.3A, B). Even under electron microscopic observation, it is not clear whether there are somatic cells that can be clearly distinguished from germline cells in the ovary (Figure 2.3C). The most peripheral oogonium-like cell mass is next to oogonia, whereas oocytes are inside the ovary. Similarly, in the testes, spermatogonia and spermatocytes are observed at the periphery, whereas spermatids and sperm are inside the testis (Figure 2.3E). In the present study, we found supporting cells surrounding the spermatocytes, similar to Sertoli cells (Ts in Figure 2.3F). Such cells may release hormones to facilitate spermatogenesis.

Neoblasts and germline cells contain electron-dense cytoplasmic structures referred to as chromatoid bodies (CBs) [47, 93, 99]. CBs are large cytoplasmic ribonucleoprotein (RNP) granules [100], morphologically similar to structures present in the germline of many organisms such as the polar granules in *Drosophila melanogaster* [101, 102]. In the somatic cell line, CBs gradually decrease in number and size during cell differentiation and disappear completely in differentiated cells [84, 103, 104]. In the present study, electron microscopic studies of the gonads of sexual worms revealed that almost all oogonia and spermatogonia contained several CBs (Note that a CB is located near the nuclear membrane, with mitochondria in the oogonium (Figure 2.3D)). The number of germline cells with CBs dramatically decreased in the more differentiated cells, oocytes, and spermatocytes. In testes, spermatids with CBs were not recognized.

Piwi proteins, which play important roles in small non-coding RNA (piRNA)-mediated RNA silencing in epigenetic regulation and repression of transposable elements, are essential for germline stem cell self-renewal, germline development and gametogenesis in diverse organisms [105, 106]. In planarians, *piwi* homologs are expressed in not only germline cells, but also in neoblasts, and are required for stem cell regulation [107–109]. Studies in diverse organisms have shown that a number of RNA-binding proteins often considered to be restricted to germline functions, such as Piwi and Vasa, operate in pluripotent progenitors and stem cells of many metazoans [100, 110]. Rouhana et al., [100] proposed a "germline multipotency program" present in pluripotent stem cells and which generates somatic cell types.

CBs seem to be a center for RNP components involved in pluripotency [100, 111–113]. Therefore, the relevant RNP components in CBs of germline cells, as well as neoblasts, are detectable, including DjCBC-1 protein, a DEAD-box RNA helicase of the RCK/p54/Me31B family [84], Spoltud-1 protein, a Tudor domain-containing protein [112], SMEDWI-3, a Piwi protein and PRMT5, a Type II protein arginine methyltransferase [100], and *gH4* mRNA, germinal histone H4 [114]. Interestingly, transcripts of a *nanos* homolog are likely to accumulate in the CBs of the germline cells, but not in neoblasts [93]. *Nanos* function in planarians is required for maintenance of the germline, as described in Section 2.5.2 in detail [95, 96].

2.4.3   PLANARIAN NURSE GLANDS

In some orders of Turbellarians, worms have particular nurse glands, referred to as yolk glands. Their eggs are called ectolecithal eggs (cocoons), which have several fertilized eggs and numerous yolk gland cells [1, 2]. In planarians, oocytes released from the ovaries are cross-fertilized with allosperm within the seminal receptacles. When fertilized eggs are transferred to the copulatory apparatus via the oviducts, yolk gland cells are also released from yolk glands to the oviducts, and then form a cocoon within the copulatory apparatus (see the cartoon of a stage 5 worm in Figure 2.2). Eventually, planarians lay a cocoon on a substrate such as a stone. Most yolk cells form a syncytium called the yolk-cell syncytium or the vitelline syncytium [115]. Embryonic development in cocoon proceeds, with nutrients being obtained by ingesting components of the yolk-cell syncytium and by cannibalizing siblings [116–118]. Several juveniles eventually hatch from a single cocoon [119].

Under light microscopy, with hematoxylin-eosin stained tissues of the sexual worms, yolk glands were strongly stained with eosin (Figure 2.3B). In freshwater planarians, some genes expressed in yolk glands have been reported. In *D. ryukyuensis*, a novel gene, *Dryg,* is expressed from primordial cells to matured tissues (Figure 2.4) [98]. This strong expression was recognized in cocoons within a day of deposition: these contain numerous intact yolk gland cells [120]. However, the biological activity of the protein product has not been examined yet. In *S. mediterranea*, *Semd-CPEB1*, which is a homolog of Cytoplasmic Polyadenylation Element Binding protein (CPEB) involved in oogenesis and spermatogenesis in various animals, is expressed in both ovaries and yolk glands [121]. Rouhana et al., [122] identified four important genes required for capsule shell maturation and tanning by transcriptome analysis, comparing intact worms with worms knocked-down for *Semd-CPEB1*. One of the four genes was named *Smed-tanning factor-1*. Interestingly, this study revealed that the tanning process of capsule shells is required for embryogenesis in cocoons.

In the present study, electron microscopic study of the yolk glands of sexual worms revealed that, morphologically, the glands contained four distinct granules (Figure 2.3G). In higher magnification views of yolk glands, numerous glycogen granules were recognized in the cytosol (Figure 2.3H).

2.5   MAINTENANCE OF ACQUIRED SEXUALITY

2.5.1   THE POINT-OF-NO-RETURN DURING SEXUAL INDUCTION

Experimental sexual induction has a "point-of-no-return" between stages 2 and 3. The worms at stages 1 and 2 return to being asexual if the feeding with sexual planarians (the administration of the sex-inducing substances) is stopped. In contrast, the worms from stages 3 onward keep developing sexual organs, even if the administration of the sex-inducing substances is stopped [79]. Thus, we can recognize a reversible phase and an irreversible phase in terms of the point-of-no-return, by external observation (Figure 2.2B). Around stages 2 and 3, how is the dormancy of postembryonic reproductive development broken down by the administration of the sex-inducing substances, and how is the acquired sexuality maintained without the administration of the sex-inducing substances? Indeed, the OH strain of *D. ryukyuensis* is an ideal model organism to investigate the reproductive organ(s) that are responsible for the acquisition and maintenance of sexuality.

Planarians can undergo "degrowth" under starvation, since their body size is homeostatically regulated by "cell turnover" from neoblasts [65]. In the present study, we carried out starvation to recovery treatment of worms after the point-of-no-return (Figure 2.5). When the worms after the point-of-no-return were starved for 5 weeks, they appeared to be asexual worms. Externally, a pair of ovaries and a copulatory apparatus were not recognized in the starved worms (Figure 2.5A). However, whole-mount *in situ* hybridization for *Dr-nanos* revealed that the starved animals were likely at stage 3 of sexual induction (Figures 2.2A and 2.5B). As we expected, all of the starved worms became fully sexual under our laboratory conditions for maintenance in which the OH worms were fed chicken liver once a week (Figure 2.5A). It is noted that the testicular PGCs, which differentiated from neoblasts because of the effects of the sex-inducing substances, were maintained without the administration of the sex-inducing substances, once worms overcame the point-of-no-return.

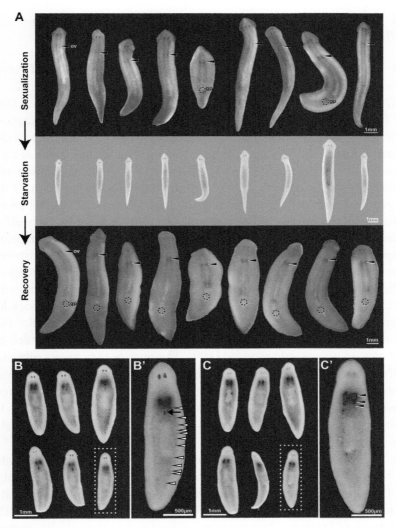

FIGURE 2.5 Acquired sexuality is maintained even under extreme starvation conditions. (A) Worms after stage 3 onwards (after the point-of-no-return) were starved for 5 weeks. Starved worms in which reproductive organs are externally invisible recovered at stage 5 under our laboratory conditions for maintenance. ov, an ovary; gp, a genital pore. (B, C) Expression patterns of *Dr-nanos* in the starved worms were similar to worms at stage 3. (B,' C') High magnification of a worm surrounded by a dashed line in (B, C), respectively. The signals of *Dr-nanos* were recognized in both cell masses of ovarian PGCs (black arrowheads) and in testicular PGCs (white arrowheads). We found the ectopic signal of *Dr-nanos* (an arrow in B') in the dorsal midline at the prepharyngeal region of all the starved worms. Such an ectopic signal was not detected in sexual worms (Figure 2.4).

2.5.2   *TESTIS PLAYS AN IMPORTANT ROLE IN MAINTAINING ACQUIRED SEXUALITY*

In *D. ryukyuensis*, four *piwi* genes have been reported [109]. *Drpiwi-1*, *Drpiwi-2*, and *Drpiwi-3* correspond to *Smediwi-1*, *Smediwi-2*, and *Smediwi-3* in *S. mediterranea* [107, 108] and *Djpiwi A*, *Djpiwi B*, and *Djpiwi C* in *D. japonica* [113]. *Drpiwi-1*, *Drpiwi-2*, and *Drpiwi-3* are expressed in not only germline cells, but also neoblasts. *Drpiwi-2* and *Drpiwi-3* are required for neoblast function in the same manner in *S. mediterranea* and *D. japonica* [109]. Therefore, RNAi knockdown of these genes caused worm lethality. This indicates the difficulties in examining the function of these *piwi* genes in germ cell development. Although *Drpiwi-1* (Drpiwi-1 protein) and its orthologs in *S. mediterranea* and *D. japonica* are used as neoblasts' markers [84, 109, 123, 124], the function of neoblasts is still unclear. In *D. japonica*, it is expected that DjPiwiA protein ubiquitously distributed in the cytoplasm of neoblasts is co-localized with DjPiwiC protein for localization of *histone* mRNA, and for transposable element repression via the production of piRNA [113]. In *D. ryukyuensis*, by RNAi knockdown of *Drpiwi-1* during sexual induction, the differentiation of testicular PGCs from neoblasts was completely inhibited, and the number of germline cells in the ovaries was greatly reduced [109]. The ovaries may be not eliminated by RNAi knockdown of *Drpiwi-1*, since ovarian PGCs originally existed in asexual worms (Figure 2.4). This result suggested that Drpiwi-1 protein is positively involved in the maintenance of gamete stem cells rather than neoblasts. Surprisingly, most *Drpiwi-1* (RNAi)-sexualized worms lacking testes returned to the asexual state.

As described in section 2.4.2, *nanos* transcripts seem to localize at the CBs in PGCs (putative gamete stem cells) in planarians [93]. Wang et al., [95] clearly showed that *nanos* function in planarians is required for maintenance of gamete stem cells. In *D. ryukyuensis*, the phenotype in RNAi knockdown of *Dr-nanos* during sexual induction was very similar to that of *Drpiwi-1* [96]. Taking these results together, we conclude that the acquisition of sexuality is associated with the determination of testicular PGCs from neoblasts in *D. ryukyuensis*. Additionally, quantitative reverse-transcription polymerase chain reaction analysis in *Dr-nanos* (RNAi)-sexualized worms revealed that expression of the yolk gland marker *Dryg* was lower than in controls (although this effect was not statistically significant due to large variation in the controls) [119]. The suppression of yolk gland differentiation may be associated with a smaller (immature) ovary and/or the absence of testis in the knockdown worms, because *Dr-nanos* is not expressed in yolk glands [98] (Figure 2.4). This result suggested that yolk glands are also associated with sexuality.

2.5.3 NEUROPEPTIDES AND LIPOPHILIC HORMONES IN PLANARIAN REPRODUCTIVE DEVELOPMENT

The neuropeptide NPY-8, which is specifically associated with testicular differentiation in *S. mediterranea*, is a notable discovery [52, 53]. Transcripts of *npy-8* that encode an NPY superfamily member in *S. mediterranea,* have been detected in a variety of cells within the central and peripheral nervous system, and within a dorsal population of cells associated with the lobes of the testes in sexual worms [52, 53]. Because RNAi knockdown of *npy-8* in sexually mature worms results in regression of the testes, the NPY-8 peptide in cells surrounding the lobes of the testes may act in a neuroendocrine fashion to influence the development of the testes and to maintain their proper function. RNAi knockdown of *npy-8* also triggered the regression of the other reproductive organs, being consistent with the results of RNAi knockdown of *nanos* and *piwi* genes [95, 96, 109]. Moreover, RNAi knockdown of a planarian central nervous system-specific gene, *pc2*, which is a homolog of pro-hormone convertase PC2 [125] and which seems to be associated with the maturation of NPY-8 peptide, also caused regression of testicular development in *S. mediterranea* [51]. As described in section 2.4.2, in planarians without a blood circulatory system, it is reasonable that various environmental stimuli should affect the regulation of the development of reproductive organs via the central nervous system.

On the other hand, a nuclear hormone receptor, NHR-1, which is involved in the development of reproductive organs in *S. mediterranea*, is another notable discovery [57]. *nhr-1* is expressed in somatic reproductive organs. Interestingly, RNAi knockdown of *nhr-1* caused repression of the development of both gonads, as well as of somatic reproductive organs. Usually, ligands for nuclear hormone receptors are lipophilic (hydrophobic) chemical compounds, such as sex steroid hormones in mammalians. Therefore, this result suggests the involvement of lipophilic (hydrophobic) hormones in planarian reproductive development. Because *S. mediterranea* seems to lack the capacity to switch reproductive modes [62], research on *npy-8* and *nhr-1* in *D. ryukyuensis* should be carried out to further clarify the molecular machinery involved in planarian reproductive switching.

2.6   SEX-INDUCING SUBSTANCES

2.6.1   *HYDROPHILIC FRACTIONS, FRS. M0 AND M10 SHOW STRONG SEX-INDUCING ACTIVITY TOWARD THE OH STRAIN*

Asexual OH strain worms were not sexualized by being fed with conspecific asexual worms [79]. Thus, sex-inducing substances may be absent or occur in very small quantities in asexual worms. The sexualized worms, by being fed with *B. brunnea* which were collected from a natural habitat, were maintained for over a year on a common diet of the chicken liver that has no sex-inducing activity. Subsequently, the asexual worms were fully sexualized by being fed with the sexualized worms [126]. Thus, the sex-inducing substances may be *de novo*-synthesized and/or specifically concentrated compounds derived from chicken liver in the sexualized worms.

We homogenized 4 g of sexually-mature planarians in phosphate-buffered saline and then obtained the cytosolic fraction of the supernatant (Supernatant-2) and two fractions of the precipitates (Precipitate-1 and -2) after a two-step centrifugation (Figure 2.6) [127]. Compounds that are more hydrophilic must be extracted into the cytosolic fraction, whereas compounds that are more hydrophobic must be contained in the precipitates. The cytosolic fraction was applied to a commercial octadecylsilane (ODS) column and eluted stepwise by changing the methanol concentration of the eluent (0, 10, and 100% (v/v)) (Figure 2.6). Using the assay system for sexual induction, we obtained hydrophilic fractions with strong sex-inducing activity toward asexual worms of *D. ryukyuensis* (the OH strain) from *B. brunnea* and conspecific sexual worms, referred to as Frs. M0 and M10. So far, we indicated that the crucial sex-inducing substances contained in the Frs. M0 and M10 that are required to overcome the point-of-no-return are hydrophilic, heat-stable, and putative low-MW compounds (<500), but are not peptides [80, 119, 127]. Neuropeptides and lipophilic hormones that we introduced in section 2.5.3 are not consistent with these features, and are not likely to be the substances that act as the crucial sex-inducing substances in our experimental sexual induction.

2.6.2   *PHYLOGENETIC RANGE OF SPECIES WITH CRUCIAL SEX-INDUCING ACTIVITY TOWARD OH STRAIN WORMS*

Recently, to further estimate phylogenetic relationships of species containing the crucial sex-inducing substance(s), a comprehensive comparison of sex-inducing activity toward asexual worms of *D. ryukyuensis* (the OH strain) was carried out using four Turbellarian species, freshwater planarians *D. ryukyuensis* (Tricladida, Continenticola, Dugesiidae) and *B. brunnea* (Tricladida, Continenticola, Dendrocoelidae), a land planarians *Bipalium nobile* (Tricladida, Continenticola, Bipaliidae), and a marine flatworm *Thysanozoon brocchii* (Polycladida) as sources of the sex-inducing substance(s). As an out-group for the assay, a slug, *Ambigolimax valentianus* (Mollusca), a natural food source for *Bi. nobile* was also used. To examine the potency of their sex-inducing activity toward asexual worms of *D. ryukyuensis*, we compared sex-inducing activity in three fractions (Frs. M0, M10 and M100 in Figure 2.6) from the five species [80]. After the feeding assay, external observations were carried out to estimate sex-inducing activity. By external observation, test worms with only a pair of ovaries (stages 1–2), those with a copulatory apparatus (stage 3), and those with a genital pore (stages 4–5) were identified (Figure 2.2).

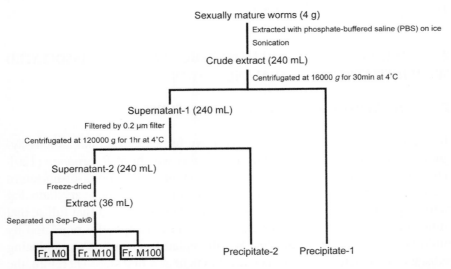

FIGURE 2.6 Procedure for fractionation of sexually mature worms (Fr., fraction. Three fractions used for the bioassay are highlighted in the box). (Adapted with permission from Nakagawa et al., [80]. http://creativecommons.org/licenses/by/4.0/).

Because the fraction with the lowest dry weight was Fr. M10 from *B. brunnea*, which showed strong sex-inducing activity [127], we set the standard dose of each sample for the bioassay at the dry weight of Fr. M10 from *B. brunnea* to compare sex-inducing activity. Figure 2.7 shows sex-inducing activity in the three cytosolic fractions toward asexual worms of *D. ryukyuensis*. The sex-inducing activity toward the asexual worms of conspecific worms was always lower than that of *B. brunnea* and *Bi. nobile*. In particular, the Fr. M0 from the land planarian *Bi. nobile* showed the highest sex-inducing activity among four Turbellarian species (Figure 2.7). Molecular phylogenetic analysis of freshwater and terrestrial planarians has suggested that in terms of phylogenetic distance, freshwater planarians in the family Dugesiidae and land planarians in the family Bipaliidae are more closely related than freshwater planarians of the family Dendrocoelidae [128]. The ability of the land planarian *Bi. nobile* (Bipaliidae) to produce strong sex-inducing activity in asexual worms of *D. ryukyuensis* (Dugesiidae) may be consistent with the aforementioned phylogenetic relationship. In contrast, insufficient sex-inducing activity to overcome the point-of-no-return was found in the fractions of the marine flatworm *T. brocchii* (Figure 2.7). These results suggest that there might be a common compound or a functional analog as the crucial hydrophilic sex-inducing substance in Tricladida, but not in Polycladida.

2.7   CHANGE IN TRYPTOPHAN METABOLISM MAY BE ASSOCIATED WITH INITIATION OF SEXUAL INDUCTION

2.7.1   *D-TRYPTOPHAN: A SEX-INDUCING SUBSTANCE*

To isolate the sex-inducing substance(s), we further carried out a bioassay-guided fractionation for Fr. M0 from 12 g wet weight of *B. brunnea* [120]. The crucial sex-inducing activity to overcome the point-of-no-return gradually decreased according to the purification step, but ovary-inducing activity persisted after the purification steps. This suggested that full sexual induction is orchestrated by several steps, and is temporally regulated by multiple sex-inducing substances. We attempted to identify a sex-inducing substance associated with ovarian development as a first step. Therefore, the bioassay-guided fractionation of the sex-inducing substance was carried out by estimating the ovary-inducing activity at each purification step.

This bioassay-guided fractionation identified L-tryptophan (Trp) on the basis of electrospray ionization–mass spectrometry, circular dichroism, and nuclear magnetic resonance spectroscopy. Originally masked by a large amount of L-Trp, D-Trp was detected by reverse-phase high-performance liquid chromatography (HPLC). We confirmed that D-Trp and L-Trp induced ovarian development in asexual worms, although the ovaries remained immature. In other words, D-Trp, and L-Trp allowed the asexual worms to induce until stage 2 (Figure 2.2). The ovary-inducing activity of D-Trp in asexual individuals was 500 times more potent than that of L-Trp [120]. This is the first report describing a role for an intrinsic D-amino acid in postembryonic germ cell development.

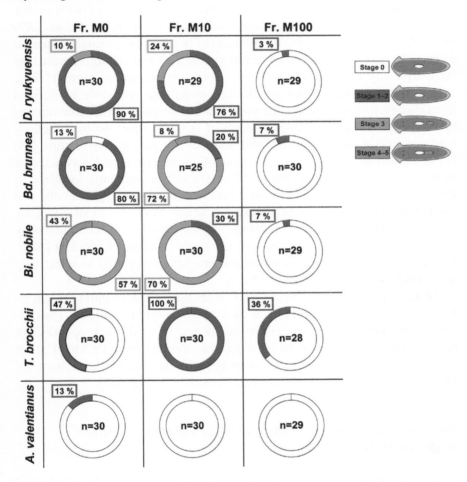

FIGURE 2.7 Comparison of sex-inducing activity toward asexual *D. ryukyuensis* worms. For assays of the three fractions from the cytosolic fraction, the proportions of test worms at the three stages of sexual induction identified by external observation were expressed as a doughnut chart. The number of test worms after the bioassay is shown in the center of the doughnut chart. The illustrations of worms in the figure correspond to those in Figure 2.2. The percentages of worms at each of the three stages of sexual induction are shown in the box with colored lines. (Adapted with permission from Nakagawa et al., [80]. http://creativecommons.org/licenses/by/4.0/).

We estimated Trp content in 4 g of wet weight of asexual and sexual worms of *D. ryukyuensis* using reverse-phase HPLC. The Trp level was much higher in the sexual worms than in the asexual animals (approximately 25-fold per wet weight). Furthermore, we performed diastereomeric derivatization of Trp from asexual and sexual worms, followed by reverse-phase HPLC. We found that the asexual worms contained D-Trp and L-Trp in a ratio of 0.2% (w/w), whereas the sexual worms contained D-Trp and L-Trp in a ratio of 1.4% (w/w). Based on these results, the amounts of D-Trp in the asexual and sexual worms (4 g wet weight) were estimated to be 0.6 μg and 100 μg, respectively (approximately 170-fold per wet weight in the sexual/asexual worms) [120].

L-Trp is an essential amino acid that, in animals, can only be derived from dietary proteins. To confirm whether the sexual worms have organs/tissues capable of pooling large amounts of Trp, we carried out immunofluorescence staining using a polyclonal antibody for free DL-Trp (Figure 2.8) [120]. In sexual worms, a strong immunofluorescence signal was detected from the yolk gland, indicating that free Trp is predominantly localized in these glands. Because fresh cocoons still have intact yolk gland cells, we compared the amounts of proteinogenic amino acids in asexual worms, sexual worms, and in fresh cocoons of *D. ryukyuensis*. We found that only Trp is prominently incorporated into the fresh cocoons (Table 2.1) [120]. These results indicate that sexual planarians have the ability to selectively incorporate and pool Trp in yolk glands.

In gastropod mollusks, including *A. valentianus*, the tripeptide L-Asn-D-Trp-L-Phe-NH$_2$ (NdWFamide) acts as a neuropeptide [129–133]. Thus, *A. valentianus* must contain free D-Trp as a degradant of this neuropeptide. In the fractionation procedure, D-Trp is recovered primarily in Fr. M0 [120]. It may be reasoned that a few test worms fed with the test food containing Fr. M0 of *A. valentianus* developed a pair of ovaries (Figure 2.7).

2.7.2   REGULATION OF SEXUAL INDUCTION BY PLANARIAN D-AMINO ACID OXIDASE (DAO)

Most amino acids exist as L-isomers within living organisms [134]. Natural proteins are exclusively built from L-amino acids. However, the development of enantioselective and other highly sensitive analytical methods have revealed that various D-amino acids naturally exist, both as free and protein-bound forms such as the tripeptide L-Asn-D-Trp-L-Phe-NH$_2$ (NdWFamide) in *A. valentianus* [129–133]. D-Amino acids play a key role in the regulation of many biological processes in living organisms [135, 136].

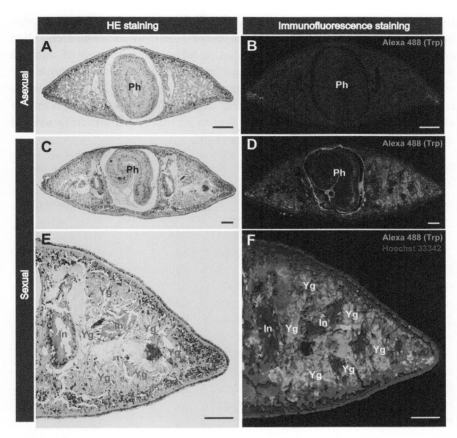

FIGURE 2.8 Immunofluorescence staining of asexual and sexual worms using free DL-Trp antibody. Transverse sections of asexual and sexual worms across the middle of the body (around the pharynx) were prepared. (A, C, E) Sections were stained with hematoxylin and eosin. (B, D, F) Immunofluorescence micrographs of asexual and sexual worms. Signals were visualized using Alexa 488 (green) fluorophore. (F) Nuclei were counterstained with Hoechst 33342 (blue), and a merged image is shown. The same settings were used to acquire all images. (A, B) Asexual worms. (C–F) Sexual worms. (E, F) High magnification of (C, D). In, intestine; Ph, pharynx; Yg, yolk gland. Scale bar, 100 μm. (Adapted with permission from Kobayashi et al., [120] . https://creativecommons.org/licenses/by/4.0/).

The regional distribution of D-amino acids is dependent on the expression of amino acid racemases and D-amino acid degrading enzymes. In other words, the activity of these enzymes may be correlated with the physiological activity of D-amino acids. In the mammalian brain, glial cells produce D-serine (D-Ser) from L-Ser via the action of serine racemase, and degrade

D-Ser by D-amino acid oxidase (DAO) [137]. The distribution of serine racemase in the brain is similar to that of D-Ser [138]. In addition, free D-Ser content is lower in the cerebellum compared to the cerebrum of mice in which DAO activity was much higher in the cerebellum compared to the cerebrum [139]. Although a tryptophan-specific racemase has not yet been isolated in animals, DAO orthologs that degrade neutral and basic D-amino acids (including D-Trp) have been found to be highly conserved in several species, ranging from yeast to humans [140].

TABLE 2.1   Relative Amounts of Each Proteinogenic Amino Acid in Asexual Worms, Sexual Worms, and Cocoons

Amino acid	Relative area			Sexual/Asexual	Cocoons/Asexual
	Asexual	**Sexual**	**Cocoons**		
Trp	1.40E-02	2.30E-01	6.10E-01	16.4	43.6
Tyr	2.20E-02	1.60E-01	6.50E-02	7.3	3.0
Cys	1.10E-03	3.00E-03	N.D.*	2.7	N.D.*
Gln	4.10E-02	7.30E-02	1.20E-02	1.8	0.3
Asp	1.30E-01	2.00E-01	9.50E-02	1.5	0.7
Thr	2.60E-02	3.60E-02	4.20E-02	1.4	1.6
Ala	1.30E-01	1.80E-01	2.60E-02	1.4	0.2
Asn	3.20E-02	4.10E-02	2.30E-02	1.3	0.7
Pro	2.40E-02	3.00E-02	1.20E-02	1.3	0.5
Ile	6.60E-02	7.60E-02	1.70E-02	1.2	0.3
His	9.00E-02	1.00E-01	7.80E-02	1.1	0.9
Phe	4.80E-02	5.20E-02	8.10E-02	1.1	1.7
Lys	8.90E-02	9.40E-02	4.60E-02	1.1	0.5
Ser	5.50E-02	5.70E-02	3.50E-02	1.0	0.6
Gly	2.10E-02	2.20E-02	6.10E-03	1.0	0.3
Leu	9.10E-02	9.20E-02	1.80E-02	1.0	0.2
Val	8.30E-02	8.10E-02	1.50E-02	1.0	0.2
Arg	8.50E-02	7.60E-02	2.40E-02	0.9	0.3
Glu	2.40E-01	1.90E-01	2.70E-02	0.8	0.1
Met	1.60E-02	9.30E-03	1.60E-03	0.6	0.1

*Not detected.

(Reproduced with permission from Kobayashi et al., [120]. https://creativecommons.org/licenses/by/4.0/).

We isolated a *DAO* ortholog gene, *Dr-DAO* from *D. ryukyuensis*. Dr-DAO recombinant protein degrades neutral, basic, and acidic D-amino acids *in vitro* [141]. In *D. ryukyuensis*, homogenates of asexual worm bodies had higher

DAO activity than those of sexual worms. Because the DAO activity was hardly detectable in *Dr-DAO* (RNAi)- asexual worms, the Dr-DAO protein would be responsible for planarian DAO activity *in vivo*. As expected, the expression of *Dr-DAO* in fresh cocoons was hardly detected, suggesting that D-Trp in yolk glands escapes degradation by Dr-DAO. This is one of the reasons the sexual worms contained much larger quantities of D-Trp than the asexual ones (approximately 170-fold per wet weight) [120]. In contrast, D-Trp may be produced by an unidentified tryptophan-specific racemase in planarians. A large amount of L-Trp, pooled in yolk glands (Figure 2.8 and Table 2.1), is the ideal substrate for supplying D-Trp via the tryptophan-specific racemase.

In asexual worms, *Dr-DAO* is highly expressed in the parenchyma. Thus, the concentration of D-amino acids in asexual worms may be maintained at comparatively lower levels. Surprisingly, RNAi knockdown of *Dr-DAO* in asexual worms results in the formation of immature ovaries in the absence of D-Trp administration [141]. Dr-DAO protein may repress early ovarian development through the degradation of D-Trp in asexual worms, with increased concentrations of D-Trp due to *Dr-DAO* knockdown inducing ovarian development. Here we propose a model of the initiation of sexual induction (ovarian development) by changes in Trp metabolism (Figure 2.9). In nature, environmental stimuli may lead to accumulation of L-Trp from foods, repression of Dr-DAO, and/or activation of an unidentified tryptophan-specific racemase in asexual worms, resulting in ovarian PGC proliferation and differentiation caused by an increase in D-Trp. The change in Trp metabolism might be essential for the initiation of sexual induction in *D. ryukyuensis*.

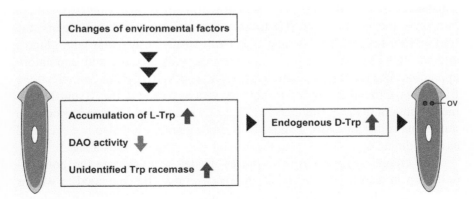

FIGURE 2.9 Changes in tryptophan metabolism may contribute to initiation of sexual induction. See the text for a detailed explanation. The illustrations of worms in the figure correspond to those in Figure 2.2.

Because the ovary-inducing activity by D-Trp was 500 times more potent than that by L-Trp, D-Trp must act as a principal bioactive compound in terms of ovarian development. In human leukocytes, D-Trp acts as a chemoattractant factor through a human niacin receptor [142]. D-Trp may function via an unknown receptor, the niacin receptor, as in ovarian development in planarians. The accumulation of L-Trp during sexual induction might lead to an increase in metabolites other than D-Trp, produced in metabolic pathways derived from L-Trp. Trp metabolites other than D-Trp may also influence ovary development. This novel, but common, mechanism of ovarian development induced by DL-Trp in metazoans may be elucidated in the near future.

2.8 PERSPECTIVES

D-Trp induces the development of immature ovaries, but cannot induce the same in other reproductive organs [120]. Other chemical compounds contained in Frs. M0 and M10 are required to trigger full sexual induction. There is much debate regarding the identity of the organs or tissues responsible for producing the crucial sex-inducing substances. As described in section 2.5, the testes and the nervous system are feasible as candidate organs/tissues producing the crucial sex-inducing substances. However, to date, the yolk gland has not been a candidate for the origin of the crucial sex-inducing substances.

Previously, to examine the distribution of the crucial sex-inducing substances in Frs. M0 and M10, sexual worms of *D. ryukyuensis* were cut into three pieces: the head (H) fragment containing no sexual organs; the middle (M) fragment containing a pair of ovaries, testes, and yolk glands; and the tail (T) fragment containing testes, yolk glands, and copulatory apparatus. The test worms were fed Frs. M0 and M10 derived from the three fragments. Crucial sex-inducing activity was observed in the M and T fragments, but not in the H fragments [117, 127]. The M and T fragments contain numerous testes and yolk glands. Thus, yolk glands were also not excluded as a candidate organ producing the crucial sex-inducing substances.

The crucial sex-inducing substances needed to overcome the point-of-no-return in asexual worms of *D. ryukyuensis* may be contained in worms of Tricladida, but not in those of Polycladida (Figure 2.7) [80]. An anatomically crucial difference between Tricladida and Polycladida is the presence or absence of yolk glands. Immediately after the point-of-no-return (stage 3), primordial yolk glands, as well as testicular PGCs, emerged in *D. ryukyuensis*

(Figure 2.2). In addition, a large quantity of L-Trp is incorporated and pooled in the yolk glands, resulting in the accumulation of D-Trp that is involved in ovarian development of asexual worms as a sex-inducing substance (Figure 2.8 and Table 2.1) [120]. Motivated by these findings, we fed the asexual worms of *D. ryukyuensis* (the OH worms) with fresh cocoons of *D. ryukyuensis* and *B. brunnea* containing numerous yolk gland cells, resulting in full sexual induction [80]. We concluded that the crucial sex-inducing substances for asexual worms of *D. ryukyuensis* is present in, at least, yolk glands. We also found that the crucial sex-inducing activity of the cocoons was significantly lower than that of individual worms (statistical significance was calculated by Chi-square test) (Figure 2.10). As described in section 2.7.1, full sexual induction may be orchestrated by several steps and also temporally regulated by multiple sex-inducing substances. This suggests that sexual planarians contain crucial sex-inducing substances or sex-inducing substances such as D-Trp which act as enhancers for sexual induction somewhere other than the yolk glands.

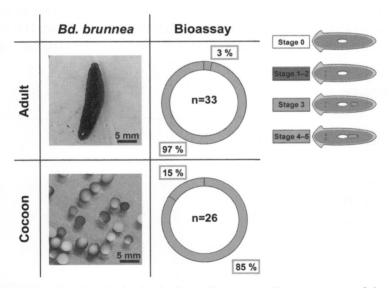

FIGURE 2.10 Sexual induction by feeding with cocoons. Test worms were fed minced sexually mature worms and cocoons of *Bdellocephala brunnea* daily for 4 weeks. The figure contains two photographs of sexually mature worms and cocoons of *B. brunnea*. Test worms at the three stages of sexual induction identified by external observation are expressed as a doughnut chart. The illustrations of worms in the figure correspond to those in Figure 2.2. The number of test worms after the bioassay is shown in the center of a doughnut chart. The percentages of worms at each of the three stages of sexual induction are shown in the box drawn with a colored line. (Adapted with permission from Nakagawa et al., [80]. https://creativecommons.org/licenses/by/4.0/).

Under natural conditions, freshwater planarians are expected to control sex-inducing substances for the development of their reproductive organs across seasons, probably producing endogenous substances or utilizing externally taken-up substances. Once the signaling pathway for sexual induction is activated, various types of molecules are expected to operate, such as transcription factors, neuropeptides, including NPY-8 [51, 52], lipophilic hormones for NHR-1 [57] and, probably, sex steroids as in mammals. The crucial sex-inducing substances we have targeted so far are able to activate all necessary endocrine systems to induce postembryonic reproductive development. In the future, the crucial sex-inducing substances may be identified using the bioassay system introduced in the present review.

ACKNOWLEDGMENTS

This study was supported in part by a Grant-in-Aid for Scientific Research (Nos. 26114501 [KK], 15K07121 [KK], 16H01249 [KK], and 18K19318 [KK]) from the Ministry of Science, Culture, Sports, and Education, Japan.

KEYWORDS

- **asexual reproduction**
- **Dugesia ryukyuensis**
- **germ cells**
- **planarian**
- **sex-inducing substance**
- **sexual induction**
- **sexual reproduction**

REFERENCES

1. Egger, B., Lapraz, F., Tomiczek, B., Müller, S., Dessimoz, C., Girstmair, J., et al., (2015). A transcriptomic-phylogenomic analysis of the evolutionary relationships of flatworms. *Curr. Biol., 25*(10), 1347–1353. https://doi.org/10.1016/j.cub.2015.03.034 (Accessed on 22 August 2019).

2. Littlewood, D. T., & Waeschenbach, A., (2015). Evolution: a turn up for the worms. *Curr. Biol., 25*(11), 457–460. https://doi.org/10.1016/j.cub.2015.04.012 (Accessed on 22 August 2019).

3. Egger, B., Steinke, D., Tarui, H., De Mulder, K., Arendt, D., Borgonie, G., et al., (2009). To be or not to be a flatworm: The acoel controversy. *PLoS One., 4*(5), e5502. https://doi.org/10.1371/journal.pone.0005502 (Accessed on 22 August 2019).

4. Inoue, T., (2017). Functional specification of a primitive bilaterian brain in planarians. In: Shigeno, S., Murakami, Y., & Nomura, T., (eds.), *Brain Evolution by Design. Diversity and Commonality in Animals* (pp. 79–100). Springer: Tokyo. https://doi.org/10.1007/978–4–431–56469–0_4 (Accessed on 22 August 2019).

5. Palmberg, I., (1990). Stem cells in microturbellarians. An autoradiographic and immunocytochemical study. *Protoplasma, 158*(3), 109–120. https://doi.org/10.1007/BF01323123 (Accessed on 22 August 2019).

6. Palmberg, I., & Reuter, M., (1983). Asexual reproduction in *Microstomum lineare* (Turbellaria). I. An autoradiographic and ultrastructural study. *Int. J. Invertebr. Reprod., 6*(4), 197–206. https://doi.org/101080/01651269198310510044 (Accessed on 23 August 2019).

7. Bode, A., Salvenmoser, W., Nimeth, K., Mahlknecht, M., Adamski, Z., Rieger, R. M., Peter, R., & Ladurner, P., (2006). Immunogold-labeled S-phase neoblasts, total neoblast number, their distribution, and evidence for arrested neoblasts in *Macrostomum lignano* (Platyhelminthes, Rhabditophora). *Cell Tissue Res., 325*(3), 577–587. https://doi.org/10.1007/s00441–006–0196–2 (Accessed on 22 August 2019).

8. Okano, D., Ishida, S., Ishiguro, S., & Kobayashi, K., (2015). Light and electron microscopic studies of the intestinal epithelium in *Notoplana humilis* (Platyhelminthes, Polycladida): The contribution of mesodermal/gastrodermal neoblasts to intestinal regeneration. *Cell Tissue Res., 362*(3), 529–540. https://doi.org/10.1007/s00441–015–2221–9 (Accessed on 22 August 2019).

9. Baguñà, J., Saló, E., Romero, R., Garcia-Fernàndez, J., Bueno, D., Muñoz-Marmol, A. M., Bayascas-Ramirez, J. R., & Casali, A., (1994). Regeneration and pattern formation in planarians: Cells, molecules and genes. *Zool. Sci., 11*(6), 781–795.

10. Agata, K., & Watanabe, K., (1999). Molecular and cellular aspects of planarian regeneration. *Semin. Cell Dev. Biol., 10*(4), 377–383. https://doi.org/10.1006/scdb.1999.0324 (Accessed on 22 August 2019).

11. Newmark, P. A., & Sánchez, A. A., (2000). Bromodeoxyuridine specifically labels the regenerative stem cells of planarians. *Dev. Biol., 220*(2), 142–153. https://doi.org/10.1006/dbio.2000.9645 (Accessed on 22 August 2019).

12. Saló, E., & Baguñà, J., (2002). Regeneration in planarians and other worms: New findings, new tools, and new perspectives. *J. Exp. Zool., 292*(6), 528–539. https://doi.org/10.1002/jez.90001 (Accessed on 22 August 2019).

13. Agata, K., Tanaka, T., Kobayashi, C., Kato, K., & Saitoh, Y., (2003). Intercalary regeneration in planarians. *Dev. Dyn., 226*(2), 308–316. https://doi.org/10.1002/dvdy.10249 (Accessed on 22 August 2019).

14. Reddien, P. W., & Sánchez, A. A., (2004). Fundamentals of planarian regeneration. *Annu. Rev. Cell Dev. Biol., 20*, 725–757. https://doi.org/10.1146/annurev.cellbio.20.010403.095114 (Accessed on 22 August 2019).

15. Orii, H., Sakurai, T., & Watanabe, K., (2005). Distribution of the stem cells (neoblasts) in the planarian *Dugesia japonica. Dev. Genes Evol., 215*(3), 143–157. https://doi.org/10.1007/s00427–004–0460-y (Accessed on 22 August 2019).

16. Sánchez, A. A., & Tsonis, P. A., (2006). Bridging the regeneration gap: Genetic insights from diverse animal models. *Nat. Rev. Genet., 7*(11), 873–884. doi: 10.1038/nrg1923.

17. Umesono, Y., & Agata, K., (2009). Evolution and regeneration of the planarian central nervous system. *Dev. Growth Differ, 51*(3), 185–195. https://doi.org/10.1111/j.1440–169X.2009.01099.x (Accessed on 22 August 2019).

18. Umesono, Y., Tasaki, J., Nishimura, K., Inoue, T., & Agata, K., (2011). Regeneration in an evolutionarily primitive brain--the planarian *Dugesia japonica* model. *Eur. J. Neurosci., 34*(6), 863–869. https://doi.org/10.1111/j.1460–9568.2011.07819.x (Accessed on 22 August 2019).

19. King, R. S., & Newmark, P. A., (2012). The cell biology of regeneration. *J. Cell Biol., 196*(5), 553–562. doi: 10.1083/jcb.201105099.

20. Liu, S. Y., Selck, C., Friedrich, B., Lutz, R., Vila-Farre, M., Dahl, A., Brandl, H., Lakshmanaperumal, N., Henry, I., & Rink, J. C., (2013). Reactivating head regrowth in a regeneration-deficient planarian species. *Nature, 500*(7460), 81–84. doi: 10.1038/nature12414.

21. Sikes, J. M., & Newmark, P. A., (2013). Restoration of anterior regeneration in a planarian with limited regenerative ability. *Nature, 500*(7460), 77–80. doi: 10.1038/nature12403.

22. Elliott, S. A., & Sánchez, A. A., (2013). The history and enduring contributions of planarians to the study of animal regeneration. *Wiley Interdiscip. Rev. Dev. Biol., 2*(3), 301–326. https://doi.org/10.1002/wdev.82 (Accessed on 22 August 2019).

23. Rink, J. C., (2013). Stem cell systems and regeneration in planaria. *Dev. Genes Evol., 223*(1/2), 67–84. https://doi.org/10.1007/s00427–012–0426–4 (Accessed on 22 August 2019).

24. Best, J. B., Goodman, A. B., & Pigeon, A., (1969). Fissioning in planarians: Control by the brain. *Science., 164*(3879), 565–566. doi: 10.1126/science.164.3879.565.

25. Morita, M., (1990). Photoperiod and melatonin control of planarian asexual reproduction. In: Hoshi, M., & Yamashita, O., (eds.), *Advances in Invertebrate Reproduction 5* (pp. 33–36). Elsevier: Amsterdam.

26. Malinowski, P. T., Cochet-Escartin, O., Kaj, K. J., Ronan, E., Groisman, A., Diamond, P. H., & Collins, E. S., (2017). Mechanics dictate where and how freshwater planarians fission. *Proc. Natl. Acad. Sci. USA, 114*(41), 10888–10893. https://doi.org/10.1073/pnas.1700762114 (Accessed on 22 August 2019).

27. Petersen, C. P., & Reddien, P. W., (2008). Smed-betacatenin-1 is required for anteroposterior blastema polarity in planarian regeneration. *Science, 319*(5861), 327–330. doi: 10.1126/science.1149943.

28. Gurley, K. A., Rink, J. C., & Sánchez, A. A., (2008). Beta-catenin defines head versus tail identity during planarian regeneration and homeostasis. *Science, 319*(5861), 323–327. doi: 10.1126/science., 1150029.

29. Iglesias, M., Gomez-Skarmeta, J. L., Saló, E., & Adell, T., (2008). Silencing of smed-betacatenin1 generates radial-like hypercephalized planarians. *Development, 135*(7), 1215–1221. doi: 10.1242/dev.020289.

30. Rink, J. C., Gurley, K. A., Elliott, S. A., & Sánchez, A. A., (2009). Planarian Hh signaling regulates regeneration polarity and links Hh pathway evolution to cilia. *Science, 326*(5958), 1406–1410. doi: 10.1126/science.1178712.

31. Petersen, C. P., & Reddien, P. W., (2009). Wnt signaling and the polarity of the primary body axis. *Cell, 139*(6), 1056–1068. https://doi.org/10.1016/j.cell.2009.11.035 (Accessed on 22 August 2019).

32. Yazawa, S., Umesono, Y., Hayashi, T., Tarui, H., & Agata, K., (2009). Planarian hedgehog/patched establishes anterior-posterior polarity by regulating Wnt signaling. *Proc. Natl. Acad. Sci. USA., 106*(52), 22329–22334. https://doi.org/10.1073/pnas.09074 64106 (Accessed on 22 August 2019).

33. Adell, T., Saló, E., Boutros, M., & Bartscherer, K., (2009). Smed-Evi/Wntless is required for beta-catenin-dependent and independent processes during planarian regeneration. *Development, 136*(6), 905–910. doi: 10.1242/dev., 033761.

34. Petersen, C. P., & Reddien, P. W., (2011). Polarized notum activation at wounds inhibits Wnt function to promote planarian head regeneration. *Science, 332*(6031), 852–855. doi: 10.1126/science.1202143.

35. Umesono, Y., Tasaki, J., Nishimura, Y., Hrouda, M., Kawaguchi, E., Yazawa, S., Nishimura, O., Hosoda, K., Inoue, T., & Agata, K., (2013). The molecular logic for planarian regeneration along the anterior-posterior axis. *Nature, 500*(7460), 73–76. doi: 10.1038/nature12359.

36. Roberts-Galbraith, R. H., & Newmark, P. A., (2013). Follistatin antagonizes activin signaling and acts with notum to direct planarian head regeneration. *Proc. Natl. Acad. Sci. USA., 110*(4), 1363–1368. https://doi.org/10.1073/pnas.1214053110 (Accessed on 22 August 2019).

37. Reuter, H., Marz, M., Vogg, M. C., Eccles, D., Grifol-Boldu, L., Wehner, D., Owlarn, S., Adell, T., Weidinger, G., & Bartscherer, K., (2015). Beta-catenin-dependent control of positional Information along the AP body axis in planarians involves a tee-shirt family member. *Cell Rep., 10*(2), 253–265. https://doi.org/10.1016/j.celrep.2014.12.018 (Accessed on 22 August 2019).

38. Umesono, Y., (2018). Postembryonic axis formation in planarians. In: Kobayashi, K., Kitano, T., Iwao, Y., & Kondo, M., (eds.), *Reproductive and Developmental Strategies. Diversity and Commonality in Animals* (pp. 743–761). Springer: Tokyo. https://doi. org/10.1007/978–4–431–56609–0_33 (Accessed on 22 August 2019).

39. Hosoda, K., Motoishi, M., Kunimoto, T., Nishimura, O., Hwang, B., Kobayashi, S., Yazawa, S., Mochii, M., Agata, K., & Umesono, Y., (2018). Role of MEKK1 in the anterior-posterior patterning during planarian regeneration. *Dev. Growth Differ., 60*(3), 341–353. https://doi.org/10.1111/dgd.12541 (Accessed on 22 August 2019).

40. Reuter, M., & Kreshchenko, N., (2004). Flatworm asexual multiplication implicates stem cells and regeneration. *Can. J. Zool., 82*(2), 334–356. https://doi.org/10.1139/ z03–219 (Accessed on 22 August 2019).

41. Itoh, M. T., Shinozawa, T., & Sumi, Y., (1999). Circadian rhythms of melatonin-synthe-sizing enzyme activities and melatonin levels in planarians. *Brain Res., 830*(1), 165–173. https://doi.org/10.1016/S0006–8993(99)01418–3 (Accessed on 22 August 2019).

42. Itoh, M. T., & Igarashi, J., (2000). Circadian rhythm of serotonin levels in planarians. *Neuroreport., 11*(3), 473–476.

43. Nishimura, K., Kitamura, Y., Inoue, T., Umesono, Y., Sano, S., Yoshimoto, K., Inden, M., Takata, K., Taniguchi, T., Shimohama, S., & Agata, K., (2007a). Reconstruction of dopaminergic neural network and locomotion function in planarian regenerates. *Dev. Neurobiol., 67*(8), 1059–1078. https://doi.org/10.1002/dneu.20377 (Accessed on 22 August 2019).

44. Nishimura, K., Kitamura, Y., Inoue, T., Umesono, Y., Yoshimoto, K., Takeuchi, K., Tani-guchi, T., & Agata, K., (2007b). Identification and distribution of tryptophan hydroxy-lase (TPH)-positive neurons in the planarian *Dugesia japonica*. *Neurosci. Res., 59*(1), 101–106. https://doi.org/10.1016/j.neures.2007.05.014 (Accessed on 22 August 2019).

45. Nishimura, K., Unemura, K., Tsushima, J., Yamauchi, Y., Otomo, J., Taniguchi, T., Kaneko, S., Agata, K., & Kitamura, Y., (2009). Identification of a novel planarian G-protein-coupled receptor that responds to serotonin in *Xenopus laevis* oocytes. *Biol. Pharm. Bull., 32*(10), 1672–1677. https://doi.org/10.1248/bpb.32.1672 (Accessed on 22 August 2019).

46. Saitoh, O., Yuruzume, E., & Nakata, H., (1996). Identification of planarian serotonin receptor by ligand binding and PCR studies. *Neuroreport., 8*(1), 173–178.

47. Morita, M., & Best, J. B., (1984). Electron microscopic studies of planarian regeneration. I. V., Cell division of neoblasts in *Dugesia dorotocephala. J. Exp. Zool., 229*(3), 425–436. https://doi.org/10.1002/jez.1402290310 (Accessed on 22 August 2019).

48. Yoshizawa, Y., Wakabayashi, K., & Shinozawa, T., (1991). Inhibition of planarian regeneration by melatonin. Hydrobiologia, *227*(1), 31–40. https://doi.org/10.1007/BF00027578 (Accessed on 22 August 2019).

49. Kreshchenko, N. D., Terenina, N. B., & Kuchin, A. V., (2018). Some details of muscles innervations by FMRF-like nerve elements in planarian *Girardia tigrina. Zoomorphology, 137*(2), 231–240. https://doi.org/10.1007/s00435–017–0392–5 (Accessed on 22 August 2019).

50. Baguñà, J., Saló, E., & Romero, R., (1989). Effects of activators and antagonists of the neuropeptides substance P and substance K on cell proliferation in planarians. *Int. J. Dev. Biol., 33*, 261–264.

51. Collins, J. J. III., Hou, X., Romanova, E. V., Lambrus, B. G., Miller, C. M., Saberi, A., Sweedler, J. V., & Newmark, P. A., (2010). Genome-wide analysis reveal a role for peptide hormones in planarian germline development. *PLoS Biol., 8*(10), e1000509. https://doi.org/10.1371/journal.pbio.1000509 (Accessed on 22 August 2019).

52. Saberi, A., Jamal, A., Beets, I., Schoofs, L., & Newmark, P. A., (2016). GPCRs direct germline development and somatic gonad function in planarians. *PLoS Biol., 14*(5), e1002457. https://doi.org/10.1371/journal.pbio.1002457 (Accessed on 22 August 2019).

53. Grasso, M., & Benazzi, M., (1973). Genetic and physiologic control of fissioning and sexuality in planarians. *J. Embryol. Exp. Morphol., 30*(2), 317–328.

54. Miller, C. M., & Newmark, P. A., (2012). An insulin-like peptide regulates size and adult stem cells in planarians. *Int. J. Dev. Biol., 56*(1–3), 75–82. doi: 10.1387/ijdb.113443cm.

55. Fukushima, M., Funabiki, I., Hashizume, T., Osada, K., Yoshida, W., & Ishida, S., (2008). Detection and changes in levels of testosterone during spermatogenesis in the freshwater planarian *Bdellocephala brunnea. Zool. Sci., 25*(7), 760–765. https://doi.org/10.2108/zsj.25.760 (Accessed on 22 August 2019).

56. Miyashita, H., Nakagawa, H., Kobayashi, K., Hoshi, M., & Matsumoto, M., (2011). Effects of 17ß-estradiol and bisphenol A on the formation of reproductive organs in planarian. *Biol. Bull., 220*(1), 47–56. https://doi.org/10.1086/BBLv220n1p47 (Accessed on 22 August 2019).

57. Tharp, M. E., Collins, J. J. III., & Newmark, P. A., (2014). A lophotrochozoan-specific nuclear hormone receptor is receptor is required for reproductive system development in the planarian. *Dev. Biol., 396*(1), 150–157. https://doi.org/10.1016/j.ydbio.2014.09.024 (Accessed on 22 August 2019).

58. Extavour, C. G., & Akam, M., (2003). Mechanisms of germ cell specification across the metazoans: Epigenesis and preformation. *Development, 130*(24), 5869–5884. doi: 10.1242/dev.00804.

59. Solana, J., (2013). Closing the circle of germline and stem cells: the primordial stem cell hypothesis. *Evo. Devo., 4*(1), 2. https://doi.org/10.1186/2041–9139–4–2 (Accessed on 22 August 2019).

60. Saló, E., & Baguñà, J., (1985). Cell movement in intact and regenerating planarians. Quantitation using chromosomal, nuclear and cytoplasmic markers. *J. Embryol. Exp. Morphol., 89*, 57–70.

61. Newmark, P. A., & Sánchez, A. A., (2002). Not your father's planarian: A classic model enters the era of functional genomics. *Nat. Rev. Genet., 3*(3), 210–219. doi: 10.1038/nrg759.

62. Wenemoser, D., & Reddien, P. W., (2010). Planarian regeneration involves distinct stem cell responses to wounds and tissue absence. *Dev. Biol., 344*(2), 979–991. https://doi.org/10.1016/j.ydbio.2010.06.017 (Accessed on 22 August 2019).

63. Wagner, D. E., Wang, I. E., & Reddien, P. W., (2011). Clonogenic neoblasts are pluripotent adult stem cells that underlie planarian regeneration. *Science, 332*(6031), 811–816. doi: 10.1126/science.1203983.

64. Shibata, N., Hayashi, T., Fukumura, R., Fujii, J., Kudome-Takamatsu, T., Nishimura, O., Sano, S., Son, F., Suzuki, N., Araki, R., Abe, M., & Agata, K., (2012). Comprehensive gene expression analyses in pluripotent stem cells of a planarian, *Dugesia japonica. Int. J. Dev. Biol., 56*(1–3), 93–102. doi: 10.1387/ijdb.113434ns.

65. González-Estévez, C., Felix, D. A., Rodriguez-Esteban, G., & Aboobaker, A. A., (2012). Decreased neoblast progeny and increased cell death during starvation-induced planarian degrowth. *Int. J. Dev. Biol., 56(1–3), 83–91.*

66. Kobayashi, K., Arioka, S., & Hoshi, M., (2002a). Seasonal changes in the sexualization of the planarian *Dugesia ryukyuensis. Zool. Sci., 19*(11), 1267–1278. https://doi.org/10.2108/zsj.19.1267 (Accessed on 22 August 2019).

67. Jenkins, M. M., (1967). Aspects of planarian biology and behavior. In: Corning, W. C., & Ratner, S. C., (eds.), *Chemistry of Learning* (pp. 117–143). Plenum Press: New York.

68. Vowinckel, C., (1970). The role of illumination and temperature in the control of sexual reproduction in the planarian *Dugesia tigrina* (Girard). *Biol. Bull., 138*(1), 77–87. https://www.jstor.org/stable/1540293 (Accessed on 22 August 2019).

69. Vowinckel, C., & Marsden, J. R., (1971a). Reproduction of *Dugesia tigrina* under short-day and long-day conditions at different temperatures. I. Sexually derived individuals. *J. Embryol. Exp. Morphol., 26*(3), 587–598.

70. Vowinckel, C., & Marsden, J. R., (1971b). Reproduction of *Dugesia tigrina* under short-day and long-day conditions at different temperatures. I. I., Asexually derived individuals. *J. Embryol. Exp. Morphol., 26*(3), 599–609.

71. Benazzi, M., (1974). Fissioning in planarians from a genetic standpoint. In: *Biology of the Turbellaria* (pp. 476–492). Riser, N. W., Morse, M. P., Eds., McGraw-Hill: New York.

72. Curtis, W., (1902). The life history, the normal fission and the reproductive organs of *Planaria maculata. Proc. Boston. Soc. Nat. Hist., 30*, 515–559.

73. Curtis, W. C., & Schulze, L. M., (1924). Formative cells of planarians. *Anat. Rec., 29*, 105.

74. Kenk, R., (1941). Induction of sexuality in the asexual from of *Dugesia tigrina. J. Exp. Zool., 87*(1), 55–69. https://doi.org/10.1002/jez.1400870105 (Accessed on 22 August 2019).

75. Benazzi, M., & Grasso, M., (1977). Comparative research on the sexualisation of fissiparous planarians treated with substances contained in sexual planarians. *Monitore Zool. Ital., 11*(1/2), 9–19.

76. Sakurai, T., (1981). Sexual induction by feeding in an asexual strain of the fresh-water planarian, *Dugesia japonica japonica. Annot. Zool. Jap., 54*, 103–112.

77. Teshirogi, W., (1986). On the origin of neoblasts in freshwater planarians (Turbellaria). In: Tyler, S., (ed.), *Advances in the Biology of Turbellarians and Related Platyhelminthes, Developments in Hydrobiology* (Vol. 32, pp. 207–216). Springer: Dordrecht. https://doi.org/10.1007/978–94–009–4810–5_29 (Accessed on 22 August 2019).

78. Hauser, J., (1987). Sexualization of *Dugesia anderlani* by feeding. *Acta Biologica Leopoldensia., 9*(1), 111–128.

79. Kobayashi, K., Koyanagi, R., Matsumoto, M., Cebrera, P. J., & Hoshi, M., (1999). Switching from asexual to sexual reproduction in the planarian *Dugesia ryukyuensis*: Bioassay system and basic description of sexualizing process. *Zool. Sci., 16*(2), 291–298. https://doi.org/10.2108/zsj.16.291 (Accessed on 22 August 2019).

80. Nakagawa, H., Sekii, K., Maezawa, T., Kitamura, M., Miyashita, S., Abukawa, M., Matsumoto, M., & Kobayashi, K., (2018). A comprehensive comparison of sex-inducing activity in asexual worms of the planarian *Dugesia ryukyuensis*: The crucial sex-inducing substance appears to be present in yolk glands in Tricladida. *Zool. Lett., 4*, 14. https://doi.org/10.1186/s40851–018–0096–9 (Accessed on 22 August 2019).

81. Inoue, T., Kumamoto, H., Okamoto, K., Umesono, Y., Sakai, M., Sánchez, A. A., & Agata, K., (2004). Morphological and functional recovery of the planarian photosensing system during head regeneration. *Zool. Sci., 21*(3), 275–283. https://doi.org/10.2108/zsj.21.275 (Accessed on 22 August 2019).

82. Inoue, T., Yamashita, T., & Agata, K., (2014). Thermosensory signaling by TRPM is processed by brain serotonergic neurons to produce planarian thermotaxis. *J. Neurosci., 34*(47), 15701–15714. https://doi.org/10.1523/JNEUROSCI.5379–13.2014 (Accessed on 22 August 2019).

83. Inoue, T., Hoshino, H., Yamashita, T., Shimoyama, S., & Agata, K., (2015). Planarian shows decision-making behavior in response to multiple stimuli by integrative brain function. *Zool. Lett., 1*, 7. https://doi.org/10.1186/s40851–014–0010-z (Accessed on 22 August 2019).

84. Yoshida-Kashikawa, M., Shibata, N., Takechi, K., & Agata, K., (2007). DjCBC-1, a conserved DEAD box RNA helicase of the RCK/p54/Me31B family, is a component of RNA-protein complexes in planarian stem cells and neurons. *Dev. Dyn., 236*(12), 3436–3450. https://doi.org/10.1002/dvdy.21375 (Accessed on 22 August 2019).

85. Nakagawa, T., Sharma, M., Nabeshima, Y., Braun, R. E., & Yoshida, S., (2010). Functional hierarchy and reversibility within the murine spermatogenic stem cell compartment. *Science, 328*(5974), 62–67. doi: 10.1126/science.1182868.

86. Nakamura, S., Kobayashi, K., Nishimura, T., Higashijima, S., & Tanaka, M., (2010). Identification of germline stem cells in the ovary of the teleost medaka. *Science, 328* (5985), 1561–1563. doi: 10.1126/science.1185473.

87. Kobayashi, S., Yamada, M., Asaoka, M., & Kitamura, T., (1996). Essential role of the posterior morphogen nanos for germline development in *Drosophila. Nature, 380* (6576), 708–711. doi: 10.1038/380708a0.

88. Mochizuki, K., Sano, H., Kobayashi, S., Nishimiya-Fujisawa, C., & Fujisawa, T., (2000). Expression and evolutionary conservation of nanos-related genes in Hydra. *Dev. Genes. Evol., 210*(12), 591–602. https://doi.org/10.1007/s004270000105 (Accessed on 22 August 2019).

89. Köprunner, M., Thisse, C., Thisse, B., & Raz, E., (2001). A zebrafish nanos-related gene is essential for the development of primordial germ cells. *Genes. Dev., 15*(21), 2877–2885. doi: 10.1101/gad.212401.

90. Suzuki, A., Tsuda, M., & Saga, Y., (2007). Functional redundancy among Nanos proteins and a distinct role of Nanos2 during male germ cell development. *Development, 134*(1), 77–83. doi: 10.1242/dev.02697.

91. Juliano, C., Yajima, M., & Wessel, G., (2010). *Nanos* functions to maintain the fate of the small micromere lineage in the sea urchin embryo. *Dev. Biol., 337*(2), 220–232. https://doi.org/10.1016/j.ydbio.2009.10.030.

92. Lai, F., Singh, A., & King, M. L., (2012). Xenopus Nanos1 is required to prevent endoderm gene expression and apoptosis in primordial germ cells. *Development, 139*(8), 1476–1486. doi: 10.1242/dev.079608.

93. Sato, K., Shibata, N., Orii, H., Amikura, R., Sakurai, T., Agata, K., Kobayashi, S., & Watanabe, K., (2006). *Identification and origin of the germline stem cells as revealed by the expression of nanos-related gene in planarians. Dev. Growth Differ, 48*(9), 615–628. https://doi.org/10.1111/j.1440–169X.2006.00897.x (Accessed on 22 August 2019).

94. Handberg-Thorsager, M., & Saló, E., (2007). The planarian *nanos*-like gene Smednos is expressed in germline and eye precursor cells during development and regeneration. *Dev. Genes Evol., 217*(5), 403–411. https://doi.org/10.1007/s00427–007–0146–3 (Accessed on 22 August 2019).

95. Wang, Y., Zayas, R. M., Guo, T., & Newmark, P. A., (2007). *Nanos function is essential for development and regeneration of planarian germ cells. Proc. Natl. Acad. Sci. USA., 104*(14), 5901–5906. https://doi.org/10.1073/pnas.0609708104 (Accessed on 22 August 2019).

96. Nakagawa, H., Ishizu, H., Chinone, A., Kobayashi, K., & Matsumoto, M., (2012a). The *Dr-nanos* gene is essential for germ cell specification in the planarian *Dugesia ryukyuensis. Int. J. Dev. Biol., 56*(1–3), 165–171. doi: 10.1387/ijdb.113433hn.

97. Wang, Y., Stary, J. M., Wilhelm, J. E., & Newmark, P. A., (2010). A functional genomic screen in planarians identifies novel regulators of germ cell development. *Genes Dev., 24*(18), 2081–2092. doi: 10.1101/gad.1951010.

98. Hase, S., Kobayashi, K., Koyanagi, R., Hoshi, M., & Matsumoto, M., (2003). Transcriptional pattern of a novel gene, expressed specifically after the point-of-no-return during sexualization, in planaria. *Dev. Genes Evol., 212*(12), 585–592. https://doi.org/10.1007/s00427–002–0288–2 (Accessed on 22 August 2019).

99. Coward, S. J., (1974). Chromatoid bodies in somatic cells of the planarian: Observations on their behavior during mitosis. *Anat. Rec., 180*(3), 533–545. https://doi.org/10.1002/ar.1091800312 (Accessed on 22 August 2019).

100. Rouhana, L., Vieira, A. P., Roberts-Galbraith, R. H., & Newmark, P. A., (2012). PRMT5 and the role of symmetrical dimethylarginine in chromatoid bodies of planarian stem cells. *Development., 139*(6), 1083–1094. doi: 10.1242/dev.076182.

101. Eddy, E. M., (1975). Germ plasm and differentiation of the germ line. *Int. Rev. Cytol., 43*, 229–280. https://doi.org/10.1016/S0074–7696(08)60070–4 (Accessed on 22 August 2019).

102. Kobayashi, S., Amikura, R., & Okada, M., (1993). Presence of mitochondrial large ribosomal RNA outside mitochondria in germ plasm of *Drosophila melanogaster. Science., 260*(5113), 1521–1524. doi: 10.1126/science.7684857.

103. Hori, I., (1982). An ultrastructural study of the chromatoid body in planarian regenerative cells. *J. Electron Microsc., 31*(1), 63–72. https://doi.org/10.1093/oxfordjournals.jmicro.a050338 (Accessed on 22 August 2019).

104. Auladell, C., Garcia-Valero, J., & Baguñà, J., (1993). Ultrastructural localization of RNA in the chromatoid bodies of undifferenti- ated cells (neoblasts) in planarians by the RNase–gold complex technique. *J. Morphol., 216*(3), 319–326. https://doi.org/10.1002/jmor.1052160307 (Accessed on 22 August 2019).

105. Ishizu, H., Siomi, H., & Siomi, M. C., (2012). Biology of PIWI-interacting RNAs: New insights into biogenesis and function inside and outside of germlines. *Genes Dev., 26*(21), 2361–2373. doi: 10.1101/gad.203786. 112.

106. Yamashiro, H., & Siomi, M. C., (2018). PIWI-interacting RNA in *Drosophila*: Biogenesis, transposon regulation, and beyond. *Chem. Rev., 118*(8), 4404–4421 doi: 10.1021/acs.chemrev.7b00393.

107. Reddien, P. W., Oviedo, N. J., Jennings, J. R., Jenkin, J. C., & Sánchez, A. A., (2005). SMEDWI-2 is a PIWI-like protein that regulates planarian stem cells. *Science, 301*(5752), 1327–1330. doi: 10.1126/science.1116110.

108. Palakodeti, D., Smielewska, M., Lu, Y. C., Yeo, G. W., & Graveley, B. R., (2008). The PIWI proteins SMEDWI-2 and SMEDWI-3 are required for stem cell function and piRNA expression in planarians. *RNA, 14*(6), 1174–1186. doi: 10.1261/rna.1085008.

109. Nakagawa, H., Ishizu, H., Hasegawa, R., Kobayashi, K., & Matsumoto, M., (2012b). *Drpiwi-1* is essential for germline cell formation during sexualization of the planarian *Dugesia ryukyuensis*. *Dev. Biol., 361*(1), 167–176. https://doi.org/10.1016/j.ydbio.2011.10.014 (Accessed on 22 August 2019).

110. Juliano, C., & Wessel, G., (2010). Developmental biology. Versatile germline genes. *Science, 329*(5992), 640–641. doi: 10.1126/science.1194037.

111. Shibata, N., Umesono, Y., Orii, H., Sakurai, T., Watanabe, K., & Agata, K., (1999). Expression of vasa(vas)-related genes in germline cells and totipotent somatic stem cells of planarians. *Dev. Biol., 206*(1), 73–87. https://doi.org/10.1006/dbio.1998.9130 (Accessed on 22 August 2019).

112. Solana, J., Lasko, P., & Romero, R., (2009). Spoltud-1 is a chromatoid body component required for planarian long-term stem cell self-renewal. *Dev. Biol., 328*(2), 410–421. https://doi.org/10.1016/j.ydbio.2009.01.043 (Accessed on 22 August 2019).

113. Kashima, M., Kumagai, N., Agata, K., & Shibata, N., (2016). Heterogeneity of chromatoid bodies in adult pluripotent stem cells of planarian *Dugesia japonica*. *Dev. Growth Differ, 58*(2), 225–237. https://doi.org/10.1111/dgd.12268 (Accessed on 22 August 2019).

114. Rouhana, L., Weiss, J. A., King, R. S., & Newmark, P. A., (2014). PIWI homologs mediate histone H4 mRNA localization to planarian chromatoid bodies. *Development., 141*(13), 2592–2601. doi: 10.1242/dev.101618.

115. Thomas, M. B., (1986). Embryology of the turbellaria and its phylogenetic significance. *Hydrobiologia., 132*(1), 105–115. https://doi.org/10.1007/BF00046236 (Accessed on 22 August 2019).

116. Sakurai, T., (1991). An electron-microscopic study of syncytium formation during early embryonic development of the freshwater planarian *Bdellocephala brunnea*. *Hydrobiologia., 227*(1) 113–118. https://doi.org/10.1007/BF00027590 (Accessed on 22 August 2019).

117. Cardona, A., Hartenstein, V., & Romero, R., (2005). The embryonic development of the triclad *Schmidtea polychroa*. *Dev. Genes Evol., 215*(3), 109–131. https://doi.org/10.1007/s00427–004–0455–8 (Accessed on 22 August 2019).

118. Harrath, H., Sluys, R., Zghal, F., & Tekaya, S., (2009). First report of adelphophagy in flatworms during the embryonic development of the planarian *Schmidtea mediterranea* (Benazzi, Baguñà, Ballester, Puccinelli & Del Papa, 1975) (Platyhelminthes, Tricladida). *Invert. Reprod. Develop., 53*(3), 117–124. https://doi.org/10.1080/07924259.2009.9652 297 (Accessed on 22 August 2019).

119. Maezawa, T., Sekii, K., Ishikawa, M., Okamoto, H., & Kobayashi, K., (2018). Reproductive strategies in planarians: Insights gained from the bioassay system for sexual induction in asexual Dugesia ryukyuensis worms. In: Kobayashi, K., Kitano, T., Iwao, Y., & Kondo, M., (eds.), *Reproductive and Developmental Strategies. Diversity and Commonality in Animals* (pp. 175–201). Springer: Tokyo. https://doi.org/10.1007/978–4–431–56609–0_9 (Accessed on 22 August 2019).

120. Kobayashi, K., Maezawa, T., Tanaka, H., Onuki, H., Horiguchi, Y., Hirota, H., Ishida, T., Horiike, K., Agata, Y., Aoki, M., Hoshi, M., & Matsumoto, M., (2017). The identification of d-tryptophan as a bioactive substance for postembryonic ovarian development in the planarian *Dugesia ryukyuensis. Sci. Rep., 7*, 45175. https://www.nature.com/articles/srep45175 (Accessed on 22 August 2019).

121. Steiner, J. K., Tasaki, J., & Rouhana, L., (2016). Germline defects caused by *Smedboule* RNA-interference reveal that egg capsule deposition occurs independently of fertilization, ovulation, mating, or the presence of gametes in planarian flatworms. *PLoS Genet., 12*(5), e1006030. https://doi.org/10.1371/journal.pgen.1006030 (Accessed on 22 August 2019).

122. Rouhana, L., Tasaki, J., Saberi, A., & Newmark, P. A., (2017). Genetic dissection of the planarian reproductive system through characterization of *Schmidtea mediterranea* CPEB homologs. *Dev. Biol., 426*(1), 43–55. https://doi.org/10.1016/j.ydbio.2017.04.008 (Accessed on 22 August 2019).

123. Rouhana, L., Shibata, N., Nishimura, O., & Agata, K., (2010). Different requirements for conserved post-transcriptional regulators in planarian regeneration and stem cell maintenance. *Dev. Biol., 341*(2), 429–443. https://doi.org/10.1016/j.ydbio.2010.02.037.

124. Shibata, N., Rouhana, L., & Agata, K., (2010). Cellular and molecular dissection of pluripotent adult somatic stem cells in planarians. *Dev. Growth Differ, 52*(1), 27–41. https://doi.org/10.1111/j.1440–169X.2009.01155.x (Accessed on 22 August 2019).

125. Agata, K., Soejima, Y., Kato, K., Kobayashi, C., Umesono, Y., & Watanabe, K., (1998). Structure of the planarian central nervous system (CNS) revealed by neuronal cell markers. *Zool. Sci., 15*(3), 433–440. https://doi.org/10.2108/zsj.15.433 (Accessed on 22 August 2019).

126. Kobayashi, K., Arioka, S., Hase, S., & Hoshi, M., (2002b). Signification of the sexualizing substance produced by the sexualized planarians. *Zool. Sci., 19*(6), 667–672. https://doi.org/10.2108/zsj.19.667 (Accessed on 22 August 2019).

127. Kobayashi, K., & Hoshi, M., (2011). Sex-inducing effect of a hydrophilic fraction on reproductive switching in the planarian *Dugesia ryukyuensis* (Seriata, Tricladida). *Front. Zool., 8*(23). https://doi.org/10.1186/1742–9994–8–23 (Accessed on 22 August 2019).

128. Alvarez-Presas, M., Baguñà, J., & Riutort, M., (2008). Molecular phylogeny of land and freshwater planarians (Tricladida, Platyhelminthes): from freshwater to land and back. *Mol. Phylogenet. Evol., 47*(2), 555–568. https://doi.org/10.1016/j.ympev.2008.01.032 (Accessed on 22 August 2019).

129. Matsuo, R., Kobayashi, S., Morishita, F., & Ito, E., (2011). Expression of Asn-D-Trp-Phe-NH2 in the brain of the terrestrial slug *Limax valentianus*. *Comp. Biochem. Physiol. B., 160*(2/3), 89–93. https://doi.org/10.1016/j.cbpb.2011.06.007 (Accessed on 22 August 2019).

130. Morishita, F., Nakanishi, Y., Kaku, S., Furukawa, Y., Ohta, S., Hirata, T., Ohtani, M., Fujisawa, Y., Muneoka, Y., & Matsushima, O., (1997). A novel D-amino-acid-containing peptide isolated from *Aplysia* heart. *Biochem. Biophys. Res. Commun., 240*(2), 354–358. https://doi.org/10.1006/bbrc.1997.7659 (Accessed on 22 August 2019).

131. Morishita, F., Sasaki, K., Kanemaru, K., Nakanishi, Y., Matsushima, O., & Furukawa, Y., (2001). NdWFamide: A novel excitatory peptide involved in cardiovascular regulation of *Aplysia*. *Peptides., 22*(2), 183–189. https://doi.org/10.1016/S0196–9781(00)00375–2 (Accessed on 22 August 2019).

132. Morishita, F., Minakata, H., Sasaki, K., Tada, K., Furukawa, Y., Matsushima, O., Mukai, S. T., & Saleuddin, A. S., (2003a). Distribution and function of an Aplysia cardioexcitatory peptide, NdWFamide, in pulmonate snails. *Peptides., 24*(10), 1533–1544. https://doi.org/10.1016/j.peptides.2003.07.017 (Accessed on 22 August 2019).

133. Morishita, F., Nakanishi, Y., Sasaki, K., Kanemaru, K., Furukawa, Y., & Matsushima, O., (2003b). Distribution of the *Aplysia* cardioexcitatory peptide, NdWFamide, in the central and peripheral nervous systems of *Aplysia*. *Cell Tissue Res., 312*(1), 95–111. https://doi.org/10.1007/s00441–003–0707–3 (Accessed on 22 August 2019).

134. Hamase, K., (2015). Recent advances on D-amino acid research. *J. Pharm. Biomed. Anal., 116*, 1.

135. Nishikawa, T., (2005). Metabolism and functional roles of endogenous D-serine in mammalian brains. *Biol. Pharm. Bull., 28*(9), 1561–1565. https://doi.org/10.1248/bpb.28.1561 (Accessed on 22 August 2019).

136. Khoronenkova, S. V., & Tishkov, V. I., (2008). D-amino acid oxidase: Physiological role and applications. *Biochemistry (Mosc)., 73*, 1511–1518. doi: 10.1134/S0006297908130105.

137. Tanaka, H., Yamamoto, A., Ishida, T., & Horiike, K., (2007). Simultaneous measurement of D-serine dehydratase and D-amino acid oxidase activities by the detection of 2-oxo-acid formation with reverse-phase high-performance liquid chromatography. *Anal. Biochem., 362*(1), 83–88. https://doi.org/10.1016/j.ab.2006.12.025 (Accessed on 22 August 2019).

138. Panatier, A., Theodosis, D. T., Mothet, J. P., Touquet, B., Pollegioni, L., Poulain, D. A., & Oliet, S. H., (2006). Glia-derived D-serine controls NMDA receptor activity and synaptic memory. *Cell, 125*(4), 775–784. https://doi.org/10.1016/j.cell.2006.02.051 (Accessed on 22 August 2019).

139. Nagata, Y., (1992). Involvement of D-amino acid oxidase in elimination of D-serine in mouse brain. *Experientia., 48*, 753–755. https://doi.org/10.1007/BF02124295 (Accessed on 22 August 2019).

140. Pollegioni, L., Piubelli, L., Sacchi, S., Pilone, M. S., & Molla, G., (2007). Physiological functions of D-amino acid oxidases: From yeast to humans. *Cell Mol. Life Sci., 64*(11), 1373–1394. https://doi.org/10.1007/s00018–007–6558–4 (Accessed on 22 August 2019).

141. Maezawa, T., Tanaka, H., Nakagawa, H., Ono, M., Aoki, M., Matsumoto, M., Ishida, T., Horiike, K., & Kobayashi, K., (2014). Planarian D-amino acid oxidase is involved in ovarian development during sexual induction. *Mech. Dev., 132*, 69–78. https://doi.org/10.1016/j.mod.2013.12.003 (Accessed on 22 August 2019).

142. Irukayama-Tomobe, Y., Tanaka, H., Yokomizo, T., Hashidate-Yoshida, T., Yanagisawa, M., & Sakurai, T., (2009). Aromatic D-amino acids act as chemoattractant factors for human leukocytes through a G protein-coupled receptor, GPR109B. *Proc. Natl. Acad. Sci. USA., 106*(10), 3930–3934. https://doi.org/10.1073/pnas.0811844106 (Accessed on 22 August 2019).

143. Best, J. B., Abelein, M., Kreutzer, E., & Pigon, P., (1975). Cephalic mechanism for social control of fissioning in planarians III: Central nervous system centers of facilitation and inhibition. *J. Comp. Physiol. Psychol., 89*(8), 923–932.

144. Yoshida, S., Sukeno, M., & Nabeshima, Y., (2007). A vasculature-associated niche for undifferentiated spermatogonia in the mouse testis. *Science, 317*(5845), 1722–1726. doi: 10.1126/science.1144885.

CHAPTER 3

Fine Tuning of Behaviors Through Neuropeptide Signaling in *Caenorhabditis elegans*

WILLIAM G. BENDENA and IAN D. CHIN-SANG

Department of Biology, Queen's University, Kingston, ON, K7L 3N6, Canada

3.1 INTRODUCTION

All animals respond to changes in the environment through alterations of neural circuit signaling that lead to alterations in behavior and metabolism. The complexity of any given neural circuit in mammalian systems undermines the ability to understand how such a circuit is regulated. The roundworm *Caenorhabditis elegans* offers an excellent model organism to analyze neural circuit function as, unlike mammalian systems, the connectome has been determined [1]. *C. elegans* is amenable to genetic analysis as it has a short generation time with each self-fertilizing hermaphrodite producing about 300 progeny. Large-scale production of several million animals is possible in a day. The hermaphrodite has 959 somatic cells that form different organs and tissues including muscle, hypodermis, intestine, reproductive organs and glands and nervous system. The hermaphrodite nervous system consists of 300 neurons whose identity and connections have been defined by morphology and ablation studies. *C. elegans* is also transparent such that fluorescent dyes and proteins can be used to visualize gene expression, calcium dynamics, and fat metabolism *in vivo*. Genetic manipulations, such as RNA interference (RNAi), can also be targeted to individual neurons then studied in the context of behavioral changes in the whole organism. *C. elegans* release small synaptic vesicles at presynaptic sites that carry most of the classical small-molecule neurotransmitters including acetylcholine, γ-aminobutyric acid (GABA), the biogenic amines octopamine, tyramine, dopamine, and serotonin (5-hydroxytryptamine) and glutamate. These function in conjunction with neuropeptides that are produced as precursor proteins that are packed with processing enzymes and stored in large dense-core vesicles in the trans-Golgi network. Release of neuropeptide-containing dense core vesicles is not restricted to nerve terminals and can occur at both axons and dendrites. Recent state-of-the-art liquid chromatography/mass spectrometry analysis identified 203 mature neuropeptides with 131 confirmed at the sequence level [2]. These are classified into three groups, the insulin-like peptides (40insulin precursors), the FMRFamide-like peptides (31 FLP precursors) and the neuropeptide-like peptides (59 NLP precursors) [2]. Several recent reviews summarize the synthesis, processing, and function of *C. elegans* neuropeptides [3–7]

In *C. elegans*, there are over 1100 G-protein coupled receptors (GPCRs), and the few that have been de-orphaned and serve as neuropeptide receptors have been reviewed [8, 9]. In this review standard nomenclature in the *C. elegans* field displays gene names in non-capitalized italicized letters (e.g., *nlp-1*) whereas the corresponding protein is capitalized (e.g., NLP-1). Each neuron is also given a capital letter designation (e.g., AIB). The physical location and connectome of all *C. elegans* neurons (e.g., AIB) can be found at: www.wormatlas.org.

This review highlights some recent advances in identifying neuropeptides that modulate behaviors.

3.2 LOCOMOTION

C. elegans display sinusoidal movements that vary in undulation frequency and amplitude to regulate speed. Animals have a bias for moving in the forward direction, especially when searching for distant food sources. Animals can reorient direction through reversals or sharp turns (omega turns) and create short distance or area restricted searches when food is nearby. The forward versus reverse movement is sensitive to metabolic state and cues from the environment.

What signals are created by the presence of food? Sensory neurons detect food cues and produce dopamine. Dopamine through interaction with its receptor in interneuron DVA activates the secretion of NLP-12 (Figure 3.1). NLP-12 interacts with a cholecystokinin receptor (CKR-2) to potentiate acetylcholine release at neuromuscular junctions. This interaction adjusts locomotory speed and undulation amplitude that results in area restricted search [10, 11]. NLP-12 may work with serotonin, which is secreted by NSM and HSN neurons, to promote restricted area search by inhibiting serotonin-gated chloride channels expressed in neurons AIY, ASI, and RIF [12, 13].

The interneuron AVK opposes the action of DVA by producing FLP-1 that increases locomotory speed, undulation amplitude, and forward locomotion [11]. Octopamine similarly stimulates search behavior [14]. FLP-1 is one of many neuropeptides that appear to enhance forward movement. PDF-1 secretion from neurons AVB, PVP, and SIAV can stimulate forward movement by interaction with its receptor, PDFR-1, in neurons AIY, RIM, and RIA [13, 15] (Figure 3.1). The neuropeptide receptor, NPR-9 in interneuron AIB, functions with glutamate signaling to stimulate forward movement. Overexpression of *npr-9* causes the animal to hyper-roam in the forward direction accompanied with a loss in the ability to sense food. The ligand for NPR-9 is still unknown [16, 17].

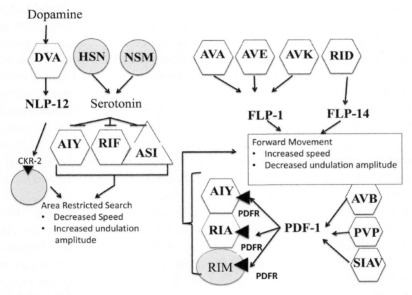

FIGURE 3.1 Area restricted search is outlined on the left-hand side of the figure. Dopamine interacts with its receptor on stretch neuron DVA in response to food cues resulting in the release of NLP-12 that interacts with the cholecystokinin receptor (CKR-2) on motorneurons (grey). Serotonin producing motorneurons HSN and NSM inhibit downstream neurons. Together these signaling pathways promote restricted area search by increasing transmission at cholinergic synapsis. The right-hand side of the figure outlines forward movement that is promoted through the action of at least three neuropeptides: FLP-1, FLP-14, and PDF-1.

The participation of FMRFamide-like peptides in locomotion has been recently reviewed ([5, 7]. The anatomy and wiring of motor neurons were recently reviewed [18]. In adults backward and forward movement are regulated by motorneurons that regulate the contraction of muscle cells in the body wall. There are eight classes of motorneurons; four classes innervating the ventral muscles and four innervating dorsal muscles. Muscle contraction is regulated by the action of dorsal (DA and DB) and ventral (VA, VB, VC, and AS) neurons. These neurons are cholinergic as acetylcholine is secreted as a neurotransmitter. D-type neurons (DD and VD) are GABAergic and inner-vate the dorsal and ventral muscles and regulate muscle relaxation. The DD neuron expresses both *flp-11* and *flp-13*, whereas the VD neurons only express *flp-11* [19]. FLP-11 and FLP-13 are expressed from different neurons during development and stress-induced sleep (see below) and may coordinate activity to muscle contraction. Sensory neurons regulate command interneurons AVA, AVD, and AVE that control VA and DA neurons for backward movement, as well as command interneurons PVC and AVB that regulate VB and DB for forwarding movement. *flp-1* is expressed in command interneurons AVA and AVE (and other head neurons), and peptide secretion modulates acetylcho-line production to promote sinusoidal movement [20]. FLP-1 may work with FLP-18, produced from B-type cholinergic motor neurons, to balance excita-tion-inhibition of muscle contraction and thus regulate the locomotory circuit. FLP-18 interacts with receptors NPR-1, NPR-4, and NPR-5. Muscle-specific expression of NPR-5 suggests that FLP-18 may work directly on muscle to regulate contraction [21]. Neuron RID produces FLP-14 and augments forward locomotion but has not been directly linked to a nutritional state. The receptor for FLP-14 is unknown [22]. RID also expresses *ins-17* and *nlp-34*, but their role in locomotion is not as yet defined [22].

Another form of motility in *C. elegans* is swimming or thrashing. Move-ment in liquid media was once thought to be undirected, hence the clas-sification "thrashing." *C. elegans* uses swimming to efficiently "chemotax" to more favorable environments. Swimming and crawling behaviors are controlled by different muscles; unlike crawling, swimming requires sensory input [23]. A survey of thrashing rate in *flp* mutants indicated that *flp* gene expression in multiple neurons outside of locomotory motor neurons affect thrashing. FLP-1, FLP-4, and FLP-9 stimulate swimming rate whereas FLP-3, FLP-10, FLP-21, and FLP-18 act inhibitory [19].

3.3 EGG LAYING

Egg-laying behavior in *C. elegans* switches between an active state where approximately five eggs are laid over a few minutes and a prolonged quies-cent period of 20 minutes where no eggs are laid. This timing sequence may disperse the eggs so that upon hatch less competition for food occurs among progeny. Egg-laying involves the release of serotonin from two hermaphrodite specific neurons (HSNs) and acetylcholine release from two ventral C motor (VC4 and VC5) neurons (Figure 3.2). HSNs and VCs both synapse with vm2 vulval muscle cells. The release of serotonin from HSNs increases the excitability of the vulval muscles whereas acetylcholine release from VC neurons (VC1-3 and VC6) slows locomotion during the period of vulval muscle contraction [24] (Figure 3.2).

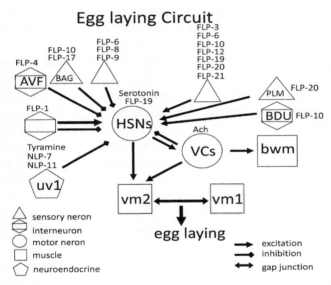

FIGURE 3.2 The *C. elegans* egg-laying circuit. The egg-laying circuit consists of two motor neurons: serotonergic/peptidergic (FLP-19) HSNs and two cholinergic (Ach) VCs (VC4 and VC5) which synapses onto the vulval muscle 2 (vm2) to cause muscle contraction and egg release. The VC neurons receive input from HSN. In addition, the VCs (VC1-3 and VC6) synapse to body wall muscles (BWM) to slow locomotion during egg-laying. The HSN neurons receive input from the VCs, AVF, and BDU interneurons, and mechanosensory PLM neurons. Neuropeptides have modulatory effects on the egg-laying circuit. FLP-1 is expressed in interneurons that are postsynaptic to sensory neurons and therefore functions to regulate the egg-laying circuit in response to sensory cues (e.g., food). The mechanosensory neuron PLM and BDU interneurons synapse directly onto the HSN neurons and express FLP-20 and FLP-10, respectively. The uv1 cells express tyramine as well as NLP-7 and NLP-11 neuropeptides, and inhibit serotonin release from HSN. Loss of function and overexpression studies show that FLPs can have excitatory (FLPs-1, 3, 6, 10, 12, 19, 20, 21) or inhibitory (FLPs-1, 4, 6, 8, 9, 10, 17) functions in the egg-laying circuit. Some neuropeptides (FLPs-1, 6 and 10) have both excitatory and inhibitory effects, suggesting that different peptides processed from each gene can have different roles in the egg-laying circuit or that overexpression of peptides can activate additional circuits regulating egg-laying.

Egg laying is also dependent on sensory inputs such that egg-laying ceases in the absence of food [14, 25]. The HSNs express the FLP-19 precursor that is thought to work in concert with serotonin to facilitate active egg-laying and the downregulation of egg-laying in the absence of food [26]. Both loss and overexpression of FLP-1 peptides decrease egg-laying rates. Different FLP-1 peptides may thus have differential effects on the egg-laying circuit, or high levels of FLP-1 may cause the activation of additional circuits to decrease egg laying [27]. FLP-10 and FLP-17 are expressed from the BAG sensory neurons and mediate the inhibition of HSNs via the GPCR EGL-6 [28]. Using loss of function mutants and transgenic overexpression lines, FLP-3, 10, 12, 19, 20, and 21 were found to promote egg-laying, while FLP-4, 6, 8, and 9 inhibited egg-laying [19].

Although not essential for basal egg-laying activity, four neuroendocrine cells (uv1) mechanically monitor the passage of eggs through the vulva and inhibit egg-laying by secretion of tyramine, which acts via the LGC-55 tyramine gated channel of HSNs [24, 29]. The uv1 cells, in addition to tyramine, produce neuropeptides NLP-7 and FLP-11 that may work with tyramine to switch egg-laying to an inactive state by reducing serotonin signaling from HSN [30]. The overall view of neuromodulation of egg-laying and reproduction is still incomplete as several orphan GPCRs have been implicated in affecting either the rate of egg-laying or the brood size [31]. In addition, the different peptides produced may have opposite effects on the egg-laying circuit as FLP-1, 6 and 10 exhibit both activation and inhibitory effects on egg-laying.

3.4 NEUROMODULATION THROUGH MECHANOSENSORY NEURONS

Neuropeptide expression in mechanosensory neurons plays a role in memory retention. The length of memory retention is related to how training occurs. Training periods with breaks between training sessions produce longer memory retention than training that is given in a mass block without breaks. *C. elegans* responds by reversing when the petri dish that they are growing in is tapped. Continuous tapping results in habituation with the size of reversals becoming smaller. Memory retention, 12 but not 24 hours after mass training, is dependent on *flp-20* expression, which increases the number of presynaptic vesicles, specifically in mechanosensory neurons [32]. This is a distinct mechanism from long term memory that results from spaced training that relies on the transcription factor cAMP response element-binding protein

(CREB) and glutamate signaling. FLP-20 expression in the mechanosensory circuit also modulates other behavioral responses.

An example of cross-modal plasticity is the ability to respond to the loss of one sense by strengthening the acuity of other senses. For example, in humans hearing and/or tactile sensation enhancement has been noted in blind individuals. In a *C. elegans* model, a mutation *(mec-4)* that lacks a DEG/ENaC channel subunit required in touch receptor neurons for sensing gentle touch, was found to have elevated the ability to sense odors. This enhancement of the chemosensory/olfactory circuit was due to reduced mechanosensory expression of the neuropeptide FLP-20 that normally suppresses the olfactory circuit. Three peptides are produced from processing of the FLP-20 precursor, AMMRFamide, AVFRMamide, and SVFRLamide [33]. FLP-20 acting on an unknown receptor then strengthens the glutamatergic synaptic connection between a chemosensory (odor-sensing) neuron (AWC) and interneuron AIY resulting in a long-lasting enhancement of the olfactory circuit [34].

3.5 BEHAVIORAL CHOICE

C. elegans in the wild forage through a mixture of compostable plant and fruit material that contains bacteria and yeast and thus is faced with decisions or behavioral choices to select optimal food sources [35]. Neuronal circuits that regulate choice depend on changes in the activities between coupled sensory neurons and interneurons and/or the strength of synaptic connections [36]. The type and concentration of neuropeptides combined with classical neurotransmitters provide the flexibility required to create decisions and alter them based on environmental influences.

The recognition of the food odor and stimulation of turning during the local search is due to glutamate signaling from neuron AWC [37, 38] Multiple glutamate receptor subunits, presumably in interneurons, are necessary for food odor recognition. AWC co-transmits glutamate with a buccalin-related neuropeptide, NLP-1, that reduces local search and facilitates adaptation to the odor. NLP-1 is the ligand for the neuropeptide F-like receptor NPR-11 [38]. Specific NLP-1:NPR-11 interaction in interneuron AIA results in the release of insulin-1 (INS-1) from AIA, which feedbacks directly or indirectly to suppress AWC activity which dampens behavioral responses to odors. Surprisingly, a mutation in the only known insulin receptor *(daf-2)* did not show the same locomotory phenotype as an *ins-1* mutant, suggesting that another insulin receptor may exist. The timing of the release of NLP-1 and INS-1 with respect to odor preferences is unclear. Glutamate released from AWC acts to inhibit AIA through a glutamate-gated chloride channel. The NLP-1/INS-1 feedback loop regulates the response to one type of food-derived cue rather than regulating food preference. Innate food odor recognition and preference have been used as a model for decision-making. When challenged to choose between the standard *Escherichia coli* OP50 laboratory strain and a pathogenic bacteria *Pseudomonas aeruginosa* PA14, a non-pathogen exposed *C. elegans* exhibits an olfactory preference for *P. aeruginosa,* but after animals have ingested *P. aeruginosa* they reduce their olfactory preference for this pathogen [37, 39]. Three of the 12 amphid sensory neurons, AWB, AWC, and ADL are required for this recognition [37, 38, 40]. In neurons, AWB and AWC recognition are accomplished by G-protein signaling via Gα-protein (ODR-3), guanylate cyclases (ODR-1 and DAF-11) and cGMP gated channels (TAX2/TAX4) [37]. Sensory neuron AWB produces the neuropeptide NLP-9 that interacts with a somatostatin-like receptor NPR-18 that is also required for food odor preference [37]. The choice for *P. aeruginosa* also includes the activity of the ADL sensory neuron. Inhibition of a GPCR, SRH-220, in ADL results in the release of FLP-4 from ADL, which interacts with receptor NPR-4 in interneuron AIB [40]. This ligand: receptor interaction activates a signaling pathway that may lead to histone H3K5 trimethylation, presumably which alters transcriptional activities [40]. How the NLP-9:NPR-18 interaction integrates with NLP-1:NPR-11 and/or FLP-4:NPR-4 interactions is unclear.

   D. melanogaster is attracted to ripening/fermenting fruit that produces a high CO_2 environment. However, *Drosophila* has specialized antennal neurons and a specific olfactory neuronal circuit that detects and creates avoidance to CO_2 The choice to feed is aided by the recognition that select odorants from the food source reduce avoidance behavior [41]. In contrast, mosquitoes are normally attracted to CO_2, and prolonged exposure to odorants can reduce this response, which confuses the choice to feed [42]. The inhibition of CO_2 sensing neurons in *C. elegans* can either result in attraction or repulsion to CO_2 levels dependent on whether the animal was previously exposed to ambient or high CO_2 levels. Animals raised in ambient CO_2 are repulsed by high CO_2, whereas those raised in high CO_2 are attracted. BAG sensory neurons mediate CO_2 avoidance and attraction through a combination of neuropeptides and glutamate. BAG neurons produce FLP-17 that together with glutamate mediates CO_2 attraction [43]. In animals raised in ambient CO_2, NLP-1 reduces CO_2 avoidance, whereas FLP-27 enhances the response. It is unclear if NLP-1 is acting on the same receptor (NPR-11) in AIA as was found in response to odors. In animals raised at high CO_2,

FLP-16 reduces CO_2 attraction, and FLP-27 again enhances the response [43]. The identity and neuronal location of neuropeptide receptors mediating these responses have not been determined. In the absence of a CO_2 sensing pathway expression of *flp-19* is reduced based on monitoring of a *flp-19* transcriptional reporter [44]. Thus multiple neuropeptides appear to function in this regulatory circuit, which encompasses choice with experience.

Generally, food leaving behavior is noted when overcrowding occurs, or food becomes limiting. A novel observation identified that mature hermaphrodites leave the food when there is an abundance of progeny. The larvae produce an ascaroside pheromone that promotes signaling of an oxytocin-like neuropeptide, nematocin, in the hermaphrodite that drives the animal to leave the food. This has been suggested to be a mechanism to ensure the new generation has ample food supply [45].

3.6 THE NEUROPEPTIDE RECEPTOR NPR-1 IS A HUB FOR NUMEROUS BEHAVIORS

Neuropeptide Y (NPY) is a 36 amino acid peptide. In vertebrates, NPY functions through multiple receptor subtypes, and dysfunctional regulation can lead to stress and anxiety disorders, drug/alcohol use disorders. NPY levels are associated with suppression of responsiveness to adverse stimuli and in promotion of food search and acquisition under adverse conditions [46, 47]. Two natural alleles of the *C. elegans* short NPF-like receptor [48], NPR-1, differing by a single amino acid at position 215 exhibit "solitary" versus "social" behavioral phenotypes. The N2 Bristol strain, used in most labs as a wild type strain, expresses *npr-1* 215Valine (*npr-1* 215V). *npr-1* 215V animals have reduced locomotory activity and disperse over the bacterial lawn as solitary individuals. The Hawaiian isolate CB4856 express *npr-1* 215 Phenylalanine (*npr-1* 215F). These animals display a "bordering" phenotype as they move rapidly on bacterial food to areas where the food is thickest, and burrow. The *npr-1* 215F animals form social clumps of intertwined animals [49]. Like *npr-1 215F, npr-1* loss-of-function *(lf)* mutants display social behavior and elevated roaming speeds at all stages of development [50]. On food, aggregation, and lawn bordering activity is used by social animals to avoid and escape from atmospheric O_2 level. A hypothesis is that worms feeding on compostable material perceive the 21% surface O_2 level as a threat and prefer the interior where the O_2 level is lower [51]. NPR-1 is expressed in O_2 –sensing neurons AQR, PQR, and URX as well as post-synaptic interneurons RMG and URX ([52, 53]. At

21% O_2, O_2-sensing neuron URX activates RMG to release neuropeptides that in turn act to promote enhanced locomotion in search of lower O_2 levels. NPR-1 215V signals to block neuropeptide release from RMG [51]. NPR-1 suppresses aggregation and bordering by inhibiting the expression/activity of two O_2-binding soluble guanylate cyclases, GCY-35, and GCY-36. 21% O_2 activates GCY-35 and GCY-36 in O_2 sensing neurons that in turn activate a cGMP-gated ion channel (TAX-2/TAX-4) and L-type Ca^{2+} channels [54–57].

FLP-21 and FLP-18 are ligands for NPR-1 [58, 59]. Deletion of *flp-21* has limited effect on increasing aggregation and bordering in *npr-1* 215V animals but increases locomotor activity. In the absence of *flp-21* expression, *npr-1* 215 Faggregation is enhanced, and the amplitude of O_2-evoked behavioral switch is increased [51, 59]. Several neuropeptides, including *flp-5, 11, 17, 21*, and *pdf-1* are regulated at different percentages of O_2 in different oxygen-sensing neurons [51].

When animals are acclimated to 21% O_2 on food, NPR-1 215V promotes avoidance of high levels of CO_2, whereas the NPR-1 215F-bearing animal only exhibits a weak avoidance to CO_2 [60]. Indeed, an increase in CO_2 leads to a burst of turning in wild type worms; however, the *npr-1 215F* strain does not respond. NPR-1 inhibits the O_2 sensing neuron URX to control CO_2 avoidance [52]. The activity of NPR-1, therefore, serves to integrate inputs from O_2 and CO_2 sensing pathways and generate an appropriate response with respect to the availability of food [61–63].

NPR-1 participates in behavioral adaptation/acute tolerance to ethanol exposure. NPR-1 regulation of tolerance appears to follow a pathway that is distinct from that of social behavior [64]. For example, ethanol consumption in wild type *C. elegans* leads to intoxication that creates a hyperexcitation phenotype followed by inhibition of locomotor activity and egg-laying. NPR-1 215F animals exhibit acute ethanol tolerance leading to decreased intoxication and faster recovery times. NPR-1 is negatively regulating acute tolerance. Expression of *npr-1* in neurons AQR, PQR, and URX can partially rescue social behavior but does not change the response to ethanol [64]. Yet activation of these three O_2-sensing neurons appears to be required for acute ethanol tolerance in *npr-1* worms [65]. Regulation of social behaviors requires the ligand FLP-21. FLP-21 does not appear to act in acute ethanol tolerance, suggesting that FLP-18 or another neuropeptide may serve as the ligand [64]. Neuropeptides appear to function as important mediators of neuroplasticity that regulates alcohol tolerance. Animals that are chronically exposed to ethanol, then subjected to ethanol removal exhibit withdrawal symptoms. The activity of a calcium-activated potassium channel, SLO-1, modulates ethanol tolerance, and withdrawal [66, 67]. Neuropeptides appear to function in ethanol withdrawal as a mutant, *egl-3,* defective in processing neuropeptide precursors showed altered behavior, but *npr-1* did not [68]. NPR-1 independent pathways appear to function in differing aspects of ethanol-induced behaviors.

NPR-1 also participates in regulating behavior in response to food availability and nutritional state. *npr-1* and *npr-2* expression in ASH sensory neurons promote adaptation to chemical repellents such as copper, glycerol, and low pH, but only in the absence of food [69]. FLP-18 and FLP-21 activate NPR-1 [59] and FLP-21 activates NPR-2 in mammalian expression assays [69]. *flp-18* or *flp-21* mutant analysis indicated that neither FLP-18 or FLP-21 peptides individually are interacting with NPR-1 or NPR-2 nor participating in ASH adaptation phenotypes. This would suggest that either an unknown neuropeptide is being expressed during times of food deprivation or possibly both peptides are necessary for an adaptation response. Dopamine works in opposition to NPR-1 and NPR-2, decreasing adaptation when animals are on food [70]. Dopamine binds to the DOP-1 receptor in AUA, which then down-regulates neuropeptide release to decrease ASH adaptation on food. Thus food can alter a behavioral state through the interaction of monoamine and neuropeptide modulators [69]. ASH also mediates the aversive response to octanol that requires neuropeptide expression. Octopamine modulates this aversive response through octopamine receptors OCTR-1 and SER-3 that act antagonistically in ASH. A third octopamine receptor, SER-6, inhibits aversive responses by regulating the release of *nlp-6,* 7, 8, 9 from sensory neurons. Based on the (over) expression of these peptides in receptor null mutant background and measuring reduced or aversive response loss, NPR-15 (functioning in AWC neurons) was identified as the receptor for NLP-8 peptides and NPR-19 (in neuron ASER) as the receptor for NLP-9 peptides. Octopamine inhibits ASH signaling directly but regulates the release of neuropeptides from sensory neurons that modulate the ASH aversive signals [71].

3.7 ANTI-MICROBIAL PEPTIDES AND NEURAL DEGENERATION

As *C. elegans* forages through a natural habitat, it encounters numerous diverse microorganisms [72]. Two strategies are used when pathogenic microorganisms are encountered; avoidance and/or launching an innate immune response. These two responses may work together and share signaling pathways. Typically, microbes attach to the intestine after ingestion or enter externally through the epidermis. Nematodes produce several types of anti-microbial peptides/proteins, including lysozymes, anti-bacterial factors (ABFs), Caenopores (saponin-like proteins), caenacins (CNCs), and NLPs. The type of infecting organism (specific Gram-positive/Gram-negative bacteria or fungus) and the route of infection (epidermis versus intestine) dictates which defense genes are upregulated (see [73, 74] for recent reviews).

There are six *nlp* genes (*nlp-27–31* and *nlp-34*) that are clustered on chromosome V, termed the *nlp-29* cluster. Translation and proteolytic processing of each precursor would release 25–27 mature peptides that have C-terminal sequences YGGYamide or YGGWamide [33]. CNCs are structurally related to NLPs with both being rich in glycine and aromatic amino acids. CNCs also form a cluster on chromosome V (*cnc-1–5 and cnc-11*), termed the *cnc-2* cluster. Not all of the peptides expressed from these clusters has been confirmed as a required component of the immune response. The obligate fungal endoparasite *Drechmeriaconiospora* infects *C. elegans* by penetrating through the cuticle and epidermis. This infective agent induces the expression of genes of the *nlp-29* cluster and *cnc-2* cluster with the exception of *cnc-3*. Several genes, *nlp-29, 30, 31* and *cnc-2* are expressed in the epidermis upon infection [75, 76]. In addition to fungal infection, *nlp-29* is induced by wounding of the epidermis and osmotic shock. However, the highest degree of induction occurs after infection [77]. Both infection and wounding appear to induce signals that activate a p38 MAP kinase signaling pathway, which controls the expression of the *nlp-29* cluster in the epidermis [78]. CNC induction appears to be regulated by a separate pathway where a neuronally derived TGF-β homolog, DBL-1 activates a signal transduction pathway in the epidermis [76].

Neurosecretion of *nlp-29* from the epidermis, rather than upregulation noted to combat infection, interacts with a receptor NPR-12 in neurons that leads to dendrite degeneration via the autophagy pathway [79]. Fungal infection results in dendrite degeneration that appears similar to the degeneration that occurs during aging.

In Alzheimer's disease, amyloid β peptides have been suggested to be anti-microbial peptides, which suggests that peptide accumulation may be associated with neurodegeneration during aging [80]. The expression in *C. elegans* motor neurons of mutant DNA/RNA binding proteins responsible for amyotrophic lateral sclerosis promoted the release of unknown secreted molecules that activated the expression of the innate immune response and *nlp-29* expression in intestinal and hypodermal cells. The same pathway activated in motor neurons led to degeneration, which supports the activity of signaling pathways possibly linking microbial defense to degeneration [81].

Neuropeptides FLP-18 and FLP-21 are ligands for a NPR-1. *npr-1 (lf)* mutantsor animals with a low-activity isoform of NPR-1 (NPR215F) have increased susceptibility to infection by pathogenic bacteria and exhibit oxygen avoidance behavior that results in animals aggregating into clumps and preferring the thickest border (most hypoxic) area of the food. This aerotaxis behavior is suppressed by a mutation in a guanylate cyclase (*gcy-35*) that functions as a molecular oxygen sensor [54]. Wildtype (N2) animals do not display oxygen avoidance when grown on standard *E. coli strains* (OP50) but do so when grown on non-pathogenic mucoid strains of *E. coli. C. elegans* that has previously ingested *P. aeruginosa* learns to avoid it thereafter. However, the pathogen avoidance behavior is lost when mucoid strains of *P. aeruginosa* are encountered. Components of the mucoid exopolysaccharide matrix impair the NPR-1–dependent regulation of oxygen sensation inhibiting oxygen and pathogen avoidance [82].

3.8 NEUROPEPTIDE REGULATION OF MALE BEHAVIOR

C. elegans males have two competing biological drives, feeding, and repro-duction. Unlike self-fertilizing hermaphrodites that can produce self-progeny or mate as females, males must mate in order to reproduce. Well-fed males, in the absence of a hermaphrodite mate, exhibit exploratory behavior and will leave a food source. The drive to explore is suppressed when a hermaphrodite is present. If the males are starved, the exploratory or mate-searching behavior is suppressed [83]. Hermaphrodites will leave a food source under conditions of food depletion or high animal density. The neuropeptide receptor, *npr-1,* dramatically increases the food leaving behavior of hermaphrodites but not of males. The pigment dispersing factor receptor (PDFR-1) and its ligand PDF-1act in the non-sex specific chemosensory neuron ASJ to stimulate the expression of *daf-7,* a TGFβ neuroendocrine modulator in males but not that of hermaphrodites [84, 85]. The dimorphic sexual behavior is due to the expression of *pdfr-1* in the ASJ neuron in males but not in the ASJ neuron of the hermaphrodite. Thus, PDF-1:PDFR-1 interaction drives *daf-7* expression only in males to stimulate mate-searching behavior [85]. Whether all three splice variants of *pdfr-1* are involved was not addressed [86]. The ligand *pdf-2* is not involved in this behavior [84, 87]. When a male makes contact with a hermaphrodite mate searching behavior is suppressed by the activity of male ray neurons and mate-sensation that inhibit *pdf-1/pdfr* signaling [84, 88]. In the hermaphrodite, PDF signaling has other roles, such as promoting normal roaming behavior [13]. Male search behavior was also impaired in nematocin and nematocin receptor (*ntr-1*) mutants. Nematocin is related to mammalian vasopressin and oxytocin peptides that are implicated in reproductive behaviors [89]. Wild type males normally contact and select a hermaphrodite by tail touch. Once selected, the vulva is located by several sliding turns around her body. Sperm is then transferred. Tail touch recognition of a hermaphrodite by a nematocin mutant required several attempts, and the time for locating and maintaining contact with the vulva was extended. Nematocin receptors are located in male-specific neurons and muscles of the copulatory organ where the loss of NTR-1 results in hermaphrodite contact defects, whereas loss of NTR-2 causes loss of reproductive efficiency [90]. A mutational assay that monitored male turning behavior during mating demonstrated that glutamate in combination with FMRF-amide-like peptides FLP-8, FLP-10, FLP-12, and FLP-20 co-regulate the behavior [91]. These three neuropeptide systems have been examined in isolation and how they function together to create the male mating ritual will be of interest.

3.9   NEUROPEPTIDES REGULATING GROWTH

Thyrotropin-releasing hormone (TRH) produced by the mammalian hypothalamus is a prime regulator of the hypothalamus-pituitary-thyroid axis, essential for the regulation of growth and metabolism. Although thought to be restricted to deuterostomes, an ancestral TRH receptor system was identified in *C. elegans* [92]. Phylogenetic analysis recognized that an orphan GPCR named NMUR-4 (C30F12.6), based on similarity to a mammalian neuromedin receptor is, in fact, orthologous to the vertebrate thyrotropin-releasing hormone receptor (TRHR) family. TRHR-1 was localized to pharyngeal muscles. Two ELFamides (C30H6.10) produced upon precursor processing were found to activate TRHR-1 *in vitro* using a mammalian expression assay. ELFamides (*trh-1*) are expressed in motor neurons M4 and M5 that innervate the pharynx. Despite the localization, manipulation of *trh-1* showed no defects in food intake, pharyngeal pumping, and locomotion or reproductive timing/maturation. Instead, the *trh-1*/TRHR signaling pathway altered growth and reduced offspring production by 17%.

3.10   NEURAL REGULATION OF QUIESCENCE

There is an increasing need to understand how sleep is regulated as numerous sleep-disorders impact negatively on human health. The signaling pathways regulating sleep-like states in all species are poorly understood but appear to be conserved in evolution. The complexity of neural and molecular components governing sleep in mammalian systems has led to researchers to look for parallels in non-mammalian genetically tractable species such as zebrafish (*Danio rerio*), fruit flies (*Drosophila melanogaster*) and the nematode worm (*Caenorhabditis elegans*). These three species have been identified as having a sleep-like state by behavioral rather than electrophysiological criteria. Behavioral criteria that define sleep include: a) sleep is reversible as sleeping animals will awaken in response to a stimulus: b) sleep exhibits an elevated arousal threshold in that there is no or limited response to a stimulus when asleep that they would normally respond to when awake: c) sleep exhibits homeostasis whereby animals deprived of sleep will show a tendency to fall asleep and will have a longer recovery sleep period: d) sleep is associated with the circadian clock or clock genes [93]. In this latter point, normally the timing of sleep is coupled to a circadian clock mechanism, but mice and *Drosophila* with mutated clock genes still sleep. Behavioral quiescence in *C. elegans* results in animals stopping their normal foraging, locomotory behavior, feeding, and defecation. There are five different conditions whereby *C. elegans* enters a sleep-like state (Figure 3.3). The first is known as 'lethargus' or developmentally timed sleep (DTS) that is a 2–3-hour quiescent period that occurs between each of the four larval molts. This period is when ecdysis occurs with the shedding of cuticle. During lethargus, animals stop moving and feeding [93]. Second, animals can enter a sleep-like quiescence after a period of stress. This is referred to as sickness sleep, or stress-induced sleep (SIS) [94]. Third, a sleep-like/quiescent state can be observed when animals are swimming in liquid media [95]. Fourth, like mammals, *C. elegans* enter a sleep-like state associated with satiation after a meal [96]. Finally, animals under harsh environmental conditions or starvation can enter and remain in a quiescent dauer state that is an alternative third larval stage ([94, 97] for reviews).

3.11 DEVELOPMENTALLY TIMED SLEEP (DTS)

The 'period' gene product PER was identified in *D. melanogaster* as a critical regulator of circadian rhythm [98]. In *C. elegans*, the PER ortholog is encoded by the *lin-42* gene. Inactivation of *lin-42* creates arrhythmic molts and continuous abnormal epidermal stem cell dynamics. Overexpression of *lin-42* produced anachronistic larval molts and a lethargus state in adults [99]. Oscillations of *lin-42* mRNA are synchronized to the molting cycles

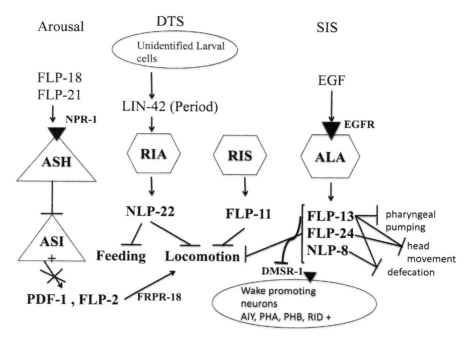

FIGURE 3.3   Developmentally Timed Sleep (DTS) between each larval molt is timed by the production of the Period protein LIN-42, which activates interneuron RIA to produce the neuropeptide NLP-22. Similarly, the neuropeptide FLP-11 is released from the interneuron RIS contributing to the DTS pathway. Stress-Induced Sleep (SIS) is initiated via the release of epidermal growth factor (EGF) which acts on the ALA interneuron that in turn releases at least three peptides, i.e., FLP-13, FLP-24, and NLP-8. The receptor for FLP-13 has been identified as a dromyosuppressin receptor (DMSR-1). Each of the neuropeptides has a negative effect on phenotypes associated with wakefulness. During periods of quiescence, a pathway that promotes arousal or wakefulness is repressed. During quiescence, neuropeptides FLP-18 and FLP-21 interact with receptor NPR-1 in sensory ASH neurons amongst others. This interaction creates an inhibitory signal in ASI (and potentially other neurons) to prevent the production of PDF-1 and FLP-2. When FLP-2 is released, it interacts with its receptor FRPR-18 to promote arousal. FRPR-18 is found in ASI, AIY, DVA, and RIM. Activation of FRPR-18 by FLP-2 enhances the production of PDF-1, which in turn enhances the production of FLP-2.

and lethargus cycles [100]. Unidentified larval cells produce LIN-42, which directly or indirectly acts on paired interneurons RIA to upregulate the expression of *nlp-22* [101]. The expression of *nlp-22* cycles in phase with *lin-42* with peak expression in the middle of active larval stages and low expression during periods of lethargus. Adults do not express *nlp-22* mRNA [101]. Overexpression of *nlp-22* induced feeding and locomotory quiescence in the adult stage. The overexpression of *nlp-22* reduced responsiveness to chemical or optic stimulation that supports the induction of a sleep state. The NLP-22 sequence, SIAIGRAGFRP-NH$_2$, is structurally similar to human Neuromedin S [101]. NLP-22 signaling decreases Protein Kinase A activity in an unknown cell type to promote quiescence or inhibit wakefulness. Expression of *nlp-22* in neuron RIA acts directly on the pharyngeal muscles to inhibit feeding although other regulators may operate since feeding quiescence is not abolished in *nlp-22* mutants. Locomotion may be inhibited through cholinergic neurons [102]. The receptor for *nlp-22* is still unknown.

During sleep, 75% of neurons become inactive relative to periods of wakefulness. Specific head neurons including RIS remain active during quiescent periods [103]. RIS expresses Υ-aminobutyric acid (GABA) along with neuropeptides that contribute to the DTS pathway [104]. RIS expresses a transcription factor LIM6 that regulates both GABA expression and the expression of a transcription factor of the AP2 family, named APTF-1. Loss of function hemizygous mutation of a human AP2 family member TFAP$_2$β results in an insomnia sleepwalking state known as Char Syndrome [105]. An ortholog of this same protein in *Drosophila* adults TfAP-2 abolished night sleep but did not affect day sleep [106]. APTF-1 regulates the expression of the neuropeptide FLP-11 during periods of sleep and wakefulness. The *flp-11* gene specifies a precursor protein that is processed into 4 peptides pQPKARSGYIRF-NH$_2$, AMRNALVRF-NH$_2$, ASGGMRNALVRF-NH$_2$, and NGAPQPFVRF-NH$_2$ [33]. At the onset of sleep, RIS depolarizes and releases FLP-11 that then induce sleep [107]. Three GPCRs, FRPR3, NPR-4, and NPR-22 are known to be activated by FLP-11 peptides *in vitro* [8]. Deletion of these GPCRs individually had only slight effects in reducing quiescence while the maximal reduction was achieved by deleting all three receptors. The alterations in phenotypes displayed by the GPCR deletions suggest that other unknown receptors may be necessary to exert a full FLP-11 phenotype [107].

After lethargus or the quiescent period, signals promoting an awake state become expressed presumably regulated by cell-intrinsic developmental clock genes. This results in arousal that is defined as a state of heightened responsiveness to sensory cues and elevated motor activities [108, 109]. These elevated motor activities are repressed by the action of neuropeptides FLP-18 and FLP-21 interacting with receptor NPR-1 [110]. NPR-1 inhibits the RMG central sensory circuit [53, 110]. Deletion of *npr-1* abolishes locomotion quiescence creating arousal during lethargus, but deletion of *flp-18* and *flp-21* only reduces this phenotype [110]. This might suggest that additional neuropeptides are involved in regulating NPR-1. Sensory neurons of the RMG circuit arouse locomotion through the expression and secretion of glutamate and neuropeptides termed pigment dispersing factor, PDF-1 [110, 111] and FLP-2 [109]. NPR-1 inhibits PDF-1 and FLP-2 secretion during lethargus thereby maintaining locomotion quiescence. PDF-1, FLP-2and the FLP-2 receptor FRPR18 is coexpressed in ASI sensory neurons where they jointly promote arousal through a reciprocal positive feedback. In this model PDF-1 increases FLP-2 secretion and in turn, FLP-2 interacts with FRPR-18 to stimulate PDF-1 secretion [109]. PDF-1 interacts with the PDF receptor located in mechanosensory neurons and body wall muscles (BWM) [111]. FRPR-18 is an ortholog of the hypocretin/orexin receptor that has been associated with narcolepsy in mammals [112].

3.12   STRESS-INDUCED SLEEP (SIS)

SIS has also been referred to as 'sickness sleep' as acute illness produces cellular stress. Under stress conditions, animals stop moving and eating, with no or limited response to small stimuli but a normal response to more severe stimuli [113]. Numerous conditions can elicit this response in *C. elegans*. These include high or low-temperature shocks, bacterial pathogens, bacterial pore-forming toxins, ethanol shock, UV light, and osmotic shock [94, 113]. Epidermal growth factor (EGF) signaling is necessary to induce SIS [113]. In *C. elegans,* the EGF ligand (LIN-3) interacts with the EGF receptor (LET-23) expressed in interneuron ALA. EGF interaction with its receptor depolarizes ALA resulting in the secretion of neuropeptideFLP-13. The FLP-13 precursor produces seven peptides upon processing. FLP-13 induces feeding, defection, and locomotion quiescence through interaction with a DroMyoSuppressin-like receptor DMSR-1which is localized diffusely to specific neuronal cell membranes, not synapses [114]. FLP-13 binds to DMSR-1which then inhibits wake-promoting neurons. *dmsr-1* is expressed in numerous neurons and produces large defects in body movement with no effect on feeding quiescence when mutated. Animals with a mutation in *flp-13* and *dmsr*-1 show greater defects in quiescence than animals with either mutation alone. This result may imply that DMSR-1 is activated by other neuropeptides or that FLP-13 activates other receptors [114]. In *C. elegans*, there are 15 additional DMSR-like receptors [8]. In contrast to DTS, FLP-13 acts upstream of pharyngeal motor neurons to induce feeding quiescence [102]. FLP-13 interacts with a second receptor, FRPR4, but a mutation in this receptor did not inhibit FLP-13 induction of SIS [115]. Deep-sequencing of ALA transcripts revealed the production of 23 *flps* and 25 *nlps* [116]. *flp-24, nlp-8,* and *nlp-7* were the most highly expressed and further tested for roles in SIS. Overexpression of *nlp-8* inhibited defecation and locomotion, *flp-24* overexpression inhibited locomotion and *flp-13* overexpression inhibited locomotion, defecation, and pharyngeal pumping [116]. These results suggest a model where multiple neuromodulators work together to control various phenotypes associated with the SIS response.

3.13   SLEEP IN RESPONSE TO FEEDING STATE

Satiety occurs in all organisms and is another form of quiescence whereby a sleep-like/drowsy state is induced after feeding. When starved *C. elegans* are refed, the pharynx's pumping rate, locomotion, and egg-laying are accelerated. Pharynx pumping and locomotion are suppressed as satiety sets in to prevent excessive feeding [96]. Satiety is promoted by the action of the insulin receptor (DAF-2) and TGFβ (DAF-7). As in mammals where the neuropeptide cholecystokinin is involved in the satiety response [117], satiety quiescence in worms requires peptidergic signaling rather than chemical transmitters such as dopamine, acetylcholine or serotonin [96]. Two neuropeptides termed Luqins, based on their original isolation from left upper quadrant (LUQ) neurons in *Aplysia*, were identified in *C. elegans* as regulators of the satiety response. These luqins PALLSRY-NH$_2$ and AVLPRY-NH$_2$, termed LURY1-1 and LURY1-2, are RYamides [118]. Upon overexpression of luqin peptides pharyngeal pumping and roaming decrease. Feeding suppression was linked to an increase in the adult life span. NPR-22 was identified as the receptor for LURY peptides. An *npr-22* mutant showed a similar phenotype to a *lury-1* mutant whereby pharyngeal pumping was elevated after refeeding, but slower satiety suppression of pumping was found compared to wild type [118]. This peptide family may work in concert with insulin signaling.

Starvation itself is a stress that can also lead to a sleep-like state. In starvation-induced sleep, brain dynamics are downregulated in association with energy deprivation. Starvation-induced sleep utilizes the insulin pathway and requires *daf-16,* known as FOXO in mammalian systems. FOXO/DAF-16 is a transcription factor whose entry into the nucleus is dependent on the state of insulin signaling. Elevated *daf-16* or low *daf-2* (Insulin receptor) signaling is required for elevated sleep upon starvation [119, 120].

3.14   THE STRESS-INDUCED DAUER STATE

Under harsh environmental conditions such as starvation, elevated temperature, or crowded conditions, first or second stage larvae (L1 or L2) can enter into a non-reproductive developmental arrest known as dauer [121]. Dauer animals have a specialized cuticle, and feeding is arrested. Using whole animal RNA sequencing comparing dauer and reproductive animals, 8042 genes were differentially regulated [122]. During the dauer commitment period, GPCR gene expression is elevated. One elevated transcript was that of the GPCR, NPR-11, which interacts with FLP peptides processed from precursors produced by five genes: *flp-1, 5, 14, 18,* and *21* [5]. Enhanced expression was found in putative neuropeptide GCPRs, including *ckr-1, frpr-7, -19, npr-17, npr-31,* C01F1.4, F13H6.5, Y37E11AL.1, and Y70D2A.1. During dauer commitment, the expression of 31 *flp* genes was coordinately upregulated. The most significant increases were in 28 of the 31 *flp* genes, *flp-1, 2, 4–9, 11–22, 24–28,* and *32–34*. Upregulation was found for seven of the 40 insulin genes *ins-1, 17–18, 24, 28,* and *30* and *daf-28*. This agrees with *ins-1* and *ins-18* overexpression, enhancing dauer arrest by inhibiting the DAF-2 insulin receptor [123]. Select insulin peptides such as INS-7 and INS-35 are degraded within the lumen during dauer so not all insulins would be expected to increase [124]. INS-4 and INS-6 were not identified in this screen but are known to act functionally redundantly with DAF-28 to suppress dauer formation [125]. 25 of the 47 *nlp* genes were found to be upregulated (*nlp-1–3, 6, 8–15, 17–18, 21, 35, 37, 38, 40–42,* and *47, ntc-1, pdf-1,* and *snet-1* [122]. The mutational analysis suggested that FLP-2, 6, 18, and 34 expressions opposed the effects of dauer entry whereas expression of FLP-11, 17, 21, 25, and 26 increased dauer entry [122]. Dauer animals perform nictation, which is a behavior where *C. elegans* stand on their tails and wave their heads with the purpose of facilitating dispersal [126]. Downregulation of insulin signaling during dauer enhances nictation [127] as does FLP-10 and 17 co-expression in CO_2 sensing BAG neurons [5]. FLP-10 and 17 act synergistically with acetylcholine to promote nictation behavior. This co-expression also switches non-dauer repulsive behavior to CO_2 to dauer CO_2 attraction [122].

3.15   SUMMARY

Genomic analysis suggests that over a hundred GPCRs exist that have the potential to interact with neuropeptide modulators [8]. Only a small number of neuropeptides have been matched to their cognate receptors. Information on the neuronal location of these GPCRs is also limited. Many GPCRs are expressed in numerous neurons, and many have been shown to regulate differing functions. Neuropeptides themselves appear to be promiscuous with many interacting with numerous GPCR types. Each neuropeptide precursor has the potential to be processed into numerous similar but different peptides. Whether all peptides within any given precursor perform the same function is still unclear. *C. elegans* offers a model whereby how modulators function in neural circuits can be uncovered using both genetic and molecular tools. Several of the behavioral analysis cited in this review demonstrates the close connection between multiple neuropeptide signals combined with classical chemical neurotransmitters. The use of *C. elegans* to uncover the interaction of several modulators will be limited only by the availability of measurable phenotypes.

KEYWORDS

- **anti-bacterial factors**
- **cAMP response element-binding protein**
- **cholecystokinin receptor**
- **G-protein coupled receptors**
- **hermaphrodite specific neurons**
- **γ-aminobutyric acid**

REFERENCES

1. White, J. G., Southgate, E., Thomson, J. N., & Brenner, S., (1986). The structure of the nervous system of the nematode *Caenorhabditis elegans*. Philosophical transactions of the Royal Society of London. *Series, B., Biological Sciences, 314*, 1–340.
2. Van Bael, S., Zels, S., Boonen, K., Beets, I., Schoofs, L., & Temmerman, L., (2018). A *Caenorhabditis elegans* mass spectrometric resource for neuropeptidomics. *Journal of The American Society for Mass Spectrometry*, doi: 10.1007/s13361–13017–11856-z.
3. Husson, S. J., Mertens, I., Janssen, T., Lindemans, M., & Schoofs, L., (2007). Neuropeptidergic signaling in the nematode *Caenorhabditis elegans*. *Progress in Neurobiology, 82*, 33–55.

4. Kim, K., & Li, C., (2004). Expression and regulation of an FMRFamide-related neuropeptide gene family in *Caenorhabditis elegans*. *Journal of Comparative Neurology, 475,* 540–550.

5. Li, C., & Kim, K., (2014). Family of FLP peptides in *Caenorhabditis elegans* and related nematodes. *Frontiers in Endocrinology, 5,* doi: 10.3389/fendo.2014.00150.

6. Li, C., & Kim, K., (2008). Neuropeptides. *WormBook,* 1–36.

7. Peymen, K., Watteyne, J., Frooninckx, L., Schoofs, L., & Beets, I., (2014). The FMRFamide-like peptide family in nematodes. *Frontiers in Endocrinology, 5,* doi: 10.3389/fendo.2014.00090.

8. Frooninckx, L., Van Rompay, L., Temmerman, L., Van Sinay, E., Beets, I., Janssen, T., Husson, S. J., & Schoofs, L., (2012). Neuropeptide GPCRs in, *C. elegans. Frontiers in Endocrinology, 3,* doi: 10.3389/fendo.2012.00167.

9. Bendena, W. G., Campbell, J., Zara, L., Tobe, S. S., & Chin-Sang, I. D., (2012). Select neuropeptides and their G-protein coupled receptors in *Caenorhabditis elegans* and *Drosophila melanogaster. Front Endocrinol., 3,* doi: 10.3389/fendo.2012.00093.

10. Hu, Z., Pym, E. C., Babu, K., Vashlishan, M. A. B., & Kaplan, J. M., (2011). A neuro-peptide-mediated stretch response links muscle contraction to changes in neurotrans-mitter release. *Neuron, 71,* 92–102.

11. Hums, I., Riedl, J., Mende, F., Kato, S., Kaplan, H. S., Latham, R., Sonntag, M., Traunmuller, L., & Zimmer, M., (2016). Regulation of two motor patterns enables the gradual adjustment of locomotion strategy in *Caenorhabditis elegans. eLife, 5,* doi: 10.7554/eLife.14116.

12. Hardaker, L. A., Singer, E., Kerr, R., Zhou, G., & Schafer, W. R., (2001). Sero-tonin modulates locomotory behavior and coordinates egg-laying and movement in *Caenorhabditis elegans. Journal of Neurobiology, 49,* 303–313.

13. Flavell, S. W., Pokala, N., Macosko, E. Z., Albrecht, D. R., Larsch, J., & Bargmann, C. I., (2013). Serotonin and the neuropeptide PDF initiate and extend opposing behavioral states in, *C. elegans. Cell, 154,* 1023–1035.

14. Horvitz, H. R., Chalfie, M., Trent, C., Sulston, J. E., & Evans, P. D., (1982). Serotonin and octopamine in the nematode *Caenorhabditis elegans. Science (New York, N.Y.), 216,* 1012–1014.

15. Meelkop, E., Temmerman, L., Janssen, T., Suetens, N., Beets, I., Van Rompay, L., Shanmugam, N., Husson, S. J., & Schoofs, L., (2012). PDF receptor signaling in *Caenorhabditis elegans* modulates locomotion and egg-laying. *Molecular and Cellular Endocrinology, 361,* 232–240.

16. Bendena, W. G., Boudreau, J. R., Papanicolaou, T., Maltby, M., Tobe, S. S., & Chin-Sang, I. D., (2008). A *Caenorhabditis elegans* allatostatin/galanin-like receptor NPR-9 inhibits local search behavior in response to feeding cues. *Proc. Natl. Acad. Sci. USA., 105,* 1339–1342.

17. Campbell, J. C., Polan-Couillard, L. F., Chin-Sang, I. D., & Bendena, W. G., (2016). NPR-9, a galanin-Like G-protein coupled receptor, and GLR-1 regulate interneuronal circuitry underlying multisensory integration of environmental cues in *Caenorhabditis elegans. PLoS Genetics, 12,* doi: 10.1371/journal.pgen.1006050.

18. Zhen, M., & Samuel, A. D., (2015). *C. elegans* locomotion: Small circuits, complex functions. *Current Opinion in Neurobiology, 33,* 117–126.

19. Chang, Y. J., Burton, T., Ha, L., Huang, Z., Olajubelo, A., & Li, C., (2015). Modulation of locomotion and reproduction by FLP neuropeptides in the nematode *Caenorhabditis elegans. PLoS One, 10,* e0135164.

20. Nelson, L. S., Rosoff, M. L., & Li, C., (1998). Disruption of a neuropeptide gene, *flp-1*, causes multiple behavioral defects in *Caenorhabditis elegans*. *Science (New York, N.Y.), 281*, 1686–1690.

21. Stawicki, T. M., Takayanagi-Kiya, S., Zhou, K., & Jin, Y., (2013). Neuropeptides function in a homeostatic manner to modulate excitation-inhibition imbalance in, *C. elegans*. *PLoS Genetics, 9*, e1003472.

22. Lim, M. A., Chitturi, J., Laskova, V., Meng, J., Findeis, D., Wiekenberg, A., Mulcahy, B., Luo, L. J., Li, Y., Lu, Y. N., et al., (2016). Neuroendocrine modulation sustains the, *C. elegans* forward motor state. *eLife,5*, doi: 10.7554/eLife.19887.

23. Pierce-Shimomura, J. T., Chen, B. L., Mun, J. J., Ho, R., Sarkis, R., & McIntire, S. L., (2008). Genetic analysis of crawling and swimming locomotory patterns in, *C. elegans. Proc. Natl. Acad. Sci. USA., 105*, 20982–209877.

24. Collins, K. M., Bode, A., Fernandez, R. W., Tanis, J. E., Brewer, J. C., Creamer, M. S., & Koelle, M. R., (2016). Activity of the, *C. elegans* egg-laying behavior circuit is controlled by competing activation and feedback inhibition. *eLife, 5*, doi:10.7554/eLife.21126.

25. Hobson, R. J., Hapiak, V. M., Xiao, H., Buehrer, K. L., Komuniecki, P. R., & Komuniecki, R. W., (2006). SER-7, a *Caenorhabditis elegans* 5-HT7-like receptor, is essential for the 5-HT stimulation of pharyngeal pumping and egg laying. *Genetics, 172*, 159–169.

26. Waggoner, L. E., Hardaker, L. A., Golik, S., & Schafer, W. R., (2000). Effect of a neuropeptide gene on behavioral states in *Caenorhabditis elegans* egg-laying. *Genetics, 154*, 1181–1192.

27. Buntschuh, I., Raps, D. A., Joseph, I., Reid, C., Chait, A., Totanes, R., Sawh, M., & Li, C., (2018). FLP-1 neuropeptides modulate sensory and motor circuits in the nematode *Caenorhabditis elegans*. *PLoS One, 13*, e0189320.

28. Ringstad, N., & Horvitz, H. R., (2008). FMRFamide neuropeptides and acetylcholine synergistically inhibit egg-laying by, *C. elegans*. *Nature Neuroscience, 11*, 1168–1176.

29. Alkema, M. J., Hunter-Ensor, M., Ringstad, N., & Horvitz, H. R., (2005). Tyramine functions independently of octopamine in the *Caenorhabditis elegans* nervous system. *Neuron, 46*, 247–260.

30. Banerjee, N., Bhattacharya, R., Gorczyca, M., Collins, K. M., & Francis, M. M., (2017). Local neuropeptide signaling modulates serotonergic transmission to shape the temporal organization of, *C. elegans* egg-laying behavior. *PLoS Genetics, 13*, e1006697.

31. Keating, C. D., Kriek, N., Daniels, M., Ashcroft, N. R., Hopper, N. A., Siney, E. J., Holden-Dye, L., & Burke, J. F., (2003). Whole-genome analysis of 60 G protein-coupled receptors in *Caenorhabditis elegans* by gene knockout with RNAi. *Current Biology, 13*, 1715–1720.

32. Li, C., Timbers, T. A., Rose, J. K., Bozorgmehr, T., McEwan, A., & Rankin, C. H., (2013). The FMRFamide-related neuropeptide FLP-20 is required in the mechanosensory neurons during memory for massed training in, *C. elegans. Learning and Memory, 20*, 103–108.

33. Nathoo, A. N., Moeller, R. A., Westlund, B. A., & Hart, A. C., (2001). Identification of neuropeptide-like protein gene families in *Caenorhabditis elegans* and other species. *Proceedings of the National Academy of Sciences USA, 98*, 14000–14005.

34. Rabinowitch, I., Laurent, P., Zhao, B., Walker, D., Beets, I., Schoofs, L., Bai, J., Schafer, W. R., & Treinin, M., (2016). Neuropeptide-driven cross-modal plasticity following sensory loss in *Caenorhabditis elegans. Plos Biology, 14*, doi: 10.1371/journal.pbio.1002348.

35. Schulenburg, H., & Felix, M. A., (2017). The natural biotic environment of caenorhabditis elegans. *Genetics, 206*, 55–86.
36. Faumont, S., Lindsay, T. H., & Lockery, S. R., (2012). Neuronal microcircuits for decision making in, *C. elegans. Current Opinion in Neurobiology, 22*, 580–591.
37. Harris, G., Shen, Y., Ha, H., Donato, A., Wallis, S., Zhang, X., & Zhang, Y., (2014). Dissecting the signaling mechanisms underlying recognition and preference of food odors. *The Journal of Neuroscience: The Official Journal of the Society for Neuroscience, 34*, 9389–9403.
38. Chalasani, S. H., Kato, S., Albrecht, D. R., Nakagawa, T., Abbott, L. F., & Bargmann, C. I., (2010). Neuropeptide feedback modifies odor-evoked dynamics in *Caenorhabditis elegans* olfactory neurons. *Nat. Neurosci., 13*, 615–621.
39. Ha, H. I., Hendricks, M., Shen, Y., Gabel, C. V., Fang-Yen, C., Qin, Y., Colón-Ramos, D., Shen, K., Samuel, A. D. T., & Zhang, Y., (2010). Functional organization of a neural network for aversive olfactory learning in *Caenorhabditis elegans. Neuron, 68*, 1173–1186.
40. Yu, Y., Zhi, L., Guan, X., Wang, D., & Wang, D., (2016). FLP-4 neuropeptide and its receptor in a neuronal circuit regulate preference choice through functions of ASH-2 trithorax complex in *Caenorhabditis elegans. Sci. Rep., 6*, doi: 10:1038/srep21485.
41. Turner, S. L., & Ray, A., (2009). Modification of CO_2 avoidance in *Drosophila* by inhibitory odorants. *Nature, 461*, 277–281.
42. Turner, S. L., Li, N., Guda, T., Githure, J., Cardé, R. T., & Ray, A., (2011). Ultra-prolonged activation of CO_2 sensing neurons disorients mosquitoes. *Nature, 474*, 87–91.
43. Guillermin, M. L., Carrillo, M. A., & Hallem, E. A., (2017). A single set of interneurons drives opposite behaviors in, *C. elegans. Current Biology, 27*, 2630–2639.
44. Rojo, R. T., Petersen, J. G., & Pocock, R., (2017). Control of neuropeptide expression by parallel activity-dependent pathways in caenorhabditis elegans. *Sci. Rep., 7*, 38734.
45. Scott, E., Hudson, A., Feist, E., Calahorro, F., Dillon, J., De Freitas, R., Wand, M., Schoofs, L., O'Connor, V., & Holden-Dye, L., (2017). An oxytocin-dependent social interaction between larvae and adult, *C. elegans. Scientific Reports, 7*, doi: 10.1038/s41598–41017–09350–41597.
46. Thorsell, A., & Heilig, M., (2002). Diverse functions of neuropeptide Y revealed using genetically modified animals. *Neuropeptides, 36*, 182–193.
47. Thorsell, A., & Mathe, A. A., (2017). Neuropeptide Y in alcohol addiction and affective disorders. *Front Endocrinol., 8*, doi: 10.3389/fendo.2017.00178.
48. Mirabeau, O., & Joly, J. S., (2013). Molecular evolution of peptidergic signaling systems in bilaterians. *Proc. Natl. Acad. Sci. USA., 110*, 2028–2037.
49. De Bono, M., & Bargmann, C. I., (1998). Natural variation in a neuropeptide Y receptor homolog modifies social behavior and food response in, *C. elegans. Cell, 94*, 679–689.
50. Stern, S., Kirst, C., & Bargmann, C. I., (2017). Neuromodulatory control of long-term behavioral patterns and individuality across development. *Cell, 171*, 1649–1662.
51. Laurent, P., Soltesz, Z., Nelson, G. M., Chen, C., Arellano-Carbajal, F., Levy, E., & De Bono, M., (2015). Decoding a neural circuit controlling global animal state in, *C. elegans. eLife, 4*, doi: 10.7554/eLife.04241.
52. Carrillo, M. A., Guillermin, M. L., Rengarajan, S., Okubo, R. P., & Hallem, E. A., (2013). O_2 sensing neurons control CO_2 response in, *C. elegans. The Journal of Neuroscience: The Official Journal of the Society for Neuroscience, 33*, 9675–9683.

53. Macosko, E. Z., Pokala, N., Feinberg, E. H., Chalasani, S. H., Butcher, R. A., Clardy, J., & Bargmann, C. I., (2009). A hub-and-spoke circuit drives pheromone attraction and social behaviour in, *C. elegans*. *Nature, 458*, 1171–1175.

54. Cheung, B. H., Arellano-Carbajal, F., Rybicki, I., & De Bono, M., (2004). Soluble guanylate cyclases act in neurons exposed to the body fluid to promote, *C. elegans* aggregation behavior. *Current Biology: CB, 14*, 1105–1111.

55. Gray, J. M., Hill, J. J., & Bargmann, C. I., (2005). A circuit for navigation in *Caenorhabditis elegans*. *Proceedings of the National Academy of Sciences USA, 102*, 3184–3191.

56. Zimmer, M., Gray, J. M., Pokala, N., Chang, A. J., Karow, D. S., Marletta, M. A., Hudson, M. L., Morton, D. B., Chronis, N., & Bargmann, C. I., (2009). Neurons detect increases and decreases in oxygen levels using distinct guanylate cyclases. *Neuron, 61*, 865–879.

57. Abergel, Z., Chatterjee, A. K., Zuckerman, B., & Gross, E., (2016). Regulation of neuronal oxygen responses in, *C. elegans* is mediated through interactions between globin 5 and the H-NOX domains of soluble guanylate cyclases. *The Journal of Neuroscience: The Official Journal of the Society for Neuroscience, 36*, 963–978.

58. Kubiak, T. M., Larsen, M. J., Nulf, S. C., Zantello, M. R., Burton, K. J., Bowman, J. W., Modric, T., & Lowery, D. E., (2003). Differential activation of "social" and "solitary" variants of the *Caenorhabditis elegans* G protein-coupled receptor NPR-1 by its cognate ligand AF9. *The Journal of Biological Chemistry, 278*, 33724–33729.

59. Rogers, C., Reale, V., Kim, K., Chatwin, H., Li, C., Evans, P., & De Bono, M., (2003). Inhibition of *Caenorhabditis elegans* social feeding by FMRFamide-related peptide activation of NPR-1. *Nature Neuroscience, 6*, 1178–1185.

60. Fenk, L. A., & De Bono, M., (2017). Memory of recent oxygen experience switches pheromone valence in Caenorhabditis elegans. *Proc. Natl. Acad. Sci. USA, 114*, 4195–4200.

61. Bretscher, A. J., Busch, K. E., & De Bono, M., (2008). A carbon dioxide avoidance behavior is integrated with responses to ambient oxygen and food in *Caenorhabditis elegans*. *Proc. Natl. Acad. Sci. USA, 105*, 8044–8049.

62. Chang, A. J., & Bargmann, C. I., (2008). Hypoxia and the HIF-1 transcriptional pathway reorganize a neuronal circuit for oxygen-dependent behavior in *Caenorhabditis elegans*. *Proc. Natl. Acad. Sci. USA, 105*, 7321–7326.

63. Hallem, E. A., & Sternberg, P. W., (2008). Acute carbon dioxide avoidance in *Caenorhabditis elegans*. *Proc. Natl. Acad. Sci. USA, 105*, 8038–8043.

64. Davies, A. G., Bettinger, J. C., Thiele, T. R., Judy, M. E., & McIntire, S. L., (2004). Natural variation in the npr-1 gene modifies ethanol responses of wild strains of, *C. elegans*. *Neuron, 42*, 731–743.

65. Chen, Y. H., Ge, C. L., Wang, H., Ge, M. H., He, Q. Q., Zhang, Y., Tian, W., & Wu, Z. X., (2018). GCY-35/GCY-36-TAX-2/TAX-4 signaling in O_2 sensory neurons mediates acute functional ethanol tolerance in *Caenorhabditis elegans*. *Sci. Rep., 8*, doi: 10.1038/s41598–41018–20477-z.

66. Bettinger, J. C., & Davies, A. G., (2014). The role of the BK channel in ethanol response behaviors: Evidence from model organism and human studies. *Front Physiol., 5*, doi: 10.3389/fphys.2014.00346.

67. Scott, L. L., Davis, S. J., Yen, R. C., Ordemann, G. J., Nordquist, S. K., Bannai, D., & Pierce, J. T., (2017). Behavioral deficits following withdrawal from chronic ethanol are influenced by SLO channel function in *Caenorhabditis elegans*. *Genetics, 206*, 1445–1458.

68. Mitchell, P., Mould, R., Dillon, J., Glautier, S., Andrianakis, I., James, C., Pugh, A., Holden-Dye, L., & O'Connor, V., (2010). A differential role for neuropeptides in acute and chronic adaptive responses to alcohol: Behavioral and genetic analysis in *Caenorhabditis elegans*. *PLoS One, 5*, e10422.

69. Ezcurra, M., Walker, D. S., Beets, I., Swoboda, P., & Schafer, W. R., (2016). Neuropeptidergic signaling and active feeding state inhibit nociception in *Caenorhabditis elegans*. *Journal of Neuroscience, 36*, 3157–3169.

70. Ezcurra, M., Tanizawa, Y., Swoboda, P., & Schafer, W. R., (2011). Food sensitizes, C. elegans avoidance behaviours through acute dopamine signaling. *The EMBO Journal, 30*, 1110–1122.

71. Mills, H., Wragg, R., Hapiak, V., Castelletto, M., Zahratka, J., Harris, G., Summers, P., Korchnak, A., Law, W., Bamber, B., et al., (2012). Monoamines and neuropeptides interact to inhibit aversive behaviour in *Caenorhabditis elegans*. *The EMBO Journal, 31*, 667–678.

72. Berg, M., Stenuit, B., Ho, J., Wang, A., Parke, C., Knight, M., Alvarez-Cohen, L., & Shapira, M., (2016). Assembly of the *Caenorhabditis elegans* gut microbiota from diverse soil microbial environments. *The Multidisciplinary Journal of Microbial Ecology, 10*, doi: 10.1038/ismej.2015.1253.

73. Dierking, K., Yang, W., & Schulenburg, H., (2016). Antimicrobial effectors in the nematode *Caenorhabditis elegans*: An outgroup to the Arthropoda. Philosophical transactions of the Royal Society of London. *Series, B., Biological Sciences, 371*, doi: 10.1098/rstb.2015.0299.

74. Ewbank, J. J., & Pujol, N., (2016). Local and long-range activation of innate immunity by infection and damage in, *C. elegans*. *Current Opinion in Immunology, 38*, 1–7.

75. Couillault, C., Pujol, N., Reboul, J., Sabatier, L., Guichou, J. F., Kohara, Y., & Ewbank, J. J., (2004). TLR-independent control of innate immunity in *Caenorhabditis elegans* by the TIR domain adaptor protein TIR-1, an ortholog of human SARM., *Nature immunology, 5*, 488–494.

76. Zugasti, O., & Ewbank, J. J., (2009). Neuroimmune regulation of antimicrobial peptide expression by a noncanonical TGF-beta signaling pathway in *Caenorhabditis elegans* epidermis. *Nature Immunology, 10*, 249–256.

77. Pujol, N., Zugasti, O., Wong, D., Couillault, C., Kurz, C. L., Schulenburg, H., & Ewbank, J. J., (2008). Anti-fungal innate immunity in, *C. elegans* is enhanced by evolutionary diversification of antimicrobial peptides. *PLoS Pathogens, 4*, e1000105.

78. Pujol, N., Cypowyj, S., Ziegler, K., Millet, A., Astrain, A., Goncharov, A., Jin, Y., Chisholm, A. D., & Ewbank, J. J., (2008). Distinct innate immune responses to infection and wounding in the, *C. elegans* epidermis. *Current Biology: CB, 18*, 481–489.

79. E, L., Zhou, T., Koh, S., Chuang, M., Sharma, R., Pujol, N., Chisholm, A. D., Eroglu, C., Matsunami, H., & Yan, D., (2018). An antimicrobial peptide and its neuronal receptor regulate dendrite degeneration in aging and infection. *Neuron, 97*, 125–138.

80. Kumar, D. K., Choi, S. H., Washicosky, K. J., Eimer, W. A., Tucker, S., Ghofrani, J., Lefkowitz, A., McColl, G., Goldstein, L. E., Tanzi, R. E., et al., (2016). Amyloid-beta peptide protects against microbial infection in mouse and worm models of Alzheimer's disease. *Science Translational Medicine, 8*, doi: 10.1126/scitranslmed.aaf1059.

81. Veriepe, J., Fossouo, L., & Parker, J. A., (2015). Neurodegeneration in, *C. elegans* models of ALS requires TIR-1/Sarm1 immune pathway activation in neurons. *Nature Communications, 6*, doi: 10.1038/ncomms8319.

82. Reddy, K. C., Hunter, R. C., Bhatla, N., Newman, D. K., & Kim, D. H., (2011). *Caenorhabditis elegans* NPR-1-mediated behaviors are suppressed in the presence of mucoid bacteria. *Proc. Natl. Acad. Sci. USA, 108*, 12887–12892.

83. Lipton, J., Kleemann, G., Ghosh, R., Lints, R., & Emmons, S. W., (2004). Mate searching in *Caenorhabditis elegans*: A genetic model for sex drive in a simple invertebrate. *Journal of Neuroscience, 24*, 7427–7434.

84. Barrios, A., Ghosh, R., Fang, C., Emmons, S. W., & Barr, M. M., (2012). PDF-1 neuropeptide signaling modulates a neural circuit for mate-searching behavior in, *C. elegans. Nat. Neurosci., 15*, 1675–1682.

85. Hilbert, Z. A., & Kim, D. H., (2018). PDF-1 neuropeptide signaling regulates sexually dimorphic gene expression in shared sensory neurons of, C. elegans. *eLife, 7*.

86. Janssen, T., Husson, S. J., Lindemans, M., Mertens, I., Rademakers, S., Donck, K. V., Geysen, J., Jansen, G., & Schoofs, L., (2008). Functional characterization of three G protein-coupled receptors for pigment dispersing factors in *Caenorhabditis elegans*. *Journal of Biological Chemistry, 283*, 15241–15249.

87. Janssen, T., Husson, S. J., Meelkop, E., Temmerman, L., Lindemans, M., Verstraelen, K., Rademakers, S., Mertens, I., Nitabach, M., Jansen, G., et al., (2009). Discovery and characterization of a conserved pigment dispersing factor-like neuropeptide pathway in *Caenorhabditis elegans. Journal of Neurochemistry, 111*, 228–241.

88. Barrios, A., Nurrish, S., & Emmons, S. W., (2008). Sensory regulation of, *C. elegans* male mate-searching behavior. *Current Biology: CB, 18*, 1865–1871.

89. Beets, I., Temmerman, L., Janssen, T., & Schoofs, L., (2013). Ancient neuromodulation by vasopressin/oxytocin-related peptides. *Worm, 2*, e24246.

90. Garrison, J. L., Macosko, E. Z., Bernstein, S., Pokala, N., Albrecht, D. R., & Bargmann, C. I., (2012). Oxytocin/vasopressin-related peptides have an ancient role in reproductive behavior. *Science (New York, N.Y.), 338*, 540–543.

91. Liu, T., Kim, K., Li, C., & Barr, M. M., (2007). FMRFamide-like neuropeptides and mechanosensory touch receptor neurons regulate male sexual turning behavior in *Caenorhabditis elegans. Journal of Neuroscience, 27*, 7174–7182.

92. Van Sinay, E., Mirabeau, O., Depuydt, G., Van Hiel, M. B., Peymen, K., Watteyne, J., Zels, S., Schoofs, L., & Beets, I., (2017). Evolutionarily conserved TRH neuropeptide pathway regulates growth in *Caenorhabditis elegans. Proceedings of the National Academy of Sciences USA, 114*, 4065–4074.

93. Raizen, D. M., Zimmerman, J. E., Maycock, M. H., Ta, U. D., You, Y. J., Sundaram, M. V., & Pack, A. I., (2008). Lethargus is a *Caenorhabditis elegans* sleep-like state. *Nature, 451*, 569–572.

94. Davis, K. C., & Raizen, D. M., (2017). A mechanism for sickness sleep: Lessons from invertebrates. *Journal of Physiology, 595*, 5415–5424.

95. Ghosh, R., & Emmons, S. W., (2010). Calcineurin and protein kinase G regulate, *C. elegans* behavioral quiescence during locomotion in liquid. *BMC Genetics, 11*, doi: 10.1186/1471–2156–1111–1187.

96. You, Y. J., Kim, J., Raizen, D. M., & Avery, L., (2008). Insulin, cGMP, and TGF-beta signals regulate food intake and quiescence in, *C. elegans*: A model for satiety. *Cell Metabol., 7*, 249–257.

97. Moosavi, M., & Hatam, G. R., (2018). The sleep in *Caenorhabditis elegans*: What we know until now. *Molecular Neurobiology, 55*, 879–889.

98. Konopka, R. J., & Benzer, S., (1971). Clock mutants of Drosophila melanogaster. *Proc. Natl. Acad. Sci. USA, 68*, 2112–2116.

99. Monsalve, Gabriela, C., Van Buskirk, C., & Frand, A. R., (2011). LIN-42/PERIOD controls cyclical and developmental progression of, *C. elegans* molts. *Current Biology, 21*, 2033–2045.
100. Jeon, M., Gardner, H. F., Miller, E. A., Deshler, J., & Rougvie, A. E., (1999). Similarity of the, *C. elegans* developmental timing protein LIN-42 to circadian rhythm proteins. *Science (New York, N.Y.), 286*, 1141–1145.
101. Nelson, M. D., Trojanowski, N. F., George-Raizen, J. B., Smith, C. J., Yu, C. C., Fang-Yen, C., & Raizen, D. M., (2013). The neuropeptide NLP-22 regulates a sleep-like state in Caenorhabditis elegans. *Nature Communications, 4*, 2846.
102. Trojanowski, N. F., & Raizen, D. M., (2016). Call it worm sleep. *Trends in Neurosciences, 39*, 54–62.
103. Nichols, A. L. A., Eichler, T., Latham, R., & Zimmer, M., (2017). A global brain state underlies, *C. elegans* sleep behavior. *Science (New York, N.Y.), 356*, doi: 10.1126/science. aam6851.
104. Turek, M., Lewandrowski, I., & Bringmann, H., (2013). An AP2 transcription factor is required for a sleep-active neuron to induce sleep-like quiescence in, *C. elegans. Current Biology: CB, 23*, 2215–2223.
105. Mani, A., Radhakrishnan, J., Farhi, A., Carew, K. S., Warnes, C. A., Nelson-Williams, C., Day, R. W., Pober, B., State, M. W., & Lifton, R. P., (2005). Syndromic patent ductus arteriosus: Evidence for haploinsufficient TFAP2B mutations and identification of a linked sleep disorder. *Proceedings of the National Academy of Sciences USA, 102*, 2975–2979.
106. Kucherenko, M. M., Ilangovan, V., Herzig, B., Shcherbata, H. R., & Bringmann, H., (2016). TfAP-2 is required for night sleep in *Drosophila. BMC Neuroscience, 17.*
107. Turek, M., Besseling, J., Spies, J. P., Konig, S., & Bringmann, H., (2016). Sleep-active neuron specification and sleep induction require FLP-11 neuropeptides to systemically induce sleep. *eLife, 5*, doi: 10.7554/eLife.12499.
108. Pfaff, D., Ribeiro, A., Matthews, J., & Kow, L. M., (2008). Concepts and mechanisms of generalized central nervous system arousal. *Annals of the New York Academy of Sciences, 1129*, 11–25.
109. Chen, D., Taylor, K. P., Hall, Q., & Kaplan, J. M., (2016). The neuropeptides FLP-2 and PDF-1 act in concert to arouse *Caenorhabditis elegans* locomotion. *Genetics, 204*, 1151–1159.
110. Choi, S., Chatzigeorgiou, M., Taylor, K. P., Schafer, W. R., & Kaplan, J. M., (2013). Analysis of NPR-1 reveals a circuit mechanism for behavioral quiescence in, *C. elegans. Neuron, 78*, 869–880.
111. Choi, S., Taylor, K. P., Chatzigeorgiou, M., Hu, Z., Schafer, W. R., & Kaplan, J. M., (2015). Sensory neurons arouse, *C. elegans* locomotion via both glutamate and neuropeptide release. *PLoS Genetics, 11*, e1005359.
112. Leonard, C. S., & Kukkonen, J. P., (2014). Orexin/hypocretin receptor signaling: A functional perspective. *British Journal of Pharmacology, 171*, 294–313.
113. Hill, A. J., Mansfield, R., Lopez, J. M., Raizen, D. M., & Van Buskirk, C., (2014). Cellular stress induces a protective sleep-like state in, *C. elegans. Current Biology, 24*, 2399–2405.
114. Iannacone, M. J., Beets, I., Lopes, L. E., Churgin, M. A., Fang-Yen, C., Nelson, M. D., Schoofs, L., & Raizen, D. M., (2017). The RFamide receptor DMSR-1 regulates stress-induced sleep in, *C. elegans. eLife, 6*, doi: 10.7554/eLife.19837.

115. Nelson, M. D., Janssen, T., York, N., Lee, K. H., Schoofs, L., & Raizen, D. M., (2015). FRPR-4 is a G-protein coupled neuropeptide receptor that regulates behavioral quiescence and posture in *Caenorhabditis elegans*. *PloS One, 10*, doi: 10.1371/journal. pone.0142938.

116. Nath, R. D., Chow, E. S., Wang, H., Schwarz, E. M., & Sternberg, P. W., (2016). *C. elegans* stress-induced sleep emerges from the collective action of multiple neuropeptides. *Current Biology: CB, 26*, 2446–2455.

117. Antin, J., Gibbs, J., Holt, J., Young, R. C., & Smith, G. P., (1975). Cholecystokinin elicits the complete behavioral sequence of satiety in rats. *Journal of Comparative and Physiological Psychology, 89*, 784–790.

118. Ohno, H., Yoshida, M., Sato, T., Kato, J., Miyazato, M., Kojima, M., Ida, T., & Iino, Y., (2017). Luqin-like RYamide peptides regulate food-evoked responses in, *C. elegans*. *eLife, 6*, doi: 10.7554/eLife.28877.

119. Skora, S., Mende, F., & Zimmer, M., (2018). Energy scarcity promotes a brain-wide sleep state modulated by insulin signaling in, *C. elegans*. *Cell Reports, 22*, 953–966.

120. McCloskey, R. J., Fouad, A. D., Churgin, M. A., & Fang-Yen, C., (2017). Food responsiveness regulates episodic behavioral states in *Caenorhabditis elegans*. *Journal of Neurophysiology, 117*, 1911–1934.

121. Hu, P. J., (2007). Dauer. *WormBook*, 1–19.

122. Lee, J. S., Shih, P. Y., Schaedel, O. N., Quintero-Cadena, P., Rogers, A. K., & Sternberg, P. W., (2017). FMRFamide-like peptides expand the behavioral repertoire of a densely connected nervous system. *Proc. Natl. Acad. Sci. USA, 114*, 10726–10735.

123. Pierce, S. B., Costa, M., Wisotzkey, R., Devadhar, S., Homburger, S. A., Buchman, A. R., Ferguson, K. C., Heller, J., Platt, D. M., Pasquinelli, A. A., et al., (2001). Regulation of DAF-2 receptor signaling by human insulin and ins-1, a member of the unusually large and diverse, *C. elegans* insulin gene family. *Genes and Development, 15*, 672–686.

124. Matsunaga, Y., Honda, Y., Honda, S., Iwasaki, T., Qadota, H., Benian, G. M., & Kawano, T., (2016). Diapause is associated with a change in the polarity of secretion of insulin-like peptides. *Nature Communications, 7*, doi: 10.1038/ncomms10573.

125. Hung, W. L., Wang, Y., Chitturi, J., & Zhen, M., (2014). A *Caenorhabditis elegans* developmental decision requires insulin signaling-mediated neuron-intestine communication. *Development, 141*, 1767–1779.

126. Lee, H., Choi, M. K., Lee, D., Kim, H. S., Hwang, H., Kim, H., Park, S., Paik, Y. K., & Lee, J., (2011). Nictation, a dispersal behavior of the nematode *Caenorhabditis elegans*, is regulated by IL2 neurons. *Nat. Neurosci., 15*, 107–112.

127. Lee, D., Lee, H., Kim, N., Lim, D. S., & Lee, J., (2017). Regulation of a hitchhiking behavior by neuronal insulin and TGF-beta signaling in the nematode *Caenorhabditis elegans*. *Biochemical and Biophysical Research Communications, 484*, 323–330.

CHAPTER 4

Annelids Neuro-Endrocrino-Immune Response

MICHEL SALZET

INSERM U1192–Laboratoire Protéomique, Réponse Inflammatoire et Spectrométrie de Masse (PRISM), Université de Lille 1, F-59000 Lille, France

4.1 INTRODUCTION

Hormonal processes along with enzymatic processing similar to that found in vertebrates occur in annelids. Amino acid sequence determination of annelids precursor gene products reveals the presence of the respective peptides that exhibit high sequence identity to their mammalian counterparts. Furthermore, these neuropeptides exert similar physiological function in annelids than the ones found in vertebrates. In this respect, the high conservation in the course of evolution of these molecules families reflects their importance. Nevertheless, some specific neuropeptides to annelids or invertebrates have also been in these animals. Another major point that can be also mentioned is the presence of a true neuroimmune system in annelids nervous system involved not only in immune response modulation but also in neurorepair and neuroregeneration showing that neuroendocrine and neuroimmune system coexist and participate in brain physiological processes.

Annelids neural tissues do not contain anatomical correlates of hypothalamus or pituitary. However, they possess localized ganglionic regions rich in mammalian-like neuroendocrine signaling molecules [1]. These molecules appear free in the animals hemolymph, demonstrating distant signaling *via* several target tissues, including immune cells [1]. Thus, the fact that a classical closed conduit system to carry signaling molecules does not exist in Annelids should not detract from an endocrine presence because the baseline functioning of the system depends on the distance between the origin of a signal molecule and its target tissue/receptor.

4.2 NEUROENDOCRINE PEPTIDES FAMILIES IN ANNELIDS WHICH ARE SIMILAR TO VERTEBRATES

Among the neuropeptides which have been identified by genomes and EST library in silico screening in invertebrates (Table 4.1), at least 40 were isolated, so far, in annelids (Tables 4.1–4.3), the most part already sequenced are related to the ones previously isolated in vertebrates [1–4]. More recently some EST databanks of annelids have been carried out by several groups and have opened the door of the discovering of novel neuropeptide families like the insulin-related peptides, RFamide, Wamides, GnRH/angiotensins, the oxytocin (OT)/vasopressin (VP), the myotropic peptides [3, 4]. With the exception to the myotropic peptides, which are more specific to the annelids, the other molecules families have also been identified in vertebrates in which they play crucial roles as neurohormones. These data further strengthen the existence of a neuroendocrine system in the annelids.

4.2.1 ANGIOTENSIN-LIKE PEPTIDES

Biochemical identification of a "central" angiotensin II (AII)-like peptide in the leech *Erpobdella octoculata* was demonstrated and found to be amidated [6–9]. This constituted the first characterization of an angiotensin-like peptide in an invertebrate, demonstrating its conservation during the course of evolution. An identification of the proteins immunoreactive to anti-AII was found both in brain extracts and *in vitro* translated brain RNA products [6]. The pro-AII precursor detected in the brain extracts possesses a *ca* 19 kDa molecular mass and is also a "multiple hormone precursor" as it is also recognized by two other antisera: a polyclonal γ-MSH and a specific mono-clonal antibody directed against leech neurons, Tt159 [10]. Furthermore, we found in leeches a *ca*. 11-kDa peptide with a sequence of DRVYIHPFHL-LXWG, which exhibits a 78.5% sequence identity to the N-terminus of the angiotensinogen and a 100% sequence identity to AI [7].

Biosynthesis study of leech AII revealed the existence of a renin- [11] and angiotensin-converting like enzyme (ACE)-like enzymes implicated in its catabolism [13]. Leech ACE has been cloned, and an approximative 2 kb cDNA has been predicted to encode a 616-amino-acid soluble enzyme containing

TABLE 4.1 Phyletic Distribution of Metazoan pNP and Neuropeptide GPCR Families (A) Phyletic distribution of metazoan pNP families. The families that are part of the CC are shown in red. Pigment-dispersing factor (Pdf), leucokinin, thyrotropin-releasing hormone (TRH), and parathyroid hormone (PTH) may be ancestral bilaterian, motilin, melanin-concentrating hormone (MCH), and endothelin ancestral chordate, based on GPCR distribution. (B) Phyletic distribution of metazoan class A and class B neuropeptide GPCR families. Class B GPCRs are indicated as (B). Ancestral bilaterian (*), protostome (+), deuterostome (o) and chordate (-) families are indicated (With permission from Gáspár Jékely published in *PNAS* (2013) *110* (21), 8702–8707).

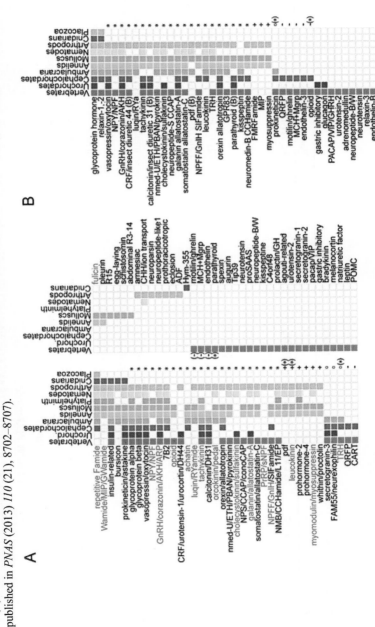

TABLE 4.2 Characterized Annelid Neuropeptides from Leeches

Species	Sequence	Name
Theromyzon tessulatum	SYVMEHFRWDKFGRKIKRRPIKVYPNGAEDE SAEAFPLE	ACTH-like Angiotensin I Angiotensin II
	DRVYIHPFHLLXWG DRVYIHPF RVYIHPF IPEPYVWD	Angiotensin III
	FMRF-amide FM(O)RF-amide FLRF-amide GDPFLRFamide PLG	LORF (leech osmoregulator factor)
	YGGFL YGGFM YGGFLRKYPK	FMRF-amide FMRF-amide sulfoxyde
	YVMGHFRWDKFamide	FLRF-amide GDPFLRF-amide MIF-1
	GSGVSNGGTEMIQLSHRERQRYWAQDNLRR	Leucine-enkephalin Méthionine-enkephalin
	RFLEKamide	α neoendorphin MSH-like peptide Leech Egg Laying Hormone
Erpobdella octoculata	DRVYIHPF-amide CFIRNCPKG-amide	Angiotensin II-amide Lysine-conopressin FMRF-amide GDPFLRF-amide
	FMRF-amide; FM(O)RF-amide GDPFLRF-amide	FLRF-amide
	FLRF-amide IPEPYVWD; IPEPYVWD-amide	LORF
Hirudo medicinalis	FMRF-amide, FM(O) RF-amide FLRF-amide	FMRF-amide FLRF-amide
	AMGMLRM-amide, GVSFLRMG, SLDMLRMG, AVSMLRMG, AVSLLRMG	Myomoduline-like peptide Hirudotocin
	CFIRNCPLG-amide	
Hirudo nipponia Whitmania pigra	WRLRSDETVRGTRAKCEGEWAIHACLCLGGN-amide	Leech excitatory peptide

TABLE 4.3 Characterized Annelids Neuropeptides Other Annelids

Species	Sequence	Name
Perinereis vancaurica	AMGMLRMamide WVGDVQ ATWLDT WMVGDVQ FYEGDVPY	Myomodulin-CARP Esophagus regulation peptides
Nereis diversicolor	FMRF-amide, FM(O)RF-amide FTRF-amide	FMRF-amide FTFR-amide
Nereis virens	FMRF-amide FLRF-amide	FMRF-amide FLRF-amide
Eisenia fetida	CFVRNCPTGamide APKCSGRWAIHSCGGNG GKCAGQWAIHACAGGNG RPKCAGRWAIHSCGGGNG	Annetocin GGNG1 GGNG2 GGNG3

a single active site, named *Tt*ACE (*T. tessulatum* ACE) [14]. Surprisingly, its primary sequence shows greater similarity to vertebrates than to invertebrates. Stable *in vitro* expression of *Tt*ACE in transfected Chinese- hamster ovary cells revealed that the leech enzyme is a functional metalloprotease. As in mammals, this 79k daglycosylated enzyme functions as a dipeptidyl carboxypeptidase capable of hydrolyzing angiotensin I to AII. However, a weak chloride inhibitory effect and acetylated *N*-acetyl-SDKP (Ac SDAcKP) hydrolysis reveal that *Tt*ACE activity resembles that of the N-domain of mammalian ACE [14]. *In situ* hybridization shows that its cellular distribution is restricted to epithelial midgut cells [14]. EST screening reveals that both membrane and soluble somatic forms of ACE are present in leeches. In the polychaete *Neanthes virens,* a dipeptidyl carboxypeptidase resembling mammalian ACE has also been demonstrated [15, 16]. The purified enzyme was homogeneous by SDS-PAGE, with a molecular mass of 71 kDa by SDS-PAGE and 69 kDa by gel filtration, indicating that it is monomeric like the leech *Tt*ACE [14]. The isoelectric point was 4.5, and the optimum pH for the activity was 8.0. It showed a specific activity of 466.8 U/mg [16].

Experiments conducted on the biological activity of the AIIamide established that this peptide is involved in the control of leeches' hydric balance exerting a diuretic effect [17]. Similarly, in the Polychaeta *Nereis diversicolor*, injections of polyclonal antisera against AII provoked a partial inhibition of the increase in body weight in animals exposed to the hypoosmotic medium. In a subsequent test, injections of synthetic AII-amide and, to a lesser extent AII enhanced the increase in body weight and, therefore, strengthened the importance of these peptides in the neuroendocrine control of *Nereis* osmoregulation [18]. In clamworm *Perinereis* sp., angiotensin III as well as AII enhanced an increase in body weight under a hypo-osmotic condition and suppressed a decrease in body weight under a hyper-osmotic condition. When clamworms were treated with tetrachloroaurate (III) after angiotensin-treatment, these enhancing and suppressive effects of the angiotensins under hypo- and hyper-osmotic conditions were inhibited. In contrast, when clamworms were pretreated with tetrachloroaurate (III) before angiotensin-treatment, these effects of angiotensins were not inhibited. Since tetrachloroaurate (III) is a representative blocker of aquaporins, these results indicate that angiotensin III, as well as AII, regulates water flow through aquaporins in clamworms [19, 20]. In leeches, the fact that AIIamide injections at different doses in *T. tessulatum* suggest the existence of two different types of receptors, one at high and the other one at low affinities towards AIIamide, we focused our interest on the identification of the leech AIIamide receptors. Binding experiments on *T. tessulatum* brain membranes with mono [^{125}I] AIIamide reveals 70% specific binding and an IC_{50} of 10 nM [21]. In addition, biochemical studies using commercial anti-AT1 receptor reveal the existence of a specific protein at a molecular weight of 140 kDa [21], which is confirmed by EST analysis where AT1 and AT2 like receptors have been detected with27 and 32% sequence identity, respectively. Immunocytochemical studies performed at the level of the brain confirmed the presence of labeling in neurons and glial cells to anti-AII, anti-leech renin, anti-leech ACE, and anti-AT1 [21]. Leech coelomocytes are also immunoreactive to polyclonal antisera raised against the *T. tessulatum* ACE and leech brain AII and a commercial anti-AT1 receptor. Biochemically, renin, ACE, and AT1-like receptor were identified in the leech immune cells [21]. Moreover, leechAII (10^{-6} M) alone does not initiate nitric oxide (NO) release in invertebrate immunocytes but does only after pre-exposing the cells to IL-1 (15.9+/–2.6 nM; P<0.005 vs. 1.1 nM when AII is added alone). Similar results were obtained with human leukocytes (14.5+/–2.7 nM; P<0.005 IL-1+AII vs. 0.9 nM when AII is added alone). Immunocytochemical studies performed at the structural and ultrastructural levels confirmed the presence in same immune cells all the molecules of the renin-angiotensin system (RAS) in leeches as epitopes to IL-1-like protein and IL-1-like receptor. These data report coaction between cytokines like substances and neuropeptides in an immune process and the involvement of the RAS in the modulation of the immune response [21]. Complementary to these findings, evidences of natural ACE inhibitors have been demonstrated in leeches [22]. In fact, the leech osmoregulator Factor (IPEPYVWD, see the LORF section), a neuropeptide found in both the central nervous system and sex ganglia of leeches is involved in water retention control through inhibition of ACE with an IC50 of 19.8 µM for rabbit ACE. Its cleavage product the IPEP is a better competitive inhibitor with an IC50 of 11.5 µM. Competition assay using p-[3H] benzoyl glycyl glycylglycine and insect ACE established that LORF and IPEP are natural inhibitors for invertebrate ACE. 54% of insect ACE activity is inhibited with 50 µM d'IPEP and 35% with 25 µM de LORF like [22]. Taken together, these data strongly suggest the existence of a complete RAS in annelids, which was conserved in the course of evolution.

4.2.2   OXYTOCIN (OT)/VASOPRESSIN (VP) PEPTIDES

The peptides of the VP/OT family have been discovered throughout the animal kingdom. They are alike, sharing at least five of nine residues and a disulfide-linked ring structure, which puts severe constraints on conformational flexibility. In vertebrates, gene duplication gave rise to two distinct families, *i.e.,* the VP one and the OT one (Table 4.4). The differential binding of VP and OT to their respective receptors is largely due to the amino acid residue in position 8, i.e., a basic one in VP- related peptides and a neutral one in OT-related peptides. In the leech *Erpobdella octoculata*, a VP related peptide; the lysine-conopressin (CFIRNCPKG) has been isolated [23]. However, EST screening in *Hirudo medicinalis* has revealed a peptide related to lysine-conopressin but belonging to the OT family (CFIRNCPLG), named Hirudotocin (Salzet, Unpublished data). In Oligochaeta, a peptide related to this family (CFVRNCPTG), the annnetocin has been discovered [24]. The proannetocin showed 37.4–45.8% amino acid homology to other prohormones. In the neurophysin domain, 14 cysteines and amino acid residues essential for association of a neurophysin with a VP/OT superfamily peptide were conserved, suggesting that the *Eisenia neurophysin* can bind to annetocin (Figure 4.1). Furthermore, *in situ* hybridization experiments demonstrated that the annetocin gene is expressed exclusively in neurons of the central nervous system predicted to be involved in the regulation of reproductive behavior [25]. Similar results are obtained for the Hirudotocin (Salzet, unpublished data). Bioinformatic EST libraries and genomes screening of annelids such like leeches (*Helobdella robusta, Hirudo medicinalis*), polychaete (*Capitella teleta,* Perinerneris, *Alvinella pompejana*) and Oligochaeta (*Lumbricus terrestritris*) have confirmed the presence of peptides of OT/VP in Annelids (Table 4.4) [4, 5].

TABLE 4.4 Oxytocin/Vasopressin Peptides Discovered in Invertebrates

Species	Sequence	Name
Erpobdella octoculata	CFIRNCPKG-amide	Lysine-Conopressin
Hirudo medicinalis	CFIRNCPLG-amide	Hirudotocin
Eisenia Fetida	CFVRNCPTG-amide	Annetocin
Conus Georgaphus	CFIRNCPKG-amide	Lysine-Conopressin
Conus Striatis	CIIRNCPRG-amide	Arginine-Conopressin
Octopus	CYFRNCPIG-amide	Cephalotocin Octopressin
Octopus	CFWTSCPIG-amide	Diuretic peptide
Locusta migratoria	CLITNCPRG-amide	

These peptides act on osmoregulation via nephridia [23–24]. Lysine-conopressin inhibits the Na+ amiloride dependent transitory current before to highly stimulate it on *Hirudo medicinalis* stomach or tegument preparation [17]. Moreover, lysine-conopressin induces egg-laying in earthworm like OT does invertebrates. This confirms that through the course of evolution, the OT/VP peptides family has conserved its function on both osmoregulation and reproduction. Annetocin receptor (AnR) has been cloned. The deduced precursor displays high sequence similarity with OT/VP receptors. Genomic analysis of the AnR gene revealed that the intron-inserted position is conserved between the AnR gene and the mammalian OT/VP receptor genes (Figure 4.1). These results indicate that AnR and mammalian OT/VP receptors share a common ancestor gene. Administration of annetocin to the AnR expressed in Xenopus oocytes induced a calcium-dependent signal transduction. Reverse transcriptase-PCR analysis and in situ hybridization showed that the AnR gene is expressed specifically in the nephridia located in the clitellum region, although the nephridia are distributed throughout the worm body. This result suggests that annetocin induces egg-laying behavior through its action on the nephridia. This is the first description concerning the functional correlation between an invertebrate OT/VP-related peptide and egg-laying behavior as in mollusk [26]. In leeches, a1.7 kb cDNA for an AVP-related receptor has been cloned from the leech *Theromyzon tessulatum* [27]. The open reading frame encodes a 435-amino acid transmembrane protein that displays seven segments of hydrophobic amino acids, typical of G-protein-coupled receptors. The overall predicted protein exhibits about 30% amino-acid identities to other invertebrates, as well as vertebrate, AVP/OT receptor family members, and displays conserved characteristic features belonging to the AVP/OT receptor superfamily. RT-PCR expression experiments showed that mRNA is expressed in the genital tract, the ovary, and the brain. The receptor expression is stage-specific, showing a weak expression after the two first blood meals, increasing dramatically after the last blood meal during the period of sexual maturation and disappearing after egg laying. Thus, the leech AVP-related receptor may mediate reproductive functions. When expressed in COS-7 cells, the receptor binds ligands with the following rank order of potency: AVP = Arg-vasotocin > Arg-conopressin > mesotocin = OT = Lys-conopressin = isotocin > annetocin. This shows an AVP-like pharmacological profile. The transfected receptor mediates AVP-induced accumulation of inositol phosphates, indicating that the leech AVP-related receptor is functional [27].

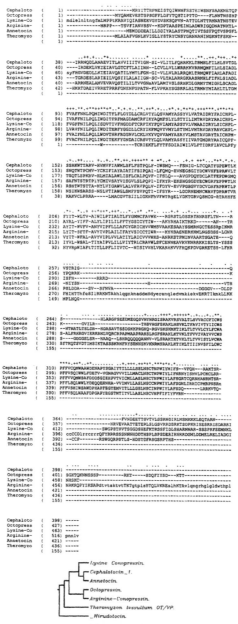

FIGURE 4.1 Sequence alignment of OT/VP receptor in Invertebrates using Mulatlin software. (http://npsa-pbil.ibcp.fr/cgi-bin/npsa_automat.pl?page=/NPSA/npsa_multalin.html): cephalotocin R and octopressin R [82, 83], lysine, and arginine-conopressin R [84], annetocin R [4], leech OT/VP R [27] and Hirudotocin R sequence are from leech EST (Salzet, unpublished data).

Taken together, these data demonstrate the characterization of AVP/OT superfamily receptor in annelids, which are considered the most distant group of coelomate metazoans possessing a functional AVP/OT-related endocrine system [4, 27].

4.2.3 OPIOIDS

Enkephalins have been isolated in Annelids [28] and its precursor the proenkephalin in the leech *Theromyzon tessulatum* [29, 30]. Bioinformatic EST libraries and genomes screening of *Platynereis dumerilii* (annelid) and *Lottia gigantea* and *Haliotis asinina* (mollusks) confirmed the presence of opioid precursor (Figure 4.2) [5]. Such a precursor has been also isolated in the venom of the Asian scorpion *Buthusmartensii Karsch* (BmK). An amidated peptide containing an enkephalin-like sequence has been identified [32]. BmK-YA is encoded by a precursor that displays a signal sequence and contains four copies of BmK-YA sequences and four of His4-BmK-YA, all flanked by single amino acid residues. BmK-YA can activate mammalian opioid receptors with selectivity for the δ subtype [32]. In leech proenkephalin contains Met- and Leu-enkephalins and Met-enkephalin-Arg-Gly-Leu and Met-enkephalin-Arg-Phe flanked by dibasic amino acid residues, which are targets for proteolytic enzymes suchlike prohormone convertases present in leeches ([29]. Specific receptors for these peptides have been characterized in neural and in immune systems [29]. Enkephalin derived peptides seem to be implicated in innate immune response [31]. In fact, lipopolysaccharides (LPS) injection into to coelomic fluid of the leech *Theromyzon tessulatum* stimulates the release of proenkephalin A (PEA)-derived peptides as determined by immunoprecipitation and Western blot analyses [33]. This release occurs in the first 15 min after LPS exposure and yields a 5.3-kDa peptide fragment corresponding to the C-terminal part of the precursor. This fragment is then cleaved to free an antibacterial peptide related to mammals arginine phenylalanine extended enkelytin: the peptide B. These PEA processing peptides were characterized using a combination of techniques including reversed-phase HPLC, microsequencing, and mass spectrometry. The isolated invertebrate peptide B presents a high sequence homology with the bovine's and the same activity against Gram + bacteria. Titrations revealed the simultaneous appearance of methionine-enkephalin (ME) and peptide B in invertebrates after stimulation by LPS (in a dose-dependent manner), surgical trauma or electrical stimulations to neural tissues of the mussel [33]. Furthermore, peptide B processing in vitro yields ME arginine phenylalanine (MERF), which exhibits via the delta receptors, immunocyte excitatory properties, i.e., movement, and conformational changes, but no antibacterial activity. We surmise that this unified response to the various stimuli is a survival strategy for the organism by providing immediate antibacterial activity and immunocyte stimulation, thereby reducing any immune latency period needed for an adequate immune response [30, 33].

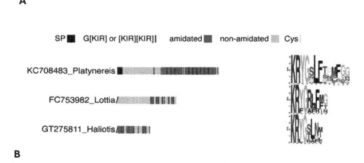

FIGURE 4.2   Structure of lophotrochozoan and scorpion opioid pNPs. A) Schematic structure of the Platynereisdumerilii (annelid) and Lottia gigantea and Haliotisasinina (mollusks) opioid pNPs. Signal peptides are shown in blue, peptides with a C-teminal Gly in green, dibasic cleavage sites in red, and Cys residues in yellow. The sequence logos show the conservation of residues in the predicted mature peptides (With permission from Gáspár Jékely published in PNAS (2013) 110 (21), 8702–8707). B) Amino acid sequence of the Bm K-YA precursor. The signal peptide is underlined. The amino acid in grey indicates cleavage sites (R, D). The glycine residues expected to serve as amide donors are shown in italic. The sequences of Bm K-YA and His 4-BmK-YA (italic) are framed in bold (from Ref. [32], © 2012 Zhang et al.)

Taken together, the results from leeches now demonstrate, those proen-kephalin precursors and many of the derived bioactive peptides, i.e., α-MSH, neoendorphin, and ACTH [35–38], which are important in mammalian neuro-endocrine signaling, are present in invertebrates. This adds to the growing body of evidence that a neuroendocrine apparatus is also present in simple animals.

4.3   NEUROENDOCRINE PEPTIDE FAMILIES IN ANNELIDS SIMILAR TO MOLLUSCSONES

Annelids and Molluscs are both lophochotrozoans, and in this context, several families of peptides have been well conserved in these two groups as reflected by the genomes screening [4]. The most represented one is the Rxamides family. However other molecules families have been detected in leech and lymnaea or Aplysia but are also now discovered in vertebrates like the RFamide peptide family.

4.3.1 RXAMIDE PEPTIDES

Several related RXamides have been found in annelids [39–46]. A myomodulin-like peptide, GMGALRLamide, has been purified and sequenced from the medicinal leech nerve cords [45]. Myomodulin-like immunoreactivity has recently been found to be present in a set of leech neurons, including Leydig neurons [45, 46]. The glial responses to Leydig neuron stimulation persisted in high-divalent cation saline, when polysynaptic pathways are suppressed, indicating that the effects on the glial cell were direct [42]. The glial responses to myomodulin A application persisted in high-Mg^{2+}/low-Ca^{2+} saline, when chemical synaptic transmission is suppressed, indicating a direct effect of myomodulin A on the glial membrane (42). The glial hyperpolarization evoked by myomodulin A was dose-dependent (EC50 = 50 nM) and accompanied by a membrane conductance increase of approximately 25%. Ion substitution experiments indicated that myomodulin A triggered a Ca^{2+}-independent K^+ conductance (42). Moreover, synthetic leech myomodulin-like peptide [46] showed identical neuronal modulation effect on the giant leech Retzius cell comparable to that by the synthetic *Aplysia myomodulin* APMGMLRLamide [47–49]. This neural and muscular modulation has been shown to be important for shaping and modifying behavior. Experiments focused on the Retzius cell (R) revealed that the myomodulin-like peptide increased the excitability of the R cell such that the cell fires more action potentials with a shorter latency to the first action potential. This effect is mediated by the activation of a Na^+-mediated inward current near the cell resting membrane potential [44]. EST library scanning reveals the presence of amyomodulin precursor sharing several peptides of this family bracketed by di-basic residues like GVSFLRMG, SLDMLRMG, AVSMLRMG, AVSLLRMG (Salzet, unpublished data) and presenting 57% sequence homology with Lymnaea or *Aplysia myomodulin* precursor [49, 50].

In polychaete annelids, a heptapeptide and AMGMLRMamide, termed Pev myomodulin, was isolated from *Perinereis vancaurica* using the esophagus of the animal as the bioassay system [43]. The sequence of the annelid peptide is highly homologous with those of the myomodulin-CARP-family peptides found in mollusks. The annelid peptide is regarded as a member of the myomodulin-CARP family, though all the molluscan peptides have a Leu-NH$_2$ at their C-termini. The annelid peptide showed a potent contractile action on the esophagus of the annelid. The peptide may be an excitatory neuromediator involved in the regulation of the esophagus. Among various myomodulin- CARP-family peptides and their analogs, the annelid peptide showed the most potent contractile action on the esophagus [43]. Replacement of the C-terminal Met-NH$_2$ of the annelid peptide with a Leu-NH$_2$ decreased its contractile potency, while replacement of the C-terminal Leu-NH$_2$ of myomodulin and CARP with a Met-NH2 increased their potency. The C-terminal Met-NH$_2$ of the annelid peptide seems to be important, but not essential, for exhibiting its contractile activity on the esophagus [43].

4.3.2   RF-AMIDE PEPTIDES

FRaP family is present in the polychaetes *Nereis virens* and *Nereis diversicolor* [51–53]. In this last species, two other RFamide peptides (FM(O) RFamide and FTRFamide) have been isolated [51–53]. Pharmacological data suggest that RFamide peptides are involved in the control of heartbeat and body wall tone in the polychaete *Sabellastarte magnifica* [54, 55] and in the oligochaete *Eisenia fetida* [56]. In the earthworm *Eisenia fetida*, FMRF-like peptides are co-localized with serotonin, suggesting a role as neuromodulators influencing serotoninergic neurons [56, 57]. In a *Lumbricus terrestris*, FMRF-like peptides seem to be involved in both central integratory processes, neuromuscular regulation, and sensory processes [56, 57]. In Hirudinae, anti-FMRFamide immunoreactivity is found in cell bodies and neuronal processes of the central nervous system [58–63]. In the segmental ganglia (SG) of the ventral nerve cord, this immunoreactivity is localized in hear texcitatory (HE) motor neurons, heart accessory (HA) modulatory neurons and several motor neurons innervating the longitudinal and mediodorsoventral muscles [64–66]. Among the 21 segmental ganglia (SG1-SG21) of the ventral nerve cord of leeches, SG5 and SG6 that innervate the sexual organs are designated as sex SG. These sex ganglia contain, as compared to the non-sex ganglia, an additional population of neurons immunostained with anti-FMRFamide in *Hirudo medicinalis* [67]. Furthermore, two RFamide peptides (FMRFamide and FLRFamide) were identified in *Hirudo medicinalis* [60, 68]. These peptides increase the strength and accelerate the rate of myogenic contractions as well as inducing myogenic contractions inquiescent hearts [65, 66]. Besides these tetrapeptides, we characterized an extended form of FLRFamide, the GDPFLRFamide from sex ganglia extracts of *E. octoculata* [60]. In *T. tessulatum* presence of RFamide peptides in neurosecretory granules in fibers of the neurohae malaria suggests that at least one of the characterized peptides is secreted into the dorsal vessel. The brain could exert a neuroendocrine control of certain functions via RFamide peptides. Taking into account a previous study showing a loss of weight of *T. tessulatum* after a GDPFLRFamide injection and an increase of weight after a FMRFamide injection [60], we surmise that GDPFLRFamide may act as adiuretic hormone and FMRFamide as an anti-diuretic hormone. Electrophysiological experiments confirmed our speculation [17]. The anti-diuretic effect of FMRFamide seemed not due to a direct action on the caecal epithelium.

Nevertheless, the control of the hydric balance might be also exerted directly on the nephridia. Indeed, Wenning et al., demonstrated that the nephridial nerve cells, which innervate the nephridia and contact the urine forming cells, contain RFamide peptide(s) in *H. medicinalis* [61, 69]. Furthermore, these authors showed that FMRFamide leads the hyperpolarization and decreases the rate of firing of the nephridial nerve cells, suggesting auto-regulation of peptide release.

4.3.3 LEECH EGG-LAYING HORMONE

In leeches, egg-laying may be under the control of a leech egg-laying hormone (L-ELH) [70]. In *Eisenia fetida*, although that the OT-VP related peptide, annetocin, is known to potentiate the pulsatory contractions in the bladder-shaking movement of the nephridia, indicating an involved of osmoregulation through nephridial function, this peptide is also implicated in egg-laying behaviors [26]. In fact, annetocin induced a series of egg-laying-related behaviors in the earthworms. These stereotyped behaviors consisted of well- defined rotatory movements, characteristic body-shape changes, and mucous secretion from the clitellum [26]. Each of these behaviors is known to be associated with the formation of the cocoon in which eggs are deposited. In fact, some of the earthworms injected with annetocin (> 5 nmol) laid eggs. Such egg-laying-related behaviors except for oviposition were also induced by OT, but not by Arg-VP. Furthermore, annetocin also induced these egg-laying-like behaviors in the leech *Whitmania pingra*, but not in the polychaete *Perinereisvancaurica*. These results suggest that annetocin plays some key role in triggering stereotyped egg-laying behaviors in terrestrial or fresh-water annelids that have the clitella [26].

4.4 ARE SOME NEUROENDOCRINE SIGNALING MOLECULES SPECIFIC TO ANNELIDS?

The history of neurobiology demonstrates the significance of the invertebrate nervous system as a valuable model. The giant axon of the squid and crayfish neuromuscular junction stands out in this regard. This field of scientific endeavor also stands out for its demonstration of the conservation of signaling molecules and their functions during evolution. Probably the first peptide found in invertebrates that later was also found in mammals is the *Hydra* head activator peptide [71–73]. In *Hydra,* this peptide modulates morphogenesis, cellular growth, and differentiation. In mammals, this peptide sequence proved to be unique in that it differed from that of any known peptide. Furthermore, the investigators surmise that this peptide may have the same functions in mammals. We propose here to extend these several families specifically found in annelids and not yet found in mammals.

4.5 LEECH NEUROIMMUNE SYSTEM

Innate immunity plays a major role as the first defense against microbes. Using in silico strategy from medicinal leech central nervous system EST, effectors of the innate response include pattern recognition receptors (PRR), proteolytic cascades and peptides/proteins with antimicrobial properties were tracked not in the peripheral system but in leech CNS [74].

4.5.1 NEURORAL HMTLR

Analysis of the *Hirudo* transcriptome database reveals the presence of putative homologs of all the factors reported to play critical roles in human TLR pathways (Figure 4.3). This stands in sharp contrast to other invertebrates, such as insects and nematodes, for which the PRR pathways thus far appear to be much simpler, with many components missing (Figure 4.3). All the identified leech putative homologs indeed play similar functional roles remains to be shown by further analysis, but their presence in the transcriptome database adds support to the hypothesis that Lophotrochozoan genetic programs are more closely related to those of vertebrates than are those of Ecdysozoans.

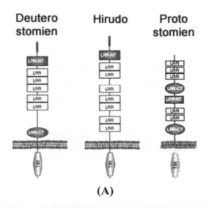

(A)

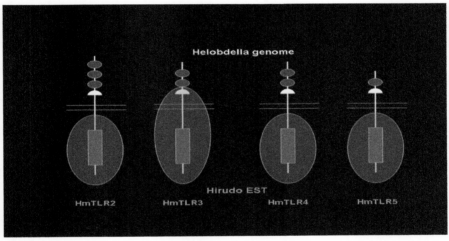

(B)

FIGURE 4.3 A) Structure comparison of HmTLR1 and Protostomian and Deurterostomian TLRs [74] B) Putative Medicinal leech HmTLR based on the Hirudo EST and completed with TLR sequence found in Hellobdella genome. (Reprinted with permission from Ref. [75]. © 2017 Elsevier.)

Five medicinal leech Toll-like receptors (TLRs) have been detected, one has already been fully characterized [74, 75]. The first *Hm*TLR characterized is presented in Figure 4.3A. TLRs in general, share similarities in their extracellular Leucine-Rich Repeat (LRR) and their intracellular Toll/IL-1 Receptor (TIR) domains. The TIR domain plays a central role in TLR signaling. All TLRs contain a cytoplasmic TIR domain, which, upon activation, acts as a scaffold to recruit adaptor proteins. It is well established that the differential recruitment of adaptors to TLRs provides a significant amount of specificity to the TLR-signaling pathways. Among these adapter proteins, MyD88 and TRIF are now considered as the signaling ones, and hence the TLR pathways can be categorized as MyD88-dependent and TRIF-dependent. The LRR domain is an extracellular domain implicated in the detection of pathogens. Based on the organization of the extra-cellular LRR array, two types of TLRs have been described. Vertebrate TLRs have an array of LRRs capped by cysteine-rich domains located at the N- and C-terminal LRR domains (LRRNT and LRRCT, respectively). By contrast, most of invertebrate TLRs also contain LRRNT and LRRCT domains, but instead of capping the LRR array, these are located within the array in tandem orientation. Interestingly, leech *HmTLR1* presents the originality to exhibit an array of LRRs capped by one LRRNT only shares sequence similarity with mouse TLR3 and similarly HmTLR3 present sequence homologies based on the Leucine-rich domain and the TIR domain with TLR7/TLR9. Based on EST medicinal leech sequence and Helobdella genome, four other HmTLR has been detected and presented in Figure 4.3B. Their complete characterization is now in progress. We recently demonstrated that *Hm*TLR1 is localized in both neurons and microglia and expressed upon septic trauma (Figure 4.4A). *Hm*TLR1 is co-expressed with a cytokine related to the endothelial monocyte-activating polypeptide (*Hm*EMAPII) sharing chemoattractive activity [76] (Figure 4.4A). We also demonstrated that *Hm*TLR1 presents an intracellular localization as can be seen in (Figure 4.4B) where it is co-localized with the Early Endosome Antigen-1 (EEA1) [77]. Thus result reflects an endosomal localization of *Hm*TLR1 in leech neurons. The complete intracellular trafficking upon pathogen challenges has to be undertaken; nevertheless, these data represent the first evidence of an intracellular TLR in an invertebrate and confirms that the similarity between *Hm*TLR1 and Human TLR3. Similarly*Hm*TLR3 also presents an intracellular localization due to its similarity with human TLR7/TLR9.

4.5.2   *HMTLR SIGNALING PATHWAYS*

Analysis of the *Hirudo* transcriptome database reveals the presence of putative homologs of all the factors reported to play critical roles in human TLR pathways. This stands in sharp contrast to other invertebrates, such as insects and nematodes, for which the PRR pathways thus far appear

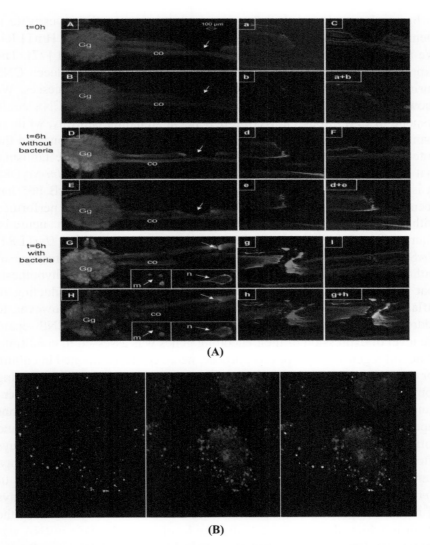

FIGURE 4.4 A: Co-appearance of *Hm*p43/EMAPII (Green) and *HmTLR*1 (Red) in the injured CNS incubated or not for 6 h with killed bacteria. Double staining of injured nerve cords was performed at t = 0 and t = 6 h post-axotomy using the fluorescent nuclear dye Hoechst 33258 (*C, F, and I*) with both the anti-p43/EMAPII (*A, D,* and *G*) and the anti-*HmTLR*1 (*B, E,* and *H*) polyclonal Ab. Immunodetection was performed using green-labeled anti-EMAPII and red labeled anti-*HmTLR*1 secondary Ab. The results demonstrate an accumulation of *Hm*-EMAPII (*G, g*) and *HmTLR*1 (*H, h*) at the lesion site 6 h after axotomy in the presence of bacteria that B: Neuronal localization of HmTLR1 at the endosomal level. Red labeling is HmTLR1 and green the endosomal makers EEA1. (Reprinted with permission from Ref. [76]. © 2009. The American Asssociation of Immunologists, Inc.)

to be much simpler, with many components missing. We investigate if members of the MyD88 family are present an associate to HmTLR1. We report the characterization of *Hm*-MyD88 and *Hm*-SARM [77]. The expression of their encoding gene was strongly regulated in leech CNS during CNS repair, suggesting their involvement in both processes. We showed for the first time that differentiated neurons of the CNS could respond to LPS through a MyD88-dependent signaling pathway, while in mammals, studies describing the direct effect of LPS on neurons and the outcomes of such treatment are scarce and controversial. We established that this PAMP induced the relocalization of *Hm*-TLR1 and *Hm*-MyD88 in isolated neurons (Figure 4.5). Besides *Hm*MyD88, *Hm*NFκB P65 has been also cloned (Figure 4.6) and immunocytochemical studies performed with specific antibodies shown a cellular localization in neurons. Interestingly, using real-time PCR analyses, we observed that *Hm*TLR1, *Hm*SOCs, *Hm*NFkB P105 expression is stimulated in the presence of pathogens in the course of time. These data allow drawing the hypothesis that leech *Hm*TLR1 stimulation induces NFκB pathway conducting to inflammasome stimulation through SOCs like in mammals. However, to address if *Hm*-MyD88 and *Hm*-SARM could be involved in CNS repair, the level of expression of their encoding genes was assessed in conditions of neural regeneration. For this purpose, nerve cords maintained in culture for up to eight days under sterile conditions were lesioned at T0 by cutting completely through half of a connective nerve that links two adjacent ganglia. The time intervals were chosen in reference to the observations reported by Müller, who demonstrated that synaptic connections and normal functions of axotomized leech neurons were restored eight days after injury [48]. Cultures were stopped at different time post axotomy. Our real-time quantitative RT-PCR experiment showed that the level of *Hm*-MyD88 mRNA decreased progressively during the first three days of repair and returned to the basal level on the fourth day and until the repair is achieved. This suggests that the level of *Hm*-MyD88 mRNA was maintained low either to avoid an over inflammation deleterious for the nerve cord or to prevent *Hm*-MyD88 from abolishing the initiation of CNS repair. In contrast, the levels of *Hm*-SARM transcripts increased as soon as six hours and remained elevated all along the repair process. SARM is known to inhibit the TLR signaling pathways specifically via MyD88 and/or TRIF and/or TRAF6. In leech brain, *Hm*-SARM seems to control the complex dynamics of neurite elongation and retraction as well as the elimination of misdirected sprouts observed during axon growth [77].

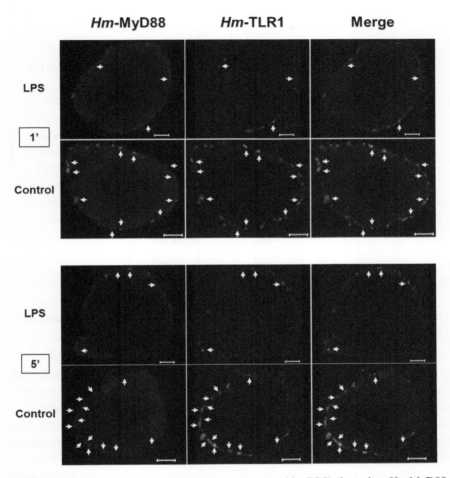

FIGURE 4.5 Neurons could respond to lipopolysaccharide (LPS) through a *Hm*-MyD88 dependent signaling pathway. Dissociated neurons were incubated with 100 ng/mL LPS for 1 and 5 minutes and double-labeled with anti-*Hm*-MyD88 (red) and anti-*Hm*-TLR1 (green). Untreated neurons served as controls. Arrows indicate regions where *Hm*-MyD88 and *Hm*-TLR1 show partial co-localization. Scale bar = 5μm. (Reprinted with permission from Ref. [75]. © 2017 Elsevier.)

4.5.3 HMTLRS EFFECTORS

Proteomic studies performed with CNS infected with a cocktail of bacteria have led to the identification of more than 100 proteins with molecular masses ranged between 12 kDa and 70 kDa which are over-expressed at the transcription level relative to controls. The sites of expression of the genes

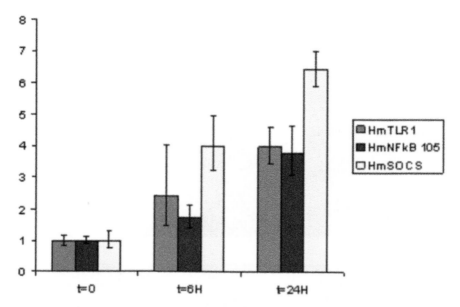

FIGURE 4.6   Real-time PCR analyses of the gene expression of molecules associated with TLR signaling pathway in leech CNS after bacteria challenge

corresponding to these biochemically-identified proteins are currently being identified by *in situ* hybridizations. Some are present uniquely in glial cells, like hemerythrin [78], others in both nervous and immune cells, like the antimicrobial peptides [79]. Interestingly, parallel biochemical studies focused on antimicrobial peptides present in the leech CNS and expressed in the course of infection or trauma have led two novel antimicrobial peptides, *Hm* lumbricin, and neuromacin [80]. Neuromacin and *Hm*-lumbricin exert bactericidal activities against Gram-positive bacteria without any hemolytic properties [80]. We have observed that in addition to exert antimicrobial activities, *Hm*-lumbricin, and neuromacin have regenerative effects on the leech CNS [80]. The capacity of both peptides to promote the regeneration of the leech nerve cord was tested *ex vivo* by adding the neuromacin and/or *Hm*-lumbricin antibody(ies) to axotomized nerve cords in the presence of killed bacteria [80]. Due to the presence of bacteria, the reconnection process should have started two days post-axotomy. It appeared that the presence of antibodies in the culture medium blocked the regeneration process since no reconnection was observed even seven days post-axotomy [80]. These observations were corroborated by the data obtained by adding native neuromacin to axotomized nerve cords under aseptic conditions. Nerve repair was evident sooner in the presence of neuromacin, reconnection starting in less than one day instead of four without an exogenous contribution in neuromacin. The participation of endogenous neuromacin and *Hm*-lumbricin in the neural repair is sustained by the accumulation of both peptides at the lesion site upon bacterial challenge of injured nerve cords [80]. Further investigations based on single-cell RT-PCR analysis and on immunohistochemical analysis of a model, developed by our group, of leech CNS almost completely devoid of microglial cells allowed us to conclude that the presence of both *Hm*-lumbricin and neuromacin at the axotomized site implicates peptide production by neurons and by the microglial cells recruited at the lesion site [80]. However, silencing studies have not allowed connecting such antimicrobial peptides with *Hm*TLR1 receptor (Figure 4.7).

By contrast, a cytokine sharing microglia chemoattractant activity recently characterized by our group in the medicinal leech, e.g., *Hm*EMAPII [76], shown based on RNAi silencing qPCR, western blot experiments (Figure 4.8) and biological tests an association to *Hm*TLR1. *Hm*EMAPII is processed from *Hm*p43 like mammals EMAP II. We hypothesized that *Hm*EMAPII could exert a chemoattractant effect on microglial cells as mammalian EMAPII does on monocytes. The chemoattractive effect of *Hm*EMAPII is blocked when an anti-human EMAPII antibody underscoring for the first time the ability of EMAPII to exert chemotactic effect toward microglial cells through CXCR3 [76]. These data points out that *Hm*TLR1 is linked to leech EMAPII and gives for the first time an immune function to a TLR in a non-ecdysozoan model, i.e., in an invertebrate model different from *C. elegans* and *D. melanogaster*.

Altogether, these data reflect that the medicinal leech express *Hm*TLR related to mammals TLR. The two first receptors characterized shown an intracellular localization and HmTLR1 is linked to the cytokine related to EMAPII which exerts chemoattraction effect after brain trauma or upon pathogen challenge in conjunction with other PRR such like NOD-like receptor [81] or RLR-like receptor (Figures 4.8 and 4.9). All these data confirm for the first time the presence of a complete TLR-signaling-Cytokine pathway implicated in immune response in medicinal leech nervous system. Such complex is also present in mammals reflecting a co-evolution between the medicinal leech and its host mammals again. It has also to be noted that such mechanism conservation is in line with a common origin of nervous system centralization between annelids and vertebrates. Moreover, we report here the conservation in leech brain of the TLR signaling (Figure 4.10).

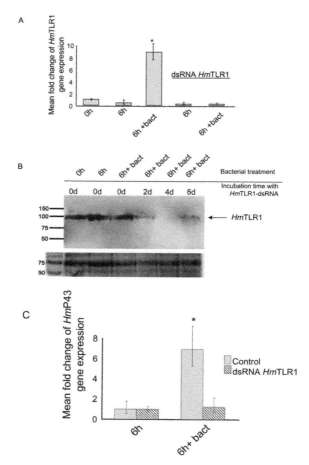

FIGURE 4.7   Impact of the *HmTLR*1 gene silencing on the induction of the *Hm*P43/EMAPII gene in the leech CNS incubated with bacteria. (A) The efficiency of the knock-down was quantified by measuring the level of *HmTLR*1 expression in nerve cords incubated with or without dsRNA. Data are expressed as relative levels comparatively to the basal level of expression measured in nerve cords processed immediately after sampling (0h). *HmTLR*1expression was quantified after 6h of culture without bacteria (6h) or with a mix of heat-killed Gram+ and Gram- bacteria (6h + bact), revealing an induction of *HmTLR*1 gene under septic conditions. The bacterial induction of *HmTLR*1 is significantly reduced when the CNS is incubated for 4 days with *HmTLR*1-dsRNA (B) Western blot analyses of *HmTLR*1 protein level in the same conditions as for Figure 4.1. Best protein extinction is observed at 4 days (4d) of incubation with specific dsRNA confirming the efficiency of the knockdown and the specificity of the anti *HmTLR*1 antibody. d, day (C) *HmTLR*1 gene silencing abolished the bacterial gene induction of *Hm*EMAPII observed in control (without dsRNA *HmTLR*1), indicating a role of the *HmTLR*1 in the gene regulation of this cytokine in the leech CNS under septic conditions. (Reprinted with permission from Ref. [76]. © 2009. The American Asssociation of Immunologists, Inc.)

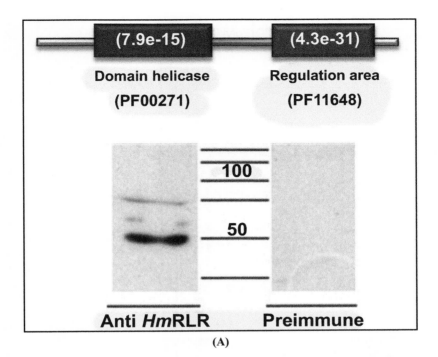

(A)

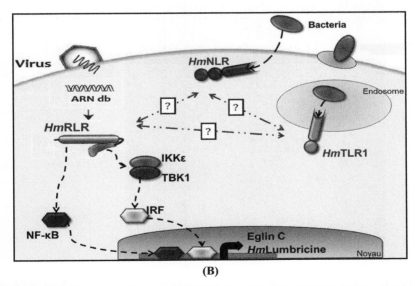

(B)

FIGURE 4.8 A) Structure of *Hirudo* RIG-I-like receptors (HmRLR) B) scheme of the cross-talk between HmTLR1 and HmRLR after bacterial and viral infection and the involvement of the Eglin-C as an anti-viral effector in conjunction with the anti-lubricin antibacterial peptide.

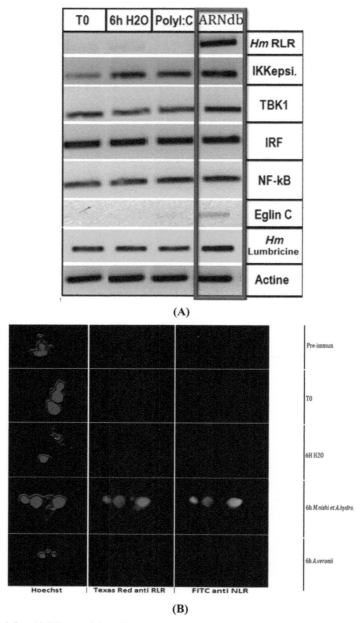

(A)

(B)

FIGURE 4.9 A) Western blot of HmRLR and the signaling pathway involving IKKb and Eglin C after either bacteria or RNA dB stimulation. B) Time course expression of Hm NLR and Hm RLR after infection with *Aeromonas veronii*, or Aeromonas hydra.

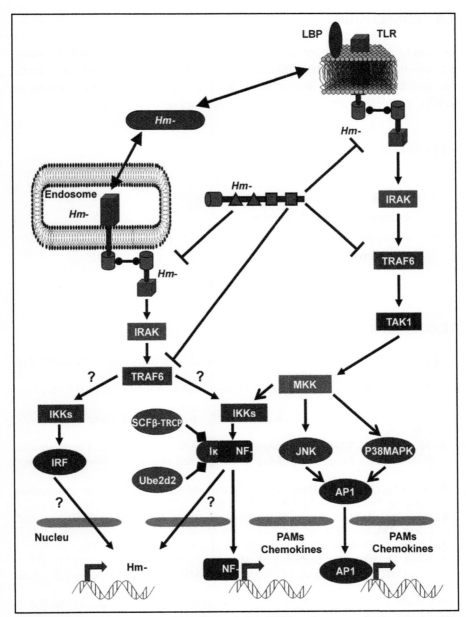

FIGURE 4.10 The putative TLR4-like and *Hm*-TLR1 pathways in leech neurons. The survey of *H. medicinalis* databases and our experiments [30, 33] pointed out that leech possesses (i) the main components of the canonical TLR4/LPS pathway and (ii) the molecules recruited by endosomal TLRs. "?" indicates that stimulation of *Hm*-TLR1 may activate an IRF or a NF-κB signaling.

4.6 CONCLUSION

In conclusion, it is more and more clear that annelids present a neuro-encrino-immune response close to what can be found in vertebrates, leads the idea that evolution forces the neuroendocrine and immune systems to evolve to an integrated response which appears since annelids.

KEYWORDS

- **annelid**
- **hormono-enzymatic system**
- **neuroimmune**
- **neuropeptides**

REFERENCES

1. Wang, W. Z., Emes, R. D., Christoffers, K., Verrall, J., & Blackshaw, S. E., (2005). *Cell Mol. Neurobiol., 25*, 427–440.
2. Kawada, T., Kanda, A., Minakata, H., Matsushima, O., & Satake, H., (2004). *Biochem. J., 382*, 231–237.
3. Voronezhskaya, E. E., Tsitrin, E. B., & Nezlin, L. P., (2003). *J. Comp. Neurol., 455*, 299–309.
4. Veenstra, J. A., (2011). Neuropeptide evolution: Neurohormones and neuropeptides predicted from the genomes of *Capitella teleta* and *Helobdellarobusta. Gen. Comp. Endocrinol., 171*(2), 160–175.
5. Jékely, G., (2013). Global view of the evolution and diversity of metazoan neuropeptide signaling. *Proc. Natl. Acad. USA, 110*(21), 8702–8707.
6. Salzet, M., Bulet, P., Wattez, C., Verger-Bocquet, M., & Malecha, J., (1995). *J. Biol. Chem., 270*, 1575–1582.
7. Laurent, V., Bulet, P., & Salzet, M., (1995). *Neurosci. Lett., 190*, 175–178.
8. Salzet, M., Wattez, C., Baert, J. L., & Malecha, J., (1993). *Brain Res., 631*, 247–255.
9. Salzet, M., Verger-Bocquet, M., Wattez, C., & Malecha, J., (1992). *Comp. Biochem. Physiol. A., 101*, 83–90.
10. Chopin, V., Bilfinger, T. V., Stefano, G. B., Matias, I., & Salzet, M., (1997). *Eur. J. Biochem., 249*, 733–738.
11. Laurent, V., & Salzet, M., (1995). *Peptides, 16*, 1351–1358.
12. Laurent, V., & Salzet, M., (1996). *Peptides, 17*, 737–745.
13. Laurent, V., & Salzet, M., (1996). *FEBS Lett., 384*, 123–127.

14. Riviere, G., Michaud, A., Deloffre, L., Vandenbulcke, F., Levoye, A., Breton, C., Corvol, P., Salzet, M., & Vieau, D., (2004). *Biochem. J., 382*, 565–573.

15. Kawamura, T., Oda, T., & Muramatsu, T., (2000). *Comp Biochem. Physiol. B. Biochem. Mol. Biol., 126*, 29–37.

16. Kawamura, T., Kikuno, K., Oda, T., & Muramatsu, T., (2000). *Biosci. Biotechnol. Biochem., 64*, 2193–2200.

17. Milde, H., Weber, W. M., Salzet, M., & Clauss, W., (2001). *J. Exp. Biol., 204*, 1509–1517.

18. Fewou, J., & Dhainaut-Courtois, N., (1995). *Biol. Cell, 85*, 21–33.

19. Satou, R., Nakagawa, T., Ido, H., Tomomatsu, M., Suzuki, F., & Nakamura, Y., (2005). *Biosci. Biotechnol. Biochem., 69*, 1221–1225.

20. Satou, R., Nakagawa, T., Ido, H., Tomomatsu, M., Suzuki, F., & Nakamura, Y., (2005). *Peptides, 26*, 2452–2457.

21. Salzet, M., & Verger-Bocquet, M., (2001). *Brain Res. Mol. Brain Res., 94*, 137–147.

22. Deloffre, L., Sautiere, P. E., Huybrechts, R., Hens, K., Vieau, D., & Salzet, M., (2004). *Eur. J. Biochem., 271*, 2101–2106.

23. Salzet, M., Bulet, P., Van Dorsselaer, A., & Malecha, J., (1993). *Eur. J. Biochem., 217*, 897–903.

24. Oumi, T., Ukena, K., Matsushima, O., Ikeda, T., Fujita, T., Minakata, H., & Nomoto, K., (1994). *Biochem. Biophys. Res. Commun., 198*, 393–399.

25. Oumi, T., Ukena, K., Matsushima, O., Ikeda, T., Fujita, T., Minakata, H., & Nomoto, K., (1996). *J. Exp. Zool., 276*, 151–156.

26. Van Kesteren, R. E., Tensen, C. P., Smi, A. B., Van Minnen, J., Van Soest, K., Kits, K. S., Meyerhof, W., Richter, D., Van Heerikhuizen, H., & Vreugdenhil, E., (1995). A novel G protein-coupled receptor mediating both vasopressin- and oxytocin-like functions of Lys-conopressin in Lymnaeastagnalis. *Neuron, 15,* 897–908.

27. Levoye, A., Mouillac, B., Riviere, G., Vieau, D., Salzet, M., & Breton, C., (2005). *Theromyzon tessulatum. J. Endocrinol., 184*, 277–289.

28. Salzet, M., Bulet, P., Verger-Bocquet, M., & Malecha, J., (1995). *FEBS Lett., 357*, 187–191.

29. Salzet, M., & Stefano, G. B., (1997). *Brain Res., 768*, 224–232.

30. Fimiani, C., Arcuri, E., Santoni, A., Rialas, C. M., Bilfinger, T. V., Peter, D., Salzet, B., & Stefano, G. B., (1999). *Cancer Lett., 146*, 45–51.

31. Salzet, M., (2001). *Neuro Endocrinol. Lett., 22*, 467–474.

32. Zhang, Y., Xu, Y., Wang, Z., Zhang, X., Liang, X., & Civelli, O., (2012). BmK-YA, an enkephalin-like peptide in scorpion venom. *PLoS One, 7*(7), e40417.

33. Tasiemski, A., Verger-Bocquet, M., Cadet, M., Goumon, Y., Metz-Boutigue, M. H., Aunis, D., Stefano, G. B., & Salzet, M., (2000). *Brain Res. Mol. Brain Res., 76*, 237–252.

34. Bilfinger, T. V., Salzet, M., Fimiani, C., Deutsch, D. G., Tramu, G., & Stefano, G. B., (1998). *Int. J. Cardiol., 64*(1), 15–22.

35. Salzet, M., Verger-Bocquet, M., Bulet, P., Beauvillain, J. C., & Malecha, J., (1996). *J. Biol. Chem., 271*, 13191–13196.

36. Salzet, M., Wattez, C., Bulet, P., & Malecha, J., (1994). *FEBS Lett., 348*, 102–106.

37. Salzet, M., & Stefano, G., (1997). *Mol. Brain Res., 52*, 46–52.

38. Salzet, M., Salzet-Raveillon, B., Cocquerelle, C., Verger-Bocquet, M., Pryor, S. C., Rialas, C. M., Laurent, V., & Stefano, G. B., (1997). *J. Immunol., 159*, 5400–5411.

39. Britz, F. C., & Deitmer, J. W., (2002). *Peptides, 23*, 2117–2125.

40. Britz, F. C., Hirth, I. C., & Deitmer, J. W., (2004). *Eur. J. Neurosci., 19*, 983–992.

41. Keating, H. H., & Sahley, C. L., (1996). *J. Neurobiol., 30*, 374–384.
42. Schmidt, J., & Deitmer, J. W., (1999). *Eur. J. Neurosci., 11*, 3125–3133.
43. Takahashi, T., Matsushima, O., Morishita, F., Fujimoto, M., Ikeda, T., Minakata, H., & Nomoto, K., (1994). *Zoolog. Sci., 11*, 33–38.
44. Tobin, A. E., & Calabrese, R. L., (2005). *J. Neurophysiol., 94*, 3938–3950.
45. Wang, Y., Price, D. A., & Sahley, C. L., (1998). *Peptides, 19*, 487–493.
46. Wang, Y., Strong, J. A., & Sahley, C. L., (1999). *J. Neurophysiol., 82*, 216–225.
47. Li, K. W., Van Golen, F. A., Van Minnen, J., Van Veelen, P. A., Van Der Greef, J., & Geraerts, W. P., (1994). *Mol. Brain Res., 25*, 355–358.
48. Lopez, V., Wickham, L., & Desgroseillers, L., (1993). *DNA Cell Biol., 12*, 53–61.
49. Santama, N., Wheeler, C. H., Burke, J. F., & Benjamin, P. R., (1994). *J. Comp. Neurol., 342*, 335–351.
50. Miller, M. W., Beushausen, S., Vitek, A., Stamm, S., Kupfermann, I., Brosius, J., & Weiss, K. R., (1993). *J. Neurosci., 13*, 3358–3367.
51. Krajniak, K. G., & Greenberg, M. J., (1992). *Comp. Biochem. Physiol. C., 101*, 93–100.
52. Baratte, B., Gras-Masse, H., Ricart, G., Bulet, P., & Dhainaut-Courtois, N., (1991). Isolation and characterization of authentic Phe-Met-Arg-Phe-NH2 and the novel Phe-Thr-Arg-Phe-NH2 peptide from Nereis diversicolor. *Eur. J. Biochem., 198*, 627–633.
53. Krajniak, K. G., & Price, D. A., (1990). Authentic FMRFamide is present in the polychaete Nereis virens. *Peptides, 11*, 75–77.
54. Diaz-Miranda, L., De Motta, G. E., & Garcia-Arraras, J. E., (1992). *J. Exp. Zool., 263*, 54–67.
55. Diaz-Miranda, L., Escalona De Motta, G., & Garcia-Arraras, J. E., (1991). *Cell Tissue Res., 266*, 209–217.
56. Banvolgyi, T., Barna, J., Csoknya, M., Hamori, J., & Elekes, K., (2000). *Acta Biol. Hung., 51*, 409–416.
57. Gershon, T. R., Baker, M. W., Nitabach, M., Wu, P., & Macagno, E. R., (1998). *J. Neurosci., 18*, 2991–3002.
58. O'Gara, B. A., Brown, P. L., Dlugosch, D., Kandiel, A., Ku, J. W., Geier, J. K., Henggeler, N. C., Abbasi, A., & Kounalakis, N., (1999). *Invert Neurosci., 4*, 41–53.
59. Calabrese, R. L., Nadim, F., & Olsen, O. H., (1995). *J. Neurobiol., 27*, 390–402.
60. Salzet, M., Bulet, P., Wattez, C., & Malecha, J., (1994). *Eur. J. Biochem., 221*, 269–275.
61. Wenning, A., Cahill, M. A., Hoeger, U., & Calabrese, R. L., (1993). *J. Exp. Biol., 182*, 81–96.
62. Walker, R. J., Holden-Dye, L., & Franks, C. J., (1993). *Comp. Biochem. Physiol. C., 106*, 49–58.
63. Norris, B. J., & Calabrese, R. L., (1990). *J. Comp. Physiol. [A], 167*, 211–224.
64. Li, C., & Calabrese, R. L., (1987). *J. Neurosci., 7*, 595–603.
65. Kuhlman, J. R., Li, C., & Calabrese, R. L., (1985). *J. Neurosci., 5*, 2301–2309.
66. Kuhlman, J. R., Li, C., & Calabrese, R. L., (1985). *J. Neurosci., 5*, 2310–2317.
67. Salzet, M., Wattez, C., Verger-Bocquet, M., Beauvillain, J. C., & Malecha, J., (1993). *Brain Res., 601*, 173–184.
68. Evans, B. D., Pohl, J., Kartsonis, N. A., & Calabrese, R. L., (1991). *Peptides, 12*, 897–908.
69. Wenning, A., & Calabrese, R. L., (1995). *J. Exp. Biol., 198*, 1405–1415.
70. Salzet, M., Verger-Bocquet, M., Vandenbulcke, F., & Van Minnen, J., (1997). *Mol. Brain Res., 49*, 211–221.
71. Schaller, H. C., (1976). *Cell Differ., 5*, 13–25.

72. Schaller, H. C., (1976). *Cell Differ., 5*, 1–11.

73. Schaller, H. C., & Bodenmuller, H., (1981). *Proc. Natl. Acad. Sci. USA, 78*, 7000–7004.

74. Eduardo, R. M., Terry, G., Lee, E., Vineet, B., Marcelo, B. S., Todd, S., Thomas, C., Corinne, D. S., Patrick, W., Aurélie, T., & Michel, S., (2010). Construction of a medicinal leech transcriptome database and its application to the dentification of leech homologs of neural and innate immune genes. *BMC Genomics, 11*, 407.

75. Tasiemski, A., & Salzet, M., (2017). Neuro-immune lessons from an annelid: The medicinal leech. *Dev. Comp. Immunol., 66*, 33–42.

76. Schikorski, D., Cuvillier-Hot, V., Boidin-Wichlacz, C., Slommiany, C., Salzet, M., & Tasiemski, A., (2009). Deciphering the immune function and the regulation by a TLR of the cytokine EMAPII in the lesioned CNS using the leech as a model. *J. Immunol., 183*(11), 7119–7128.

77. Rodet, F., Tasiemski, A., Boidin-Wichlacz, C., Van Camp, C., Vuillaume, C., Slomianny, C., & Salzet, M., (2015). Hm-MyD88 and Hm-SARM: Two key regulators of the neuroimmune system and neural repair in the medicinal leech. *Sci. Rep., 5*, 9624.

78. Vergote, D., Macagno, E., Salzet, M., & Sautière, P. E., (2006). Proteomic analysis of leech brain after bacterial infection. *Proteomics, 6*(17), 4817–4825.

79. Schikorski, D., Cuvillier, V., Hot, M. L., Macagno, E., & Salzet, M. A., (2008). Tasiemski. The medicinal leech as a model for studying the antimicrobial response of the central nervous system. *J. immunol., 181*(2), 1083–1095.

80. Céline, B. W., David, V., Christian, S., Nathalie, J., Michel, S., & Aurélie, T., (2012). Morphology and function of leech circulating blood cells in immunity and neural repair of the damaged CNS. *Cell Mol Life Sci., 69*(10), 1717–1731.

81. Cuvillier-Hot, V., Boidin-Wichlacz, C., Slomianny, C., Salzet, M., & Tasiemski, A., (2011). Characterization and immune function of two intracellular sensors, HmTLR1 and HmNLR, in the injured CNS of an invertebrate. *Dev. Comp. Immunol., 35*(2), 214–226.

CHAPTER 5

Neuropeptide Signaling in Echinoderms: From "Physiologic Activity of Nerve Extracts" to Neuropeptidomics and Beyond

MAURICE R. ELPHICK

Queen Mary University of London, School of Biological and Chemical Sciences, Mile End Road, London E1 4NS, UK, Tel.: +44(0) 20 7882 6664, Fax: +44(0) 20 7882 7732, E-mail: m.r.elphick@qmul.ac.uk

5.1 INTRODUCTION

An abstract with a mysteriously vague title–"Physiologic activity of nerve extracts"–was published in 1959 on pages 407 and 408 of volume 117 of *The Biological Bulletin* [1]. The authors of the abstract, A.B. Chaet and R.A. McConnaughy from the Department of Biology at The American University in Washington, DC, reported the observation that intracoelomic injection of an extract of radial nerve cords from the starfish *Asterias forbesi* triggers spawning in reproductively mature starfish. The active component was named gamete-shedding substance (GSS), and subsequently, Chaet and colleagues demonstrated that GSS is a polypeptide [2]. Thus, the 1959 abstract heralded the beginnings of research on neuropeptides in echinoderms.

Chaet's rationale for selecting starfish as an experimental model was the recognition that these animals had largely been neglected as a source of eggs and sperm for studies in reproductive and developmental biology. He attributed this to the difficulty in obtaining mature and synchronized gametes from starfish. In sea urchins, mature gametes can be obtained by simply injecting ripe animals with potassium chloride solution, but this method does not work in starfish [3]. Therefore, Chaet investigated whether an endogenous gonadotropic substance present in the nerve cords would trigger the release of mature gametes–and thus GSS was discovered. Furthermore, this was an important novel contribution to the emerging field of neuroendocrinology, which had been pioneered by Enrst and Berta Scharrer in the 1940s [4].

So what is the rationale for using echinoderms as experimental animals for neuroendocrinology? A primary justification is a phylogenetic position that echinoderms occupy in the animal kingdom. Bilaterian animals are subdivided into two major clades: the deuterostomes and the protostomes (Figure 5.1). The vertebrates are deuterostomes, whereas the majority of invertebrates are protostomes, including arthropods (e.g., *Drosophila melanogaster* and other insects), nematodes (e.g., *Caenorhabditis elegans*), mollusks, and annelids. There are, however, some invertebrate deuterostomes, including the chordate sub-phyla that are closely related to vertebrates–the urochordates and cephalochordates–and the ambulacrarians–hemichordates and echinoderms [5–7]. Neuroendocrinologists have generally focused on vertebrates and selected protostomes (arthropods, mollusks, and more recently *C. elegans*), largely for practical reasons and/or for applied research [8, 9]. Until recently, however, the lack of data from invertebrate deuterostomes has hindered efforts to reconstruct the evolutionary history of neuroendocrine systems and to determine orthologous relationships between neuropeptide signaling systems in vertebrates and in protostomes. But with the availability of transcriptomic/genomic sequence data from echinoderms and other invertebrate deuterostomes, important new insights into the evolution of neuropeptide signaling systems are now being obtained [10–13], as discussed in more detail below.

Echinoderms are not, however, solely of interest in providing "missing pieces" in the "jigsaw puzzle" of neuropeptide evolution. These animals are also of intrinsic interest from a physiological perspective. Echinoderms are unique amongst the Bilateria in exhibiting radial symmetry (typically penta-radial) as adult animals. Consequently, echinoderms do not have an anterior brain, but instead, they have a central nervous system comprising a circum-oral nerve ring and five or more radial nerve cords [14, 15]. Therefore, it is of interest to investigate how neuropeptide signaling systems function to regulate physiological processes and behavior in the context of a radial body plan. Furthermore, there are many fascinating aspects of echinoderm physiology. Perhaps most notable is the mutable collagenous tissue (MCT) of echinoderms that changes its mechanical state rapidly under the control of the nervous system [16], with neuropeptides having been identified as candidate regulators of MCT [17]. Another intriguing property of echinoderms is the ability to autotomize and then regenerate body parts; for example, the arms of starfish and brittle stars and the visceral organs of sea cucumbers [18]. Again there is evidence that neuropeptides are involved in regulating these processes [19].

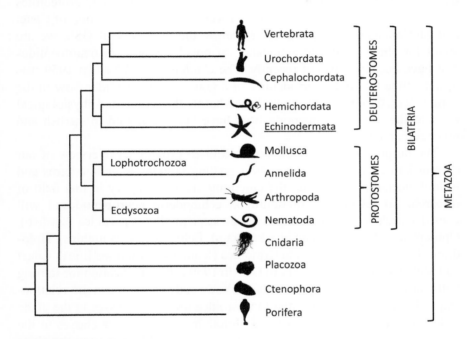

FIGURE 5.1   Animal phylogeny.
Figure 5.1 shows the phylogenetic position of the phylum Echinodermata with respect to other selected animal phyla and sub-phyla. The Metazoa comprise bilaterian phyla and non-bilaterian phyla. The bilaterians comprise two super-phyla: the deuterostomes, which include chordates and echinoderms, and the protostomes, which include ecdysozoans (e.g., the arthropod *Drosophila melanogaster* and the nematode *Caenorhabditis elegans*) and lophotrochozoans (e.g., the mollusk *Aplysia californica*). The non-bilaterians include phyla that lack nervous systems (Porifera, Placozoa) and phyla that have nervous systems (Ctenophora, Cnidaria). Note that the branch lengths in the tree are arbitrary. (Reprinted with permission from Ref. [10]. https://creativecommons.org/licenses/by/3.0/).

The first neuropeptides to be sequenced in an echinoderm were two peptides isolated from the nerve cords of the starfish species *Asterias rubens* and *A. forbesi*, which were named SALMFamide-1 (S1) and SALMFamide-2 (S2) [20]. In keeping with the experimental approaches widely employed at the time of their discovery, these peptides were isolated on account of their cross-reactivity with antibodies to a molluscan FMRFamide-like peptide. Thus, nothing was known about the physiological roles of S1 and S2 in starfish when they were first discovered. The same is true of the plethora of putative neuropeptides that are now being discovered in echinoderms using the modern techniques of transcriptomics, genomics, and proteomics [21–25]. So as we pass the sixtieth anniversary of the publication of Chaet and McConnaughy's [1] abstract reporting the discovery of GSS, we are spoilt for choice with the abundance of novel echinoderm neuropeptides that have been discovered recently. The challenge for Chaet in 1959 was to determine the molecular identity of a gonadotropic peptide. Now in the "omics" era, the challenge is to find out what are the actions and physiological roles of the many neuropeptides that have been identified in starfish and other echinoderms.

The purpose of this review is two-fold–to provide an overview of our current knowledge of neuropeptide signaling systems in echinoderms and to look ahead in identifying the emerging areas of inquiry in this field of research. To review the literature on echinoderm neuropeptides, I will consider each of the five extant classes sequentially: Asteroidea (starfish), Ophiuroidea (brittle stars and basket stars), Echinoidea (sea urchins), Holothuroidea (sea cucumbers) and Crinoidea (feather stars and sea lilies). Chaet and McConnaughy's discovery of GSS in 1959 is the justification for starting with starfish, and the sequence that follows reflects phylogenetic relationships. Thus, the Asteroidea and Ophiuroidea are sister classes in the clade Asterozoa, and the Echinoidea and Holothuroidea are sister classes in the clade Echinozoa. Collectively, the Asterozoa and Echinozoa form the clade Eleutherozoa, with the Crinoidea occupying a basal position with respect to the Eleutherozoa (Figure 5.2) [26, 27].

5.2   NEUROPEPTIDE SIGNALING SYSTEMS IN THE ASTEROIDEA

1.   *From GSS to RGP: The Long Road to Identification of a starfish Gonadotropic Neuropeptide*

The discovery of GSS or gonad stimulating substance (GSS) [1] heralded the beginning of a program of research that continues to this day. Chaet proceeded to investigate the properties of GSS in the 1960s and summarized his findings in two review articles published in 1966 [2, 28]. Subsequently, the Japanese scientists H. Kanatani and H. Shirai became the leading researchers on GSS [29, 30], progressing with investigations of the chemical nature of

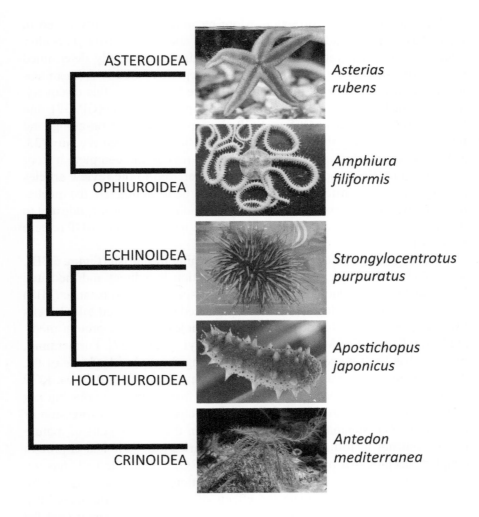

FIGURE 5.2 Phylogenetic relationships of extant echinoderm classes.
The phylum Echinodermata comprises five extant classes: Asteroidea (starfish), Ophiuroidea (brittle stars), Echinoidea (sea urchins), Holothuroidea (sea cucumbers) and Crinoidea (feather stars and sea lilies). The phylogenetic relationships of the five classes are shown on the left, based on the findings of [26, 27]. Photographs of species belonging to each class are illustrated and include *Asterias rubens* (taken by Ray Crundwell), *Amphiura filiformis* (taken by Paola Oliveri), *Strongylocentrotus purpuratus* (taken by Maurice Elphick), *Apostichopus japonicus* (taken by Ding Kui) and *Antedon mediterranea* (taken by Dario Fassini).

GSS and its mechanism of action. More recently, M. Mita, also based in Japan, has been the leading researcher on GSS and in 2009, fifty years after GSS was first discovered, Mita, and colleagues successfully determined the structure of GSS as a heterodimeric polypeptide related to the mammalian reproductive hormone relaxin [31] (Figure 5.3A). With this discovery, Mita renamed GSS as a relaxin-like gonadotropic peptide or RGP [32], and henceforth I will refer to GSS as RGP. Research on RGP is on-going, and reviews summarizing the latest findings have been published recently [33, 34]. It is beyond the scope of this chapter to review the complete history of research on RGP and for this readers are referred to the review articles highlighted above. Instead, the focus here will be to summarize the mechanisms by which RGP exerts its effects as a gonadotropic neuropeptide and to discuss what is known about the physiological mechanisms of RGP release in starfish.

RGP triggers gamete maturation and spawning in starfish (Figure 5.3B). The effect of RGP in causing gamete maturation is indirect and mediated by the maturation-inducing substance or meiosis-inducing substance (MIS) 1-methyl-adenine (1-MeAde), which is produced by associated follicle cells [36]. RGP triggers 1-MeAde production in follicle cells via G-protein-mediated stimulation of cAMP synthesis by adenylyl cyclase [37]. Furthermore, progress has been made recently in elucidating the molecular basis of the development of gonadal responsiveness to the effects of RGP. Thus, RGP triggers 1-MeAde production in follicle cells from mature ovaries but not from young ovaries, and this is explained by changes in the expression of a G-protein $G_s\alpha$-subunit, which is expressed in the follicle cells of mature ovaries but not in follicle cells of young ovaries [38, 39]. The G-protein coupled receptor that mediates the effect of RGP on follicle cells has yet to be identified, but it is likely that it is evolutionarily related to G-protein coupled receptors that mediate the effects of relaxins in vertebrates [40]. Thus, the discovery of the RGP receptor represents an important goal for future research on RGP.

Whilst our knowledge of the molecular mechanisms by which RGP exerts its gonadotropic effects in starfish has advanced significantly, as discussed above, we still know very little about the mechanisms by which RGP is released physiologically. The concentration of RGP in the coelomic fluid of starfish increases prior to spawning, indicating that RGP acts as a hormone [41, 42]. Because RGP was originally isolated from the radial nerve cords of starfish, there has been an assumption that the radial nerve cords are the physiological source of RGP in the context of its role as a gonadotropin. With the molecular identification of RGP, it has become feasible to investigate its pattern of expression in starfish. Analysis of the expression of the RGP precursor in the starfish *A. rubens* using mRNA *in situ* hybridization methods revealed the presence of transcripts in cells located in the ectoneural region of the radial nerve cords and circumoral nerve ring, consistent with original isolation of RGP from this tissue. Furthermore, a dense population of RGP-expressing cells was also revealed at the tips of the arms in close association with sensory organs–the terminal tentacle and the optic cushion [35] (Figure 5.3C–F). The detection of RGP-expressing cells in the arm tips is intriguing because it is suggestive of a mechanism by which environmental cues for spawning, such as changes in day length, temperature, and/or release of gametes by conspecifics, could be detected by sensory cells and relayed to nearby RGP-expressing cells. A key question that remains to be addressed is whether the RGP-expressing cells in the arm tips have processes that project to sites where RGP could be released directly into the coelomic fluid. Recently, antibodies to RGP were generated and used to quantify RGP expression in tissues/organs of the starfish *Patiria pectinifera* using radioimmunoassays and enzyme-linked immunosorbent assays (ELISAs). Consistent with the distribution of RGP precursor transcripts, RGP was detected in the radial nerve cords (1.54 ± 0.09 pmol/mg) and the circumoral nerve rings (0.87 ± 0.04 pmol/mg) but not in other cells/tissues/organs analyzed, which included the pyloric stomach, pyloric caeca, tube feet, ovaries, testes, and ovarian follicle cells [43, 44]. With the development and characterization of antibodies to RGP, there now exist opportunities to use these antibodies to examine the distribution of RGP in starfish using immunohistochemical methods and to identify environmental cues that trigger the release of RGP into the coelomic fluid.

Interestingly, analysis of transcriptome sequence data has revealed the presence of a second relaxin-like peptide precursor in *A. rubens* (ArRLPP2) [23], and subsequently, an ortholog of ArRLPP2 was discovered in the crown-of-thorns starfish *Acanthaster planci* [45]. Currently, nothing is known about the physiological roles of this peptide in starfish, and therefore, this will be an important question to address in the future. Thus, does the second relaxin-like peptide in starfish also act as gonadotropic hormone-like RGP or does it have other functions?

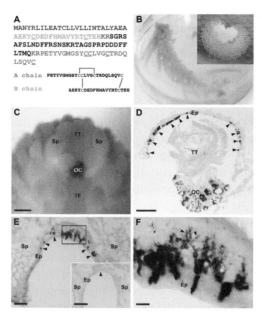

FIGURE 5.3 Relaxin-like gonadotropic peptide (RGP) in the starfish *Asterias rubens*. (A) Sequence of the *A. rubens* RGP (ArRGP) precursor protein. The N-terminal signal peptide is shown in blue, dibasic cleavage sites are shown in green and the A chain and B chain peptides are shown in pink and orange, respectively. The A chain and B chain dimerize to form mature RGP, which has two interchain disulfide bridges and a single intrachain disulfide bridge in the A chain. (B) Ovary dissected from a female specimen of *A. rubens*; the inset shows the effect of synthetic ArRGP in triggering the release of eggs from an ovary fragment *in vitro*. (C) Photograph of a living specimen of *A. rubens* showing the arm tip region viewed under a microscope. The most prominent feature is the pigmented optic cushion, which is located at the base of the terminal tentacle. The terminal tentacle and optic cushion are bounded on each side by spines and rows of tube feet can be seen adjacent to the optic cushion. (D) Localization of ArRGP precursor expression in a transverse section of the arm tip region of *A. rubens*, using mRNA *in situ* hybridization methods with antisense probes. Stained cells expressing ArRGP precursor transcripts (arrowheads) can be seen in the body wall epithelium lining a cavity that surrounds the terminal tentacle and the pigmented optic cushion. (E) Localization of ArRGP precursor expression in a transverse section of the distal region of the arm tip beyond the terminal tentacle in *A. rubens*, using mRNA *in situ* hybridization methods with antisense probes. Stained cells (arrowheads and rectangle) can be seen in the body wall epithelium at the base of two adjacent spines; the region highlighted with a rectangle is shown in panel F. The inset shows absence of staining (arrowhead) in a section of the arm tip adjacent to the section shown in the main panel, which was incubated with sense probes instead of the anti-sense probes, demonstrating the specificity of staining observed with anti-sense probes. (F) Detail of the region highlighted with a rectangle in panel E, showing stained cells with processes (arrowheads) at high magnification. Abbreviations: Ep, epithelium of body wall; OC, optic cushion; TF, tube foot; Sp, spine; TT, terminal tentacle. Scale bars: C, 400 μm; D, 100 μm; E, 50 μm; E inset, 100 μm; F, 10 μm. (Adapted from figures shown in Ref. [35]. © 2016 The Authors The Journal of Comparative Neurology Published by Wiley Periodicals, Inc.)

2. SALMFamides: The First Echinoderm Neuropeptides to be Sequenced

A detailed review of twenty-five years of research on SALMFamides was published in 2014 [46], and it would be superfluous to replicate that here. Therefore, here, I will summarize key discoveries and then go on to briefly review a few papers that have been published since 2014.

Two peptides that were found to be immunoreactive with antibodies to an FMRFamide-like peptide were isolated from the starfish species *A. rubens* and *A. forbesi* and identified as structurally related peptides–the octapeptide GFNSALMFamide and the dodecapeptide SGPYSFNSGLTFamide. The C-terminal pentapeptide of the octapeptide, SALMFamide, was coined as a name for the peptides and the octapeptide was named SALMFamide-1 (S1) and the dodecapeptide was named SALMFamide-2 (S2) [20, 47]. Investigation of the expression of S1 and S2 in *A. rubens* using immunohistochemistry revealed widespread patterns of neuronal expression in larval and adult animals [48–51]. Furthermore, examination of the pharmacological actions of S1 and S2 in *A. rubens* revealed that both peptides act as muscle relaxants [52–54]. At the level of whole-animal behavior, starfish feed by everting their stomach out of their mouth and over the digestible parts of prey (e.g., mussels) and, interestingly, both S1 and S2 cause relaxation and eversion of the stomach in *A. rubens*. Therefore, SALMFamides may be involved in the neural mechanisms that control extra-oral feeding in starfish [54, 55].

Transcriptome sequencing has revealed that S1 and S2 are derived from different precursor proteins in *A. rubens* [23] and the larval expression pattern of transcripts encoding these precursors has been reported [56]. S1 is derived from a precursor that comprises six other related peptides that have in common with S1 a C-terminal LxFamide motif (where x is variable) and hence the S1 precursor is referred to as an L-type SALMFamide precursor (Figure 5.4). Interestingly, some of the other peptides derived from the S1 precursor have an Amino Terminal Cu(II), Ni(II) Binding (ATCUN) motif, and it has been shown that these peptides can bind Cu(II) and form metal linked dimers; however, the functional significance of this property of S1 precursor-derived SALMFamides remains to be determined [57]. Like S1, S2 has a C-terminal LxFamide motif, but it is atypical of the precursor it is derived from, which comprises seven SALMFamides with a C-terminal FxFamide motif and hence is referred to as an F-type SALM-Famide precursor (Figure 5.4). The functional significance of the presence of the L-type SALMFamide S2 in an F-type SALMFamide precursor is not known, but it appears to be an evolutionarily ancient characteristic because it is also observed in orthologous F-type SALMFamide precursors in other starfish and in other echinoderms [58, 59] (Figure 5.4). Further insights into the functional significance of the occurrence of the neuropeptide "cocktails" derived from SALMFamide precursors could be obtained if the receptors that mediate their effects are identified. Therefore, the discovery of SALMFamide receptors is a key objective for the future.

3. Transcriptomic/Genomic Identification of Neuropeptide Precursors in Starfish: From Genes to "Physiologic Activity"

The fifty years that separate the discovery of GSS and its molecular identification as a relaxin-type peptide are indicative of how challenging the purification and molecular identification of bioactive peptides can be [1, 31]. Now in an age of high-throughput transcriptome/genome sequencing, it is far easier to first identify transcripts/genes encoding putative neuropeptide precursors and then proceed toward the molecular and functional characterization of the associated neuropeptides. Thus, sequencing of the neural transcriptome of *A. rubens* recently enabled identification of forty neuropeptide precursors [23]. Similarly, analysis of transcriptomic and genomic sequence data from the crown-of-thorns starfish *Acanthaster planci* enabled identification of orthologs of the precursors found in *A. rubens* as well as several other novel candidate neuropeptide precursors. In addition, mass spectroscopic analysis of *A. planci* tissue extracts enabled determination of the structures of some of the neuropeptides derived from the precursor proteins [45]. Identification of neuropeptide precursors in starfish has provided a valuable new resource for neuropeptide research in these animals. Furthermore, in some cases, the discovery of starfish neuropeptide precursors has provided important insights into the evolution of neuropeptide signaling systems, as discussed below.

Several of the neuropeptide precursors identified in *A. rubens* were the first to be identified outside the phylum Chordata; for example, a precursor of kisspeptin-type neuropeptides and a precursor of a melanin-concentrating hormone (MCH)-type neuropeptide. Thus, new insights into the evolutionary history of kisspeptin-type and MCH-type signaling were obtained [13, 23]. Another important finding to emerge from *A. rubens* neural transcriptome data was the identification of two precursors of gonadotropin-releasing hormone-related peptides–ArGnRH1 and ArGnRH2 [23]. Analysis of *A. rubens* transcriptome sequence data also revealed an ortholog of vertebrate GnRH-type receptors (ArGnRHR) and an ortholog of insect corazonin

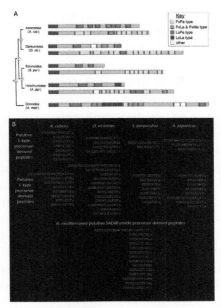

FIGURE 5.4 The occurrence and properties of SALMFamide precursors and SALMFamide peptides in species representing each of the five extant echinoderm classes. (A) SALMFamide precursors are shown in a phylogenetic diagram in accordance with phylogeny shown in Figure 5.2, with crinoids basal to the Echinozoa (Holothuroidea and Echinoidea) and the Asterozoa (Asteroidea and Ophiuroidea). The estimated divergence times for the nodes (labeled with numbers in pentagons) according to [26] are: 1. 501–542 Ma, 2. 482–421 Ma, 3. At least 479 Ma, 4. 464–485 Ma. *A. rub* is the starfish *Asterias rubens* (Asteroidea), *O. vic* is the brittle star *Ophionotus victoriae* (Ophiuroidea), *S. pur* is the sea urchin *Strongylocentrotus purpuratus* (Echinoidea), *A. jap* is the sea cucumber *Apostichopus japonicus* (Holothuroidea), and *A. med* is the feather star *Antedon mediterranea* (Crinoidea). Signal peptides are shown in blue and dibasic or monobasic cleavage sites are shown in green. L-type SALMFamides with a C-terminal LxFamide motif or with an L-type-like motif (e.g., IxFamide) are shown in red. F-type SALMFamides with a FxFamide motif or with an F-type-like motif (e.g., YxFamide) are shown in yellow. SALMFamides with a FxLamide-type motif are shown in orange, and SALMFamides with LxLamide-type motif are shown in dark red. Peptides that do not conform with any of the four color-coded categories are shown in white (e.g., GVPPYVVKVTYamide in *A. japonicus* and SRLPFHSGLMQamide in *O. victoriae*). The diagram shows how in a presumed ancestral-type precursor in crinoids the majority of the putative peptides have a FxLamide-type motif or a LxLamide-type motif and there is only one L-type SALMFamide and one F-type SALMFamide. However, as a consequence of specialization following a presumed duplication of the ancestral-type gene in a common ancestor of the Echinozoa and Asterozoa, two types of SALMFamide precursor have evolved: one that is predominantly comprised of L-type SALMFamides (red) and another that is exclusively or predominantly comprised of F-type SALMFamides (yellow). (B) C-terminal alignments of SALMFamide neuropeptides derived from the precursor proteins shown in A. The C-terminal regions of each peptide are color-coded according to the key shown in A. This figure is an adapted version of Figure 4 from Ref. [59], (© 2015 Elphick, Semmens, Blowes, Levine, Lowe, Arnone and Clark), with sequence data from *Asterias rubens* [23] replacing sequence data from *Patiria miniata*.

receptors (ArCRZR), which are closely related to GnRH receptors. Functional characterization of ArGnRHR and ArCRZR revealed that ArGnRH1 is the ligand for ArGnRHR and ArGnRH2 is the ligand for ArCRZR. Therefore, ArGnRH1 was renamed ArGnRH, and ArGnRH2 was renamed ArCRZ (Figure 5.5A, B). ArGnRHR and ArCRZR were the first neuropeptide receptors to be pharmacologically characterized in starfish. Furthermore, the discovery of ArCRZ was of broader significance because it is the first ligand for a corazonin-type receptor to be identified in a deuterostome. Thus, it was established that the evolutionary origin of paralogous GnRH-type and CRZ-type signaling pathways could be traced back to the common ancestor of protostomes and deuterostomes, but with subsequent loss of CRZ-type signaling in some taxa (e.g., vertebrates, and nematodes) [60, 61].

Molecular identification of novel neuropeptide signaling systems in *A. rubens* has provided a basis for investigation of their physiological roles in starfish. Here I will highlight the progress that has been made so far in the form of published outputs, starting with the GnRH-type and CRZ-type signaling systems introduced above. Analysis of the expression of the ArGnRH and ArCRZ precursors in *A. rubens* using immunohistochemistry and/or mRNA *in situ* hybridization revealed that both precursors are widely expressed but with differences in their patterns of expression. Informed by these findings, the *in vitro* pharmacological effects of ArGnRH of ArCRZ on neuromuscular systems were examined. Both ArGnRH and ArCRZ were found to be myoactive, causing contraction of apical muscle, tube foot, and cardiac stomach preparations. However, ArGnRH was more potent/effective than ArCRZ in its effect on cardiac stomach preparations, whereas ArCRZ was more potent/effective than ArGnRH in its effect on apical muscle and tube foot preparations [62]. As this was the first study to compare the expression and bioactivity of paralogous GnRH-type and CRZ-type neuropeptides in a deuterostome, new insights into the evolution of neuropeptide function in the animal kingdom were obtained from experimental studies on starfish.

Another starfish neuropeptide that causes contraction of cardiac stomach preparations from *A. rubens* is the amidated pentapeptide NGFFYamide, which is an ortholog of neuropeptide-S (NPS) in tetrapod vertebrates and crustacean cardioactive peptide (CCAP) in protostomes [63, 64] (Figure 5.5C), as discussed below in more detail with respect to NGFFFamide, a sea urchin homolog of NGFFYamide. Investigation of the *in vivo* effects of NGFFYamide revealed that it causes reversal of magnesium chloride-induced eversion of the cardiac stomach in *A. rubens* [64]. Thus, NGFFYamide may act physiologically to trigger cardiac stomach retraction in starfish. It is noteworthy that ArGnRH was found not to cause reversal of magnesium

A. GnRH

```
A.rub  pQIHYKNPGWGPGa
O.vic  pQLHSR-MRWEPGa
S.pur  pQVHHRFSGWRPGa
```

C. NG peptide (NPS/CCAP)

```
A.rub     NGFFYa
O.vic(1)  NGFFYa
O.vic(2)  NGFFFa
S.pur     NGFFFa
H.sca1    NGIWYa
H.sca2    NGIWFa
```

E. Luqin

```
A.rub   EEKTRFPKFMRWa
O.vic   pQGFNRDGPAKFMRWa
S.pur   GKPHKFMRWa
H.sca   KPYKFMRWa
```

B. Corazonin

```
A.rub  HNTFTMGGQNRWKAGa
O.vic  HNTFSFKGSNRWNA-a
S.pur  HNTFSFKGRSRYFP-a
H.sca  HNTYSMKGKYRWRA-a
```

D. Vasopressin/Oxytocin

```
A.rub  CLVQDCPEGa
O.vic  CLVSDCPEGa
S.pur  CFISNCPKGa
H.sca  CFVTNCLLGa
```

F. TRH

```
A.rub   pQWYTa
O.vic   pQFSAa
S.pur   pQYPGa
A.jap   pQYFAa
```

G. Calcitonin

```
A.rub     NGESRGCSG-FGGCGVLTIGHNAAMRMLAESNSP-F-GASGPa
O.vic(1)  S-GNGGCAG-FTGCAQLAAGQNALRNFMHSNRASLFTGASGPa
O.vic(2)  N-GNGGCAG-FTGCAQLAAGQSALQAMIHSGRASLF-GSGGPa
S.pur     ---SKGCGS-FSGCMQMEVAKNRVAALLRNSNAHLF-GLNGPa
A.jap(1)  -----SCSNKFAGCAHMKVANAVLKQNSRGQQQFKF-GSAGPa
A.jap(2)  --RVGGCGD-FSGCASLKAGRDLVRAMLRPSK---F-GSGGPa
```

H. Orexin

```
A.rub1  SNADSA-CCARTFRC-NLRSDCTCMVREILCRDPSEGMLNSa
A.rub2  ----NA-CC-RGT-CHDIPPGCNCPYKSYLCGELN--ALTMa
O.vic1  ---DRA-CCRLTTGC-QLRTDCLCVAKEVMCRDPSVGLLNMa
O.vic2  --pQKQSCCRVK-GC-SIPPDCDCPLKQELCKDVTKGILSMa
S.pur1  ---DRA-CCKRTVGC-NLRSDCTCRIREITCTDPSLGLQNYa
S.pur2  --pQSP-CCRRAKGC-SFPPGCHCPLKMSFCGDPSRGLQIVa
H.sca1  ---DRR-CCQRTRVC-KIPSDCTCVTKELVCKYHVRNNIHIa
H.sca2  --pQMG-CCSRVVDC-NIPAGCFCPLKKSMCRDGARRHFISa
```

FIGURE 5.5 Sequences of echinoderm representatives of selected neuropeptide families. Neuropeptides identified in species from four echinoderm classes are shown: Class Asteroidea (A. rub, *Asterias rubens*), Class Ophiuroidea (O. vic, *Ophionotus victoriae*), Class Echinoidea (S. pur, *Strongylocentrotus purpuratus*) and Class Holothuroidea (H. sca, *Holothuria scabra*; A. jap, *Apostichopus japonicus*). Amino acid residues that are conserved in the majority of neuropeptides are underlined; post-translational modifications include amidation (a) and conversion of glutamine to pyroglutamate (pQ). Numbers in parentheses indicate that multiple peptides are derived from the same precursor protein. Numbers without parentheses indicate that related peptides are derived from different precursor proteins. The sequences shown are taken from the following publications: [23–25]. Note that for the TRH-type peptides only the sequences of the most abundant peptide in each precursor are shown. Note also that the numbering of the *Holothuria scabra* orexins is based on sequence similarity with the two orexin types identified in other echinoderms and is different from the numbering in [24].

chloride-induced eversion of the cardiac stomach in *A. rubens* [62]. Thus, although both NGFFYamide and ArGnRH cause dose-dependent contraction of *in vitro* cardiac stomach preparations from *A. rubens*, only NGFFYamide triggers stomach retraction *in vivo*. This may, at least in part, reflect the fact that NGFFYamide is more potent than ArGnRH as a cardiac stomach contractant *in vitro*. However, differences in the patterns of expression of ArGnRH and NGFFYamide in the cardiac stomach may also be relevant here, and therefore investigation of the expression pattern of NGFFYamide in *A. rubens* (and other starfish) represents an important objective for future work.

4.   Discovery of Pedal Peptide/Orcokinin (PP/OK)-Type Neuropeptides as Muscle Relaxants in Starfish

As highlighted above, the first neuropeptides to be identified in starfish were the SALMFamide-type neuropeptides S1 and S2, which act as muscle relaxants [46]. Recently, other neuropeptides that act as muscle relaxants in starfish have been identified. Employing the use of the apical muscle as a bioassay system for myoactive peptides, extracts of the starfish *P. pectinifera* were found to contain a peptide that acts as a muscle relaxant. Purification of this peptide revealed that it is a hexadecapeptide with the amino-acid sequence Phe-Gly-Lys-Gly-Gly-Ala-Tyr-Asp-Pro-Leu-Ser-Ala-Gly-Phe-Thr-Asp and it was named starfish myorelaxant peptide (SMP). A cDNA encoding the SMP precursor revealed the presence of twelve copies of SMP and seven copies of other SMP-like peptides. Furthermore, comparative sequence analysis revealed that SMP and the other SMP-like peptides in *P. pectinifera* belong to a bilaterian family of neuropeptides that include molluscan neuropeptides known as pedal peptides (PPs) and arthropod neuropeptides known as orcokinins (OKs) [65]. SMP also acts as a relaxant of tube foot and cardiac stomach preparations from *P. pectinifera* [65], and analysis of the effects of an SMP-like peptide in *A. rubens* has revealed that it likewise causes dose-dependent relaxation of apical muscle, tube foot and cardiac stomach preparations from this species [66]. Furthermore, analysis of the distribution of SMP-type peptides in *A. rubens* using mRNA *in situ* hybridization and immunohistochemistry has revealed expression in the cell bodies and axonal processes of neurons that innervate muscles [66]. Thus, combining the findings from *P. pectinifera* and *A. rubens*, it appears that SMP-type peptides may act as inhibitory neuromuscular transmitters or modulators throughout the class Asteroidea.

Interestingly, a second precursor of PP/OK-type neuropeptides has been identified in *A. rubens* [23], which is now referred to as *A. rubens* pedal peptide-like neuropeptide precursor 2 (ArPPLNP2) so as to distinguish it from the SMP precursor, which is also referred to as ArPPLNP1 [66]. Investigation of the actions of a peptide derived from ArPPLNP2 has revealed that it also causes relaxation of cardiac stomach preparations, but it has no effect on apical muscle and tube foot preparations [67]. The pattern of expression of ArPPLN2 in *A. rubens* is similar to that of the SMP precursor (ArPPLNP1); however, there are differences in the expression patterns of ArPPLNP1-derived and ArPPLNP2-derived neuropeptides in *A. rubens*, consistent with the differences in bioactivity [67].

The discovery and functional characterization of PP/OK-type neuropeptides in starfish have provided important new insights into the comparative physiology of this family of neuropeptides. Hitherto, the pharmacological actions of PP/OK-type neuropeptides had been characterized only in protostomes, with excitatory effects on muscle preparations being a common theme [68, 69]. Thus, this contrasts with inhibitory effects of PP/OK-type neuropeptides on starfish neuromuscular systems that have been discovered recently. It would be interesting, therefore, to investigate the physiological effects of PP/OK-type neuropeptides in other echinoderms to determine if inhibitory effects on muscle systems are common features of these neuropeptides throughout the phylum Echinodermata.

Surprisingly, nothing is known about the molecular identity of the receptors that mediate the effects of PP/OK-type neuropeptides in any bilaterian. Therefore, if the receptors that mediate the effects of PP/OK-type neuropeptides in starfish or other echinoderms were to be identified, this would have a broad impact in providing insights into the mechanisms by which PP/OK-type neuropeptides exert their effects and the relationships of PP/OK-type neuropeptides with other types of neuropeptides.

5. The Starfish Enterprise: Novel Neuropeptides in Search of a Mission

Only a handful of the novel neuropeptides that have been identified in starfish based upon analysis of genome/transcriptome sequence data [23, 45] have been functionally characterized thus far, as illustrated in the examples discussed above. Therefore, there are numerous opportunities ahead to investigate neuropeptide function in starfish. Representatives of many bilaterian neuropeptide families (vasopressin/oxytocin, calcitonin, thyrotropin-releasing hormone (TRH), orexin, and others [23]; Figure 5.5) have yet to be functionally characterized in starfish. Furthermore, there are other starfish neuropeptides that do not appear to belong to any of the known bilaterian neuropeptide families. For example, "AN peptides," which are characterized by an N-terminal Ala-Asn motif and were first identified in sea urchins [22] but have subsequently been found in other echinoderms, including starfish [23, 25, 45].

Investigation of the patterns of neuropeptide expression in adult starfish have provided an anatomical basis for investigation of their physiological roles, as discussed above with reference to specific studies (e.g., ArGnRH, and ArCRZ [62]). Accordingly, analysis of the anatomical expression patterns and pharmacological actions of the many other neuropeptides that have been identified recently in starfish [23, 45] may shed light on their physiological roles. However, it should be noted here that the investigation of neuropep-tide function in starfish need not be restricted to the adult stage of the life cycle. As highlighted above, with reference to SALMFamides, neuropep-tides are also expressed by neurons in the nervous systems of the bilaterally symmetrical larvae of starfish. Thus, using mRNA *in situ* hybridization tech-niques, the expression of neuropeptide precursors has been examined in the bipinnaria and brachiolaria stage larvae of *A. rubens* [56]. Eight neuropep-tide precursors were analyzed–the SALMFamide-type S1 precursor and S2 precursor, as highlighted above, and precursors of a vasopressin/oxytocin-type peptide ("asterotocin") (Figure 5.5D), NGFFYamide (Figure 5.5C), ArGnRH (Figure 5.5A), TRH-type peptides (Figure 5.5F), a calcitonin-type peptide (ArCT; Figure 5.5G)) and a corticotropin-releasing hormone-type peptide (ArCRH). Expression of the S1, S2 and NGFFYamide precursors was revealed in bipinnaria larvae, but the expression of all eight precur-sors was revealed in brachiolaria stage larvae. Furthermore, expression of the precursors was observed to be associated with the attachment complex, which enables larval attachment to a substratum prior to the metamorphic transition into pentaradially symmetrical juvenile starfish. Therefore, neuro-peptides are likely to be involved in signaling processes associated with the process of larval attachment prior to metamorphosis. Interestingly, several of the neuropeptide precursors were also found to be expressed in cells associ-ated with the ciliary bands, which mediate larval locomotion and generation of currents for feeding on plankton [56]. Recently, high-resolution methods for analyzing the currents generated by the ciliary bands of starfish larvae have been developed [70], and therefore exciting opportunities lie ahead to use these techniques to investigate the physiological roles of neuropeptides as regulators of ciliary activity in starfish larvae.

5.3 NEUROPEPTIDE SIGNALING SYSTEMS IN THE OPHIUROIDEA

1. *SALMFamide-Type Neuropeptide Signaling in Brittle Stars:*

The discovery of the SALMFamides in starfish [20] facilitated investigation of the occurrence and functions of related neuropeptides in other echinoderms, including ophiuroids. Using antibodies to S1 the anatomical distribution of S1-like immunoreactive peptides was investigated in brittle star species [71, 72]. More recently, antibodies to the sea cucumber SALMFamide neuropeptide GFSKLYFamide have also been used to for analysis neuropeptide expression in brittle stars [73]. Furthermore, the modulatory effects of S1 and S2 on light production in bioluminescent brittle stars have also been reported [74]. However, the low potency of S1 or S2 in exerting these effects probably reflects the use of starfish neuropeptides that are not native to brittle stars. Immunocytochemical investigation of SALMFamide expression in brittle stars has also been extended to larvae, with patterns of expression indicative of roles in the regulation of swimming, feeding, and gut activity [75].

Opportunities for further investigation of the physiological roles of SALM-Famides in brittle stars have been facilitated recently with the discovery of transcripts encoding SALMFamide precursors in ophiuroid species [25, 59]. As in starfish, there are two SALMFamide precursors in ophiuroid species. One precursor is orthologous to the *A. rubens* S1 precursor, comprising peptides with a LxFamide motif ("L-type") and the other precursor is orthologous to the *A. rubens* S2 precursor, largely comprising peptides with a FxFamide motif ("F-type") but also containing an L-type peptide (Figure 5.4). Availability of transcriptome sequence data from over fifty brittle star species has enabled an evolutionary analysis of SALMFamide precursor structure with reference to a molecular-based ophiuroid phylogeny, revealing examples of clade-specific gain or loss of SALMFamide neuropeptides [25]. The evolutionary and functional significance of changes in the composition of the neuropeptide "cocktails" derived from SALMFamide precursors in ophiuroids is currently unknown. As with the occurrence of SALMFamide "cocktails" in starfish and other echinoderms (see below), insights into the functional significance of evolutionary changes in neuropeptide precursor composition may emerge if the receptors that mediate the effects of SALMFamides are discovered. Furthermore, the availability of SALMFamide precursor sequences from a variety of brittle star species [25, 59] has also provided a basis for detailed investigations of the anatomical expression patterns and physiological roles of SALMFamides in ophiuroids, building upon earlier studies that relied on use of antibodies to heterologous peptides and pharmacological tests with starfish SALMFamides [71, 74].

2. Transcriptomic Identification of Other Neuropeptide Precursors in Brittle Stars

Analysis of transcriptome sequence data from ophiuroids [25] has enabled identification of orthologs of many of the neuropeptide precursors previously identified in starfish [23, 45]; for example, vasopressin/oxytocin-type, NG peptide-type, GnRH-type, and corazonin-type neuropeptide precursors (Figure 5.5). However, in some cases, multiple precursors of related neuropeptides were identified in ophiuroids. For example, a single precursor of a somatostatin-type neuropeptide was identified starfish [23], but analysis of ophiuroid sequence data revealed two precursors of somatostatin-type peptides–SS1 (an ortholog of the putative neuropeptide original identified in starfish) and SS2 (a novel peptide). Likewise, two precursors of cholecystokinin (CCK)-type peptides were identified in ophiuroids–one precursor comprising three CCK-type peptides and another comprising a single CCK-type peptide. Another notable finding was the discovery of four different corticotropin-releasing hormones (CRH)-type precursors in ophiuroid species, contrasting with the single CRH-type precursor previously found in starfish [25].

Perhaps the most important finding to emerge from the analysis of ophiuroid sequence data was the discovery of representatives of bilaterian neuropeptide families that hitherto had not been identified in starfish or echinoderms. Thus, precursors of neuropeptide-Y (NPY)-type peptides were identified in ophiuroid species, and this then facilitated the discovery of orthologous precursors in starfish species [25]. Another noteworthy finding to emerge from the analysis of ophiuroid sequence data was the identification of echinoderm homologs of the insect neuropeptide eclosion hormone (EH). Hitherto, EH-type peptides had only been identified in arthropods so the discovery of precursors of EH-type peptides in a deuterostome provided the first evidence that the evolutionary origin of this neuropeptide may date back to the common ancestor of the Bilateria.

3. Analysis of the Expression and Functions of Neuropeptides in Brittle Stars

The molecular characterization of many types of neuropeptide precursors in ophiuroids [25], as discussed above, has provided a superb resource for investigations of neuropeptide expression and function in brittle stars. Thus far, there has been no progress in this regard. However, it should be noted that recently a very detailed analysis of the anatomy of the ophiuroid nervous system has been reported, utilizing electron microscopy, molecular markers and 3D reconstructions [73], extending, and re-evaluating earlier anatomical studies [76–78] and studies that employed use of dye-filling of neurons in combination with electrophysiological recordings of neuronal activity [79]. These anatomical studies provide a valuable framework for the analysis of the patterns of neuropeptide expression in ophiuroids, as revealed by the use of techniques such as mRNA *in situ* hybridization and immunohistochemistry. In particular, it will be interesting to compare the expression patterns of orthologous neuropeptides in ophiuroids and asteroids to investigate the conservation and/or diversification of neuropeptide function in the asterozoan clade of the phylum Echinodermata. For example, are neuropeptide types that are expressed by motoneurons in the hyponeural region of the starfish nervous system (see above) likewise expressed by motoneurons in the hyponeural region of the brittle star nervous system? Furthermore, if the success of J.L.S. Cobb in making electrophysiological recordings from ophiuroid nervous systems [79] can be replicated, then the effects of neuropeptides on neuronal activity in brittle stars could be examined. Other areas of interest for further investigation are the roles of neuropeptides as candidate regulators of arm autotomy and regeneration in brittle stars [80–82] and the roles of neuropeptides in mechanisms of neural control of whole-animal locomotory behavior [83].

5.4   NEUROPEPTIDE SIGNALING SYSTEMS IN THE ECHINOIDEA

1.   *SALMFamide-Type Neuropeptide Signaling in Sea Urchins*

As with ophiuroids, the availability of antibodies to the starfish SALM-Famides S1 and S2 enabled investigation of the occurrence of related peptides in echinoids. Immunocytochemical studies on larvae of the sand dollar *Dendraster excentricus* revealed the distribution of S1-like immunoreactivity [84]. Subsequently, the distribution of SALMFamide-like immunoreactivity in larvae of the sea urchin *Psammechinus miliaris* was reported [85]. Furthermore, efforts were made to purify and sequence SALMFamides from the extracts of the sea urchin *Echinus esculentus*, employing radioimmunoassays for S1 and S2. However, only a partial N-terminal sequence (MRYH) of one purified peptide was determined [86]. As discussed in more detail below, it was the sequencing of the genome of the sea urchin *Strongylocentrotus purpuratus* that transformed opportunities for determination of the sequences of sea urchin neuropeptides [87, 88]. Analysis of *S. purpuratus* transcriptome/genome sequence data revealed two SALMFamide precursors. The first to be identified was an F-type precursor comprising seven peptides with a C-terminal FxFamide motif [86] (Figure 5.4). Then an *S. purpuratus* L-type precursor was identified that comprises two SALMFamide neuropeptides [89] (Figure 5.4), including MRLHPGLLFamide–a homolog of the partially sequenced (MRYH) peptide that was purified from *E. esculentus* [86].

Little is known about the anatomical distribution of SALMFamide-type neuropeptide expression in adult echinoids, although recently it was reported that antibodies to the sea cucumber SALMFamide neuropeptide GFSKLY-Famide label processes in the ectoneural region of the radial nerve cords in the sea urchin *Echinometra lucunter* [14]. Likewise, little is known about the physiological roles of SALMFamides in echinoids. However, *in vitro* pharmacological tests with the starfish SALMFamides S1 and S2 revealed that both peptides cause relaxation of tube foot preparations from the sea urchin *E. esculentus* [86]. Thus, SALMFamides act as muscle relaxants not only in starfish but also in other echinoderms. With the availability of the sequences of SALMFamide precursors from *S. purpuratus* and other sea urchin species, there now exist opportunities to investigate the physiological roles of SALMFamides in sea urchins more extensively, both in larval and adult animals.

2.　Transcriptomic/Genomic Identification of Neuropeptide Precursors in Sea Urchins

The genome of the sea urchin *S. purpuratus* was the first to be sequenced in an echinoderm. A large number of candidate G-protein coupled neuropeptide receptors were identified from analysis of the genome sequence data, but only a few neuropeptide precursors were identified. These included precursors of a vasopressin/oxytocin-type neuropeptide ("echinotocin"), bursicon-type neuropeptides and glycoprotein hormone-type polypeptides [87]. Subsequently, a systematic analysis of neural transcriptome sequence data from *S. purpuratus* enabled identification of transcripts encoding twenty neuropeptide precursors [22]. For example, a precursor of peptides that exhibit sequence similarity with vertebrate TRH was a noteworthy finding because this was the first TRH-type precursor to be discovered in an invertebrate. Other vertebrate neuropeptides for which homologs have been identified in *S. purpuratus* include calcitonin, CCK, GnRH, kisspeptin, orexin, MCH and somatostatin [11, 12, 22, 25] (Figure 5.5).

From a different perspective, analysis of sea urchin sequence data also enabled the discovery of homologs of neuropeptides that hitherto had only been identified in protostomes. For example, two precursors (SpPPLNP1, SpPPLNP2) of pedal peptide/orcokinin (PP/OK)-type neuropeptides–a family of neuropeptides that was first discovered in mollusks and arthropods. Thus, the discovery of PP/OK-type neuropeptides in an echinoderm revealed that the evolutionary origin of this neuropeptide family could be traced to a bilaterian common ancestor of protostomes and deuterostomes [22]. Currently, nothing is known about the physiological roles of PP/OK-type peptides in sea urchins; however, progress has been made in the functional characterization of PP/OK-type peptides in starfish, as discussed above and in [65–67]. Another example of a neuropeptide identified in sea urchins that belongs to a neuropeptide family that hitherto had only been found in protostomes is luqin (Figure 5.5E). The neuropeptide luqin was originally discovered in the mollusk *Aplysia californica* [90], and subsequently, related peptides have been identified in other protostomes, including arthropod RYamide-type neuropeptides [91, 92]. Comparative analysis of genomic sequence data enabled identification of a protein in the sea urchin *S. purpuratus* comprising a luqin/RYamide-type neuropeptide and a C-terminal domain containing two cysteine residues that are highly conserved amongst luqin/RYamide-type precursors [11]. The discovery of this precursor was consistent with the presence of luqin/RYamide-type receptors in echinoderms and hemichordates [11, 12]. Thus, the discovery of a luqin-type neuropeptide precursor and receptor in the sea urchin *S. purpuratus* established that luqin/RYamide-type signaling originated in a common ancestor of protostomes and deuterostomes, but with subsequent loss in chordates [11]. Furthermore, the discovery of the luqin/RYamide-type precursor in *S. purpuratus* facilitated the discovery of luqin/RYamide-type precursors in other echinoderms, including sea cucumbers [21], starfish [23, 45] and brittle stars [25]. However, currently, nothing is known about the physiological roles of luqin/RYamide-type neuropeptides in sea urchins or other echinoderms.

In addition to precursors of neuropeptides that are homologs of known neuropeptides from other phyla, other sea urchin precursor proteins comprise peptides that do not appear to exhibit sequence similarity with any known neuropeptides. For example, Spnp13, which comprises a peptide with the amino-acid sequence LPANLARE [22]. However, progress has been made in establishing relationships for some of the neuropeptides identified in *S. purpuratus*. For example, we now know that the sea urchin precursor originally designated as "Spnp12" [22] is the precursor of a peptide that is an ortholog of protostome corazonin-type neuropeptides [60, 61].

3. Discovery of the Sea Urchin Neuropeptide NGFFFamide and its Cognate Receptor Provides New Insights into Neuropeptide Evolution in the Bilateria

One of the most interesting discoveries to emerge from the analysis of the *S. purpuratus* genome/transcriptome sequence data was the discovery of a precursor comprising two copies of the neuropeptide NGFFFamide [93]. This precursor was discovered on account of the sequence similarity that NGFFFamide shares with the myoactive neuropeptide NGIWYamide, which had been discovered previously in the sea cucumber *Apostichopus japonicus* (see below and [94]); Figure 5.5C). Surprisingly, analysis of the sequence of the NGFFFamide precursor revealed the presence of a C-terminal neurophysin domain, which was an unexpected finding because neurophysins hitherto had only been found in the precursors of vasopressin/oxytocin-type neuropeptides (including the sea urchin "echinotocin" precursor), where they participate in intracellular transport of the mature vasopressin/oxytocin-type neuropeptides [95, 96]. The presence of a neurophysin domain in the *S. purpuratus* NGFFFamide precursor suggested a close evolutionary relationship with vasopressin/oxytocin-type precursors [93]. Subsequently, it was discovered that the NGFFFamide precursor is one of a family of precursor proteins in invertebrate deuterostomes that comprise neuropeptides that have an Asn-Gly (NG) motif ("NG peptides") and, like the NGFFFamide precursor, have a C-terminal neurophysin domain [97]. Notably, the NG peptide precursor of the cephalochordate *Branchiostoma floridae* comprises two copies of the peptide SFRNGVamide, which is identical to the N-terminal region of the vertebrate neuropeptide NPS. Thus, a relationship between NG peptides and vertebrate NPS was established, and a relationship with vasopressin/oxytocin-type signaling was again revealed because the NPS receptor is closely related to vasopressin/oxytocin-type receptors [98]. Furthermore, NPS receptors are orthologous to the receptors for CCAPs in arthropods, which exhibit some structural similarity with vasopressin/oxytocin-type peptides. Therefore, it was postulated that NG peptides are ligands for NPS/CCAP-type receptors in invertebrate deuterostomes [13, 99]. To test this hypothesis, the NPS/CCAP-type receptor from *S. purpuratus* was cloned and expressed heterologously in Chinese hamster ovary (CHO) cells so that NGFFFamide could be tested as a ligand for this receptor. Importantly, it was discovered that NGFFFamide is a potent ligand for the *S. purpuratus* NPS/CCAP-type receptor, with an EC_{50} of 0.4nM [63]. Furthermore, based on this finding, it was inferred that NGFFYamide and NGIWYamide are ligands for NPS/CCAP-type receptors in the starfish *A. rubens* and the sea cucumber *A. japonicus*, respectively [63].

The discovery that NG peptides are ligands for NPS/CCAP-type receptors in echinoderms provided key evidence in support of a scenario of neuropeptide evolution in the Bilateria. Thus, following duplication of a vasopressin/oxytocin-type signaling system in a common ancestor of the Bilateria, one copy of the system retained the ancestral features and gave rise to the highly conserved vasopressin/oxytocin-type peptides and receptors that occur throughout the Bilateria. In contrast, the other copy of the system diverged, but this took different courses in protostomes and deuterostomes. In the protostomes, the neurophysin domain was lost from the precursor protein, but the neuropeptide derived from the precursor (CCAP) retained a vasopressin/oxytocin-like feature–the presence of a disulfide bridge. In the deuterostomes, the neurophysin domain was retained (although it was subsequently lost in vertebrates and holothurians), but the neuropeptide(s) derived from the precursor (NG peptides and NPS) lost the disulfide bridge that is characteristic of vasopressin/oxytocin-type peptides [63].

The retention of a neurophysin domain in the sea urchin NGFFFamide precursor (and starfish and brittle star NG peptide precursors) is interesting because its functional significance remains unknown. As highlighted above, in vasopressin/oxytocin-type precursors the neurophysin domain binds vasopressin or oxytocin and is required for targeting of vasopressin/oxytocin-type peptides to the regulated secretory pathway [95, 96]. Therefore, it is possible that neurophysins derived from deuterostome NG peptide precursors also have this role. However, there is a 1:1 stoichiometry in the interaction between neurophysins and vasopressin/oxytocin type peptides, which reflects the occurrence of single copies of neurophysins and vasopressin/oxytocin-type peptides in the precursor proteins [95, 96]. In contrast, in the sea urchin, starfish, brittle star, and cephalochordate NG peptide precursors, there are two NG peptide copies combined with a single copy of neurophysin [63]. Therefore, if neurophysins bind NG peptides in a manner similar to the interaction between vasopressin/oxytocin-type peptides, then a 2:1 stoichiometry would be expected. However, the occurrence of an NG peptide precursor in the hemichordate *Saccoglossus kowalevskii* comprising a single neurophysin in combination with six NG peptides represents a challenge to this hypothesis. Furthermore, the loss of neurophysin in the sea cucumber NGIWYamide precursor and in vertebrate NPS precursors suggests that neurophysins are not essential for the biosynthesis of NG peptides/NPS in deuterostomes. If this is the case, then what is the function of neurophysin in NG peptide precursors that have retained neurophysins? This represents an interesting line of inquiry for further research on the sea urchin NGFFFamide precursor and other neurophysin-containing NG peptide precursors.

4. Functional Analysis of Neuropeptide Signaling Sea Urchins

Currently, very little is known about the physiological roles of neuropeptides in echinoids. To the best of my knowledge, papers reporting the effects of echinotocin (Figure 5.5D) and NGFFFamide (Figure 5.5C) in causing contraction of sea urchin tube foot and oesophagus preparations [93] and the effects of the starfish neuropeptides S1 and S2 in causing relaxation of sea urchin tube foot preparations [86] are the only studies that have examined the pharmacological actions of neuropeptides in echinoids. Furthermore, with the exception of studies reporting the presence of SALMFamide-type neuropeptides in larval [84, 85] and adult [14] echinoids (see above), little is known about the patterns of neuropeptide expression in echinoids. With the discovery of many novel neuropeptides in echinoids, as discussed above, there is tremendous scope for further work in this area of inquiry. An indication of the possibilities was a recent report revealing the expression of insulin-like peptides in the digestive system of the pluteus larvae of *S. purpuratus* [100]. Thus, insulin-like peptides may be involved in the regulation of digestive physiology in larval echinoids.

Another aspect of larval echinoid biology that is likely to be regulated by neuropeptides is the ciliary system that mediates both feeding and locomotion. Studies on other larval marine invertebrates such as the annelid *Platynereis dumerilii* and the brachiopod *Terebratalia transversa* have revealed that neuropeptides are involved in the neural control of swimming depth and predator avoidance behavior [101, 102]. The few studies that have examined neuropeptide expression in sea urchins have revealed the presence of neuropeptide immunoreactivity in fiber tracts associated with the ciliary bands of sea urchin larvae [84, 85]. Therefore, as discussed above with respect to starfish larvae, there are opportunities for studies examining the effects of neuropeptides on echinoid larval behavior.

5.5   NEUROPEPTIDE SIGNALING SYSTEMS IN THE HOLOTHUROIDEA

1. SALMFamide-Type Neuropeptide Signaling in Sea Cucumbers

The use of antibodies to a molluscan FMRFamide-like neuropeptide to monitor purification of the starfish SALMFamides S1 and S2 established a methodology that was then applied to other echinoderms. Two SALM-Famide-type peptides were purified from the sea cucumber *Holothuria glaberrima* and identified as GFSKLYFamide and SGYSVLYFamide [103]. Investigation of the distribution of GFSKLYFamide in *H. glaberrima* using immunohistochemistry revealed a widespread pattern of neuronal expression, with immunolabeled cells and/or processes detected in the radial nerve cords, body wall and digestive system [104]. Furthermore, consistent with relaxing actions of S1 and S2 in starfish, GFSKLYFamide was found to cause relaxation of intestinal and body wall longitudinal muscle preparations from *H. glaberrima* [105]. Other members of the SALMFamide neuropeptide family were discovered in the sea cucumber *Apostichopus japonicus* as part of a systematic effort to identify bioactive peptides that exert effects on neuromuscular preparations [106]. Two SALMFamides that cause relaxation of *A. japonicus* intestine preparations were purified and identified as GYSPFMFamide and FKSPFMFamide. The discovery of these two peptides revealed for the first time the occurrence of F-type (FxFamide) SALMFamides in echinoderms. More recently, the sequence of the precursor protein that GYSPFMFamide and FKSPFMFamide are derived from has been determined, revealing that it also comprises six other SALMFamide neuropeptides, including L-type (LxFamide) SALMFamides [107] (Figure 5.4). A second SALMFamide precursor in *A. japonicus* that comprises three L-type SALMFamides has also been identified [58] (Figure 5.4). Furthermore, the sequences of SALMFamide precursors have also been determined recently in *H. glaberrima* and *H. scabra* as part of a comprehensive transcriptomic analysis of neuropeptide precursors in these species [24].

With the identification of the two types of SALMFamide precursors in several sea cucumber species, each comprising multiple neuropeptides, there is scope to investigate the expression of these precursors and the actions of peptides derived from them in both larval and adult sea cucumbers. In particular, nothing is known about the expression pattern and pharmacological actions of peptides derived from the L-type SALMFamide precursor in sea cucumbers, and so this represents an interesting line of inquiry for the future.

2. Discovery of Neuropeptides That Act as Regulators of Muscle and/or Collagenous Tissue in the Sea Cucumber A. japonicus

As highlighted above, purification of components of extracts of the sea cucumber *A. japonicus* that cause relaxation of *in vitro* preparations of intestinal tissue from this species led to the identification of two F-type SALMFamide neuropeptides [106]. Furthermore, many other myoactive neuropeptides were identified using the intestine and the body wall radial longitudinal muscle (RLM) for bioassays. These included peptides that cause intestinal contraction (GLRFA, holokinin, KHTAYTGIamide) and sixteen peptides that potentiate or inhibit electrically-evoked contraction of the RLM [94, 106]. More recently, the precursor proteins that give rise to many these bioactive peptides have been identified by analysis of transcriptome sequence data [107]. Importantly, this has revealed that most of the peptides are derived from proteins with an N-terminal signal peptide, consistent with the notion that these peptides are secreted signaling molecules. One exception, however, is the peptide holokinin, which was found to be derived from collagen [107]. Since the discovery of these myoactive peptides in *A. japonicus*, in most cases, nothing more has been learned about their expression and actions in this species. However, two of the peptides (stichopin and NGIWYamide) have been investigated in more detail, as discussed below.

A peptide named stichopin was purified from extracts of *A. japonicus* on account of its effect in causing inhibition of electrically-evoked contraction of RLM preparations [94]. Stichopin is a 17-residue peptide with two cysteine residues that form a disulfide bridge, and it is derived from a 39-residue precursor protein that simply comprises an N-terminal signal peptide and the stichopin sequence [107]. Further investigation of the pharmacological actions of stichopin revealed that it also causes inhibition of acetylcholine-induced stiffening of *in vitro* preparations of the collagenous body wall of *A. japonicus* [17]. This effect of stichopin provided a basis for investigation of the expression of this peptide in *A. japonicus* using immunohistochemistry. Consistent with the effects of stichopin on collagenous body wall tissue, stichopin was found to be present in collagenous tissue associated with a variety of different organ systems. More specifically, stichopin was found to be expressed in two types of cells: neuron-like cells with processes and non-neuronal oval-shaped cells without processes. Informed by these findings, it is thought that stichopin acts as a neuropeptide derived from neurons and as a hormone derived from oval-shaped secretary cells to regulate the mechanical properties of MCT in sea cucumbers [108].

An amidated pentapeptide with the amino-acid sequence Asn-Gly-Ile-Trp-Tyr-NH_2 (NGIWYamide; Figure 5.5C) was purified from extracts of *A. japonicus* on account of its effect in causing contraction of RLM preparations [94]. Subsequently, it was found that NGIWYamide also causes stiffening of *in vitro* preparations of the collagenous body wall of *A. japonicus* [17], contraction of tentacle preparations and inhibition of spontaneous contractile activity of intestine preparations [109]. Consistent with these pharmacological actions of NGIWYamide, immunohistochemical analysis of the expression of this peptide in *A. japonicus* revealed that it is present in neuronal cell bodies and/or processes located in the radial nerve cords, circumoral nerve ring, tube feet, tentacles, intestine, and body wall dermis [109].

The sequence of the *A. japonicus* NGIWYamide precursor has been determined from analysis of transcriptomic sequence data, revealing that it comprises five copies of NGIWYamide [107]. Similarly, a precursor comprising five copies of NGIWYamide has been identified in *Holothuria glaberrima*, whereas in *Holothuria scabra* the NGIWYamide precursor comprises four copies of NGIWYamide and one copy of the structurally similar peptide NGIWFamide [24] (Figure 5.5C). As discussed above, NGIWYamide belongs to a family of neuropeptides in invertebrate deutero-stomes that are characterized by an Asn-Gly motif–"NG peptides" (Figure 5.5C) -, which include NGFFFamide in sea urchins [93], NGFFYamide in starfish [64], both NGFFFamide and NGFFYamide in brittle stars [25, 63], NGFYNamide, and NGFWNamide in the hemichordate *Saccoglossus kowalevskii* [97] and SFRNGVamide in the cephalochordate *Branchiostoma floridae* [97]. As also discussed above, a unifying feature of these peptides is that they are derived from precursor proteins that have C-terminal neuro-physin domain, which reflects an evolutionary relationship with the neuro-physin-containing precursors of vasopressin/oxytocin-type neuropeptides [97]. Interestingly, the NGIWYamide precursor in *A. japonicus* and other sea cucumber species has lost the neurophysin domain [24, 107], but the functional significance of this loss remains to be investigated.

3. Discovery of Gonadotropic Peptides in Sea Cucumbers

Sea cucumbers are used as foodstuffs in China, Japan, and other Asian countries, and consequently, they have high economic value. With the depletion of natural populations, methods for aquaculture of economically important sea cucumber species have been established [110]. However, there are challenges associated with sea cucumber aquaculture, including methods for reliably obtaining gametes. A widely used method to induce spawning is thermal and/or mechanical shocking, but there are concerns regarding the quantity and quality of viable gametes obtained using this approach [111]. Therefore, efforts to purify and identify endogenous regulators of gamete release in sea cucumbers have been initiated.

In 2009, Kato et al., [112] reported the purification of a peptide named cubifrin that triggers oocyte maturation and ovulation of ovarian tissue *in vitro* [112]. Determination of the structure of cubifrin revealed that it is identical to the neuropeptide NGIWYamide that had previously been identified in *A. japonicus* as a muscle contractant (see above and [94, 109]). Interestingly, a synthetic analog of NGIWYamide, NGLWYamide, was found to be 10–100 times more potent than NGIWYamide. Furthermore, injection of NGIWYamide or NGLWYamide into sexually mature animals induced a characteristic spawning behavior, where animals raise and shake their anterior (oral) region prior to releasing gametes from the genital pore. In addition to the identification of NGIWYamide as a gonadotropic peptide, Kato et al., also purified a second gonadotropic peptide from *A. japonicus* that was identified as the amidated heptapeptide QGLF-SGVamide. However, QGLFSGVamide was found to be much less potent than NGIWYamide.

In parallel with the discovery of NGIWYamide and QGLFSGVamide as gonadotropic peptides in *A. japonicus* [112], another paper reporting the identification of a gonadotropic peptide in *A. japonicus* was published in 2009 [113]. Radial nerve cord extracts were found to contain a peptide or peptides that induce germinal vesicle breakdown in oocytes. Efforts to purify the bioactive peptide led to partial sequencing of a 4.8 kDa polypeptide comprising the amino acid sequence AEIDDLAGNIDY. Furthermore, a partial cDNA sequence was determined that comprised an open reading frame encoding the AEIDDLAGNIDY sequence. Based on these sequence data, a 43-residue peptide was synthesized, tested for gonadotropic activity, and found to cause germinal vesicle breakdown in 50% of immature ovarian oocytes at a concentration of 6 μM. Furthermore, a synthetic peptide corresponding to the N-terminal 21 residues of the 43-residue peptide was also found to exhibit gonadotropic activity. However, BLAST analysis of *A. japonicus* transcriptome sequence data using the amino acid sequence AEIDDLAGNIDY as a query reveals that it is part of a myosin heavy chain (M.R. Elphick, unpublished data). Therefore, the physiological significance of the gonadotropic activity of the peptide identified by Katow et al., is unclear.

Further studies are now required to investigate the occurrence and properties of other gonadotropic peptides in sea cucumbers, in addition to NGIWY-amide and QGLFSGVamide. Recently, analysis of transcriptome sequence data from the sea cucumbers *Holothuria scabra* and *Holothuria glaberrima* revealed the presence of a transcript encoding a precursor of a relaxin-like peptide that is closely related to starfish RGP [24]. Therefore, it will be of interest to investigate if the RGP-like molecule derived from this precursor acts as a gonadotropic peptide in sea cucumbers.

4. Transcriptomic Identification of Neuropeptide Precursors in Sea Cucumbers

Generation of transcriptome sequence data from the sea cucumber *A. japonicus* [114] enabled identification of transcripts encoding a number of neuropeptide precursors, including precursors of TRH-type, calcitonin-type, pedal peptide-type, luqin-type, glycoprotein hormone-type and bursicon-type peptides [21] (Figure 5.5). More recently, precursors of other neuropeptides have been identified in *A. japonicus*, including CCK-type, corazonin-type, kisspeptin-type, orexin-type, pigment dispersing factor (PDF)-type, somatostatin-type, and vasopressin/oxytocin-type neuropeptides [24, 25]. Furthermore, two transcripts encoding calcitonin-type precursors were identified in *A. japonicus*–one precursor comprising two calcitonin-type peptides (AjCT1 and AjCT2; Figure 5.5G) and the second precursor comprising a single copy of AjCT2. Comparison of the sequence of these two precursors indicates that they are products of the same gene, but with inclusion or exclusion of AjCT1 determined by alternative splicing of transcripts [24, 25]. Furthermore, opportunities to identify genes encoding neuropeptide precursors and neuropeptide receptors have emerged recently with the sequencing of the genome of *A. japonicus* [115]. With the availability of this important resource, it will now be possible to identify the complete complement of neuropeptide-related genes in a sea cucumber species for the first time. Furthermore, analysis of gene structure and gene synteny may provide new insights into relationships between neuropeptides in sea cucumbers and neuropeptides in other echinoderms and neuropeptides in other phyla.

Transcriptomic identification of neuropeptide precursors has also been extended to two other sea cucumber species–*Holothuria scabra* and *Holothuria glaberrima* [24]. Furthermore, the expression of three neuropeptide precursors was investigated specifically in *H. scabra*. Thus, analysis of expression of transcripts encoding both the long isoform (including HsCT1 and HsCT2) and the short isoform (including the only HsCT2) of calcitonin-type precursors revealed that both isoforms are expressed in the circumoral nerve ring, radial nerve cords, intestine, and longitudinal body wall muscle. Similarly, a homolog of the *A. japonicus* precursor of the neuropeptide GN-19 [107] was found to be expressed in the radial nerve cords, intestine, and longitudinal body wall of *H. scabra*, whereas expression of a homolog of the *A. japonicus* precursor of the neuropeptide GLRFA [107] was found to be restricted to the circumoral nerve ring and radial nerve cords [24]. These findings provide a taster of what now needs to be accomplished–a comprehensive analysis of the expression patterns of all the neuropeptide precursors in sea cucumbers, employing the use of mRNA *in situ* hybridization and immunohistochemical methods that will reveal more specifically the cell populations that express neuropeptide precursors. This will provide an anatomical basis for investigation of the physiological roles of neuropeptides using *in vitro* and *in vivo* pharmacological methods, as has already been accomplished successfully in starfish species as discussed above. And there are several aspects of sea cucumber physiology and behavior where neuropeptides may have important regulatory roles. The importance of neuropeptides as regulators of gamete release in sea cucumbers has already been discussed above. Other processes that are likely to require complex neurochemical control mechanisms are the eversion of Cuvierian tubules and/or evisceration that occur as a defense against predation in some sea cucumber species [116, 117]. Both of these processes involve changes in the mechanical state of MCT in sea cucumbers and progress has been made recently in elucidating the mechanisms underpinning MCT mechanical adaptability [118]. However, there is still much to be learned about how changes in MCT stiffness are controlled by the nervous system. Therefore, the recent discovery of a plethora of novel neuropeptides in sea cucumbers, some of which appear to be unique to sea cucumbers or echinoderms [21, 24, 107], provides a rich resource for future studies in this fascinating area of echinoderm biology.

5.6 NEUROPEPTIDE SIGNALING SYSTEMS IN THE CRINOIDEA

1. SALMFamide-Type Neuropeptide Signaling in Feather Stars

Evidence of the occurrence of SALMFamide-type neuropeptides in crinoids was first obtained using immunohistochemical methods. Antibodies to the starfish SALMFamide neuropeptide S2 revealed immunoreactivity in neuronal somata and their processes in the brachial nerve of the feather star *Antedon bifida* [119]. More recently, sequencing of the transcriptome of the feather star *Antedon mediterranea* enabled determination of the sequence of the first SALMFamide precursor to be identified in a crinoid [59]. The *Antedon* SALMFamide precursor comprises fourteen putative neuropeptides, ranging in length from eight to twenty-four residues. Furthermore, the C-terminal regions of these peptides have a variety of motifs, including L-type (LxFamide) and F-type (FxFamide), as in other echinoderms, LxLamide, FxMamide, and others (Figure 5.4). Because just one SALMFamide precursor has been identified in *A. mediterranea* that exhibits similarity with both

L-type and F-type SALMFamide precursors in other echinoderms, it has been proposed that this precursor may represent an ancestral type that predates a gene duplication event that gave rise to the L-type SALMFamide precursors and F-type SALMFamide precursors in other echinoderm classes. However, there remains the possibility that a second SALMFamide precursor exists in *A. mediterranea* but has yet to be discovered due to sequence divergence or incomplete transcriptome coverage.

2.   *Transcriptomic Identification of Other Neuropeptide Precursors in Crinoids*

Sequencing of the transcriptome of the feather star *A. mediterranea* has enabled identification of transcripts encoding precursors of a variety of neuropeptides, including calcitonin-type, CCK-type, luqin-type, orexin-type, MCH-type and vasopressin/oxytocin-type (M.R. Elphick et al., unpublished data). Further insights into the diversity of neuropeptide precursors in crinoids will be obtained if the genome of a crinoid species is sequenced. Informed by identification of neuropeptide precursors in crinoids, there are exciting opportunities ahead to investigate the expression and functions of neuropeptides in both larval and adult crinoids. One aspect of adult crinoids that makes them of interest from a comparative perspective is that the ecto-neural and hyponeural regions of the nervous system are less prominent than in other echinoderms and it is the entoneural (apical or aboral) region of the nervous system that is predominant both anatomically and functionally [14]. Furthermore, immunohistochemical studies on *A. mediterranea* have revealed the presence of L-glutamate in the brachial nerves of the entoneural nervous system and pharmacological studies have revealed that L-glutamate induces rhythmic contractions of arm muscles and arm autotomy in this species [120, 121]. These studies are illustrative of experimental approaches that could be employed to investigate the physiological roles of neuropeptides in crinoids.

5.7   GENERAL CONCLUSIONS AND LOOKING AHEAD

At the outset of this chapter, two perspectives that make echinoderms of particular interest for research on neuropeptide signaling are highlighted. Firstly, a phylogenetic perspective, with echinoderms providing "missing pieces" in the "jigsaw puzzle" of neuropeptide evolution by virtue of their status as non-chordate deuterostomes that occupy an "intermediate" evolutionary position with respect to the well-studied vertebrates and selected protostomes (e.g., *C. elegans, D. melanogaster*). Secondly, a functional perspective, with the pentaradial symmetry and other unusual properties (e.g., MCT) of adult echinoderms providing unique contexts for gaining insights into the physiological roles of neuropeptides and the evolution of neuropeptide function in the animal kingdom. Furthermore, it will be apparent from reading this chapter that there is also a third perspective on neuropeptide signaling in echinoderms–a developmental perspective.

Here in the final concluding section of this chapter, I will summarize progress in investigating the biology of neuropeptides in echinoderms from these three perspectives. Furthermore, I will highlight some of the opportunities that lie ahead for the investigation of neuropeptide signaling in echinoderms.

1.   *Reconstructing Neuropeptide Evolution: Insights from Echinoderms*

A review article summarizing progress in this aspect of echinoderm neuropeptide biology was published recently [13], and I refer readers to this article for a more detailed overview of this topic. A unifying theme is that insight into the evolutionary origins of neuropeptide signaling systems have been obtained with the discovery of echinoderm orthologs of neuropeptide-related genes/proteins that hitherto had only been identified in vertebrates/chordates or in protostomes [10, 13]. Definitive proof of orthology has been obtained by the experimental demonstration that neuropeptides act as ligands for their candidate cognate receptors. The discovery that NGFFFamide is the ligand for the *S. purpuratus* NPS/CCAP-type receptor is an important example of this because it provided the key "missing link" between the NPS signaling system in vertebrates and the CCAP signaling system in protostomes [63]. Similarly, the discovery of a corazonin-type signaling system in *A. rubens* illustrates how research on an echinoderm species [60] has had a broad impact in changing our perspective on the evolution of a neuropeptide signaling system [61]. Are there other instances where identification of ligand-receptor partnerships in echinoderms could provide important insights into neuropeptide evolution? Here I will highlight some examples.

The first putative precursor of TRH-like peptides to be discovered in an invertebrate was identified in the sea urchin *S. purpuratus* [22], but the receptor(s) for these peptides have yet to be characterized experimentally. However, peptides that act as ligands for TRH-type receptors have been identified recently in two protostomes–the annelid *Platynereis dumerilii* and the nematode *C. elegans* [122, 123]. Interestingly, alignment of these peptides, and related peptides from other protostomes, has revealed structural similarities with the TRH-like peptides in sea urchins and other echinoderms, providing further evidence of orthology. Furthermore, investigation of the roles of TRH-type signaling in *C. elegans* has revealed an evolutionarily ancient role in the regulation of growth [123]. In this context, it will be interesting to gain insights into the physiological roles of TRH-type signaling in sea urchins and/or other echinoderms because it will provide a "missing link" between protostomes and vertebrates.

The MCH-type and kisspeptin-type precursor proteins that were identified in the starfish *A. rubens* and in other echinoderms were the first to be identified outside the phylum Chordata [23–25, 45] and therefore functional characterization of the neuropeptides derived from these precursors and the receptors they act on would provide important insights into the evolution of MCH-type and kisspeptin-type signaling. Conversely, there are a number of neuropeptide signaling systems that were discovered in protostomes but which are not present in the vertebrate/chordate lineage. These include corazonin-type and PP/OK-type neuropeptides, and functional characterization of these neuropeptides in starfish has provided new insights into the evolution of neuropeptide function in the animal kingdom [62, 66, 67]. However, the receptors that mediate the effects of PP/OK-type peptides have yet to be discovered in any animal. Therefore, this represents an area of research where echinoderms could expand significantly our knowledge of the mechanisms of neuropeptide signaling. Other bilaterian neuropeptide signaling systems that have been lost in chordates but retained in echinoderms are the luqin-type and PDF-type signaling systems. Therefore, investigations of the physiological roles of the luqin and PDF signaling systems in echinoderms are interesting research objectives for the future. It is noteworthy that PDF signaling has an important role in mediating circadian control of behavior in insects [124, 125], so it will be of particular interest to discover if this role extends to echinoderms.

2. Discovering Neuropeptide Function in Adult Echinoderms

The discovery of a gonadotropic peptide in starfish heralded the beginnings of research on neuropeptide function in adult echinoderms [1, 3]. Furthermore, the identification of this neuropeptide as a relaxin-like peptide provided evidence that relaxin-type peptides are evolutionarily ancient regulators of reproductive processes [31]. It remains to be determined, however, if other relaxin-type peptides that have been identified in starfish and in other echinoderms also act as gonadotropins. The discovery that NGIWYamide [94], an ortholog of the vertebrate neuropeptide known as NPS [63], acts as a gonadotropin in the sea cucumber *A. japonicus* [112] may be evidence of divergence in the mechanisms of neuroendocrine regulation of spawning in different echinoderm classes. In this regard, it should be noted that the common ancestor of asterozoans (starfish and brittle stars) and echinozoans (sea cucumbers and sea urchins) is estimated to have lived ~480 mya [126] and over this geological timescale, there has been huge scope for the divergence of physiological mechanisms. In this context, it will be interesting to obtain further insights into the mechanisms of neuropeptide-mediated control of gamete maturation and spawning in echinoderm species belonging to each of the five extant classes. Likewise, as highlighted elsewhere in this chapter, comparative analysis of neuropeptide function in species belonging to each of the echinoderm classes will also be of interest for the investigation of other aspects of echinoderm physiology. For example, is the action of PP/OK-type neuropeptides as muscle relaxants in starfish [65–67] applicable to other echinoderms? However, each of the five extant echinoderm classes may provide differing opportunities for investigation of neuropeptide function, as discussed below.

A unique aspect of echinoderm biology is the presence of MCT and progress in elucidating the mechanisms of MCT has largely been made through studies on sea cucumbers, which have a voluminous and easily accessible layer of MCT in their body wall [16]. Thus, the molecular identification of peptides and proteins that regulate or mediate mechanisms of MCT has been accomplished in sea cucumbers [17, 127, 128]. Building upon these important advances and recent insights into the mechanisms of MCT in the sea cucumber body wall [118], there are opportunities ahead to utilize the power of transcriptomics and proteomics to obtain more comprehensive molecular insights into how changes in the mechanical state of the sea cucumber body are regulated by the nervous system through the release of neuropeptides and other neurotransmitters/neuromodulators. In other echinoderm classes, MCT may be less abundant and accessible than in the body wall of sea cucumbers, but species from other echinoderm classes have nevertheless been valuable model systems for research on MCT. For example, the compass depressor ligament in sea urchins [129], the interossicular ligaments of crinoids [130] and the inner dermis of some starfish species [131]. Using these experimental preparations, it will be interesting to determine if the mechanisms of neuropeptidergic control of MCT are conserved between the echinoderm classes.

Another aspect of the biology of echinoderms that has attracted interest is their capacity to autotomize arms or eviscerate internal organs as a defense against predation and then regenerate the lost arms or organs [18]. There is evidence that neuropeptides may trigger arm autotomy in starfish [19], but the identity of these molecules remains to be determined, and therefore, this represents a fascinating objective for the future. Furthermore, immunohisto-chemical evidence that SALMFamide-type neuropeptides may be involved in the regulation of arm regeneration in starfish has also been reported [132]; now there is scope for studies of this kind to be extended to the many other neuropeptides that have been identified in starfish [23]. Arguably the most impressive capacity for regeneration in echinoderms is seen in brittle star species such as *Amphiura filiformis* (Figure 5.2), which frequently lose arms due to predation and then very rapidly regrow new arms [133]. For this reason, *A. filiformis* is an emerging model system in regenerative biology and insights into the molecular mechanisms of regeneration in this species have been obtained recently [80, 81]. Furthermore, as in starfish, the recent discovery of multiple neuropeptide precursors in *A. filiformis* [25] has provided an opportunity to investigate the roles of neuropeptide signaling systems in regenerative processes in this species.

With a more specific focus on starfish, an intriguing aspect of their biology is the mechanism of feeding, which in many species involves eversion of the stomach out of the mouth and over prey [134]. For example, the common European starfish *A. rubens* feeds on mussels and other bivalves and there-fore in order that the everted stomach can gain access to soft and digestible tissues, a gap between the valves of their prey has to be created by employing the pulling power of the tube feet on the underside of each arm. By way of contrast, the crown-of-thorns starfish *A. planci* feeds on coral so here the stomach simply has to be everted from the mouth so that external digestion of soft polyp tissue can commence. Investigation of the physiological roles of the first neuropeptides to be discovered in starfish, the SALMFamides S1 and S2, revealed that intracoelomic injection of these peptides induces stomach eversion in *A. rubens* but with S2 being more effective than S1 [54]. Transcriptomic identification of neuropeptide precursors in *A. rubens* [23, 64] has enabled investigation of other neuropeptides as candidate regula-tors of stomach eversion or retraction. Thus, the NG peptide NGFFYamide was identified as a potent stimulator of stomach contraction (*in vitro*) and retraction (*in vivo*) [64]. However, it has been found that neuropeptides that trigger stomach relaxation or contraction *in vitro* do not necessarily trigger stomach eversion or retraction *in vivo*. For example, GnRH-type and cora-zonin-type neuropeptides both trigger the contraction of *in vitro* preparations of the stomach from *A. rubens* but, unlike NGFFYamide, neither peptide was observed to cause stomach retraction [62]. Conversely, neuropeptides derived from two PP/OK-type precursors in *A. rubens* (ArPPLNP1, ArPPLNP2) were found to cause relaxation of stomach preparations *in vitro* but did not trigger stomach eversion when injected *in vivo* [66, 67]. It is clear that neuropeptidergic control of stomach activity in starfish is highly complex and there is much to be learned about the roles of the many neuropeptide signaling systems that are present in the starfish stomach and in other regions of the digestive system. Furthermore, gaining deeper insights into the mechanisms of neuropeptidergic regulation of feeding in starfish may have applications as part of the effort to identify ways of controlling species such as *A. planci*, which is contributing to the loss of coral on the Great Barrier Reef and on other reefs in the Indo-Pacific region [135, 136].

Finally, the discovery of multiple neuropeptides signaling systems in echinoderms provides new opportunities to gain insights into the functional anatomy of echinoderm nervous systems in the unique context of a pentaradially symmetrical body plan. Use of molecular markers has provided valuable insights into the neuroarchitecture of echinoderm nervous systems [14, 73] and with the development of a growing toolkit of neuropeptide-specific antibodies, there are exciting opportunities ahead to gain new neuroanatomical insights from the perspective of neuropeptide signaling. An example of how immunohistochemical analysis on neuropeptide expression has already provided such insights is the localization of PP/OK-type neuropeptides in the lateral motor nerves of starfish [66]. The pioneering studies of J. Eric Smith using classical histological staining methods provided remarkably detailed accounts and illustrations of the neuroanatomy of the starfish nervous system and one feature of the nervous system that Smith described for the first time was the lateral motor nerves [137–139]. The detection of neuropeptides derived from the ArPPLN1 precursor in the lateral motor nerves has provided the first evidence that these nerves contain the axons of peptidergic neurons. Furthermore, it also enabled immunohistochemical visualization of axonal processes in branches of the lateral motor nerves that innervate interfascicular muscles or project into the circular muscle layer of the coelomic lining of the arms [66]. Analysis of neuropeptide expression in *A. rubens* has also revealed that some neuropeptides (e.g., ArPPLN1, ArPPLN2, ArGnRH) are expressed in cell bodies of hyponeural motoneurons in the radial nerve cords and circumoral nerve ring, whereas other neuropeptides (e.g., ArCRZ) are not [62, 66, 67]. Looking ahead it may be possible to analyze the neuropeptide expression profile of hyponeural neurons to identify subpopulations of neurons that innervate different regions of the starfish body, which would transform our knowledge of the neuroanatomy of motor systems in starfish. Likewise, the same approach could be employed for other echinoderms, with the intriguing potential prospect of identifying sub-populations of homologous motoneurons in different echinoderm classes.

3. Developmental Analysis of Neuropeptide Function: Echinoderms as Model Systems

The sequencing of the genome of the sea urchin *S. purpuratus* was founded upon a long history of using this species as a model system in developmental biology [88]. However, very little is known about neuropeptide expression and function during the embryonic and larval development of this species and other echinoderms. Prior to the sequencing of the *S. purpuratus* genome, the use of antibodies to SALMFamides enabled immunocytochemical visualization of neuropeptides in the larvae of echinoids [84, 85]. Now with the identification of many neuropeptide precursors in *S. purpuratus* [22, 25], there are opportunities ahead to investigate neuropeptide expression and function in sea urchin larvae more comprehensively. The recent analysis of the expression of eight neuropeptide precursors in larvae of the starfish *A. rubens* using mRNA *in situ* hybridization methods has provided an indication of the insights that can be obtained [56]. Now, these studies need to be extended to a wider range of neuropeptides and to other echinoderms, employing the use of both mRNA *in situ* hybridization methods and immunocytochemistry for anatomical studies. Furthermore, pharmacological testing of the effects of neuropeptides combined with gene-knockdown (using morpholino antisense oligonucleotides) and gene-knockout (CRISPR/Cas9) techniques [140, 141] could be used to reveal the functions of neuropeptides in larvae. Thus, in echinoderms there exists a unique opportunity amongst extant bilaterian phyla to discover the physiological roles of neuropeptides in both the pre-metamorphic bilaterally symmetrical nervous systems of the larval stage and the post-metamorphic pentaradially symmetrical nervous systems of juvenile and adult animals.

And so we have come full circle. Alfred Chaet's rationale sixty years ago for using radial nerve cord extracts to induce spawning in starfish was to establish a method for obtaining eggs and sperm for studies in reproductive and developmental biology [1–3, 28]. Now in the "omics" era, we have the resources to not only trigger spawning in starfish using RGP or its downstream effector 1-MeAde [33, 34] but also to examine the expression and functions of multiple neuropeptides signaling systems during the embryonic and larval development of starfish and other echinoderms. We have come a long way since that first report of the "physiologic activity of nerve extracts," but there is so much more to be discovered.

ACKNOWLEDGMENTS

I am grateful to the Leverhulme Trust (grants RGP-2013-351 and RGP-2016-353), BBSRC (grant BB/M001644/1), China Scholarship Council, Society of Experimental Biology, CONACYT, and Queen Mary University of London (QMUL) for supporting my research on neuropeptide signaling in echinoderms. I thank the talented postdoctoral researchers and PhD, MSc, and BSc students who have worked in my research group at QMUL and who have made and continue to make so many important contributions to our knowledge and understanding of neuropeptide signaling in echinoderms. I also thank collaborators in the UK and overseas for their contributions to our research on neuropeptide signaling in echinoderms. Special thanks to Dean Semmens for commenting on the manuscript during its preparation and to Dr. Iain Wilkie (University of Glasgow) for providing meticulous, helpful feedback as a reviewer of the chapter. Thank you also to Ming Lin for his help in producing Figure 5.3 and to Paola Oliveri, Dario Fassini and Ding Kui for providing the photographs of *Amphiura filiformis*, *Antedon mediterranea*, and *Apostichopus japonicus*, respectively, in Figure 5.2. Last but not least, I dedicate this chapter to Mike Thorndyke (Professor Emeritus, Royal Swedish Academy of Sciences and Göteborg University). Mike's pioneering research on neuropeptides in urochordates and his undergraduate course in "Comparative Endocrinology" at the Royal Holloway University of London inspired me to enter this field of research over thirty years ago. The mission of *The Starfish Enterprise* is still far from complete!

KEYWORDS

- **Neuropeptide precursor**
- **G-protein coupled receptor**
- **Starfish**
- **Brittle star**
- **Sea urchin**
- **Sea cucumber**

REFERENCES

1. Chaet, A., & McConnaughy, R., (1959). Physiologic activity of nerve extracts. *Biol. Bull., 117*, 407–408.
2. Chaet, A. B., (1966a). The gamete-shedding substances of starfishes: A physiological-biochemical study. *Am. Zool., 6*, 263–271.
3. Chaet, A. B., (1964). A mechanism for obtaining mature gametes from starfish. *Biological Bulletin, 126*, 8–13.
4. Oksche, A., (1997). Ernst and Berta Scharrer–Pioneers in Neuroendocrinology. In: Korf, H. W., & Usadel, K. H., (eds.), *Neuroendocrinology: Retrospect and Perspectives* (pp. 1–4). Berlin, Heidelberg: Springer Berlin Heidelberg.
5. Dunn, C. W., Giribet, G., Edgecombe, G. D., & Hejnol, A., (2014). Animal phylogeny and its evolutionary implications. *Annual Review of Ecology, Evolution, and Systematics, 45*, 371–395.
6. Halanych, K. M., (2004). The new view of animal phylogeny. *Annual Review of Ecology, Evolution, and Systematics, 35*, 229–256.
7. Holland, P. W. H., (2011). *The Animal Kingdom: A Very Short Introduction.* Oxford, New York: Oxford University Press.
8. Schoofs, L., Loof, A. D., & Hiel, M. B. V., (2017). Neuropeptides as regulators of behavior in insects. *Annual Review of Entomology, 62*, 35–52.
9. Taghert, P. H., & Nitabach, M. N., (2012). Peptide neuromodulation in invertebrate model systems. *Neuron, 76*, 82–97.
10. Elphick, M. R., Mirabeau, O., & Larhammar, D., (2018). Evolution of neuropeptide signaling systems. *The Journal of Experimental Biology,* 221.
11. Jekely, G., (2013). Global view of the evolution and diversity of metazoan neuropeptide signaling. *Proceedings of the National Academy of Sciences of the United States of America, 110*, 8702–8707.
12. Mirabeau, O., & Joly, J. S., (2013). Molecular evolution of peptidergic signaling systems in bilaterians. *Proceedings of the National Academy of Sciences of the United States of America, 110*, 2028–2037.
13. Semmens, D. C., & Elphick, M. R., (2017). The evolution of neuropeptide signalling: Insights from echinoderms. *Brief Funct Genomics, 16*, 288–298.
14. Mashanov, V., Zueva, O., Rubilar, T., Epherra, L., & García-Arrarás, J., (2016). Echinodermata. In: Schmidt-Rhaesa, A., Harzsch, S., & Purschke, G., (eds.), *Structure and Evolution of Invertebrate Nervous Systems* (pp. 665–688). Oxford: Oxford University Press.
15. Pentreath, V. W., & Cobb, J. L., (1972). Neurobiology of echinodermata. *Biological Reviews of the Cambridge Philosophical Society, 47*, 363–392.
16. Wilkie, I. C., (2005). Mutable collagenous tissue: Overview and biotechnological perspective. *Prog. Mol. Subcell. Biol., 39*, 221–250.
17. Birenheide, R., Tamori, M., Motokawa, T., Ohtani, M., Iwakoshi, E., Muneoka, Y., Fujita, T., Minakata, H., & Nomoto, K., (1998). Peptides controlling stifness of connective tissue in sea cucumbers. *The Biological Bulletin, 194*, 253–259.
18. Wilkie, I. C., (2001). Autotomy as a prelude to regeneration in echinoderms. *Microsc. Res. Tech., 55*, 369–396.
19. Mladenov, P. V., Igdoura, S., Asotra, S., & Burke, R. D., (1989). Purification and partial characterization of an autotomy-promoting factor from the sea star *Pycnopodia helianthoides. The Biological Bulletin, 176*, 169–175.

20. Elphick, M. R., Price, D. A., Lee, T. D., & Thorndyke, M. C., (1991a). The SALMFamides: A new family of neuropeptides isolated from an echinoderm. *Proceedings. Biological Sciences / The Royal Society, 243,* 121–127.

21. Rowe, M. L., Achhala, S., & Elphick, M. R., (2014). Neuropeptides and polypeptide hormones in echinoderms: New insights from analysis of the transcriptome of the sea cucumber *Apostichopus japonicus. General and Comparative Endocrinology, 197,* 43–55.

22. Rowe, M. L., & Elphick, M. R., (2012). The neuropeptide transcriptome of a model echinoderm, the sea urchin *Strongylocentrotus purpuratus. General and Comparative Endocrinology, 179,* 331–344.

23. Semmens, D. C., Mirabeau, O., Moghul, I., Pancholi, M. R., Wurm, Y., & Elphick, M. R., (2016). Transcriptomic identification of starfish neuropeptide precursors yields new insights into neuropeptide evolution. *Open Biol., 6,* 150224.

24. Suwansa-Ard, S., Chaiyamoon, A., Talarovicova, A., Tinikul, R., Tinikul, Y., Poomtong, T., Elphick, M. R., Cummins, S. F., & Sobhon, P., (2018). Transcriptomic discovery and comparative analysis of neuropeptide precursors in sea cucumbers (Holothuroidea). *Peptides, 99,* 231–240.

25. Zandawala, M., Moghul, I., Yanez Guerra, L. A., Delroisse, J., Abylkassimova, N., Hugall, A. F., O'Hara, T. D., & Elphick, M. R., (2017). Discovery of novel representatives of bilaterian neuropeptide families and reconstruction of neuropeptide precursor evolution in ophiuroid echinoderms. *Open Biol., 7.*

26. O'Hara, T. D., Hugall, A. F., Thuy, B., & Moussalli, A., (2014). Phylogenomic resolution of the class ophiuroidea unlocks a global microfossil record. *Curr. Biol., 24,* 1874–1879.

27. Telford, M. J., Lowe, C. J., Cameron, C. B., Ortega-Martinez, O., Aronowicz, J., Oliveri, P., & Copley, R. R., (2014). Phylogenomic analysis of echinoderm class relationships supports Asterozoa. *Proceedings of the Royal Society B: Biological Sciences, 281.*

28. Chaet, A. B., (1966b). Neurochemical control of gamete release in starfish. *The Biological Bulletin, 130,* 43–58.

29. Kanatani, H., (1979). Hormones in echinoderms. In: Barrington, E., (ed.), *Hormones and Evolution* (Vol. 1, pp. 273–307). London: Academic Press.

30. Shirai, H., (1986). Gonad-stimulating and maturation-inducing substance. *Methods Cell Biol., 27,* 73–88.

31. Mita, M., Yoshikuni, M., Ohno, K., Shibata, Y., Paul-Prasanth, B., Pitchayawasin, S., Isobe, M., & Nagahama, Y., (2009). A relaxin-like peptide purified from radial nerves induces oocyte maturation and ovulation in the starfish, *Asterina pectinifera. Proc. Natl. Acad. Sci. USA, 106,* 9507–9512.

32. Mita, M., Daiya, M., Haraguchi, S., Tsutsui, K., & Nagahama, Y., (2015). A new relaxin-like gonad-stimulating peptide identified in the starfish *Asterias amurensis. General and Comparative Endocrinology, 222,* 144–149.

33. Mita, M., (2013). Relaxin-like gonad-stimulating substance in an echinoderm, the starfish: A novel relaxin system in reproduction of invertebrates. *General and Comparative Endocrinology, 181,* 241–245.

34. Mita, M., (2016). Starfish gonadotropic hormone: Relaxin-like gonad-stimulating peptides. *General and Comparative Endocrinology, 230, 231,* 166–169.

35. Lin, M., Mita, M., Egertova, M., Zampronio, C. G., Jones, A. M., & Elphick, M. R., (2017b). Cellular localization of relaxin-like gonad-stimulating peptide expression in *Asterias rubens*: New insights into neurohormonal control of spawning in starfish. *The Journal of Comparative Neurology, 525,* 1599–1617.

36. Kanatani, H., & Shirai, H., (1969). Mechanism of starfish spawning. II. Some aspects of action of a neural substance obtained from radial nerve. *The Biological Bulletin, 137,* 297–311.

37. Mita, M., & Nagahama, Y., (1991). Involvement of G-proteins and adenylate cyclase in the action of gonad-stimulating substance on starfish ovarian follicle cells. *Developmental Biology, 144,* 262–268.

38. Mita, M., Haraguchi, S., Uzawa, H., & Tsutsui, K., (2013). Contribution of de novo synthesis of galphas-proteins to 1-methyladenine production in starfish ovarian follicle cells stimulated by relaxin-like gonad-stimulating substance. *Biochemical and Biophysical Research Communications, 440,* 798–801.

39. Mita, M., Yamamoto, K., Nakamura, M., Takeshige, Y., Watanabe, M., & Nagahama, Y., (2012). Participation of Gs-proteins in the action of relaxin-like gonad-stimulating substance (GSS) for 1-methyladenine production in starfish ovarian follicle cells. *General and Comparative Endocrinology, 176,* 432–437.

40. Halls, M. L., Bathgate, R. A., Sutton, S. W., Dschietzig, T. B., & Summers, R. J., (2015). International union of basic and clinical pharmacology. X. C., V., Recent advances in the understanding of the pharmacology and biological roles of relaxin family peptide receptors 1–4, the receptors for relaxin family peptides. *Pharmacol. Rev., 67,* 389–440.

41. Kanatani, H., & Ohguri, M., (1966). Mechanism of starfish spawning. I. Distribution of active substance responsible for maturation of oocytes and shedding of gametes. *Biological Bulletin, 131,* 104–114.

42. Kanatani, H., Shirai, H., Nakanishi, K., & Kurokawa, T., (1969). Isolation and indentification on meiosis inducing substance in starfish *Asterias amurensis. Nature, 221,* 273–274.

43. Mita, M., & Katayama, H., (2017). Enzyme-linked immunosorbent assay of relaxin-like gonad-stimulating peptide in the starfish *Patiria (Asterina) pectinifera. General and Comparative Endocrinology.*

44. Yamamoto, K., Kiyomoto, M., Katayama, H., & Mita, M., (2017). Radioimmunoassay of relaxin-like gonad-stimulating peptide in the starfish *Patiria (=Asterina) pectinifera. General and Comparative Endocrinology, 243,* 84–88.

45. Smith, M. K., Wang, T., Suwansa-Ard, S., Motti, C. A., Elizur, A., Zhao, M., Rowe, M. L., Hall, M. R., Elphick, M. R., & Cummins, S. F., (2017). The neuropeptidome of the crown-of-thorns starfish, *Acanthaster planci. J. Proteomics, 165,* 61–68.

46. Elphick, M. R., (2014). SALMFamide salmagundi: The biology of a neuropeptide family in echinoderms. *General and Comparative Endocrinology, 205,* 23–35.

47. Elphick, M. R., Reeve, J. R., Jr., Burke, R. D., & Thorndyke, M. C., (1991b). Isolation of the neuropeptide SALMFamide-1 from starfish using a new antiserum. *Peptides, 12,* 455–459.

48. Moore, S. J., & Thorndyke, M. C., (1993). Immunocytochemical mapping of the novel echinoderm neuropeptide SALMFamide 1 (S1) in the starfish *Asterias rubens. Cell and Tissue Research, 274,* 605–618.

49. Moss, C., Burke, R. D., & Thorndyke, M. C., (1994). Immunocytochemical localization of the neuropeptide S1 and serotonin in larvae of the starfish *Pisaster ochraceus* and *Asterias rubens. Journal of the Marine Biological Association of the United Kingdom, 74,* 61–71.

50. Newman, S. J., Elphick, M. R., & Thorndyke, M. C., (1995a). Tissue distribution of the SALMFamide neuropeptides S1 and S2 in the starfish *Asterias rubens* using

novel monoclonal and polyclonal antibodies. 1. Nervous and locomotory systems. *Proceedings Biological Sciences / The Royal Society, 261*, 139–145.

51. Newman, S. J., Elphick, M. R., & Thorndyke, M. C., (1995b). Tissue distribution of the SALMFamide neuropeptides S1 and S2 in the starfish *Asterias rubens* using novel monoclonal and polyclonal antibodies. 2. Digestive system. *Proceedings Biological Sciences / The Royal Society, 261*, 187–192.

52. Elphick, M. R., Newman, S. J., & Thorndyke, M. C., (1995). Distribution and action of SALMFamide neuropeptides in the starfish *Asterias rubens*. *The Journal of Experimental Biology, 198*, 2519–2525.

53. Melarange, R., & Elphick, M. R., (2003). Comparative analysis of nitric oxide and SALMFamide neuropeptides as general muscle relaxants in starfish. *The Journal of Experimental Biology, 206*, 893–899.

54. Melarange, R., Potton, D. J., Thorndyke, M. C., & Elphick, M. R., (1999). SALMFamide neuropeptides cause relaxation and eversion of the cardiac stomach in starfish. *Proceedings Biological Sciences / The Royal Society, 266*, 1785–1789.

55. Elphick, M. R., & Melarange, R., (2001). Neural control of muscle relaxation in echinoderms. *The Journal of Experimental Biology, 204*, 875–885.

56. Mayorova, T. D., Tian, S., Cai, W., Semmens, D. C., Odekunle, E. A., Zandawala, M., Badi, Y., Rowe, M. L., Egertova, M., & Elphick, M. R., (2016). Localization of neuropeptide gene expression in larvae of an echinoderm, the starfish *Asterias rubens*. *Front Neurosci., 10*, 553.

57. Jones, C. E., Zandawala, M., Semmens, D. C., Anderson, S., Hanson, G. R., Janies, D. A., & Elphick, M. R., (2016). Identification of a neuropeptide precursor protein that gives rise to a "cocktail" of peptides that bind Cu(II) and generate metal-linked dimers. *Biochimica et Biophysica Acta, 1860*, 57–66.

58. Elphick, M. R., Achhala, S., & Martynyuk, N., (2013). The evolution and diversity of SALMFamide neuropeptides. *PloS One, 8*, e59076.

59. Elphick, M. R., Semmens, D. C., Blowes, L. M., Levine, J., Lowe, C. J., Arnone, M. I., & Clark, M. S., (2015). Reconstructing SALMFamide neuropeptide precursor evolution in the phylum echinodermata: Ophiuroid and crinoid sequence data provide new insights. *Front Endocrinol (Lausanne), 6*, 2.

60. Tian, S., Zandawala, M., Beets, I., Baytemur, E., Slade, S. E., Scrivens, J. H., & Elphick, M. R., (2016). Urbilaterian origin of paralogous GnRH and corazonin neuropeptide signalling pathways. *Sci. Rep., 6*, 28788.

61. Zandawala, M., Tian, S., & Elphick, M. R., (2018). The evolution and nomenclature of GnRH-type and corazonin-type neuropeptide signaling systems. *General and Comparative Endocrinology.*

62. Tian, S., Egertova, M., & Elphick, M. R., (2017). Functional characterization of paralogous gonadotropin-releasing hormone-type and corazonin-type neuropeptides in an echinoderm. *Front Endocrinol (Lausanne), 8*, 259.

63. Semmens, D. C., Beets, I., Rowe, M. L., Blowes, L. M., Oliveri, P., & Elphick, M. R., (2015). Discovery of sea urchin NGFFFamide receptor unites a bilaterian neuropeptide family. *Open Biol., 5*, 150030.

64. Semmens, D. C., Dane, R. E., Pancholi, M. R., Slade, S. E., Scrivens, J. H., & Elphick, M. R., (2013). Discovery of a novel neurophysin-associated neuropeptide that triggers cardiac stomach contraction and retraction in starfish. *The Journal of Experimental Biology, 216*, 4047–4053.

65. Kim, C. H., Kim, E. J., Go, H. J., Oh, H. Y., Lin, M., Elphick, M. R., & Park, N. G., (2016). Identification of a novel starfish neuropeptide that acts as a muscle relaxant. *J. Neurochem., 137*, 33–45.

66. Lin, M., Egertova, M., Zampronio, C. G., Jones, A. M., & Elphick, M. R., (2017a). Pedal peptide/orcokinin-type neuropeptide signaling in a deuterostome: The anatomy and pharmacology of starfish myorelaxant peptide in *Asterias rubens*. *The Journal of Comparative Neurology, 525*, 3890–3917.

67. Lin, M., Egertova, M., Zampronio, C. G., Jones, A. M., & Elphick, M. R., (2018). Functional characterization of a second pedal peptide/orcokinin-type neuropeptide signaling system in the starfish *Asterias rubens*. *The Journal of Comparative Neurology, 526*, 858–876.

68. Hall, J. D., & Lloyd, P. E., (1990). Involvement of pedal peptide in locomotion in *Aplysia:* Modulation of foot muscle contractions. *J. Neurobiol., 21*, 858–868.

69. Stangier, J., Hilbich, C., Burdzik, S., & Keller, R., (1992). Orcokinin: A novel myotropic peptide from the nervous system of the crayfish, *Orconectes limosus*. *Peptides, 13*, 859–864.

70. Gilpin, W., Prakash, V. N., & Prakash, M., (2016). Vortex arrays and ciliary tangles underlie the feeding–swimming trade-off in starfish larvae. *Nature Physics, 13*, 380.

71. De Bremaeker, N., Deheyn, D., Thorndyke, M. C., Baguet, F., & Mallefet, J., (1997). Localization of S1- and S2-like immunoreactivity in the nervous system of the brittle star *Amphipholis squamata* (Delle Chiaje., 1828). *Proceedings. Biological Sciences / The Royal Society, 264*, 667–674.

72. Ghyoot, M., Cobb, J. L., & Thorndyke, M. C., (1994). Localization of neuropeptides in the nervous system of the brittle star *Ophiura ophiura*. *Philosophical Transactions of the Royal Society of London. Series, B., Biological Sciences, 346*, 433–444.

73. Zueva, O., Khoury, M., Heinzeller, T., Mashanova, D., & Mashanov, V., (2018). The complex simplicity of the brittle star nervous system. *Frontiers in Zoology, 15*, 1.

74. Bremaeker, N. D., Baguet, F., Thorndyke, M. C., & Mallefet, J., (1999). Modulatory effects of some amino acids and neuropeptides on luminescence in the brittlestar *Amphipholis squamata*. *The Journal of Experimental Biology, 202*(13), 1785–1791.

75. Cisternas, P., & Byrne, M., (2003). Peptidergic and serotonergic immunoreactivity in the metamorphosing ophiopluteus of *Ophiactis resiliens* (Echinodermata, Ophiuroidea). *Invertebrate Biology, 122*, 177–185.

76. Cobb, J. L., & Stubbs, T. R., (1981). The giant neurone system in Ophiuroids. I. The general morphology of the radial nerve cords and circumoral nerve ring. *Cell and Tissue Research, 219*, 197–207.

77. Cobb, J. L., & Stubbs, T. R., (1982). The giant neurone system in ophiuroids. III. The detailed connections of the circumoral nerve ring. *Cell and Tissue Research, 226*, 675–687.

78. Stubbs, T. R., & Cobb, J. L., (1981). The giant neurone system in Ophiuroids. II. The hyponeural motor tracts. *Cell and Tissue Research, 220*, 373–385.

79. Cobb, J. L. S., (1985). The neurobiology of the ectoneural/hyponeural synaptic connection in an echinoderm. *The Biological Bulletin, 168*, 432–446.

80. Czarkwiani, A., Dylus, D. V., & Oliveri, P., (2013). Expression of skeletogenic genes during arm regeneration in the brittle star *Amphiura filiformis*. *Gene Expression Patterns: GEP, 13*, 464–472.

81. Czarkwiani, A., Ferrario, C., Dylus, D. V., Sugni, M., & Oliveri, P., (2016). Skeletal regeneration in the brittle star *Amphiura filiformis*. *Front Zool., 13*, 18.

82. Dylus, D. V., Czarkwiani, A., Stangberg, J., Ortega-Martinez, O., Dupont, S., & Oliveri, P., (2016). Large-scale gene expression study in the ophiuroid *Amphiura filiformis* provides insights into evolution of gene regulatory networks. *Evo. Devo., 7*, 2.

83. Kano, T., Sato, E., Ono, T., Aonuma, H., Matsuzaka, Y., & Ishiguro, A., (2017). A brittle star-like robot capable of immediately adapting to unexpected physical damage. *R. Soc. Open Sci., 4*, 171200.

84. Thorndyke, M. C., Crawford, B. D., & Burke, R. D., (1992). Localization of a SALMFamide Neuropeptide in the Larval Nervous System of the Sand Dollar *Dendraster excentricus*. *Acta Zoologica, 73*, 207–212.

85. Beer, A. J., Moss, C., & Thorndyke, M., (2001). Development of serotonin-like and SALMFamide-like immunoreactivity in the nervous system of the sea urchin *Psammechinus miliaris*. *The Biological Bulletin, 200*, 268–280.

86. Elphick, M. R., & Thorndyke, M. C., (2005). Molecular characterisation of SALMFamide neuropeptides in sea urchins. *The Journal of Experimental Biology, 208*, 4273–4282.

87. Burke, R. D., Angerer, L. M., Elphick, M. R., Humphrey, G. W., Yaguchi, S., Kiyama, T., Liang, S., Mu, X., Agca, C., Klein, W. H., et al., (2006). A genomic view of the sea urchin nervous system. *Developmental Biology, 300*, 434–460.

88. Sodergren, E., Weinstock, G. M., Davidson, E. H., Cameron, R. A., Gibbs, R. A., Angerer, R. C., Angerer, L. M., Arnone, M. I., Burgess, D. R., Burke, R. D., et al., (2006). The genome of the sea urchin *Strongylocentrotus purpuratus*. *Science, 314*, 941–952.

89. Rowe, M. L., & Elphick, M. R., (2010). Discovery of a second SALMFamide gene in the sea urchin *Strongylocentrotus purpuratus* reveals that L-type and F-type SALMFamide neuropeptides coexist in an echinoderm species. *Marine Genomics, 3*, 91–97.

90. Aloyz, R. S., & DesGroseillers, L., (1995). Processing of the L5–67 precursor peptide and characterization of LUQIN in the LUQ neurons of *Aplysia californica*. *Peptides, 16*, 331–338.

91. Collin, C., Hauser, F., Krogh-Meyer, P., Hansen, K. K., Gonzalez De Valdivia, E., Williamson, M., & Grimmelikhuijzen, C. J., (2011). Identification of the *Drosophila* and *Tribolium* receptors for the recently discovered insect RYamide neuropeptides. *Biochemical and Biophysical Research Communications, 412*, 578–583.

92. Ida, T., Takahashi, T., Tominaga, H., Sato, T., Kume, K., Ozaki, M., Hiraguchi, T., Maeda, T., Shiotani, H., Terajima, S., et al., (2011). Identification of the novel bioactive peptides dRYamide-1 and dRYamide-2, ligands for a neuropeptide Y-like receptor in *Drosophila*. *Biochemical and Biophysical Research Communications, 410*, 872–877.

93. Elphick, M. R., & Rowe, M. L., (2009). NGFFFamide and echinotocin: Structurally unrelated myoactive neuropeptides derived from neurophysin-containing precursors in sea urchins. *J. Exp. Biol., 212*, 1067–1077.

94. Iwakoshi, E., Ohtani, M., Takahashi, T., Muneoka, Y., Ikeda, T., Fujita, T., Minakata, H., & Nomoto, K., (1995). Comparative aspects of structure and action of bioactive peptides isolated from the sea cucumber *Stichopus japonicus*. In: Ohno, M., (ed.), *Peptide Chemistry* (pp. 261–264). Osaka, Japan: Protein Research Foundation.

95. De Bree, F. M., (2000). Trafficking of the vasopressin and oxytocin prohormone through the regulated secretory pathway. *J. Neuroendocrinol., 12*, 589–594.

96. De Bree, F. M., & Burbach, J. P., (1998). Structure-function relationships of the vasopressin prohormone domains. *Cell Mol. Neurobiol., 18*, 173–191.

97. Elphick, M. R., (2010). NG peptides: A novel family of neurophysin-associated neuropeptides. *Gene, 458*, 20–26.

98. Xu, Y. L., Reinscheid, R. K., Huitron-Resendiz, S., Clark, S. D., Wang, Z., Lin, S. H., Brucher, F. A., Zeng, J., Ly, N. K., Henriksen, S. J., et al., (2004). Neuropeptide S: A neuropeptide promoting arousal and anxiolytic-like effects. *Neuron, 43*, 487–497.

99. Valsalan, R., & Manoj, N., (2014). Evolutionary history of the neuropeptide S receptor/ neuropeptide S system. *General and Comparative Endocrinology, 209*, 11–20.

100. Perillo, M., & Arnone, M. I., (2014). Characterization of insulin-like peptides (ILPs) in the sea urchin *Strongylocentrotus purpuratus*: Insights on the evolution of the insulin family. *General and Comparative Endocrinology, 205*, 68–79.

101. Conzelmann, M., Offenburger, S. L., Asadulina, A., Keller, T., Munch, T. A., & Jekely, G., (2011). Neuropeptides regulate swimming depth of *Platynereis* larvae. *Proc. Natl. Acad. Sci. USA, 108*, 1174–1183.

102. Thiel, D., Bauknecht, P., Jekely, G., & Hejnol, A., (2017). An ancient FMRFamide-related peptide-receptor pair induces defense behavior in a brachiopod larva. *Open Biol., 7.*

103. Díaz-Miranda, L., Price, D. A., Greenberg, M. J., Lee, T. D., Doble, K. E., & García-Arrarás, J. E., (1992). Characterization of two novel neuropeptides from the sea cucumber *Holothuria glaberrima*. *The Biological Bulletin, 182*, 241–247.

104. Díaz-Miranda, L., Blanco, R. E., & García-Arrarás, J. E., (1995). Localization of the heptapeptide GFSKLYFamide in the sea cucumber *Holothuria glaberrima* (Echinodermata): A light and electron microscopic study. *The Journal of Comparative Neurology, 352*, 626–640.

105. Díaz-Miranda, L., & García-Arrarás, J. E., (1995). Pharmacological action of the heptapeptide GFSKLYFamide in the muscle of the sea cucumber *Holothuria glaberrima* (Echinodermata). *Comparative biochemistry and physiology. Part, C., Pharmacology, Toxicology & Endocrinology, 110*, 171–176.

106. Ohtani, M., Iwakoshi, E., Muneoka, Y., Minakata, H., & Nomoto, K., (1999). Isolation and characterization of bioactive peptides from the sea cucumber, *Stichopus japonicus*. In: Shimonishi, Y., (ed.), *Peptide Science–Present and Future* (pp. 419–420). Dordrecht, the Netherlands: Kluwer Academic Publishers.

107. Elphick, M. R., (2012). The protein precursors of peptides that affect the mechanics of connective tissue and/or muscle in the echinoderm *Apostichopus japonicus*. *PloS One., 7*, e44492.

108. Tamori, M., Saha, A. K., Matsuno, A., Noskor, S. C., Koizumi, O., Kobayakawa, Y., Nakajima, Y., & Motokawa, T., (2007). Stichopin-containing nerves and secretory cells specific to connective tissues of the sea cucumber. *Proceedings Biological Sciences / The Royal Society, 274*, 2279–2285.

109. Inoue, M., Birenheide, R., Koizumi, O., Kobayakawa, Y., Muneoka, Y., & Motokawa, T., (1999). Localization of the neuropeptide NGIWYamide in the holothurian nervous system and its effects on muscular contraction. *Proceedings of the Royal Society B-Biological Sciences, 266*, 993–1000.

110. Chen, J., (2004). Present status and prospects of sea cucumber industry in China. In: Lovatelli, A., (ed.), *Advances in Sea Cucumber Aquaculture and Management* (pp. 25–38). Rome: Food and Agriculture Organization of the United Nations.

111. Lovatelli, A., (2004). Advances in sea cucumber aquaculture and management. In: *FAO Fisheries Technical Paper* (p. 425). Rome: Food and Agriculture Organization of the United Nations.

112. Kato, S., Tsurumaru, S., Taga, M., Yamane, T., Shibata, Y., Ohno, K., Fujiwara, A., Yamano, K., & Yoshikuni, M., (2009). Neuronal peptides induce oocyte maturation and

gamete spawning of sea cucumber, *Apostichopus japonicus. Developmental Biology, 326*, 169–176.

113. Katow, H., Katow, T., & Moriyama, A., (2009). Gonad-stimulating substance-like molecule from the radial nerve of the sea cucumber. *Int. J. Dev. Biol., 53*, 483–491.

114. Du, H., Bao, Z., Hou, R., Wang, S., Su, H., Yan, J., Tian, M., Li, Y., Wei, W., Lu, W., et al., (2012). Transcriptome sequencing and characterization for the sea cucumber *Apostichopus japonicus* (Selenka, 1867). *PLoS One, 7*, e33311.

115. Zhang, X., Sun, L., Yuan, J., Sun, Y., Gao, Y., Zhang, L., Li, S., Dai, H., Hamel, J. F., Liu, C., et al., (2017). The sea cucumber genome provides insights into morphological evolution and visceral regeneration. *PLoS Biol., 15*, e2003790.

116. Byrne, M., (2001). The morphology of autotomy structures in the sea cucumber *Eupentacta quinquesemita* before and during evisceration. *The Journal of Experimental Biology, 204*, 849–863.

117. Demeuldre, M., Hennebert, E., Bonneel, M., Lengerer, B., Van Dyck, S., Wattiez, R., Ladurner, P., & Flammang, P., (2017). Mechanical adaptability of sea cucumber Cuvierian tubules involves a mutable collagenous tissue. *The Journal of Experimental Biology, 220*, 2108–2119.

118. Mo, J., Prevost, S. F., Blowes, L. M., Egertova, M., Terrill, N. J., Wang, W., Elphick, M. R., & Gupta, H. S., (2016). Interfibrillar stiffening of echinoderm mutable collagenous tissue demonstrated at the nanoscale. *Proc. Natl. Acad. Sci. USA, 113*, 6362–6371.

119. Heinzeller, T., & Welsch, U., (1994). Crinoidea. In: Harrison, F. W., & Chia, F. S., (eds.), *Microscopic Anatomy of Invertebrates: Volume 14 Echinodermata* (pp. 9–148). New York: Wiley-Liss.

120. Wilkie, I. C., Barbaglio, A., & Carnevali, M. D., (2013). The elusive role of L-glutamate as an echinoderm neurotransmitter: Evidence for its involvement in the control of crinoid arm muscles. *Zoology (Jena), 116*, 1–8.

121. Wilkie, I. C., Barbaglio, A., Maclaren, W. M., & Carnevali, M. D., (2010). Physiological and immunocytochemical evidence that glutamatergic neurotransmission is involved in the activation of arm autotomy in the featherstar *Antedon mediterranea* (Echinodermata: Crinoidea). *The Journal of Experimental Biology, 213*, 2104–2115.

122. Bauknecht, P., & Jékely, G., (2015). Large-scale combinatorial deorphanization of *Platynereis* neuropeptide GPCRs. *Cell Rep., 12*, 684–693.

123. Van Sinay, E., Mirabeau, O., Depuydt, G., Van Hiel, M. B., Peymen, K., Watteyne, J., Zels, S., Schoofs, L., & Beets, I., (2017). Evolutionarily conserved TRH neuropeptide pathway regulates growth in *Caenorhabditis elegans. Proc. Natl. Acad. Sci. USA, 114*, 4065–4074.

124. Mezan, S., Feuz, J. D., Deplancke, B., & Kadener, S., (2016). PDF signaling is an integral part of the *Drosophila* circadian molecular oscillator. *Cell Rep., 17*, 708–719.

125. Renn, S. C., Park, J. H., Rosbash, M., Hall, J. C., & Taghert, P. H., (1999). A pdf neuropeptide gene mutation and ablation of PDF neurons each cause severe abnormalities of behavioral circadian rhythms in *Drosophila. Cell, 99*, 791–802.

126. Pisani, D., Feuda, R., Peterson, K. J., & Smith, A. B., (2012). Resolving phylogenetic signal from noise when divergence is rapid: A new look at the old problem of echinoderm class relationships. *Molecular Phylogenetics and Evolution, 62*, 27–34.

127. Takehana, Y., Yamada, A., Tamori, M., & Motokawa, T., (2014). Softenin, a novel protein that softens the connective tissue of sea cucumbers through inhibiting interaction between collagen fibrils. *PloS One, 9*, e85644.

128. Tipper, J. P., Lyons-Levy, G., Atkinson, M. A., & Trotter, J. A., (2002). Purification, characterization and cloning of tensilin, the collagen-fibril binding and tissue-stiffening factor from *Cucumaria frondosa* dermis. *Matrix Biol., 21*, 625–635.

129. Ribeiro, A. R., Barbaglio, A., Benedetto, C. D., Ribeiro, C. C., Wilkie, I. C., Carnevali, M. D., & Barbosa, M. A., (2011). New insights into mutable collagenous tissue: Correlations between the microstructure and mechanical state of a sea-urchin ligament. *PloS One, 6*, e24822.

130. Motokawa, T., Shintani, O., & Birenheide, R., (2004). Contraction and stiffness changes in collagenous arm ligaments of the stalked crinoid *Metacrinus rotundus* (Echinodermata). *The Biological Bulletin, 206*, 4–12.

131. Motokawa, T., (2011). Mechanical mutability in connective tissue of starfish body wall. *The Biological Bulletin, 221*, 280–289.

132. Moss, C., Jackie, H. A., & Thorndyke, M. C., (1998). Patterns of bromodeoxyuridine incorporation and neuropeptide immunoreactivity during arm regeneration in the starfish *Asterias rubens*. *Philosophical Transactions of the Royal Society of London. Series B: Biological Sciences, 353*, 421–436.

133. Dupont, S., & Thorndyke, M. C., (2006). Growth or differentiation? Adaptive regeneration in the brittlestar *Amphiura filiformis*. *The Journal of Experimental Biology, 209*, 3873–3881.

134. Anderson, J. M., (1954). Studies on the cardiac stomach of the starfish *Asterias forbesi*. *Biol. Bull., 107*, 157–173.

135. Hall, M. R., Kocot, K. M., Baughman, K. W., Fernandez-Valverde, S. L., Gauthier, M. E. A., Hatleberg, W. L., Krishnan, A., McDougall, C., Motti, C. A., Shoguchi, E., et al., (2017). The crown-of-thorns starfish genome as a guide for biocontrol of this coral reef pest. *Nature, 544*, 231–234.

136. Leray, M., Beraud, M., Anker, A., Chancerelle, Y., & Mills, S. C., (2012). *Acanthaster planci* outbreak: Decline in coral health, coral size structure modification and consequences for obligate decapod assemblages. *PloS One, 7*.

137. Smith, J. E., (1937). On the nervous system of the starfish *Marthasterias glacialis* (L.). *Philosophical Transactions of the Royal Society of London. Series, B., Biological Sciences, 227*, 111–173.

138. Smith, J. E., (1946). The mechanics and innervation of the starfish tube foot-ampulla system. *Philosophical Transactions of the Royal Society of London. Series, B., Biological Sciences, 232*, 279–310.

139. Smith, J. E., (1950). The motor nervous system of the starfish, *Astropecten irregularis* (Pennant), with special reference to the innervation of the tube feet and ampullae. *Philosophical Transactions of the Royal Society of London. Series, B., Biological Sciences, 234*, 521–558.

140. Mellott, D. O., Thisdelle, J., & Burke, R. D., (2017). Notch signaling patterns neurogenic ectoderm and regulates the asymmetric division of neural progenitors in sea urchin embryos. *Development, 144*, 3602–3611.

141. Oulhen, N., Swartz, S. Z., Laird, J., Mascaro, A., & Wessel, G. M., (2017). Transient translational quiescence in primordial germ cells. *Development, 144*, 1201–1210.

CHAPTER 6

Endocrine Control of Gametogenesis and Spawning in Bivalves

MAKOTO OSADA[1] and TOSHIE MATSUMOTO[2]

[1]*Graduate School of Agricultural Science, Tohoku University,*
1-1 Tsutsumidori-Amamiyamachi, Sendai 981-8555, Japan,
Tel.: +81227178725, Fax: +81227178727,
E-mail: makoto.osada.a8@tohoku.ac.jp

[2]*National Research Institute of Aquaculture, 422-1 Nakatsuhamaura,*
Minami-ise, Mie 516-0193, Japan

6.1 INTRODUCTION

The reproductive process represented by gametogenesis and spawning accompanied by oocyte maturation and sperm motility activation in bivalves undergoes under the endocrine and neuroendocrine control as well as other animals. General morphology of germ cell and somatic cell supporting the development of germ cell in bivalves are described in this chapter. Endocrine and neuroendocrine factors and their signal transduction, which involved in the reproductive process are additionally reviewed based on the literatures described morphological, physiological, pharmacological, biochemical, and molecular biological aspect in reproduction.

The class name of Bivalvia is a synonym of Pelecypoda in taxonomy, which can be easily recognized because of two bisymmetric valves to cover their soft bodies. Bivalve species are commercially important organisms harvested as capture and aquaculture products in the fishery industry. The study on the reproduction of bivalve species is an essential subject for the biological evaluation of broodstock and environmental evaluation of aquaculture area. Reproduction and broad-ranging biological phenomena controlled by the nervous system in mollusks have been well-reviewed by Joosse and Geraerts [1], although the knowledge was limited to a gastropod. Very little is known of endocrine control of gametogenesis and spawning of bivalve mollusks, while the artificial seed production based on reproductive control is thirsted to improve productivity in aquaculture.

Development of eggs and sperms during gametogenesis process is a common phenomenon among oviparous organisms. In oogenesis, the primary germ cells (stem cell) undergo mitotic division and give rise to oogonia [2], and then develop into oocytes accumulating yolk materials. In the testis, the stem cells undergo a series of mitotic division with decreasing cytoplasmic volume and successively give rise to spermatogonia [2] and then undergo meiosis to form spermatozoa through spermatocytes and spermatids followed by spermiogenesis. The gametogenesis normally undergoes an annual cycle and the changes in the gonadal development have been qualitatively classified into several stages; in the scallop undifferentiating, early differentiating, growing, mature, spawning, post-spawning, and degenerating stages [3], based on the characterization of overall morphology of the gonads, but not on the quantitative analysis of the germ cells (Figure 6.1).

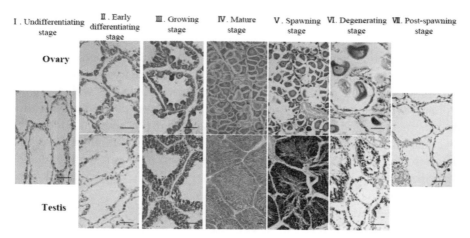

FIGURE 6.1 Morphological changes in gametogenesis of the scallop *Patinopecten yessoensis*. Bars, 50 μm. (Reprinted with permission from Ref. [4]. © 1987 Springer Nature.)

In this chapter, the histology of germ cells and cells related to gametogenesis, the mitotic process in the early development of germ cells, vitellogenesis resulting in oocyte growth and spawning based on oocyte maturation and sperm motility in bivalves are described. Most of the descriptions will be focused on bivalve reproduction from the point of view of comparative endocrinology.

6.2 HISTOLOGY OF GERM CELLS AND CELLS RELATED TO GAMETOGENESIS

Bivalves possess single gonads. The gonad most frequently takes the form of a mass, surrounding the digestive gland in the visceral region (oysters, clams). In mussels, it also invades paired mantle lobes, which run along the internal lining of the shell and ramify into the digestive gland. In scallops, the gonad is a discrete organ, and therefore the gonad index (GI), which represents the weight of the gonad as a proportion of total body weight, is used to assess gametogenic cycles.

Histological examinations are necessary to describe reproductive events concerning gamete development. Bivalve reproductive maturation is categorized into several stages; differentiation, growing or developing, ripening of gametes, spawning, and resting stages. In *Crassostrea gigas*, the gonad is a diffuse organ, which consists of numerous tubules in vaginated in a connective tissue. Expansion of the gonadal tubules in the visceral mass varies broadly with the progression of gametogenesis. At the undifferentiated stage, the gonadal area is largely filled with the vesicular connective tissue cells. Genital tubules are reduced in size and number and contain only undifferentiated cells. Sex is not distinguishable. At the developing stage, oogonia or spermatogonia develop from undifferentiated cells and divide actively. Gonoduct distributed near the surface of the gonad are lined with oogonia and oocytes or spermatogonia on the inner side and a ciliated epithelium on the outer side [5]. Genital tubules gradually enlarge, ramify, and invade the surrounding vesicular connective tissue. In females, oocytes are distributed along the inner wall of ovarian genital tubules. Vitellogenic oocytes with prominent germinal vesicles (GVs) predominate. In males, sperm can be found in the lumina of genital tubules. The vesicular connective tissue can be still observed partly. At the ripe stage, genital tubules are filled with fullgrown oocytes or sperm and occupy most of the region between the mantle epithelium and the digestive diverticula. The vesicular connective tissue can hardly be observed at this stage. At the spawning stage, genital tubules are partly empty because some of eggs or sperm are discharged during spawning. After completion of spawning, undischarged eggs or sperm and spermatocytes are re-absorbed. The vesicular connective tissue cells proliferate again.

The gonadal development of the scallop *Patinopecten yessoensis* is histologically classified into seven stages (Figure 6.1). The scallops at the undifferentiated stage show the lowest value of GI. A few oogonia and oocytes, or spermatogonia, and occasionally spermatocytes are distributed along the epithelia of germinal acini in the ovary and testis, respectively, at the early differentiating stage. The scallops at the growing stage have growing oocytes or an increased number of spermatogonia and spermatocytes, and the gonads developed to the mature stage. The disappearance of oogonia and the increased number of growing oocytes in the ovary, and the increased number of spermatogonia, spermatocytes, spermatids, and spermatozoa in the testis are observed at the mature stage. In the scallops at the spawning stage, the germinal acini are filled with fully grown oocytes in the ovary and mostly spermatozoa with spermatogonia, spermatocytes, and spermatids in the testis. A spawned trace is partially seen. After spawning, the gonad enters the post-spawning and degenerating stages. The GI continuously increases from the early differentiating stage to the spawning stage for both females and males and decreases after spawning to a similar level as that at the undifferentiating stage. Similarly to the GI profile, the oocyte diameter increases from the early differentiating stage to the spawning stage.

6.2.1 ACCESSORY CELLS IN FEMALES

In bivalves, the ovarian acinus is a simple structure containing mainly oocytes and associated accessory cells within a thin germinal epithelium. The relationship between intra-acinal accessory cells termed "follicle cells" and the successive stages of the developing oocyte were shown in *Mytilus edulis* [6]. During the early stages of development, oogonia are located along the internal wall of the acini (Figure 6.2A) and develop into oocytes, which are surrounded by a limited number of small follicle cells (Figures 6.2B and 6.3). With the development of the oocyte, the follicle cells are restricted to the stalk region, which attaches the oocyte to the acinus wall (Figure 6.2C). A few follicle cells are also observed in association with oocytes at all stages of *Crassostrea virginica* [7]. These cells of *Pecten maximus* are called "auxiliary cells" [8]. Auxiliary cells, as well as follicle cells, are closely associated with developing oocytes (Figure 6.4). The contact between the follicle cells and the developing oocyte is maintained by means of desmosome-like gap junctions. The presence of desmosome-like gap junctions indicates that the exchange of small molecules and ions may take place between follicle cells and developing oocytes. Ovarian follicle cells or auxiliary cells are the site of vitellogenin (Vtg) synthesis in *C. gigas* and *P. yessoensis* [9, 10].

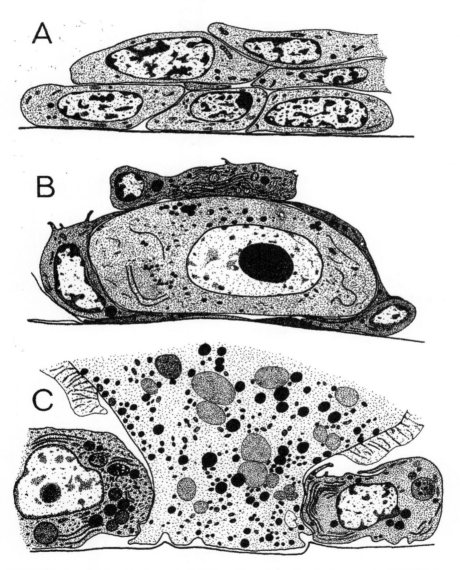

FIGURE 6.2 Relationship between follicle cells and the developing oocyte. (A) Primary oogonia. (B) Developing oocyte surrounded by small follicle cells. (C) Basal region of post-vitellogenic oocyte showing follicle cells in contact with stalk region of oocyte. (Reprinted with permission from Ref. [6]. © 1987 Springer Nature.)

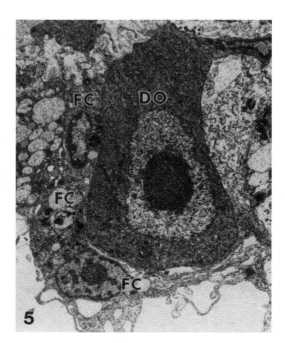

FIGURE 6.3   Developing oocyte (DO) surrounded by small follicle cells (FC) (×6500). (Reprinted with permission from Ref. [6]. © 1987 Springer Nature.)

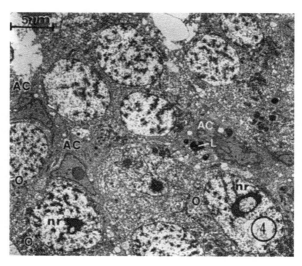

FIGURE 6.4   Auxiliary cells (AC) in close contact with the young oocytes (O); nr: nucleolar ring in oocyte. An auxiliary cell displays dense lysosomal-like organelles (L). (Reprinted with permission from Ref. [8]. © 1989 Springer Nature.)

6.2.2 ACCESSORY CELLS IN MALES

Bivalve testes contain accessory (somatic) cells, which presumably play important roles in the processes of spermatogenesis. In the ultrastructural study of spermatogenesis in *C. gigas*, somatic intragonadal cells are observed among the male germ cells [11]. With light microscopy, such cells are difficult to identify because their cytoplasm do not stain differentially from that of germ cells. Intermediate junctions or zonula adherens and phagolysosomes containing degenerating sperm cells are noticed in some somatic cells. Intragonadal somatic cells are observed in numerous species of bivalves, but the terminology used to describe these cells differs among the studies [12]. In *C. virginica*, accessory cells are closely associated with all sperm stages except the mature spermatozoon (Figure 6.5). Accessory cells appear to be assigned the term "Sertoli cells" in *M. edulis* [13], and "Sertoli-like cells" in *P. maximus* [14].

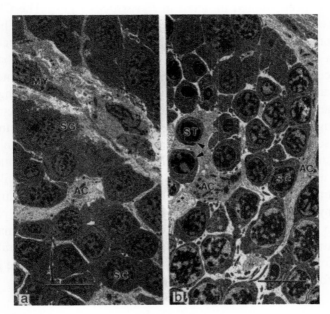

FIGURE 6.5 Accessory cells closely associated with sperm cells. (a) The accessory cell (AC) is distributed in close association with spermatogonia (SG) and spermatocytes (SC); MY: myoepithelial cell. (b) Spermatocytes (SC), spermatids (ST) and accessory cells (AC). One accessory cell surrounds two adjacent spermatids (arrowheads). Scale bar = 4.0 μm. (Reprinted with permission from Ref. [12]. © 1996 Springer Nature.)

6.2.3   ESTROGENIC CELLS IN GONADS

In *P. yessoensis*, estrogenic cells were identified immunohistochemically in the gonad. Immunoreactivities against P450 aromatase and estradiol-17β (E_2) were detected in the cells along the inside of the acinus wall of the testis [15], whereas the cells are distributed along the outside of the acinus wall in the ovary [16]. The localization of estrogen synthesizing cells in the testis is similar to that of a Sertoli cell in the testis of vertebrates. In contrast, the distribution of the same type of cells in the ovary is similar to that of Leydig cells in the testis of vertebrates.

6.2.4   GENERAL ANATOMY OF THE CENTRAL NERVOUS SYSTEM (CNS)

The central nervous system (CNS) that controls the reproduction of the bivalves has been studied in detail in *P. yessoensis* [17]. The CNS of the scallops is composed of the cerebral, pedal, and visceral ganglia. The cerebral ganglia, consisting of an anterior lobe and a posterior lobe, respectively, are located on either side of the pedal ganglion, which lies at the base of the foot. The cerebrovisceral connective exits the posteroventral corner of each cerebral ganglion and extends under the gonad to the visceral ganglion. The visceral ganglia are the largest and have the most intricate structure of the central ganglia. They are situated on the ventral side of the adductor muscle. The accessory ganglia are positioned at the point of the lateral lobes of the visceral ganglia. The width of the cerebropedal commissure and the visceral ganglia in an adult is approximately 5.5 mm and 3.4 mm, respectively.

In contrast to other bivalves, CNS of oysters is simple. It includes a pair of very tiny cerebral ganglia located at the bases of the labial palps. The coalesced visceral ganglia are at the ventral end of the visceral mass on the anteroventral border of the adductor muscle. The cerebrovisceral connectives run from the cerebral ganglia to the visceral ganglia. There is no foot and, therefore, there is no pedal ganglion.

6.2.5   NEUROSECRETORY PRODUCTS

There are both aminergic and peptidergic secretions from CNS. The existence of monoamines, including dopamine and serotonin (5-hydroxytryptamine, 5-HT) has been demonstrated in the scallop by histofluorescence methods [17]. Green fluorescent cells which presumably contained dopamine were detected in the anterior lobe of cerebral ganglion and the lateral lobe of visceral ganglion, whereas all the cells of accessory ganglion showed yellow fluorescence which suggested the presence of serotonin. In the gonadal area, the wall of the gonoduct contained both green and yellow fluorescent fibers. Green fluorescent varicose fibers were distributed in the epithelium around the gonad and along the intestinal epithelium. Moreover, the localization of 5-HT neurons in CNS and the gonad of the scallop was examined immunohistochemically [18]. In CNS, 5-HT neurons were distributed in a part of the anterior lobe and the posterior lobe of the cerebral ganglion, the pedal ganglion, and the accessory ganglion, whereas no nerve cells containing 5-HT were found in the visceral ganglion. All central ganglia, including the visceral ganglion, had numerous 5-HT varicose fibers in their neuropil. In the gonadal region, a number of bundles of 5-HT nerve fibers, most of which appeared to be ramifications of the cerebrovisceral connectives, were found to run into the gonad through the pallial epithelium of the gonad near the adductor muscle. Bundles of 5-HT nerve fibers were also observed in the subepithelial layer of the gonoduct, and the varicose fibers were distributed along the germinal epithelium. The development of 5-HT and catecholamine neuron during the larval period and their distribution has been also examined in detail using immunocytochemical and histofluorescent techniques in mussel species from the point of view of different development among trochozoan groups [19].

Several kinds of neuropeptides have been identified in bivalve mollusks. The cardioexcitatory neuropeptide FMRFamide was first isolated in the clam *Macrocallista nimbosa*, and its cardioexcitatory effect on the cardiac muscle was confirmed in other bivalves (see [20]). Presence of FMRFamide-like peptides was immunologically evidenced in the scallop *Placopecten magellanicus* through the CNS and peripheral organs of juveniles and adults [21]. Immunoreactivity of the neuropeptide, APGWamide, which functions as a neurotransmitter within the CNS of gastropods, was detected in the CNS and the axonal terminals within peripheral tissues, including gonads of bivalves [22]. Gonadotropin (GTH)-releasing hormone (GnRH) is a neuropeptide that plays key roles in the regulation of reproduction in vertebrates. The demonstration of GnRH neurons has been made in the nervous system of bivalves using immunohistochemical techniques. In the scallop, the GnRH-like neurons detected with anti-mammalian GnRH antibody were distributed sparsely in the pedal ganglia and predominantly in the anterior lobe of the cerebral ganglia of both sexes at the growing stage, whereas no GnRH nerve fibers were found in the gonad [23]. A study carried out in the nervous system of *C. gigas*, and *M. edulis* detected neurons, which react positively to antibodies raised against octopus GnRH [24]. GnRH neurons and fibers were scattered in the visceral ganglia of *C. gigas*. In *M. edulis*, both cerebral and pedal ganglia contained GnRH neurons and fibers. In addition, anti-GnRH immunoreactivity was not found in the male and female gonad of *M. edulis* in the previous study [25].

6.3 MULTIPLICATION OF GONIAL CELLS

Quantification of development of gonial cells is necessary to understand endocrine control of early development of germ cells in bivalves. The mitotic division of gonial cells has been quantified in bivalve mollusks. In blue mussel, *M. edulis*, [^{3}H] thymidine incorporation into young germinal cells, particularly spermatogonia, in the mantle tissue was demonstrated by autoradiography [26]. The [^{3}H] thymidine incorporation into dissociated cell suspensions from the mantle tissue associated with the estimation of aspartate transcarbamylase activity was applied to quantitatively evaluate the mitogenic influence of an endogenous factor on spermatogonial multiplication [26, 27]. In scallop, *P. yessoensis*, BrdU incorporation into gonial cells during early development showed that the proliferation of gonial cells in the scallop could be divided into phase I and II [4] (Figure 6.6). In phase I, oogonia, and spermatogonia show a slow proliferation. In phase II, oogonia terminate proliferation followed by development into the vitellogenic oocytes while spermatogonia show a rapid proliferation. In vertebrates, developing spermatogonia have been classified into two types: Non-proliferated type A spermatogonium as a stem cell and type B spermatogonium as a differentiated spermatogonium [28]. Osada et al., [4] proposed that spermatogonia in bivalves could be classified into a spermatogonial stem cell and a series of development of differentiated spermatogonium during early spermatogenesis on the basis of proliferation potency of spermatogonia.

Mollusks are useful to scientists involved in the study of neurophysiological responses due to their extremely large nerve cells. The effect of neural factors on gonial mitosis in mollusks was first reported in *M. edulis* of bivalve. The cerebral ganglia stimulated gonial mitosis, the reinitiation of meiosis in males and previtellogensis and vitellogenesis in females in organ culture experiment of the mussel in vitro, although the ablation of the ganglia did not affect such a reproductive function in vivo [29, 30]. Involvement of neural factors in the multiplication of germ cells has been reported also in the slipper limpet, a gastropod [31]. Mathieu et al., [32] found the occurrence of a neural factor named gonial mitosis-stimulating factor from the cerebral ganglia of *M. edulis* with a molecular mass of less than 5 kDa, which stimulated [³H] thymidine incorporation into dissociated cell suspensions of the mussel mantle tissue and existed in the hemolymph and circulatory cells. This neural factor was supposed to be primarily concerned with the GnRH of bivalve mollusks.

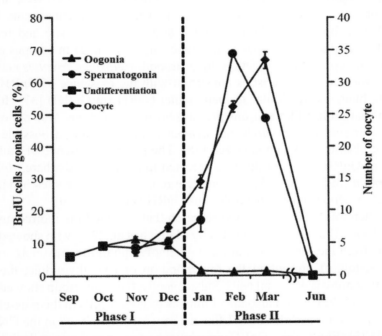

FIGURE 6.6 Changes in the percentage of BrdU immunopositive gonial cells per gonial cells in germinal acini of the ovary and testis and the number of oocyte in germinal acini of the ovary of *Patinopecten yessoensis* during sexual maturation. Each value represents the mean ± SE (*n* = 7–10). (Reprinted with permission from Ref. [23]. © 2007 Springer Nature.)

GnRH superfamily, which includes GnRH, adipokinetic hormone (AKH), corazonin (Crz), and AKH/Crz-related peptides, is almost ubiquitous throughout the bilateral animals [33]. In vertebrates, GnRH is synthesized in the hypothalamic region, and is then transported to, and acts on the pituitary to promote the release of GTH consisting of follicle-stimulating hormone, and luteinizing hormone (LH). The connection between the brain and the gonads via the pituitary is called the hypothalamus–pituitary–gonadal axis (HPG axis), which forms the basis for both the neural and endocrine regulation of reproduction in all vertebrates. Interestingly, α-mating factor, a tridecapeptide mating pheromone of yeast has been identified as a homolog of GnRH with LH-release activity from gonadotroph at high doses, suggesting a conservation of the structural and functional properties of GnRH-related peptides during evolution [34].

Since then, the existence and function of GnRH-like peptides in mollusks have been demonstrated using heterologous GnRH antibodies. In cephalopod, *Octopus vulgaris*, immunopositive GnRH-like peptides were shown to be present in the optic gland, which is a major endocrine organ [35]. GnRH was found throughout the CNS and in both the male and female reproductive ducts of octopus [36], suggesting that GnRH in octopus could have some reproductive function. In gastropod, *Hellisoma trivolvis* nervous system showed characteristics consistent with the existence of GnRH-like peptide functionally similar to mammalian GnRH (m-GnRH) [37]. It was shown that GnRH-like neurons are present in the nervous systems of the freshwater snails *H. trivolvis* and *Lymnaea stagnalis*, and a possible role for reproduction was hypothesized [38]. The possible presence of multiple forms of GnRH-like peptides was reported in the marine sea hare, *Aplysia californica* [39] and the expression were refined to specific sites inside the CNS of the *Aplysia* using heterologous GnRH antibodies [40].

As mentioned above, the first research that involved the indirect use of mollusk GnRH deeply was reported by Mathieu et al., [32], who showed that extracts from the cerebral ganglion and hemolymph of the mussel *M. edulis* were capable of promoting the incorporation of [³H] thymidine, thereby promoting mitosis of gonial cells. Subsequently, the same group showed that vertebrate GnRHs could also affect mitosis of gonial cells in the mussel and the Pacific oyster and the GnRH-like neurons were identified in the CNS of mussel [25] (Figure 6.7). GnRH signal maybe transduced into cells through the membrane receptor as a GnRH receptor ortholog was cloned from the gonad of oysters [41, 42]. And then two GnRH-related peptides (pQNYHFN-SNGWQP-NH$_2$Cg-GnRH-a; amidated undecapeptide and pQNYHFNSNG-WQPG Cg-GnRH-G; non-amidated decapeptide) were confirmed by mass spectrometry from the CNS of the Pacific oyster [24]. However, a specific affinity of the GnRH receptor orthologs to endogenous oyster GnRH-like peptides identified [24] have not yet been confirmed. In the Japanese scallop, Nakamura et al., [23] reported an occurrence of GnRH-like peptide in the scallop CNS-regulating development of the scallop gonad. The GnRH-like neurons detected with anti-m-GnRH antibody were distributed in the CNS of the scallop, whereas no immunopositive GnRH nerve fibers were found in the gonad and both the neural factor extracted from the CNS and m-GnRH strongly stimulated mitosis of spermatogonia in vitro. The reactions of the neural factor and m-GnRH were abolished by absorption with anti-m-GnRH antibody and competition with m-GnRH-specific antagonists, which interfere GnRH function on a GnRH receptor, suggested occurrence of endogenous GnRH-like peptide and GnRH receptor-like receptor. And the same mitotic activity as the neural factor and m-GnRH was found in the hemocyte lysate, but not in the serum. These results suggested that the neural factor has a similar antigenicity to m-GnRH, and the function of the factor may be mediated through m-GnRH receptor-like receptor in the testis under a neuroendocrine pathway (Figure 6.8).

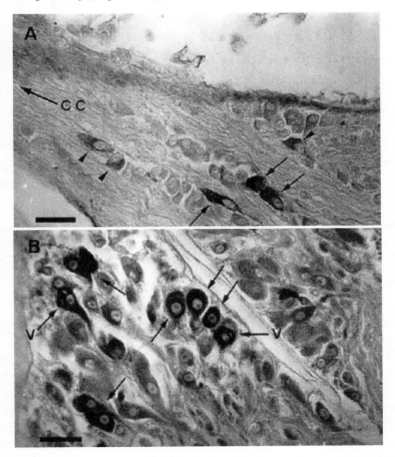

FIGURE 6.7 GnRH-like immunoreactivity in the central nervous system of *M. edulis*. (A) Intensely (arrows) and weakly (arrowheads) stained cells in the cerebral ganglia (CC, the direction toward the cerebral commissure). (B) Immunoreactive cells (arrows) in the pedal ganglia. Two stained cells present vacuoles (V). Bar, 20μm. (Reprinted with permission from Ref. [25]. © 1999 Elsevier.)

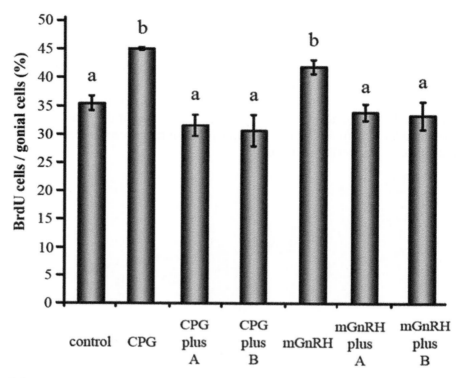

FIGURE 6.8   Effect of two types of antagonists (1×10^{-6} M each), antide (A) and [D-pGlu[1], D-Phe[2], D-Trp[3, 6]] GnRH (B), of m-GnRHon 1×10^{-6} M m-GnRH and CPG extract-induced spermatogonial proliferation of *P. yessoensis*. Each value represents the mean ± SE ($n = 3$–4). Values with different letters are significantly different ($p < 0.05$). (Reprinted with permission from Ref. [23]. © 2007 Springer Nature.)

The full cDNA sequences of GnRHs were cloned from the Japanese scallop and Pacific oyster as py-GnRH and cg-GnRH, respectively. The GnRH-like peptide sequences of both bivalve species have a high similarity to oct- and ap-GnRHs [43]. The extra dipeptide insertion after N-terminal pyro-glutamate residue was recognized in both bivalve species in common with other molluscan species (Figure 6.9) [44]. The logically predictedpQN-FHYSNGWQP-NH₂ (py-GnRH-P-NH₂), an amidated undecapeptide peptide, was synthesized and was shown to stimulate proliferation of spermatogonia in testicular tissue culture of the scallop as well as the previous culture with m-GnRH [43]. However, the peptide failed to induce LH release from quail pituitary, suggesting that molluscan GnRHs might conserve a fundamental molecular structure similar to that found in other animals, but may not bind to the vertebrate receptor [43]. The predicted pQNFHYSNGWQP-NH$_2$ (py-GnRH11AA-NH$_2$) peptide was definitely confirmed by mass spectrometry from the CNS of the scallop as an endogenous GnRH-like peptide along with pQNFHYSNGWQPG-OH (py-GnRH12AA-OH) [45] as two types of GnRHs in the Pacific oyster [24].

		Signal peptide	GnRH peptide and Cleavage sequence	
Bivalve	py-GnRH(AB486004) 1	------MSSYTQILVAQLLLAGLLVAVVS-G	QNFHYSNGWQPGKR-	GAP sequence 104
	cg-GnRH(HQ712119) 1	----MKVSPCTQVIVMVLTLG--LLCEVH-A	QNYHFSNGWQPGKR-	GAP sequence 91
Cephalopod	oct-GnRH(AB037165)1	MSATASTTSSRKMAFFIFSMLLLSLCLQTQA	QNYHFSNGWPGGKR-	GAP sequence 90
	ue-GnRH(AB447557) 1	MSTSPVTSTLRRMVFLTCAIFLLSLCMQTQA	QNYHFSNGWHPGGKR-	GAP sequence 91
Gastropod	ap-GnRH(EU204144) 1	-MACRITSATTTLFSILLLIVIAELCS---A	QNYHFSNGWYAGKKR-	GAP sequence 148
	lg-GnRH(FC805608) 1	-------MMPVPLKYFGLALTLALVTELAVG	QHYHFSNGWKSGKKR-	GAP sequence 108
	m-GnRH(EAW63591) 1	---MCLRMKPIQKLLAGLILLTWCVEG-CSS	Q--HWSYGLRPGGKR-	GAP sequence 97

	Postulated function	Ref
py-GnRH	spermatogonial proliferation, steroidogenesis	Nakamura et al., 2007; Osada and Treen, 2013
oct-GnRH	steroidogenesis, brain function	Iwakoshi–Ukena et al., 2004; Kanda et al., 2006
ap-GnRH	behavior	Tsai et al., 2010

FIGURE 6.9 Amino acid alignment of pre-pro-GnRH sequences from molluscan species deduced from their coding DNAs and their postulated functions. The alignment was created by CLUSTALW2, and conserved amino acids are shaded. The sequences containing the GnRH peptide with the cleavage site are boxed. Signal peptide and GAP regions are also indicated. EMBL/GenBank accession numbers: *Patinopecten yessoensis*: py-GnRH (AB486004), *Crassostrea gigas*: cg-GnRH (HQ712119), *Octopus vulgaris*: oct-GnRH (AB037165), *Aplysia californica*: ap-GnRH (ABW82703), *Lottia gigantea*: lg-GnRH (FC805608), *Uroteuthis edulis*: ue-GnRH (AB447557), *Homo sapiens*: m-GnRH (EAW63591). (Reprinted with permission from Ref. [44]. © 2013 Elsevier.)

In mollusks, several physiological functions for mollusk GnRH have been suggested other than spermatogonial proliferation. The immunopositive nerve fibers of oct-GnRH were identified in both CNS and peripheral organs, suggesting that the GnRH may act as a modulatory factor in controlling higher brain functions as well as a reproductive factor [46, 47]. The oct-GnRH did modulate the contraction of the heart and oviducts [46, 48]. In addition, the oct-GnRH stimulated the production of progesterone, testosterone, and E$_2$ in both testis and ovary of the octopus, suggesting the reproductive role of the oct-GnRH [47] because ovarian development in octopus was found to be associated with the fluctuation of the sex steroid hormone [49, 50]. The ap-GnRH, distributed in the central tissue, functioned as a modulator of behavioral attributes controlling parapodia, foot, and head movement, not as an acute reproductive trigger for the development of ovotestis and secretion of egg-laying hormone (ELH) [51–53]. The presence of multiple forms of GnRH in *A. californica* was suggested [39], and the nap-AKH shared common ancestry with AKH/RPHC was newly identified in addition to ap-GnRH, which inhibited feeding resulting in a reduction of body and gonadal mass [54]. In the scallop py-AKH-like gene was identified with py-AKH receptor gene, which was predominantly expressed in the CNS [55]. Bigot et al., [24] identified two kinds of GnRH by mass spectrometry in the CNS of the Pacific oyster, while each function has not been confirmed in the oyster. In vivo administration of the scallop py-GnRH into the scallop, gonad accelerated spermatogenesis associated with a significant increase in testis mass and conversely inhibited oocyte development associated with apoptosis, implying early phenotypic alteration to masculinization [56]. The py-GnRH signal to induce spermatogenesis and masculinization was supposed to be utilized via py-GnRH receptor which broadly distributed in various tissues, including the gonad as well as the CNS [55].

In view of the roles of GnRH in the reproduction of mollusks without a pituitary, it is suggested that GnRH could be involved in spermatogonial proliferation by mediating gonadal steroidogenesis [44]. Estrogen synthesizing cells have been immunologically found along the outside and inside acinar wall in the ovary and testis of the scallop, respectively, and their localization is similar to Leydig cells and Sertoli cells in the testis, respectively [15, 16]. Aromatase activity and E_2 content increased in synchronization with reproductive progress [15]. E_2 is involved in spermatogenesis in vertebrates [57, 58]. Interestingly, oct-GnRH was capable of inducing steroidogenesis of testosterone, progesterone, and estrogen [47]. It is possible that py-GnRH stimulates the estrogen synthesis that, in turn, promotes spermatogonial mitosis because the py-GnRH and E_2-induced spermatogonial proliferation were blocked by an estrogen antagonist [44]. It was suggested that spermatogonial proliferation stimulated by py-GnRH might be mediated through estrogen whose synthesis was induced by py-GnRH (Figure 6.10) [44].

Steroid production and function on mollusk reproduction have been well examined using vertebrate steroids in the past; quantification of steroids by the immunological method and HPLC, etc., identification of putative sites of steroid production, and induction of sex reversal and oogenesis have been demonstrated (see [59]. However, steroid function in mollusks is recently in contention because convincing evidence for the biosynthesis of vertebrate steroids by mollusks and the biological effects of steroids on mollusks are required [60, 61]. Series of enzymes composing steroid synthetic pathway, structure determination of endogenous steroids and corresponding receptors should be uncovered hereafter.

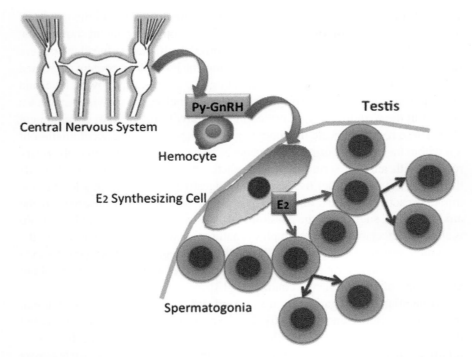

FIGURE 6.10 Transduction mechanism of py-GnRH for inducing spermatogonial proliferation in bivalve mollusks. The py-GnRH secreted from the anterior lobe of the cerebral ganglion may be transported by the hemocytes in the circulation system and received by estrogen synthesizing cells to stimulate the secretion of estradiol-17β, resulting in the induction of spermatogonial proliferation. (Reprinted with permission from Ref. [44]. © 2013 Elsevier.)

6.4 OOCYTE GROWTH

In the developing or growing stage, the oocyte rapidly increases in size. In these periods, a large amount of yolk protein is accumulated in oocytes. Vitellin (Vn) is a major yolk protein stored in yolk granules of oocytes and used as a nutrient during embryogenesis. In vertebrates such as fish, amphibians, and birds, yolk protein is synthesized from a precursor, Vtg, produced by the liver and is transported to the oocytes via the blood circulation system. In bivalves, a few biochemical studies of yolk proteins were carried out [62, 63]. Yolk protein was purified, and using a specific antiserum against the purified protein, the distribution of the protein was examined in *C. gigas* and *P. yessoensis*. The presence in the oocytes indicated the protein is a Vn. The ovarian Vn content of *C. gigas* increased as the oocyte developed and became low after spawning.

This variation agrees with the profile of the oocyte diameter and the results of histological observation of the ovaries [64]. In scallops, Vn content showed low level till the growing stage and thereafter showed the highest value at the mature stage. At the degenerating stage after spawning, the level of Vn content decreased markedly. GI value at the early differentiating stage showed the lowest level and gradually increased, reaching the highest value just before the spawning stage. Vn content in the scallop ovary increased in parallel with the ovarian development [65]. These results indicate that oocyte growth depends on the accumulation of Vn within the oocytes during vitellogenesis. Since no organ other than the ovary reacted with anti-Vn serum, it was predicted that Vn was synthesized inside the ovary, not in the other tissues such as digestive diverticula. In addition, the immunoreactivity against anti-scallop Vn antibody was also observed in the auxiliary cells, suggesting the possibility of Vtg synthesis in these cells [65]. Vn is slightly detectable in hemolymph of mature females, in particular, during the spawning season. This appears to originate from degenerated oocytes. Oocyte degeneration and resorption is not unusual in bivalves and may be brought about by a variety of environmental conditions [13].

Molecular investigations were carried out to elucidate the mechanism involved in vitellogenesis [9, 10]. The expression of Vtg mRNA indicated the maximum level at the developing or growing stage, and the level was retained during the mature stage. In RT-PCR analysis, Vtg mRNA was detected in the ovary but not in the digestive diverticula, which was consistent with the results of immunohistochemistry. In situ hybridization indicated Vtg mRNA signals were detected in the follicle cells in the ovary. Similarly, scallop Vtg mRNA expression is demonstrated in the auxiliary cells. In marine bivalves, autosynthetic yolk formation in the oocytes had been thought to be the main type of vitellogenesis, based on the morphological evidence [6, 8, 12, 63]. In oyster and other bivalve species, the ovarian acini contain only developing oocytes and associated follicle cells within a thin germinal epithelium. The functions of follicle cells in the bivalve ovary were not well understood, though they were suspected of playing some roles in oocyte nutrition. Results of RT-PCR and in situ hybridization, together with the immunocytological results support Vtg is synthesized in the auxiliary cells surrounding the vitellogenic oocyte through a heterosynthetic pathway and directly passed to oocyte. In common with bivalve Vtgs, the abalone Vtg gene is expressed in the follicle cells in the ovary [66]. In abalone species like in bivalves, follicle cells are present on the stalk of developing oocytes in the maturing ovary. A recent study in abalone showed the immunoreactivity in the follicle cells with anti-Vn antibody and indicated that transcription and translation of the Vtg gene occur in the ovarian follicle cells [67]. In addition, positive reactions with the antibody appeared first in the stalk part of the oocytes at the early phase of yolk accumulation, and the follicle cells adjacent to the stalk of these oocytes were also stained positively. These observations imply that Vn or Vtg is transported from the follicle cells to the oocyte through an extracellular space around the oocyte stalk. In the scallop, Vn immunoreactivity was observed in the oocytes and the auxiliary cells [65]. Similar systems of yolk protein transport can be expected in the oyster and the scallop.

It is clearly established that the Vtg gene expression is regulated by E_2 through estrogen receptor (ER) in oviparous vertebrates [68]. In marine bivalves, the presence of steroids was reported; the structural identity of endogenous steroids as vertebrate-type sex steroids (progesterone, androstenedione, testosterone, E_2, and estrone) has been demonstrated in the mussel by gas chromatography and mass spectrometry [69]. In oysters and scallops, E_2 is detected in the ovary, and its contents showed a synchronous profile with gametogenesis on the basis of HPLC [16]. In the ovary of scallops, estrogenic cells were demonstrated from observation of immunoreactivities against P450 aromatase and E_2, which distributed along the outside of acinus wall. In addition, aromatase activity and E_2 showed maximum values at the mature stage before spawning [15]. The Vtg synthesis terminates at the mature stage [10, 65]. These results suggest that E_2 synthesized in the estrogenic cells through P450 aromatase may be involved in the induction of Vtg synthesis. The physiological function of estrogen on vitellogenesis has also been reported in various mollusks, though it is still under scientific debate [61]. Estrogen-induced Vn synthesis has been demonstrated in the scallop [65] and oyster [64]. In vitro E_2 treatment resulted in an increase in Vn content in the ovarian tissue. Further, in scallop, in vitro ovarian tissue culture with the cerebral plus pedal ganglion (CPG) extract resulted in a greater increase in Vn content. This indicates the CPG contains a vitellogenesis promoting factor (VPF) which regulates Vtg synthesis. However, Vtg mRNA expression in the ovarian tissue cultured in vitro showed no change with VPF, while it was promoted by E_2 [10]. VPF appears to promote vitellogenesis at the level of translation (Figure 6.11). ER-like immunoreactivity was found in the growing oocyte and the auxiliary cells, in which Vn immunoreactivity and Vtg mRNA were also found [65]. These findings suggest that E_2 may be involved in the control of vitellogenesis mediated by ER.

ERs are members of the nuclear receptor (NR) superfamily, which has a number of common features, and their proteins can be divided into six domains. The DNA-binding domain (DBD) and the ligand-binding domain (LBD) are the most highly conserved among species. The DBD recognizes and binds to specific responsive elements in DNA. The LBD regulates hormone-dependent transcription of target genes, such as Vtg.

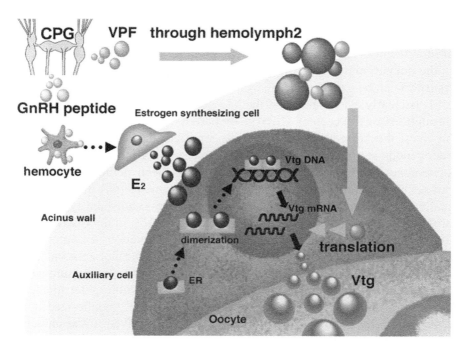

FIGURE 6.11 Central nervous system-gonadal control of vitellogenesis in scallops. Vitellogenin (Vtg) is synthesized in the auxiliary cell closely associated with the growing oocyte. The transcription of Vtg mRNA is promoted by estradiol-17β (E_2), and the translation is enhanced by vitellogenesis promoting factor (VPF) from the cerebral plus pedal ganglion (CPG). The GnRH peptide secreted from CPG may be transported by the hemocyte in the circulation system and received by estrogen synthesizing cell along the outside of the acinus wall to stimulate the secretion of E_2. It remains unknown whether vitellogenesis is mediated by estrogen receptor (ER), though ER immunoreactivity is found in the auxiliary cells.

To understand the estrogen signaling in the vitellogenesis of bivalves, the oyster ER homolog was isolated, which is highly similar to the ERs of other species [70]. In invertebrates, ER has been cloned from mollusks, *A. californica* [71], *Thais clavigera* [72], and *O. vulgaris* [73]. The amino acid sequence of oyster ER protein revealed a high identity to other mollusks ERs. The phylogenetic analysis indicated that oyster ER is most closely related to the other mollusk ERs. This indicates that the oyster ER isolated is an ortholog of Aplysia ER, snail ER, and octopus ER. The oyster ER also did not activate luciferase expression in the presence of E_2 and constitutively activated the reporter transcription. The addition of E_2 did not induce the further enhancement of the upregulation. This is consistent with the results on other mollusks ERs.

Mollusk ER ortholog is called ER because of high sequence similarity to vertebrate ERs; however, it activates transcription in the absence of ligand and does not bind steroid hormones. Annelid ER exceptionally conserved architecture of ligand pocket and responded to estrogen [74]. The X-ray crystal structure of oyster ER was determined and found that its ligand pocket is filled with bulky residues that prevent ligand occupancy [75]. The oyster genome possesses 43 putative NR sequences [76]. It contains two members of NR3, the sex steroid hormone receptor analogs, an ER homolog identified as oyster ER and an estrogen-related receptor homolog, which is constitutively activated and unlikely to bind estrogen. Additional NR3 members, which could interact with vertebrate sex steroids, were not identified. Given prior studies of vitellogenesis induced by estrogens, these imply the effects must be mediated by mechanisms other than ER activation in the presence of estrogens.

Although the function of ERs from mollusks remains unsolved, snail ER mRNA was expressed in the ovary, and cerebral ganglia [72] and octopus ER expression was observed in both sexes, with the highest levels in the ovary [73]. The oyster ER mRNA was detected in all tissues tested, with the higher expression in the ovary. The immunohistochemical localization of oyster ER using the antiserum to synthetic oyster ER peptide was found in the nuclei of follicle cells, the site of Vtg synthesis, and oocytes [70]. In *M. edulis*, two novel forms of ER-like genes have been isolated as an ER and an estrogen-related receptor (ERR), respectively, which were localized in the oocytes and follicle cells in contact with developing oocytes in the ovary and Sertoli cells in the testis, and in the ciliated cells of the gill [77]. And a significant increase in ER not ERR as well as Vtg mRNA expression was observed when mussels were exposed to estrogens during the early stage of gametogenesis [78]. We cannot rule out the possibility that E_2 might be mediated through alternative NRs and activate Vtg gene expression in the ovary. Nevertheless, the question of what roles mollusk ERs have remains unanswered.

6.5 SPAWNING

Spawning in marine invertebrates including bivalve mollusk is suggested to correlate with changes in temperature, lunar age, illumination, salinity, abundance of phytoplankton, food availability, physical shock, tidal surge, drying, and radical oxygen, and such an environmental fluctuation has been thought to be a natural cue to induce spawning mediated through endogenous regulation mechanism [79]. Blake and Sastry [80] have reported that a neurosecretion associated with stage V of the neurosecretory cycle in the CNS has been associated with spawning in the bay scallop, *Argopecten irradians*. The involvement of the cerebral ganglion in spawning has been suggested on the basis of the histological change in the ganglion associated with spawning and induction of spawning by ablation of the ganglion (see [81]). The implication of the CNS in bivalve spawning has been predicted for many years, although endogenous and specific factor controlling spawning in bivalves has not been found by this time.

The relationship between specific neurosecretory substances and spawning has been first investigated in Japanese scallop, *P. yessoensis*. Exogenous serotonin (5-HT) has been found to strongly induce spawning in the scallop in vivo and suggested to play an important role in the mechanism of the spawning in bivalves [82]. The reproducibility of 5-HT-induced spawning has been ascertained in a considerable number of other marine bivalves [83–85]. The 5-HT neuron has been immunologically identified in the pedal ganglion, cerebral ganglion, and accessory ganglion adjacent to the visceral ganglion, and 5-HT nerve fibers have been found to distribute around the gonoduct and along the outside of germinal acini of the scallop [18]. The localization of the 5-HT neuron and nerve fiber strongly supported a regulation of spawning process in the gonad by endogenous 5-HT. The UV ray-irradiated seawater is well known to induce spawning of the scallop as well as abalone, *Haliotis discus hannai* [86, 87]. The mechanism of induction of spawning with UV ray-irradiated seawater of the scallop has been pharmacologically demonstrated. The UV ray-irradiated seawater stimulates serotonergic mechanisms via dopaminergic mechanisms and induces spawning, which is modulated with prostaglandins [88].

A quantitative analysis of monoamine and prostaglandin during spawning of bivalves has been attempted to understand the roles of each endogenous substance in the spawning. In scallop, *P. yessoensis*, Osada et al., (1987) showed that significantly decreasing level of dopamine in the CNS and gonad of both genders after spawning induced by UV ray-irradiated seawater, suggesting that the release of dopamine might stimulate serotonergic mechanisms for induction of spawning. Thermal stimulation-induced spawning in other kinds of scallop, *Argopecten purpuratus*, has, in turn, resulted in changes of levels of dopamine, noradrenaline, and serotonin in the CNS, muscle, and gonad [89]. Seasonal variations in the levels of prostaglandin F2α (PGF2α) and prostaglandin E_2 (PGE$_2$) in the gonad were closely related to the reproductive cycle, suggesting that PGF2α and PGE$_2$ might be involved in the sexual maturation and spawning of the scallop [90]. PGF2α and PGE$_2$ in the gonad have been reported to significantly decrease during spawning of females and conversely increase during spawning of males. It was suggested that these PGs in females and males are suppressive neuromodulators of 5-HT-induced egg release and acceleratory neuromodulators of 5-HT-induced sperm release from the gonad tissue of scallop, respectively [94]. In *the in-vitro* experiment, PGF2α?significantly inhibited 5-HT-induced egg release from ovarian tissue and PGE2enhanced the 5-HT function, suggesting that PGF2α and PGE2play a role as a suppressive and acceleratory modulator of 5-HT-induced egg release, respectively [92]. Taken together, PGF2α particularly may be a suppressive modulator in the spawning of the females and an acceleratory modulator in the spawning of the males [91].

5-HT function to induce spawning was also modulated by steroid as well as by PGs. The response to artificial stimulation with UV-irradiated seawater tended to become higher during sexual maturation [87], suggesting that the sensitivity to 5-HT on spawning could rise, depending on maturity. Estrogen has been thought to be a stimulator on gametogenesis in bivalve because of its seasonal variation associated with gametogenesis and vitellogenesis accelerated by estrogen [10, 15, 16, 64]. The promoting effect of estrogen on 5-HT-induced egg release in the scallop was observed when the ovarian pieces of the scallop were pre-treated with estrogen before induction of egg release with 5-HT [93]. In bivalves, there have been several pharmacological characterizations of 5-HT receptors on the surface of a germ cell to transduce the 5-HT signal into the cell. The 5-HT receptor was pharmacologically characterized as a mixed profile of 5-HT$_1$/5-HT$_2$ subtypes in the oocyte membrane of scallop and 5-HT$_1$ subtype in that of oyster [94]. A unique type that showed mixed pharmacological properties [95] in the oyster, a mixed 5-HT$_1$/5-HT$_3$ type [96, 97] or a novel type that was distinct from any mammalian 5-HT receptors [98–100] in the surf clam, and an original type in zebra mussels [101] were also reported in the oocyte and sperm. Osada et al., [94] reported that the expression of the 5-HT receptor in the oocyte membrane of the scallop was induced by E_2 via a genomic action in their pharmacological experiment, suggesting that 5-HT receptors induced by estrogen distribute in the oocyte membrane and increase during oocyte growth, leading to a rising sensitivity to 5-HT involved in spawning (Figure 6.12).

The 5-HT acts as a neurohormone to directly mediate meiosis reinitiation of prophase-arrested oocytes [98, 102–107] as evidences by germinal vesicle breakdown (GVBD) [92] using isolated oocytes in *Spisula solidissima*,

Spisula sachalinensis, *C. gigas*, and *Ruditapes philippinarum*. In the scallop in which it is impossible to isolate oocyte from the ovary due to cytolysis after detachment from the germinal epithelium, GVBD oocytes were observed in paraffin section of ovarian tissue treated by 5-HT with dose-dependent manner, which was identical to a variation of egg release induced by 5-HT [108] (Figure 6.13). These results suggested that the induction of oocyte maturation is a primary role of 5-HT for their spawning. These facts suggest a distribution of 5-HT receptor on the surface of membrane of germ cell as mentioned before.

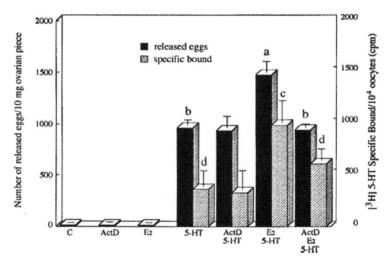

FIGURE 6.12 Effects of estradiol-17β and actinomycin D on 5-HT-induced egg release from ovarian pieces and specific bound of [³H] 5-HT to scallop oocyte membranes teased from ovarian pieces before 5-HT induction.2 × 10⁴ cells of prepared oocytes were incubated with 0.3 μM [³H] 5-HT. Each value represents the mean ± SE (*n* = 3). (a) Significantly different from (b) (*p*<0.005). (c) Significantly different from (d) (*p*<0.01). C; control, ActD; actinomycin D. (Reprinted with permission from Ref. [108]. © 2006 Elsevier.)

In mollusk, primary structures for seven 5-HT receptors have been determined by molecular cloning. Six 5-HT receptor cDNAs were cloned from CNS and reproductive system of gastropods, pond snail *L. stagnalis* and sea hare *A. californica* [109–113]. The full-length cDNA encoding a putative 5-HT receptor has been isolated from the ovary of the scallop as a 5-HT$_{py}$ [114]. The 5-HT$_{py}$ was classified into a vertebrate serotonin receptor 5-HT$_1$ subtype based on molecular architecture, homology searches, and phylogenetic analysis. The features of the absence of introns in the coding region of the gene, a relatively long third cytoplasmic loop, and a short fourth inner terminal domain (C-terminal tail) characterized the 5-HT$_{py}$ as an ancestral 5-HT receptor and a member of the 5-HT$_1$ receptor family coupled with G protein with high probability [115, 116]. A positive 5-HT$_{py}$ signal was ubiquitously observed in any peripheral tissue as well as the nervous system, interestingly in the spermatid, oocyte, and ciliated epithelium of gonoducts in both male and female gonads [114] (Figure 6.14). These results suggested that effects of 5-HT on a series of events of spawning consisting of induction of oocyte maturation, sperm motility, and transportation of mature oocyte and sperm through the ciliated epithelium of the gonoducts could be mediated by5-HT$_{py}$. The gene expression of 5-HT$_{py}$ in the ovarian tissues was significantly up-regulated by E$_2$ [114], which supported the pharmacological results that the expression of the 5-HT receptor in the oocyte membrane has been induced by E$_2$ via a genomic action leading to a rising sensitivity to 5-HT in relation to spawning, and explained a higher sensitivity to external stimuli for spawning depending on maturity [87, 94].

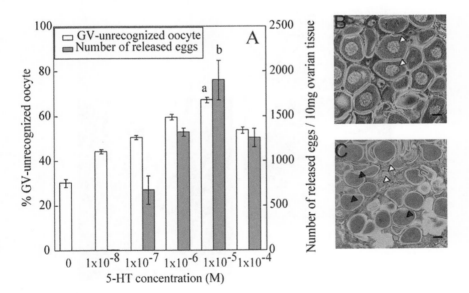

FIGURE 6.13 (A) Effect of various concentration of 5-HT on the induction of oocyte maturation and egg release of *Patinopecten yessoensis*.(B) A germinal vesicle (white arrowhead) can be seen in the section of ovarian tissue incubated without 5-HT. (C) Spindle fibers (black arrowhead) and a few germinal vesicles (white arrowhead) can be seen in the section of ovarian tissue incubated with 5-HT. [a,b] Significantly different from the others ($p<0.05$). Each value indicates the mean ±SE ($n=4$). Scale bar indicates 20 μm. (Reprinted with permission from Ref. [94]. © 2096 Elsevier.)

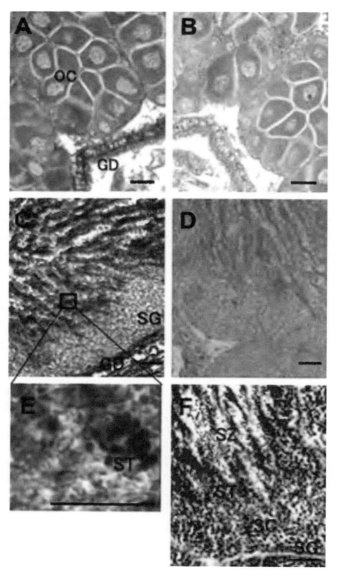

FIGURE 6.14   Localization of 5-HT$_{py}$ mRNA in the scallop ovary and testis. The ovary tissue hybridized with the antisense (A) or sense (B) DIG-labeled 5-HT$_{py}$cRNA probe. The testis tissue hybridized with the antisense (C and E) or sense (D) DIG-labeled 5-HT$_{py}$cRNA probe or was stained by hematoxylin–eosin (F). (E) is the large scale of the boxed part in C. Abbreviations: OC, oocyte; GD, gonoduct; SG, spermatogonia; SC, spermatocyte; ST, spermatid; SZ, spermatozoa. The scale bar represents 10 μm in E and 20 μm in A, B, C, D, and F. (Reprinted with permission from Ref. [114]. © 2010 Elsevier.)

After the binding of 5-HT to 5-HT receptors on the membrane of oocyte and sperm, the 5-HT signal is transmitted into the cells. Sperms are in a quiescence state just before ejaculation from the testis. The 5-HT signal transduction into the sperm mediated through 5-HT receptor was suggested to initiate 5-HT-dependent and osmolality-independent sperm motility in marine bivalve mollusks, which is associated with a K^+ efflux and Ca^{2+} influx via voltage-dependent ion channels under alkaline conditions [117]. And Na^+ influx was thought to be also important for the initiation of sperm motility probably via regulation of Ca^{2+} exchange (Figure 6.15) [117]. Oocytes of oviparous animal are generally arrested in late prophase of meiosis I just before ovulation. At this stage, the oocytes possess a developmental stage-specific nucleus, the GV corresponding to the dictyate stage of oocyte. Meiosis reinitiation is a process of oocyte maturation as evidenced by GVBD. It has been demonstrated that oocyte maturation and spawning of invertebrates are initiated by1-methyladenine, cubifrin, and serotonin in starfish, sea cucumber, and surf clam, respectively [104, 105, 118, 119]. 5-HT-induced Oocyte maturation in bivalves was suggested to be resulted from a combination of the transduction mechanisms of 5-HT signal and their cross-talk involved in 5-HT-induced oocyte maturation [120]. The 5-HT signal may reduce cyclic AMP in the oocyte cytoplasm inhibiting adenylate cyclase coupled to G_i protein. The 5-HT signal simultaneously induces conversion of phosphatidylinositol-4,5 bisphosphate (PIP2) into inositol triphosphate (IP3), and diacylglycerol (DAG) activating phospholipase C coupled to G_0protein, an increase in Ca^{2+} uptake activating membrane voltage-dependent Ca^{2+} channels coupled to G_x protein. These pathways activate protein kinases (PKA and PKC) and mitogen-activated protein kinase (MAPK), which activate maturation promoting factor (MPF) consisting of cdc2 and cyclin B and result in oocyte maturation.

Given the 5-HT functions in oocyte maturation and initiation of sperm motility, an administration of exogenous 5-HT into bivalves must be able to induce spawning in vivo. Interestingly, it has been reported that the full-grown oocytes of bivalves at the mature stage are arrested at the dictyate stage corresponding to late prophase of meiosis I, which is commonly observed in marine invertebrates [120]. The arrest state is held for 2 months after reaching the highest 5-HT-specific binding to oocyte membrane until the spawning stage [94]. Furthermore, the application of exogenous 5-HT in the induction of spawning does not always succeed, and the number of released eggs widely varied among individuals [92]. These observations indicate the occurrence of modulatory mechanisms acting on 5-HT-induced oocyte/sperm maturation and spawning by maturation-competent extracellular signals.

Oocyte maturation inhibitor and granulosa cell factor found in mammals have been reported to be heat-stable polypeptides with masses of less than 2 and 6 kDa, respectively [121–124]. In the bivalve *S. solidissima*, substances originating from oocytes that inhibit oocyte maturation induced by 5-HT have been identified as a *Spisula* factor with a mass of less than 1 kDa [125, 126] and an oocyte membrane component with a mass of more than 18 kDa [127]. However, the gonad of bivalves has no such follicle structure like vertebrates, and the roles of the *Spisula* factor in the endocrine mechanisms have not been well demonstrated. PGF2α?has bee suggested as a suppressive neuromodulator of 5-HT function in spawning of bivalves as mentioned before. In fact PGF2α certainly blocked 5-HT-induced egg release through the gonoduct from ovarian tissue, but did not inhibit 5-HT-induced oocyte maturation, suggesting that the cilioexcitatory activity of 5-HT in the gonoducts to transport mature eggs may be inhibited byPGF2α [108]. In addition to PGF2α, a novel inhibitor of 5-HT-induced egg release from ovarian tissue was found in the CNS of the scallop *P. yessoensis* of both genders. A main action of the neural factor was to arrest 5-HT-induced oocyte maturation, named "oocyte maturation arresting factor," OMAF [108] (Figure 6.16). The OMAF was a universal substance for bivalve species in both genders and thought to be transported from the CNS to the ovary through blood flow based on the occurrence of OMAF function and its identification in the hemolymph. The OMAF may prohibit 5-HT-induced oocyte maturation and sperm motility due to the interference of extracellular Ca^{2+} influx into oocytes, eventually resulting in the inhibition of spawning [108, 128] (Figure 6.17). Internal amino acid sequences of the OMAF with a molecular mass of 52 kDa were determined, and an antibody against the partial peptide of OMAF strongly amplified the 5-HT-induced release of egg and sperm due to the release from the suppressive activity of OMAF by neutralizing endogenous OMAF with the antibody. These results confirmed that the OMAF acts as an inhibitor of 5-HT-induced oocyte maturation and sperm motility [128]. Taken all together, 5-HT is an essential neurohormone for crucial events of oocyte maturation and sperm motility and following spawning of bivalves. The mode of actions of 5-HT on activation of germ cell and gonoduct regulated by OMAF and PGF2 is supposed to explain an arrest of oocyte maturation and sperm motility until spawning and simultaneous spawning in nature. Moreover, it is necessary for us to uncover a mystery as to receptor mechanisms of them and how bivalves can be released from the negative control by OMAF and PGF2α and secret 5-HT to induce simultaneous spawning.

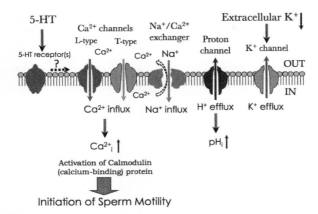

FIGURE 6.15 Ionic fluxes required for the initiation of sperm motility in marine bivalve mollusks. The stimulatory effect of serotonin (5-HT) on the initiation of sperm motility were associated with potassium (K^+) efflux, proton (H^+) efflux and calcium (Ca^{2+}) influx through a voltage-dependent K^+ channel, a proton channel and both L-type and T-type voltage-dependent Ca^{2+}, respectively. A sodium (Na^+)/Ca^{2+} exchanger regulates Na^+ influx to control intracellular Ca^{2+} during motility period. These steps stimulate Ca^{2+}-dependent or calcium-calmodulin (CaM) protein phosphatase (s) in the flagellum for the initiation of sperm motility. (Source: Republished with permission of Society for Reproduction and Fertility from Ref. [117] Alavi, S. M. H., Matsumura, N., Shiba, K., Itoh, N., Takahashi, K. G., Inaba, K., & Osada, M., (2014), Roles of extracellular ions and pH in serotonin-dependent initiation of sperm motility in marine bivalve mollusks. *Reproduction, 147,* 331-345. Permission conveyed through Copyright Clearance Center, Inc.

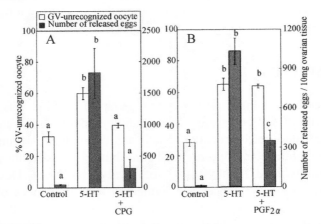

FIGURE 6.16 Effects of cerebral and pedal ganglia (CPG) extract and prostaglandin $PGF2\alpha$ on 5-HT-induced oocyte maturation and egg release of *Patinopecten yessoensis*. 1×10^{-5} M 5-HT was applied to the ovarian tissues with CPG extract (A) or 1×10^{-6} M $PGF2\alpha$(B). Values with different letters within each section A and B, and among parameters are significantly different ($p<0.05$). Each value indicates the mean ± SE (n=4). (Reprinted with permission from Ref. [94]. © 2016 Elsevier.)

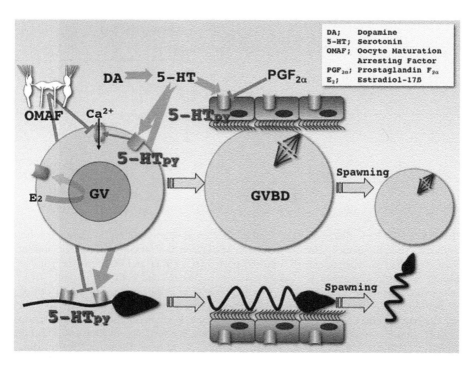

FIGURE 6.17 Illustration of transduction of extracellular signals for regulation of oocyte maturation, sperm motility, and spawning. The stimulatory effect of 5-HT on oocyte maturation and sperm motility via 5-HT receptor was negatively regulated by OMAF mediated through Ca^{2+} ion. Transportation of activated oocyte and sperm through gonoduct is regulated with 5-HT and PGF2α.

6.6 PROSPECTIVE

A genetic research focused on the animal group of mollusk was not many compared with that of other animals, including model invertebrates such as fruit fly, nematodes, etc. Recently, genetic analysis data are accumulated in mollusks, and molecular works are well contributing to realize the molecular mechanisms involved in physiology of bivalve mollusks. Development of Next Generation Sequencer accelerated to read sequences of genome and transcripts on a massive scale in any kind of organisms. Pearl oyster, *Pinctada fucata* [129], and Pacific oyster, *C. gigas* [130], genomes have been sequenced, and draft genome database has been constructed. The genome database was utilized as a platform for identification of specific genes for calcification, which is an essential phenomenon to produce an ornamental pearl, stress adaptation, shell formation, and larval development. The released draft genome of the pearl oyster has been analyzed to screen reproductive-related genes [131]. A de novo transcriptome sequencing and analysis have led us to the understanding of sex determination and differentiation, oocyte maturation, growth, and stress response [Dheilly et al., 2012 ; 132–135]. For the future, the application of genome and transcriptome database is supposed to be a strong tool for research on comprehensive understanding of the mechanism of reproduction in bivalve mollusk because it is generally difficult to isolate a specific gene on the basis of the conserved region obtained from an alignment of a few mollusk sequences or vertebrate sequences which are taxonomically far from mollusks. The biological function of genes screened from genomic and transcriptomic resource should be proven in each animal to understand their primary physiological function in bivalve mollusks.

KEYWORDS

- **Bivalvia**
- **endocrine**
- **gametogenesis**
- **gonad**
- **intragonadal somatic cells**
- **neuroendocrine**
- **spawning**

REFERENCES

1. Joosse, J., & Geraerts, W. P. M., (1983). Endocrinology. In: Saleuddin, A. S. M., & Wilbur, K. M., (eds.), *The Mollusca* (Vol. 4, pp. 317–406). Academic Press: New York.
2. Sastry, A. N., (1979). *Pelecypoda* (exclusive *ostreidae*). In: Giese, A. C., & Pearse, J. S., (eds.), *Reproduction of Marine Invertebrates* (Vol. V., pp. 113–292). Academic Press: New York, Chapter 5.
3. Osanai, K., (1975). Seasonal gonad development and sex alteration in the scallop, *Patinopecten yessoensis*. *Bull. Mar. Biol. St. Asamushi, Tohoku Univ.*, *15*, 81–88.
4. Osada, M., Nakamura, S., & Kijima, A., (2007). Quantitative analysis of the pattern of gonial proliferation during sexual maturation in the Japanese scallop *Patinopecten yessoensis*. *Fish. Sci.*, *73*, 1318–1324.

5. Eble, A. F., & Scro, R., (1996). General anatomy. In: Kennedy, V. S., Newell, R. I. E., & Eble, A. F., (eds.), *The Eastern Oyster Crassostrea virginica* (p. 734). Maryland Sea Grant College.

6. Pipe, R. K., (1987a). Oogenesis in the marine mussel *Mytilus edulis*: An ultrastructural study. *Mar. Biol., 95*, 405–414.

7. Eckelbarger, K. J., & Davis, C. V., (1996a). Ultrastructure of the gonad and gameto-genesis in the eastern oyster, *Crassostreavirginica*. I. ovary and oogenesis. *Mar. Biol., 127*, 79–87.

8. Dorange, G., & Le Pennec, M., (1989b). Ultrastructural study of oogenesis and oocytic degeneration in *pecten maximus* from the bay of St. Brieuc. *Mar. Biol., 103*, 339–348.

9. Matsumoto, T., Nakamura, A. M., Mori, K., & Kayano, T., (2003). Molecular charac-terization of a cDNA encoding putative vitellogenin from the pacific oyster *Crassostrea gigas*. *Zool. Sci., 20*, 37–42.

10. Osada, M., Harata, M., Kishida, M., & Kijima, A., (2004a). Molecular cloning and expression analysis of vitellogeninin scallop, *Patinopecten yessoensis* (Bivalvia, Mollusca). *Mol. Reprod. Dev., 67*, 273–281.

11. Franco, A., Heude, B. C., Goux, D., Sourdaine, P., & Mathieu, M., (2008). Fine structure of the early stages of spermatogenesis in the pacific oyster, *Crassostrea gigas* (Mollusca, Bivalvia). *Tissue Cell, 40*, 251–260.

12. Eckelbarger, K. J., & Davis, C. V., (1996b). Ultrastructure of the gonad and gametogenesis in the eastern oyster, *Crassostreavirginica*. I. I., testis and spermatogenesis. *Mar. Biol., 127*, 89–96.

13. Pipe, R. K., (1987b). Ultrastructural and cytochemical study on interactions between nutrient storage cells and gametogenesis in the mussel *Mytilus edulis*. *Mar. Biol., 96*, 519–528.

14. Dorange, G., & Le Pennec, M., (1989a). Ultrastructural characteristics of spermatogenesis in *pecten maximus* (Mollusca, Bivalvia) *Invert. Reprod. Dev., 15*, 109–117.

15. Osada, M., Tawarayama, H., & Mori, K., (2004b). Estrogen synthesis in relation to gonadal development of japanese scallop, *Patinopecten yessoensis*: Gonadal profile and immunolocalization of P450 aromatase and estrogen. *Comp. Biochem. Physiol., 139B*, 123–128.

16. Matsumoto, T., Osada, M., Osawa, Y., & Mori, K., (1997). Gonadal estrogen profile and immunohistochemical localization of steroidogenic enzymes in the oyster and scallop during sexual maturation. *Comp. Biochem. Physiol., 118B*, 811–817.

17. Matsutani, T., & Nomura, T., (1984). Localization of monoamines in the central nervous system and gonad of the scallop, *Patinopecten yessoensis*. *Bull. Jpn. Soc. Sci. Fish., 50*, 425–430.

18. Matsutani, T., & Nomura, T., (1986a). Serotonin-like immunoreactivity in the central nervous system and gonad of the scallop, *Patinopecten yessoensis*. *Cell Tissue Res., 244*, 515–517.

19. Voronezhskaya, E. E., Nezlin, L. P., Odintsova, N. A., Plummer, J. T., & Croll, R. P., (2008). Neuronal development in larval mussel *Mytilustrossulus* (Mollusca: Bivalvia), *Zoomorphology, 127*, 97–110.

20. Zatylny-Gaudin, C., & Favrel, P., (2014). Diversity of the RFamide peptide family in mollusks. *Front. Endocrinol., 5*, Article 178.

21. Too, C. K., & Croll, R. P., (1995). Detection of FMRFamide-like Immunoreactivities in the sea scallop *Placopecten magellanicus* by immunohistochemistry and western blot analysis. *Cell Tissue Res., 281*, 295–304.

22. Smith, S. A., Nason, J., & Croll, R. P., (1997). Detection of APGWamide-like immunoreactivity in the sea scallop, *Placopectenmagellanicus*. *Neuropeptides*, *31*, 155–165.

23. Nakamura, S., Osada, M., & Kijima, A., (2007). Involvement of GnRH neuron in the spermatogonial proliferation of the scallop, *Patinopecten yessoensiss*. *Mol. Rep. Dev.*, *74*(1), 108–115.

24. Bigot, L., Zatylny-Gaudin, C., Rodet, F., Bernay, B., Boudry, P., & Favrel, P., (2012). Characterization of GnRH-related peptides from the Pacific oyster *Crassostrea gigas*. *Peptides*, *34*, 303–310.

25. Pazos, A. J., & Mathieu, M., (1999). Effects of five natural gonadotropin-releasing hormones on cell suspensions of marine bivalve gonad: Stimulation of gonial DNA synthesis. *Gen. Comp. Endocrinol.*, *113*(1), 112–120.

26. Mathieu, M., (1987). Utilization of ATCase activity in the study of neuroendocrine control of gametogenesis in *Mytilus edulis*. *J. Exp. Zool.*, *241*, 247–252.

27. Mathieu, M., (1985). Partial characterization of aspartate transcarbamylase from the mantle of the mussel *Mytilus edulis*. *Comp. Biochem. Physiol.*, *82B*, 667–674.

28. Miura, T., Miura, C., (2001). Japanese Eel: A model for analysis of spermatogenesis. *Zool. Sci.*, *18*, 1055–1063.

29. Lubet, P., & Mathieu, M., (1982). The action of internal factors on gametogenesis in pelecypod mollusks. *Malacologia*, *22*, 131–136.

30. Mathieu, M., & Lubet, P., (1980). Analyseexpérimentaleen cultures d'organes de l'action des ganglions nerveux sur la gonadeadulte de la moule . *Bull. Soc. Zool. Fr.*, *105*, 149–153.

31. Le Gall, S., Feral, C., Lengronne, C., & Porchet, M., (1987). Partial purification of the endocrine mitogenic factor in the mollusk *Crepidulafornicata*, L. *Comp. Biochem. Physiol.*, *86B*, 393–396.

32. Mathieu, M., Lenoir, F., & Robbins, I., (1988). A Gonial mitosis-stimulating factor in cerebral gangulia and hemolymph of the marine mussel *Mytilus edulis*, L. *Gen. Comp. Endocrinol.*, *72*, 257–263.

33. Roch, G. J., Busby, E. R., & Sherwood, N. M., (2011). Evolution of GnRH: Diving deeper. *Gen. Comp. Endocrinol.*, *171*, 1–16.

34. Loumaye, E., Thorner, J., & Catt, K. J., (1982). Yeast mating pheromone activates mammalian gonadotrophs: Evolutionary conservation of a reproductive hormone? *Science*, *218*, 1323–1325.

35. Di Cosmo, A., & Di Cristo, C., (1998). Neuropeptidergic control of the optic gland of *octopus vulgaris*: FMRF-amide and GnRH immunoreactivity. *J. Comp. Neurol.*, *398*(1), 1–12.

36. Di Cristo, C., Paolucci, M., Iglesias, J., Sanchez, J., Di Cosmo, A., (2002). Presence of two neuropeptides in the fusiform ganglion and reproductive ducts of *octopus vulgaris*: FMRFamide and gonadotropin-releasing hormone (GnRH). *J. Exp. Zool.*, *292*(3), 267–276.

37. Goldberg, J. I., Garofarlo, R., Price, C. J., & Chang, J. P., (1993). Presence and biological activity of a gnrh-like factor in the nervous system of *Helisomatrivolvis*. *J. Comp. Neurol.*, *336*(4), 571–582.

38. Young, K. G., Chang, J. P., & Goldberg, J. I., (1999). Gonadotropin-releasing hormone neuronal system of the freshwater snails *Helisomatrivolvis* and *Lymnaea stagnalis*: Possible involvement in reproduction. *J. Comp. Neurol.*, *404*(4), 427–437.

39. Zhang, L., Wayne, N. L., Sherwood, N. M., Postigo, H. R., & Tsai, P. S., (2000). Biological and immunological characterization of multiple GnRH in an opisthobranch mollusk, *Aplysia californica*. *Gen. Comp. Endocrinol.*, *118*(1), 77–89.

40. Tsai, P-S., Maldonado, T. A., & Lunden, J. B., (2003). Localization of gonadotropin-releasing hormone in the central nervous system and a peripheral chemosensory organ of *Aplysia californica. Gen. Comp. Endocrinol., 130*(1), 20–28.

41. Rodet, F., Lelong, C., Dubos, M. P., Costil, K., & Favrel, P., (2005). Molecular cloning of a molluscan gonadotropin-releasing hormone receptor orthologue specifically expressed in the gonad. *Biochim. Biophys. Acta, 1730*(3), 187–195.

42. Rodet, F., Lelong, C., Dubos, M. P., & Favrel, P., (2008). Alternative splicing of a single precursor mRNA generates two subtypes of gonadotropin-releasing hormone receptor orthologues and their variants in the bivalve mollusc *Crassostrea gigas. Gene, 414*, 1–9.

43. Treen, N., Itoh, N., Miura, H., Kikuchi, I., Ueda, T., Takahashi, K. G., Ubuka, T., Yamamoto, K., Sharp, P. J., Tsutsui, K., & Osada, M., (2012). Mollusc gonadotropin-releasing hormone directly regulates gonadal functions: A primitive endocrine system controlling reproduction. *Gen. Comp. Endocrinol., 176*, 167–172.

44. Osada, M., & Treen, N., (2013). Molluscan GnRH associated with reproduction. *Gen. Comp. Endocrinol., 181*, 254–258.

45. Nagasawa, K., Osugi, T., Suzuki, I., Itoh, N., Takahashi, K. G., Satake, H., & Osada, M., (2015c). Characterization of GnRH-like peptides from the nerve ganglia of Yesso scallop, *Patinopecten yessoensis. Peptides, 71*, 202–210.

46. Iwakoshi-Ukena, E., Ukena, K., Takuwa-Kuroda, K., Kanda, A., Tsutsui, K., & Minakata, H., (2004). Expression and distribution of octopus gonadotropin-releasing hormone in the central nervous system and peripheral organs of the octopus (*Octopus vulgaris*) by *in situ* hybridization and immunohistochemistry. *J. Comp. Neurol., 477*(3), 310–323.

47. Kanda, A., Takahashi, T., Satake, H., & Minakata, H., (2006). Molecular and functional characterization of a novel gonadotropin-releasing-hormone receptor isolated from the common octopus (*Octopus vulgaris*). *Biochem. J., 395*(1), 125–135.

48. Iwakoshi, E., Hisada, M., & Minakata, H., (2000). Cardioactive peptides isolated from the brain of a japanese octopus, *Octopus minor. Peptides, 21*, 623–630.

49. Di Cosmo, A., Di Cristo, C., & Paolucci, M., (2001). Sex steroid hormone fluctuations and morphological changes of the reproductive system of the female of *Octopus vulgaris* throughout the annual cycle. *J. Exp. Zool., 289*, 33–47.

50. Di Cristo, C., (2013). Nervous control of reproduction in *octopus vulgaris*: A new model. *Invertebr. Neurosci., 13*, 27–34.

51. Sun, B., & Tsai, P. S., (2011). A gonadotropin-releasing hormone-like molecule modulates the activity of diverse central neurons in a gastropod mollusk, *Aplysia californica. Front. Endocrinol., 2*, 36.

52. Sun, B., Kavanaugh, S. I., & Tsai, P. S., (2012). Gonadotropin-releasing Hormone in protostomes: Insights from functional studies on *Aplysia californica. Gen. Comp. Endocrinol., 176*, 321–326.

53. Tsai, P. S., Sun, B., Rochester, J. R., & Wayne, N. L., (2010). Gonadotropin-releasing hormone-like molecule is not an acute reproductive activator in the gastropod, *Aplysia californica. Gen. Comp. Endocrinol., 166*(2), 280–288.

54. Johnson, J. I., Kavanaugh, S. I., Nguyen, C., & Tsai, P. S., (2014). Localization and functional characterization of a novel adipokinetic hormone in the mollusk, *Aplysia californica. PLoS One, 9*(8), e106014.

55. Nagasawa, K., Muroi, M., Thitiphuree, T., Minegishi, Y., Itoh, N., & Osada, M., (2017). Cloning of invertebrate gonadotropin-releasing hormone receptor (GnRHR)-like gene in Yesso scallop, *Patinopecten yessoensis. Agri Gene, 3*, 46–56.

56. Nagasawa, K., Oouchi, H., Itoh, N., Takahashi, K. G., & Osada, M., (2015a). *In vivo* administration of scallop GnRH-like peptide influences on gonad development in the yesso scallop, *Patinopecten yessoensis. PLoS One, 10*(6), e0129571.

57. Hess, R. A., Bunick, D., Lee, K. H., Bahr, J., Taylor, J. A., Korach, K. S., & Lubahn, D. B., (1997). A role for oestrogens in the male reproductive system. *Nature, 390*, 509–512.

58. Pierantoni, R., Cobellis, G., Meccariello, R., Cacciola, G., Chianese, R., Chioccarelli, T., & Fasano, S., (2009). Testicular gonadotropin-releasing hormone activity, progression of spermatogenesis, and sperm transport in vertebrates. *Ann. N. Y. Acad. Sci.*, 279–291.

59. Lafont, R., & Mathieu, M., (2007). Steroids in aquatic invertebrates, *Ecotoxicology, 16*, 109–130.

60. Scott, A. P., (2012). Do mollusks use vertebrate sex steroids as reproductive hormones? Part I: Critical appraisal of the evidence for the presence, biosynthesis and uptake of steroids. *Steroids, 77*, 1450–1468.

61. Scott, A. P., (2013). Do mollusks use vertebrate sex steroids as reproductive hormones? Part, I. I., Critical review of the evidence that steroids have biological effects. *Steroids, 78*, 268–281.

62. Osada, M., Unuma, T., & Mori, K., (1992b). Purification and characterization of a yolk protein from the scallop ovary. *Nippon Suisan Gakkaishi, 58*, 2283–2289.

63. Suzuki, T., Hara, A., Yamaguchi, K., & Mori, K., (1992). Purification and immunolocalization of a vitellin-like protein from the pacific oyster *Crassostrea gigas. Mar. Biol., 113*, 239–245.

64. Li, Q., Osada, M., Suzuki, T., & Mori, K., (1998). Changes in vitellin during oogenesis and effect of estradiol-17β on vitellogenesis in the pacific oyster *Crassostrea gigas. Invertebr. Reprod. Dev., 33*, 87–93.

65. Osada, M., Takamura, T., Sato, H., & Mori, K., (2003). Vitellogenin synthesis in the ovary of scallop, *Patinopecten yessoensis*: control by estradiol-17 beta and the central nervous system. *J. Exp. Zool., 299*, 172–179.

66. Matsumoto, T., Yamano, K., Kitamura, M., & Hara, A., (2008). Ovarian follicle cells are the site of vitellogenin synthesis in the pacific abalone *Haliotis discus hannai. Comp. Biochem. Physiol., 149A*, 293–298.

67. Awaji, M., Matsumoto, T., Yamano, K., Kitamura, M., & Hara, A., (2011). Immunohistochemical observations of vitellin synthesis and accumulation processes in ovary of ezoabalone *haliotis discus hannai. Fish. Sci., 77*, 191–197.

68. Polzonetti-Magni, A. M., Mosconi, G., Soverchia, L., Kikuyama, S., & Carnevali, O., (2004). Multihormonal control of vitellogenesisin lower vertebrates. *Int. Rev. Cytol., 239*, 1–45.

69. Reis-Henriques, M. A., Le Guellec, D., Remy-Martin, J. P., & Adessi, G. L., (1990). Studies of endogenous steroids from the marine mollusc *mytilus edulis*, L. by gas chromatography and mass spectrometry. *Comp. Biochem. Physiol., 95B*, 303–309.

70. Matsumoto, T., Nakamura, A. M., Mori, K., Akiyama, I., Hirose, H., & Takahashi, Y., (2007). Oyster estrogen receptor: cDNA cloning and immunolocalization. *Gen. Comp. Endocrinol., 151*, 195–201.

71. Thornton, J. W., Need, E., & Crews, D., (2003). Resurrecting the ancestral steroid receptor: Ancient origin of estrogen signaling. *Science, 301*, 1714–1717.

72. Kajiwara, M., Kuraku, S., Kurokawa, T., Kato, K., Toda, S., Hirose, H., et al., (2006). Tissue preferential expression of estrogen receptor gene in the marine snail, *Thais clavigera. Gen. Comp. Endocrinol., 148*, 315–326.

73. Keay, J., Bridgham, J. T., & Thornton, J. W., (2006). The *Octopus vulgaris* estrogen receptor is a constitutive transcriptional activator: Evolutionary and functional implications. *Endocrinology, 147,* 3861–3869.

74. Keay, J., & Thornton, J. W., (2009). Hormone-activated estrogen receptors in annelid invertebrates: Implications for evolution and endocrine disruption. *Endocrinology, 150,* 1731–1738.

75. Bridgham, J. T., Keay, J., Ortlund, E. A., & Thornton, J. W., (2014). Vestigialization of an allosteric switch: Genetic and structural mechanisms for the evolution of constitutive activity in a steroid hormone receptor. *PLoS Genet., 10,* e1004058.

76. Vogeler, S., Galloway, T. S., Lyons, B. P., & Bean, T. P., (2014). The nuclear receptor gene family in the pacific oyster, *Crassostrea gigas,* contains a novel subfamily group. *BMC Genomics, 15,* 369.

77. Nagasawa, K., Treen, N., Kondo, R., Otoki, Y., Itoh, N., Rotchell, J. M., & Osada, M., (2015b). Molecular characterization of an estrogen receptor and estrogen-related receptor and their autoregulatory capabilities in two *Mytilus* Species. *Gene, 564,* 153–159.

78. Ciocan, C. M., Cubero-Leon, E., Puinean, A. M., Hill, E. M., Minier, C., Osada, M., Fenlon, K., & Rotchell, J. M., (2010). Effects of estrogen exposure in mussels, *Mytilus edulis,* at different stages of gametogenesis. *Environ. Pollut., 158,* 2977–2984.

79. Giese, A., & Kanatani, H., (1987). Maturation and spawning. In: Giese, A. C., Pearse, J. S., & Pearse, V. B., (eds.), *Reproduction of Marine Invertebrates, Vol., IX—General Aspects: Seeking Unity in Diversity* (Vol. IX, pp. 251–329). Blackwell Scientific Publication/Boxwood Press: California, Chapter 4.

80. Blake, N. J., & Sastry, A. N., (1979). Neurosecretory regulation of oögenesis in the bay scallop, *Argopecten irradians* (Lamarck). In: Naylor, E., & Hortnoll, R. G., (eds.), *Cyclic Phenomena in Marine Plants and Animals* (pp. 181–190). Pergamon Press: New York.

81. Barber, B. J., & Blake, N., (2006). Reproductive physiology. In: Shumway, S. E., & Parsons, G. J., (eds.), *Scallops: Biology, Ecology and Aquaculture* (pp. 357–416). Elsevier: San Diego, CA, Chapter 6.

82. Matsutani, T., & Nomura, T., (1982). Induction of spawning by serotonin in the scallop *Patinopecten yessoensis* (JAY). *Mar. Biol. Lett., 3,* 353–358.

83. Braley, R. D., (1985). Serotonin-induced spawning in giant clams (Bivalvia: Tridacnidae). *Aquaculture, 47,* 321–325.

84. Gibbons, M. C., & Castagna, M., (1984). Serotonin as an inducer of spawning in six bivalve species. *Aquaculture, 40,* 189–191.

85. Tanaka, Y., & Murakoshi, M., (1985). Spawning induction of the hermaphroditic scallop, *Pectenalbicans,* by injection with serotonin. *Bull. Natl. Res. Inst. Aquacult., 7,* 9–12.

86. Kikuchi, S., & Uki, N., (1974). Technical study on artificial spawning of abalone, genus *Haliotis.* II, effect of irradiated seawater with ultraviolet rays on induction to spawn. *Bull. Tohoku Reg. Fish. Res. Lab., 33,* 79–86.

87. Uki, N., & Kikuchi, S., (1974). On the effect of irradiated seawater with ultraviolet rays on inducing spawning of the scallop, *Patinopecten yessoensis* (Jay). *Bull. Tohoku Reg. Fish. Res. Lab., 34,* 87–92 (Abstract in English).

88. Matsutani, T., & Nomura, T., (1986b). Pharmacological observations on the mechanism of spawning in the scallop *Patinopecten yessoensis. Bull. Jpn. Soc. Sci. Fish., 52,* 1589–1594.

89. Martínez, G., Saleh, F. L., Mettifogo, L., Campos, E., & Inestrosa, N., (1996). Monoamines and release of gamets by the scallop, *Argopecten purpuratus. J. Exp. Zool., 274,* 365–372.

90. Osada, M., & Nomura, T., (1990). The levels of prostaglandins associated with the reproductive cycle of the scallop, *Patinopecten yessoensis*. *Prostaglandin*, *40*, 229–239.

91. Osada, M., Nishikawa, N., & Nomura, T., (1989). Involvement of prostaglandins in the spawning of the scallop *Patinopecten yessoensis*. *Comp. Biochem. Physiol.*, *94C*, 595–601.

92. Matsutani, T., & Nomura, T., (1987). *In vitro* effects of serotonin and prostaglandins on release of eggs from the ovary of the scallop, *Patinopecten yessoyensis*. *Gen. Comp. Endocrinol.*, *67*, 111–118.

93. Osada, M., Mori, K., & Nomura, T., (1992a). *In vitro* effects of estrogen and serotonin on release of eggs from the ovary of the scallop. *Nippon Suisan Gakkaishi*, *58*(2), 223–227.

94. Osada, M., Nakata, A., Matsumuto, T., & Mori, K., (1998). Pharmacological characterization of serotonin receptor in the oocyte membrane of bivalve molluscs and its formation during oogenesis. *J. Exp. Zool.*, *281*, 124–131.

95. Kyozuka, K., Deguchi, R., Yoshida, N., & Yamashita, M., (1997). Change in intracellular Ca^{2+} is not involved in serotonin-induced meiosis reinitiation from the first prophase in oocytes of the marine bivalve *Crassostrea gigas*. *Dev. Biol.*, *182*, 33–41.

96. Bandivdekar, A. H., Segal, S. J., & Koide, S. S., (1991). Demonstration of serotonin receptors in isolated *spisula* oocyte membrane. *Invert. Reprod. Dev.*, *19*, 147–150.

97. Bandivdekar, A. H., Segal, S. J., & Koide, S. S., (1992). Binding of 5-hydroxytryptamine analogs by isolated *Spisula* Aperm Membrane. *Invert. Reprod. Dev.,* *21*, 43–46.

98. Krantic, S., Dube, F., Quirion, R., & Guirrier, P., (1991). Pharmacology of the serotonin-induced meiosis reinitiation in *Spisula solidissima* oocytes. *Dev. Biol.*, *146*, 491–498.

99. Krantic, S., Dube, F., & Guerrier, P., (1993a). Evidence for a new subtype of serotonin receptor in oocytes of the surf clam *Spisula solidissima*. *Gen. Comp. Endocrinol.*, *90*, 125–131.

100. Krantic, S., Guerrier, P., & Dube, F., (1993b). Meiosis reinitiation in surf clam oocytes is mediated via a 5-hydroxytryptamine serotonin membrane receptor and a vitellin envelope-associated high affinity binding site. *J. Biol. Chem.*, *268*, 7983–7989.

101. Fong, P. P., Wall, D. M., & Ram, J. L., (1993). Characterization of serotonin receptors in the regulation of spawning in the zebra mussel *Dreissena polymorpha* (Pallas). *J. Exp. Zool.*, *267*, 475–482.

102. Gobet, I., Durocher, Y., Leclerc, C., Moreau, M., & Guerrier, P., (1994). Reception and transduction of the serotonin signal responsible for meiosis reinitiation in oocytes of the Japanese clam *Ruditapes philippinarum*. *Dev. Biol.*, *164*, 540–549.

103. Guerrier, P., Ledlerc-David, C., & Moreau, M., (1993). Evidence for the involvement of internal calcium stores during serotonin-induced meiosis reinitiation in oocytes of the bivalve mollusk *Ruditapes philippinarum*. *Dev. Biol.*, *159*, 474–484.

104. Hirai, S., Kishimoto, T., Kadam, A. L., Kanatani, H., & Koide, S. S., (1988). Induction of spawning and oocyte maturation by 5-hydroxytryptamine in the surf clam. *J. Exp. Zool.*, *245*, 318–321.

105. Osanai, K., & Kuraishi, R., (1988). Response of oocytes to meiosis-inducing agents in pelecypods. *Bull. Mar. Biol. Stn. Asamushi, Tohoku Univ.*, *18*(2), 45–56.

106. Osanai, K., (1985). *In vitro* induction of germinal vesicle breakdown in oyster oocytes. *Bull. Mar. Biol. Stn. Asamushi, Tohoku Univ.*, *18*(1), 1–9.

107. Varaksin, A. A., Varaksina, G. S., Reunova, O. V., & Latyshev, N. A., (1992). Effect of serotonin, some fatty acids and their metabolites on reinitiation of meiotic maturation in oocytes of bivalve *Spisula sachalinensis* (Schrenk). *Comp. Biochem. Physiol.*, *101C*(3), 627–630.

108. Tanabe, T., Osada, M., Kyozuka, K., Inaba, K., & Kijima, A., (2006). A novel oocyte maturation arresting factor in the central nervous system of scallops inhibits serotonin-induced oocyte maturation and spawning of bivalve mollusks. *Gen. Comp. Endocrinol., 147*, 352–361.

109. Angers, A., Storozhuk, M. V., Duchaîne, T., Castellucci, V. F., & DesGroseillers, L., (1998). Cloning and functional expression of an *Aplysia*5-HT receptor negatively coupled to adenylate cyclase. *J. Neurosci., 18*, 5586–5593.

110. Barbas, B., Zappulla, J. P., Angers, S., Bouvier, M., Castellucci, V. F., & DesGroseillers, L., (2002). Functional characterization of a novel serotonin receptor (5-HTap$_2$) expressed in the CNS of *Aplysia californica*. *J. Neurochem., 80*, 335–345.

111. Gerhardt, C. C., Leysen, J. E., Planta, R. J., Vreugdenhil, E., & Van-Heerikhuizen, H., (1996). Functional characterization of a 5-HT$_2$ receptor cDNA Cloned from a *Lymnaea stagnalis. Eur. J. Pharmacol., 311*, 249–258.

112. Li, X. C., Giot, J. F., Kuhl, D., Hen, R., & Kandel, E. R., (1995). Cloning and characterization of two related serotonergic receptors from the brain and the reproductive system of *Aplysia* that activate phospholipase, C. *J. Neurosci., 15*, 7585–7591.

113. Sugamori, K. S., Sunahara, R. K., Guan, H. C., Bulloch, A. G., Tensen. C. P., Seeman, P., Niznik, H. B., & Van Tol. H. H., (1993). Serotonin receptor cDNA, cloned from *Lymnaea stagnalis. Proc. Natl. Acad. Sci. U.S.A., 90*, 11–15.

114. Tanabe, T., Yuan, Y., Nakamura, S., Itoh, N., Takahashi, K. G., & Osada, M., (2010). The role in spawning of a putative serotonin receptor isolated from the germ and ciliary cells of the gonoduct in the gonad of the Japanese scallop, *Patinopecten yessoensis. Gen. Comp. Endocrinol., 166*, 620–627.

115. Paul, R. A., & Mario, T., (2001). Receptor signaling and structure: Insights from serotonin-1 receptors. *Trends Endocrinol. Metabol., 12*, 453–460.

116. Tierney, A. J., (2001). Structure and function of invertebrate 5-HT receptors: A review. *Comp. Biochem. Physiol., 128A*, 791–804.

117. Alavi, S. M. H., Matsumura, N., Shiba, K., Itoh, N., Takahashi, K. G., Inaba, K., & Osada, M., (2014). Roles of extracellular ions and pH in serotonin-dependent initiation of sperm motility in marine bivalve mollusks. *Reproduction, 147*, 331–345.

118. Kanatani, H., Shirai, H., Nakanishi, K., & Kurokawa, T., (1969). Isolation and identification of meiosis-inducing substance in starfish. *Nature, 21*, 273–274.

119. Kato, S., Tsurumaru, S., Taga, M., Yamane, T., Shibata, Y., Ohno, K., Fujiwara, A., Yamano, K., & Yoshikuni, M., (2009). Neuronal peptides induce oocyte maturation and gamete spawning of sea cucumber, *Apostichopusjaponicus. Dev. Biol., 326*, 169–176.

120. Krantic, S., & Rivailler, P., (1996). Meiosis reinitiation in molluscan oocytes: A model to study the transduction of extracellular signals. *Invert. Reprod. Dev., 30*, 55–69.

121. Hillensjo, T., Brannstrom, M., Chari, S., Daume, E., Magnusson, C., & Tornell, J., (1985). Oocyte maturation as regulated by follicular factors. *Ann. N. Y. Acad. Sci., 442*, 73–79.

122. Franchimont, P., Demoulin, A., & Valcke, J. C., (1988). Endocrine, paracrine, and autocrine control of follicle development. *Horm. Metab. Res., 20*, 193–203.

123. Sato, E., & Koide, S. S., (1984). A factor from bovine granulose cells preventing oocyte maturation. *Differentiation, 26*, 59–62.

124. Tsafriri, A., & Pomerantz, S. H., (1986). Oocyte maturation inhibitor. *Clin. Endocrinol. Metab., 15*, 157–170.

125. Kadam, A. L., & Koide, S. S., (1990). Inhibition of Serotonin-induced oocyte maturation by a *spisula* factor. *J. Exp. Zool., 255*, 239–243.
126. Sato, E., Wood, H. N., Lynn, D. G., Sahni, M. K., & Koide, S. S., (1985). Meiotic arrest in oocytes regulated by a *Spisula* factor. *Biol. Bull., 169*, 334–341.
127. Sato, E., Toyoda, Y., Segal, S. J., & Koide, S. S., (1992). Oocyte membrane components preventing trypsin-induced germinal vesicle breakdown in surf clam oocyte. *J. Rep. Dev., 38*, 309–315.
128. Yuan, Y., Tanabe, T., Maekawa, F., Inaba, K., Maeda, Y., Itoh, N., Takahashi, K. G., & Osada, M., (2012). Isolation and functional characterization for oocyte maturation and sperm motility of the oocyte maturation arresting factor from the Japanese scallop, *Patinopecten yessoensis*. *Gen. Comp. Endocrinol., 179*, 350–357.
129. Takeuchi, T., et al., (2012). Draft genome of the pearl oyster *Pinctada fucata*: A platform for understanding bivalve biology. *DNA Res., 19*, 117–130.
130. Zhang, G., et al., (2012). The oyster genome reveals stress adaptation and complexity of shell formation. *Nature, 490*, 49–54.
131. Matsumoto, T., Masaoka, T., Fujiwara, A., Nakamura, Y., Satoh, N., & Awaji, M., (2013). Reproduction-related genes in the pearl oyster genome. *Zool. Sci., 30*, 826–850.
132. Ghiselli, F., Milani, L., Chang, P. L., Hedgecock, D., Davis, J. P., Nuzhdin, S. V., & Passamonti, M., (2012). Assembly of the manila clam *Ruditapes philippinarum* transcriptome provides new insights into expression bias, mitochondrial doubly uniparental inheritance and sex determination. *Mol. Biol. Evol., 29*, 771–786.
133. Hou, R., Bao, Z., Wang, S., Su, H., Li, Y., Du, H., Hu, J., Wang, S., & Hu, X., (2011). Transcriptome Sequencing and *de novo* analysis for yesso scallop (*Patinopecten yessoensis*) Using 454 GS, F. L., X., *PLos One, 6*(6), e21560.
134. Pauletto, M., Milan, M., De Sousa, J. T., Huvet, A., Joaquim, S., Matias, D., Leitão, A., Patarnello, T., Bargelloni, L., (2014). Insight into molecular features of *Venerupis decussate* oocyte: A microarray-based study. *PLoS One, 9*(12), e113925.
135. Teaniniuraitemoana, V., Huvet, A., Levy, P., Klopp, C., Lhuillier, E., Gaertner-Mazouni, N., Gueguen, Y., & Le Moullac, G., (2014). Gonad transcriptome analysis of pearl oyster *Pinctada margaritifera*: Identification of potential sex differentiation and sex determining genes. *BMC Genomics, 15*, 491.
136. Dheilly, N. M., Lelong, C., Huvet, A., Kellner, K., Dubos, M. P., Riviere, G., Boudry, P., & Favrel, P., (1998). Gametogenesis in the Pacific Oyster *Crassostrea gigas*: A microarray-based analysis identifies sex and stage specific genes. *PLoS One, 7*(5), e36353.

CHAPTER 7

Peptidergic Systems in the Pond Snail Lymnaea: From Genes to Hormones and Behavior

PAUL R. BENJAMIN and ILDIKÓ KEMENES

Sussex Neuroscience, School of Life Sciences, University of Sussex, Brighton BN1 9QG, UK

7.1   INTRODUCTION

The pond snail, *Lymnaea stagnalis* (L.) (Pulmonata, Basomatophora), is a model molluscan system used in many laboratories to study a wide range of biological problems [1, 2]. The snails are widely distributed throughout Europe, the Northern parts of the United States and parts of Asia in freshwater ponds, lakes, and rivers with rich vegetation. They are from 2 to 5 cm in shell length (Figure 7.1) and are typically found in large numbers close to the water surface feeding on floating pond weed. A major advantage for experimental work is that snails deposit their gelatinous egg-masses, each containing 100 eggs or more, on a substrate (e.g., the surface of a floating leaf) and tiny snails eventually emerge in adult form without any free-living veliger larval stage. This makes it simple to breed *Lymnaea* in the laboratory. This is extremely useful for endocrinological and molecular studies where large numbers of animals are required.

Multidisciplinary endocrinological, electrophysiological, and molecular approaches on *Lymnaea* have led to a deep understanding of peptide-mediated neurohormonal and neuromodulatory mechanisms underlying physiology and behavior. Neuropeptides control the major physiological processes of the animal, such as the cardiovascular system, digestion, reproduction, growth, and metabolism and ion and water regulation (Figure 7.2). The classic separation of neuropeptides into hormones and transmitters is blurred in *Lymnaea*. Certainly, many of the peptides are released into the

FIGURE 7.1   Photograph of a laboratory-bred *Lymnaea stagnalis*. It is about four months old.

blood and reach their targets via the cardiovascular system behaving as conventional hormones, but the same peptides may also act as neurotransmitters/modulators within the CNS to play a role in controlling behavior. A good example of this is CDCH (Caudodorsal Cell Hormone) that acts as a hormone to stimulate ovulation but also acts as a transmitter on motoneurons to organize the sequence of body movements that underlying egg-laying behavior. Release of peptides into the bloodstream may occur at specific neurohaemal sites in nerves leaving the cerebral ganglia. For example, the median lip nerves (MLN) act a unique release site for the Light Green Cells (Figure 7.2). In other peptidergic neurons, there are no specific neurohaemal organs, and peptide release into the blood occurs more generally from nerve terminals in the connective sheath surrounding the ganglia, for instance, the Yellow Cells (YCs) and the Light YCs. Many peptidergic neurons form synaptic connections with peripheral organs (e.g., heart or penis) or with other neurons in central circuits of the brain such as those controlling the heart. In these examples, peptides may act as primary transmitters but also as co-transmitters to modulate the effects of classical transmitters such as 5-HT (serotonin) or acetylcholine (ACh). So far, 22 neuropeptide genes have been cloned in *Lymnaea* and at least 100 neuropeptides identified with certainty (examples in Table 7.1). This diversity in *Lymnaea* arises from a variety of general molecular mechanisms that occur in all types of nervous systems. These mechanisms in the *Lymnaea* context will be considered next.

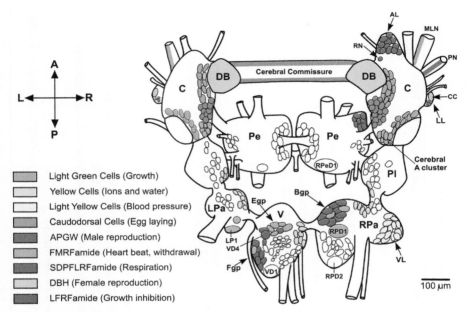

Light Green Cells (Growth)
Yellow Cells (Ions and water)
Light Yellow Cells (Blood pressure)
Caudodorsal Cells (Egg laying)
APGW (Male reproduction)
FMRFamide (Heart beat, withdrawal)
SDPFLRFamide (Respiration)
DBH (Female reproduction)
LFRFamide (Growth inhibition)

FIGURE 7.2 Map of neurohormonal peptidergic neurons in the CNS of *Lymnaea*. The cerebral commissure is the neurohaemal organ of the CDCs (pink shading). The median lip nerves (MLN) are the neurohaemal organ of the LGCs (light brown shading). The anterior lobe (AL) APGW neurons project along the right penis nerve (PN) and innervate the penis complex. Note that the cluster of Yellow Cells that lie on the hand side of the visceral ganglion are a special group of cells whose axons project along the distal processes of the intestinal nerve to innervate the pericardium and the reno-pericardial canal (see text). Abbreviations: B/Egp, B/E group; C, cerebral ganglion; CC, Canopy Cell; DB, Dorsal Body; LL, lateral lobe; LP1, left parietal 1; LPa, left parietal ganglion; Pe, pedal ganglion; Pl, pleural ganglion; RN, Ring Neuron; RPa, right parietal ganglion; RPD1/2, right parietal dorsal 1/2; RPeD1, right pedal dorsal 1; V, visceral ganglion; VD1, visceral dorsal 1; VD4, visceral dorsal 4; VL ventral lobe. (Reprinted with permission from Ref. [1].)

7.2 MOLECULAR BASIS OF PEPTIDE DIVERSITY

7.2.1 THE PRESENCE OF MULTIPLE GENE FAMILIES

A good example of this is the gene family encoding insulin related-peptides (MIPs, molluscan insulin-related peptides). Five different MIP genes (MIP I, II, III, V, VII) give rise to five highly divergent MIPs [3]. Another example is the CDCH gene that occurs in 3 variants (CDCH I-III) [4].

TABLE 7.1 Examples of Neuropeptide Hormones and Transmitters

Neuropeptides	Amino Acid Sequence
α-peptides	DMYEGLAGRCQHHPRNCPGFN (α1), DMVTTTRIGTGGLAGRCQHHPRNCPGFN (α2)
APGW	APGWamide
Caudodorsal cell hormone (CDCH)	LSITNDLRAIADSYLYDQHKLRERQEENLRRFLELamide
Calfluxin (CaFl)	RVDSADESNDDGFD
FMRFamide-related peptides (FARPs)	FMRFamide, FLRFamide, SDPFLRFamide, GDPFLRFamide
F(X) Rlamides	ASSFVRlamide, SPSSFVRlamide, PNSFLRlamide, YPMNRFIRlamide
Granularin	EPCEHNGVTYNPGDAYHKDQCTTCYCGEDSEAFCIPLQCDWPQCEDGASPVYLEDSCCPGCP
LFRFamides	NTLFRFamide, GTLLRFamide, GGSLFRFamide, TLFRFamide
Light Yellow Cell (LYC I-III) peptides	AFIVEEDDLTGYPTTIDAAMTTIRP (LYC I'), TPDKSILLNRL (LYC II')
Lys-conopressin	CFIRNCPKGamide
Lymnaea cardioactive peptide (LyCEP)	TPHWRPQGRFamide
Lymnaea inhibitory peptides (LIPs)	GAPRFVamide (LIP A), SAPRFVamide (LIP B), ARPSKFVamide (LIP C)
Lymnaea leucokinin-like peptide	PSFHSWSamide
Lymnaea neuropeptide Y (LNPY)	TEAMLTPPERPEEFKNPNELRKYLKALNEYYAIVGRPRFamide
Lymnaea tetradecapeptide	GFRANSASRVAHGYamide
Molluscan insulin- related peptide (MIP)	QGTTNIVCECCMKPCTLSELRQYCP-- (A Chain) QFSACNINDRPHRRGVCGSALADLVDFACSSSNQPAMV-- (B chain) (MIP1)
Myomodulins	PMSMLRLamide, GLQMLRLamide, SMSMLRLamide,
Ovipostatin	EKDQTPSCSPDTFEANLYCTDGSVCGKYAVDWTQNVSVVQLKSFRLVFYINQIKGFRIS- NEEECNPVDGRSNQIPIRCIPPNAVIRLKGNAFGADFFSFDVVSPSPTVTWYPEKEHVPKILKI- FIDGGRLFPGDFVFFPDLLKTDDLSLSLFDFPVTCPYFEGYDPE
Schistosomin	DNYWCPQSGEAFECFESDPNAKFCLNSGKTSVVICSKCRKKYEFCRNGLKVSKRPDY- DCGAGWESTPCTGDNSAVPAVF
Small cardioactive peptides	SGYLAFPRMamide (SCP$_A$), QNYLAFPRMamide (SCP$_B$)
Sodium influx-stimulating peptide (SIS)	SRTQSRFASYELMGTEGTECVTTKTISQICYQCATRHEDSFVQVYQ-ECCKKEMGLREYCEEI- YTELPIRSGLWQPN

7.2.2   THE ENCODING OF STRUCTURALLY-RELATED PEPTIDES ON THE SAME GENE

This is widespread in *Lymnaea* where families of peptides sharing similar amino acid motifs were found to be encoded on a single gene. There are examples where 2 (SCP_A and SCP_B, small cardioactive peptide A and B, [5]), 3 (LIPs, *Lymnaea* inhibitory peptides, [6]), 5 (myomodulins, [7]; LFRFamides, [8]) and 13 (F(X) RIamides, [9]) structurally-related peptides are encoded on the same gene. There are also examples of different peptides encoded on the same gene, for example, the CDCH1 gene where 11 structurally diverse peptides are found (see section 7.2.2).

7.2.3   ALTERNATIVE SPLICING OF A PRIMARY TRANSCRIPT ENCODING DIFFERENT PROTEIN PRECURSORS AND NEUROPEPTIDES

Alternative mRNA splicing also leads to diverse peptide expression, and this is best understood in the gene encoding FMRFamide-related peptides (FARPS) [10, 11]. The single FMRFamide gene consists of 5 exons (I-V). Two different mRNA variants are spliced from the primary transcript (Figure 7.3A). One (mRNA1) consists of exons I and II (mRNA1) the other (mRNA2) of exons I, III, IV, and V (mRNA2). Exon I encodes a highly hydrophobic sequence and an N-terminal cleavage site typical of leader sequence required for the targeting of the precursor sequences to the endoplasmic reticulum. The protein precursor 1 derived from Exon II encodes five different confirmed peptides FMRFamide, FLRFamide, EFLRIamide, (p) QFYRIamide and 'SEEPLY,' a 22 amino acid peptide (Figure 7.3A). The protein precursor 2 formed from the translation of exons III, IV, and V encodes seven different peptides, SDPFLRFamide, GDPFLRFamide, SDPYLRFamide, SKPYMRFamide, 'Acidic peptide,' a 35 amino acid peptide (P1), and a '22 amino acid' amidated peptide (P2) (Figure 7.3A). *In situ* hybridization studies using either cDNA or exon-specific oligonucleotides have revealed that the two alternatively spliced mRNA species are expressed in the CNS in a striking, differential, and mutually exclusive manner at the single neuron level (Figure 7.3B). Of the ~ 340 neurons that express the FMRFamide gene, the majority (80%) express the mRNA1, the rest mRNA 2 [12].

7.2.4   POST-TRANSLATIONAL PROCESSING OF PEPTIDES YIELDS MORE TYPES OF PEPTIDES

Post-translational truncation of peptides has been reported for MIPs, FARPs, and the LYC peptides. Direct peptide profiling by mass spectrometry of individual LYCs (Li et al., 1994a) has revealed a delicate pattern of peptide processing. The LYC peptide precursor is processed to yield three different peptides (LYC I, LYC II, and LYC III) by cleavage at dibasic sites. LYC I and LYCII are further processed at their N-termini to yield truncated mature peptides LYCP I' and LYCP II' (Table 7.1 and section 8.2). A peptidomics analysis of individual VD1/RPD2 neurons (Jimenez et al., 2006) has revealed a whole battery of different post-translational modifications including phosphorylation of an amino acid in the β peptide and a number of modified α2 peptides that include hydroxylated and glycosylated forms (see also Section 7.2).

7.3   PEPTIDE PROCESSING ENZYMES

All the sequences of peptides such as those expressed in the FMRFamide peptide precursors are flanked by mono-basic (R, arginine; K, lysine) dibasic RR, KK) or tetrabasic amino acids (RRKR, RKRR) that are cleavage sites for enzymes known as endoproteases. Examples of the mRNA sequences that code for the prohormone convertases (PCs) involved in endoproteolytic processing have been cloned for *Lymnaea*. These are LPC2 (*Lymnaea* PC2, [13]), one of a class of convertases that cleave at dibasic cleavage sites RR or KR and L*furin*1 (*Lymnaea* furin1) and L *furin* 2 that are known to cleave tetrabasic cleavage sites such as RKRR [14]. Activation of PC2 in *Lymnaea* and other organisms is regulated by the molecular chaperone protein called 7B2 that inhibits the active site [15]. A final step in peptide biosynthesis is amidation of peptides by two enzymes that are synthesized by one precursor known as the α-amidating enzyme whose gene has been cloned in *Lymnaea* [16]. This amidation protects peptides from degradative enzymes and is often essential for peptide bioactivity. These regulatory processes would be expected to act together in peptidergic neurons and Spijker et al., [17] has shown that there is a neuron-specific expression of the transcripts encoding convertases, α-amidating enzyme and the 7B2 chaperone in the egg-laying hormone-producing CDCs (Caudodorsal Cells). Furthermore, environmental stimuli that induce egg-laying (clean water treatment) cause co-regulated induction of the three transcripts and the egg-laying hormone. This

A Alternative mRNA splicing of the FMRFamide gene

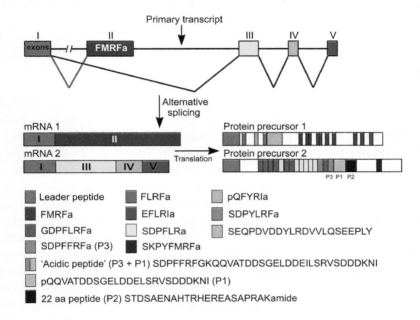

B Mutually exclusive expression of exon II and exon III

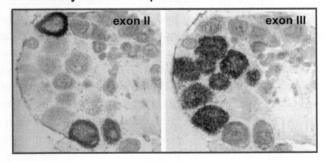

FIGURE 7.3 The *Lymnaea* FMRFamide gene: alternate mRNA spicing and mutually exclusive neuronal expression. Two mRNA variants are spliced from the primary transcript. Protein precursor 1 encodes five different peptides, including multiple copies of the tetrapeptides, FMRFamide, and FLRFamide. Post-translational processing of QFYRIamide converted Q into pQ. Protein precursor 2 encodes seven peptides including multiple copies of the heptapeptides, SDPFLRFamide and GDPFLRFamide. Post-translational cleavage of the 'acidic peptide' resulted in two further peptides P3 and P1. Only peptides that were confirmed by sequencing and mass spectrometry are included in the list of peptides. B *In situ* hybridization of the alternatively spliced transcripts shows the mutually exclusive mRNA expression at the single neuron level. The same neurons can be identified in these adjacent sections of the visceral ganglion. (Reprinted with permission from Ref. [1].)

shows that neuropeptide release and the regulation of transcript levels of both prohormones and processing enzymes are regulated in accordance with physiological demands.

7.4   NEUROPEPTIDE RECEPTORS

Rather less work has been carried out on the identification of neuropeptide receptors in *Lymnaea* and the ones that have been identified, apart from the putative molluscan insulin-related receptor (MIPR, [18]), were receptors for novel peptides that were previously unknown. The single *Lymnaea* MIPR has the typical insulin receptor features including a cysteine-rich domain, a single transmembrane domain, and a tyrosine-kinase domain. Most *Lymnaea* peptide receptors, like those from mammals, mediate their effects through one subgroup of GPCRs (G protein-coupled receptors) that are characterized by seven alpha-helical trans-membrane spanning domains and the presence of particular conserved amino acids. In *Lymnaea,* degenerate oligomers corresponding to these conserved amino acids were used in PCR analysis to probe cDNA libraries from the CNS or other tissues like the heart where the receptors were expected to be located. Orphan GPCR clones were expressed in *Xenopus* oocytes or CHO cells, and putative peptide ligands from the CNS were screened for their ability to mobilize intracellular calcium. Perhaps it is not surprising that a rather random and sometimes unexpected set of receptors were discovered using this 'shot-gun' approach. For instance, attempts to isolate the receptors of cardioexcitatory FMRFamide family from the snail heart turned up another type of -RFamide receptor whose ligand is LyCEP (*Lymnaea* cardioexcitatory peptide, Table 7.1). This LyCEP receptor is insensitive to FMRFamide [19]. Another type of FMRFamide receptor is the FMRFamide-gated sodium channel (LsFaNaC), which is a non-GPCR type receptor containing only two transmembrane regions [20]. Other GPCRs have endogenous ligands to a leukocidin-like peptide and a 39aa neuropeptide homologous to neuropeptide Y (Table 7.1). An orphan GPCR, expressed widely in the CNS, is related to vertebrate galanin and nociception/orphanin-FQ receptor families [21]. From the point of view of the molecular evolution of receptors, the neuropeptide receptors to the peptide lys-conopressin (Table 7.1) are of particular interest. A novel GPCR receptor, LSCPR1 (*Lymnaea stagnalis* conopressin receptor 1), mediates both the vasopressin- and oxytocin-like functions of lys-conopressin. Oxytocin-like functions are mediated by receptors located on central neurons that control the male reproductive organs, and vasopressin-like functions are indicated by receptors that modulate insulin release [22]. The discovery of a second lys-conopressin receptor (LSCPR2) gives a clue to the evolutionary origin of vertebrate oxytocin and a vasopressin receptors. LSCPR2 is maximally activated by both lys-conopressin and Ile-conopressin, an oxytocin-like analog of lys-conopressin, unlike LSCPR1, that only responds to lys-conopressin. Together with a study of the phylogenetic relationships of lys-conopressin receptors in *Lymnaea* and their vertebrate counterparts, these data suggest that LSCPR2 represents an ancestral receptor to the vasopressin/oxytocin receptor family in the vertebrates [23].

7.5 HEARTBEAT: MULTIPLE ROLES OF PEPTIDES IN THE CONTROL OF A MYOGENIC PERIPHERAL ORGAN

Peptidergic control of heartbeat is the best-understood example of the modulatory control of a rhythmically active muscular organ by extrinsic peptide-containing neurons. The range of general roles that peptides play as transmitters and co-transmitters are exemplified by results from the heart control system. The heart of *Lymnaea* is myogenic, but the heartbeat is controlled by a network of five different types of centrally located moto-neurons with diverse excitatory and inhibitory effects. These neurons are all located in the visceral and right parietal ganglia (Figure 7.4A) with axonal projections to the heart along the pericardial branch of the intestinal nerve. One branch of this nerve penetrates the pericardium to innervate both regions of the heart, with the highest number of branches in the auricle (Figure 7.4B). Simultaneous electrophysiological recordings of heart motoneurones and single muscle fibers show that they directly innervate individual heart muscle fibers [24, 25]. One of the exciting features of the investigation of the cardiac control system is that the peptides of individually dissected identified moto-neurons were subjected to MALDI-TOF MS (matrix-assisted laser desorption/ionization-time of flight mass spectrometric) analysis [26, 27] as well as immunocytochemistry [28] and radioimmunoassay [29] to firmly establish their peptide content. The same techniques were applied to the cardiac tissue itself to correlate central neuronal content with the peripheral release [27, 30]. A total of 28peptides were identified in heart and pericardial tissue (examples in Figure 7.4B) arising from 8 different genes indicating a rich peptidergic control of this peripheral muscular organ. The auricle is much more heavily modulated by peptides (Figure 7.4C) than the ventricle (Figure 7.4D) and of the eight peptide families found in the heart only one, the FARP family (FMRFamide, FLRFamide), has been found in the ventricle (Figure 7.4D).

7.5.1 *DIVERSE NEUROPEPTIDE TRANSMITTERS ENCODED ON THE SAME GENE HAVE DIFFERENT BUT COMBINATORIAL EFFECTS ON HEARTBEAT*

The pair of E_{he} (E heart excitor) motoneurons are unique because they allow us to study the synaptic role of the well-known cardio-excitatory peptide FMRFamide and other types of peptides that are encoded on them RNA transcript 1 of the FARP gene (Figure 7.5A). The increased frequency and amplitude produced by electrical stimulation of the E_{he} cells are mimicked by the perfusion of these peptides through the heart with each type of peptide having a different detailed effect [24, 28, 32, 33]. The absence of classical transmitters [24, 30] suggests that these gene-related peptides are primary transmitters mediating the monosynaptic effects that the E_{he} cells have on heart muscle fibers. The peptides processed from the prohormone precursor 1 are divided into three types. In addition to the RFamides (FMRFamide, FLRFamide), there are the RIamides (pQFYRIamide, EFLRIamide) and the 22 amino-peptide 'SEEPLY' (Figure 7.3A). All five peptides are found in the E_{he} neurons (Figure 7.5Ai), the pericardial nerve, and the auricle [30]. RIA (radioimmunoassay) analysis of heart perfusates following heart electrical stimulation shows that all three peptide types are released from the heart tissue in a Ca^{++}-dependent manner [30]. The RIamide and RFamide peptides are released in about equal quantities suggesting co-release, and this was confirmed by double immuno-gold staining of secretory granules in nerve terminals in the heart. This showed that both types of peptides are present in the same type of secretary granule [30]. SEEPLY appear to be released separately from different granules, and in less quantities, but is still part of the same cocktail of peptides shown to be released by heart electrical stimulation. The three types of peptides have distinct effects on heartbeat, and these are mediated by different second messenger pathways. The increases in the frequency and amplitude of heart-beat produced by FMRFamide/ FLRFamide (Figure 7.5Aii) are mediated by mobilizing the inositol phosphate pathway, and they follow the same time course [32]. The effects of the RIamide peptides on heartbeat are mediated by a separate cyclic AMP-mediated pathway that produces a more prolonged excitatory effect on heartbeat compared to the RFamides (Figure 7.5Aiii) [33]. An initial cyclic AMP-independent inhibitory response on the heartbeat to one of the RIamides, ELRFamide (Figure 7.5Aiii), is due to direct effects on heart muscle ion channels. SEEPLY has no effects on heart-beat when applied alone but prolongs the effects of FMRFamide by delaying its mobilizing effects on the

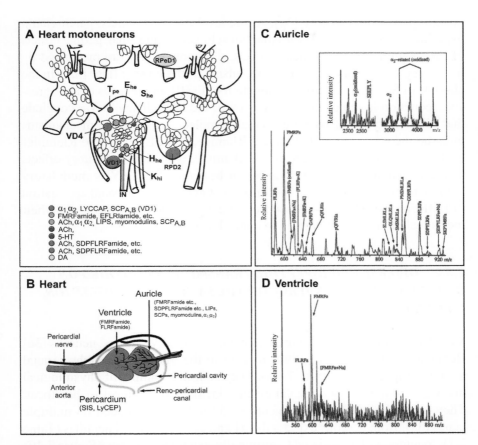

FIGURE 7.4 Heart motoneurons and peptides in the heart of *Lymnaea*. The location and neuropeptide content of heart motoneurons and interneurons in the visceral and right parietal ganglia. All five types of motoneurons project to the heart by axons that project along the intestinal nerve (IN). VD4 (visceral dorsal 1) and RPeD1 (right pedal 1) are interneurons that form part of the CPG that controls both cardiac and respiratory functions. Cell types: E_{he}, E heart excitor; H_{he}, H heart excitor; K_{hi}, K heart inhibitor; S_{he}, S heart excitor; T_{pe}, T pericardium. B The nerve in nervation of the heart and a summary of the peptide content of the heart and pericardium. The pericardial nerve is a branch of the intestinal nerve that penetrates the pericardium to innervate both the auricle and ventricle, with more extensive branching in the auricle. Note that there are more types of peptides in the auricle than the ventricle. C Analysis of neuropeptide content in the auricle by Maldi-TOF MS. Masses corresponding to FARP peptides and other peptides encoded on the FMRFamide gene (FMRFamide, FLRFamide, SDFLRFamide, GDPFLRFamide, SDPYLRFamide, pQFLRIamide, pQFYRIamide and SEEPLY), myomodulins (PMSMLRLamide, SMSMLRLamide and GLQMLRLamide), LIPs (GAPRFVamide) and the VD1/RPD2 peptides α_1 and α_2 were identified. D Analysis of peptides in the ventricle by MALDI-TOF MS. Masses corresponding to FMRFamide and FLRFamide were identified. (Reprinted with permission from Ref. [31].)

inositol phosphate pathway [32]. At the level of ion channels, patch-clamp experiments on isolated ventricular muscle cells show that gating effects of FMRFamide on non-voltage gated Ca^{2+} channels [34] are prolonged by the co-application of SEEPLY to the outside of the muscle fiber [35]. In the absence of SEEPLY, repeated application of FMRFamide leads to a progressive reduction in the opening frequency of the Ca^{2+} channels. We conclude that different types of receptor-mediated molecular and ionic mechanisms underlie the effects of FMRFamide, RIamide, and SEEPLY. The RIamides prolong the excitatory effects of FMRFamide, with early inhibitory effects acting to maintain a steady increase in beat rate. SEEPLY is modulatory helping to prevent the loss in responses that occur with repeated application of FMRFamide. Their co-release from the E_{he} motoneurons means that their individual effects are integrated to control heart function.

7.5.2 DIVERSE NEUROPEPTIDE TRANSMITTERS ENCODED ON DIFFERENT GENES CONVERGE ON THE SAME HEART-MUSCLE ION CHANNEL

VD1/RPeD2 are functionally-related giant electrically-coupled neurons [36] Benjamin and Pilkington, 1990) located in the visceral and right parietal ganglia, respectively (Figure 7.4A). They control a number of physiological processes related to pO_2 regulation that includes the control of heartbeat. They express a wide range of neuropeptides that are encoded on multiple genes [37]. Both neurons express the α1 and α2 gene-related peptides (Table 7.1), together with α2-related C-terminally-extended versions (Figure 7.4C. insert), and a novel 53 amino acid peptide LyCCAP (*Lymnaea* calcium current activating peptide, N-terminal sequence GPIGAKKF...) that is encoded on a different gene. VD1 alone expresses the two SCP peptides (SCP_A and SCP_B, Table 7.1) encoded on a third gene. Despite the structural diversity of these peptides, they all target the same ion channel. When synthetic versions of the α2, LyCCAP, and SCP_A and SCP_B peptides are applied to isolated ventricular cells, they all specifically enhance the size of the HVA (high voltage-activated) L-type Ca^{2+} current, although with varying potencies [38]. It is interesting that the post-translationally-modified mono and di-glycosylated forms of the α2 peptide are more potent than the unmodified isoform. These three types of peptides appear to be the only transmitters present that mediate the excitatory synaptic effects that VD1 makes with isolated ventricular muscle cells in culture. The HVA Ca^{2+} currents in heart ventricular muscle cells are important in generating the pacemaker

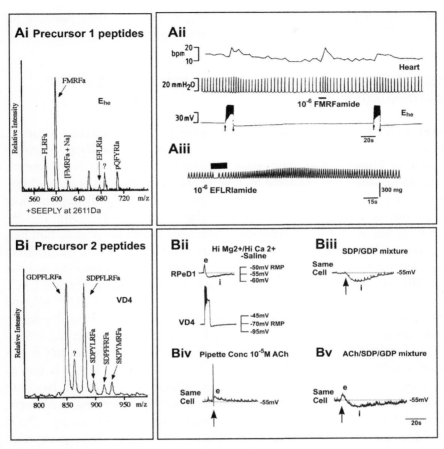

FIGURE 7.5 The synaptic function of identified neurons expressing precursor 1 and precursor two peptides of the FMRFamide gene. Ai MALDI-TOF MS analysis of dissected E_{he} heart excitor motoneuron shows the presence of precursor 1 peptides, FMRFamide, FLRFamide, EFLRIamide, pQFYRIamide, and SEEPLY but no precursor 2 peptides. Aii Perfusion of FMRFamide through the heart increases beat rate and amplitude as do bursts of induced spikes in the E_{he} cell. Neuronal bursts produce longer duration effects than peptide application as shown in the plots of instantaneous beat rate (bpm, beats per minute) in the upper trace. Aiii EFLFIamide initially inhibits spontaneous heartbeat and then increases the beat rate and amplitude, similar to FMRFamide, but over a more prolonged time period. Bi VD4 only expresses precursor 2 peptides (SDPFLRFamide, GDPFLRFamide, SDPYLRFamide, SKPYMRFamide). No precursor 1 peptides are present. Bii An evoked burst of spikes in VD4 generates a biphasic depolarizing (e) followed by hyperpolarizing synaptic response (i) in RPeD1 (right pedal D1). Its persistence in high Mg^{2+}/high Ca^{2+} saline indicates a monosynaptic connection. Biii, Application of SDPFLRFamide/GDPFLRFamide produces a hyperpolarizing response, ACh a depolarizing response (Biv), and a mixture of both a biphasic response (Bv) similar to neuronal stimulation. (Reprinted with permission from Ref. [1].)

mechanism of the myogenic heartbeat [39], so one important function for the VD1/RPeD2 peptides is to increase the myogenic beating rate.

7.5.3 NEUROPEPTIDES ACT AS CO-TRANSMITTERS IN HEART MOTONEURONS

5-HT is the classical transmitter of two types of excitatory heart motoneurons, the H_{he} (H heart excitor) and the S_{he} (S heart excitor) cells (cell body locations shown in Figure 7.4A). The main excitatory effects that these neurons have on heartbeat are mediated by this monoamine (Buckett et al., 1999c) . However, one of these neurons, the $H_{he,}$ expresses a remarkably large number of peptide co-transmitters that modulate the effects of 5-HT [30]. The H_{he} cell body has been shown to contain members of 4 different peptide families, the myomodulins, $SCP_{A,B}$, LIPs, and α1/α2 (Figure 7.4A) with 12 different peptides identified by MALDI-TOF MS spectrometry of dissected neurons [26, 30]. All these peptides have also been found in the auricle (Figure 7.4C). The SCPs have highly potent effects on the auricle (threshold 10^{-10} M), and similar to 5-HT act to increase beat rate but in a more prolonged manner. Importantly these peptides also increase heart tonus (Buckett et al., 1999a) . This increase in tonus is observed with neuronal stimulation but is not mimicked by application of 5-HT to the heart suggesting that it is one of the specific modulatory roles of SCP. The effects of mymodulin have not been tested on the *Lymnaea* heart, but in the nudibranch mollusk *Archidoris*, mymodulins act as modulators to potentiate the cardioexcitatory effects of 5-HT, preventing desensitization effects that occur following repeated application of the monoamine [40]. This is also likely to be important in *Lymnaea* because repeated electrical activation of H_{he} neurons does not result in a reduction in a heartbeat. The α peptides increase the cellular excitability of heart muscle fibers and increase beating as previously explained for the VD1/RPD1 neurons (see above). Thus the peptide co-transmitters in the H_{he} cells modulate the effects of 5-HT and in combination mediate the effects of neuronal stimulation. In other examples, the K_{hi} (K heart inhibitor) and T_{pe} (T heart pericardium) motoneurons (Figure 7.4A), ACh is the classical inhibitory transmitter on heartbeat [28]. Here peptide co-transmitters, extended peptides of the FARP family. e.g., SDPFLRamide and GDPFLRamide found in the T_{pe} motoneuron, and the auricle (Figure 7.4A, C) underlie its ability to increase cardiac tonus [24].

7.5.4   INTERNEURONAL CONTROL OF HEART PEPTIDERGIC MOTONEURONS

Several types of interneurons control spike activity in peptidergic heart motoneurons. Two of these form part of the CPG for rhythmic pneumostome movements, and so respiration and heart-beat are linked by a common control mechanism (see [1]). The most significant effect of the CPG arises from excitatory effects on the H_{he} motoneurons. This generates periodic bursting which excites the heart in a cyclical manner (Buckett et al., 1999b). At the same time, the inhibitory input to the heart provided by the K_{hi} motoneurons is suppressed, promoting further excitation to the heart (Buckett et al., 1996b). An important interneuron in the heart/respiratory CPG is VD4 (location in Figure 7.4A). ACh and FARP peptides of the extended heptapeptide type (SDPFLRFamide/GDPetc) are co-transmitters in this neuron (Figure 7.5Bi) Skingsley et al., 2003; [27, 41]. Spikes inVD4 generate biphasic excitatory/inhibitory monosynaptic responses on another CPG interneuron, RPeD1 (Figure 7.5Bii). ACh mediates an early nicotinic depolarizing response on RPeD1 (Figure 7.5Biv) and the extended RFamides a delayed hyperpolarizing response (Figure 7.5Biii). Applied together, they mimic the synaptic effects of VD4 stimulation (Figure 7.5Bv). This is an important example of a neuron that uses peptides processed frommRNA2 variant of the FMRFamide gene (Figure 7.3A) as a co-transmitter and allows direct comparison with the E_{he} motoneurons (Figure 7.5Aii) that express the alternative mRNA1.

7.6   NEUROPEPTIDERGIC CONTROL OF GUT CONTRACTIONS

Most work in *Lymnaea* has concentrated on the oesophageal region of the gut whose peristaltic contractions are controlled by peptidergic motoneurons in the buccal (feeding) ganglia. Here we focus on the paired B2 (oesophageal motoneurons) that are located in the buccal ganglia (Figure 7.6Aii). These neurons have extensive axonal projections to all regions of the pro-esophagus (Figure 7.6Ai, Aii). The B2s contain SCP_A and SCP_B [9] and a variety of the myomodulin family peptides [7, 42] as well as the classical transmitter ACh [42]. The isolated esophagus is capable of spontaneous peristaltic contractions, but its contractions are modulated by electrical activity in the B2 neurons [42]. The frequency, amplitude, and underlying tonus of these contractions are increased by B2 stimulation (Figure 7.6Aiv).

Suppression of spike activity in the B2s reduced the amplitude, frequency of contractions, and the tonus of the esophagus (Figure 7.6Aiii).

The peptides and ACh each have distinct effects on the esophagus when applied separately to the isolated esophagus but act co-operatively to mimic the effect of B2 stimulation. ACh increases the frequency and amplitude of oesophageal contractions (Figure 7.6B). Myomodulins have similar effects to ACh (Figure 7.6C) on frequency and amplitude (Figure 7.6Ci, Cii) but are slower in onset compared with ACh and involve a different signaling pathway [42]. Like ACh, the myomodulins also increase the tonus of oesophageal contractions (Figure 7.6C) mimicking another effect of B2 stimulation (Figure 7.6Aiv). Applying a single type of myomodulin peptide (e.g., PMSMLR-Lamide) at different concentrations produces dose-dependent increases in oesophageal tonus (Figure 7.6Ci). Application of mixes of mymodulins also produces dose-dependent increases in tonus (Figure 7.6Cii). Comparing the effects of mixes with single peptide applications show that the mixtures produce greater effects at each concentration (Figure 7.6Ciii). This indicates that the different types of myomodulin peptides (shown in Table 7.1) have additive effects on tonal muscular control, assuming that they are co-released from the peripheral B2 nerve terminals in the oesophageal tissue. Both SCP peptides (SCP_A and SCP_B) also increase the underlying tonus of oesophageal contractions in a dose-dependent manner (SCP_A Figure 7Dii; SCP_B Figure 7.6Diii) compared with saline alone (Figure 7.6Di) but the responses are much slower to build-up compared with the myomodulins (Figure 7.6Ci, Cii). They have no effect on the amplitude or the frequency of oesophageal contractions (Figure 7.6Dii, 6Diii).

There are other peptides in the esophagus that originate from unidentified motoneurons on the buccal ganglia; for example, *Lymnaea* tetradeca-peptide (Table 7.1) [43]. This peptide has excitatory effects on the isolated esophagus. The gaseous transmitter nitric oxide is also produced by the B2 neurons [44], but its role in controlling the gut is unknown.

7.7 NEUROENDOCRINOLOGY OF MALE AND FEMALE REPRODUCTIVE BEHAVIOR

The reproductive biology of *Lymnaea* has been intensively studied. It is a simultaneous hermaphrodite, but during mating behavior, one individual acts as the male and the other the female (Figure 7.7A). Often this is immediately followed by a reversal of sexual roles by the same pair of snails. During oviposition (Figure 7.7B3.), gelatinous egg masses, each containing 100

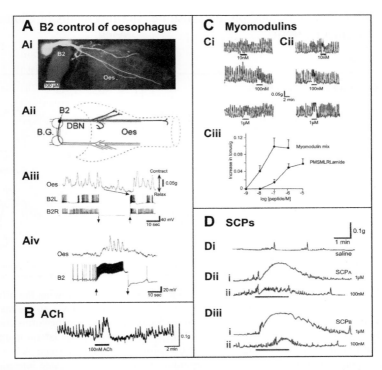

FIGURE 7.6 Neural and peptidergic modulation of gut motility. (A) Control of esophageal contractions by the paired B2 motoneuron. (Ai) Carboxyflourescein dye injection into B2 reveals branching axonal projections to the esophagus. These divide into finer terminal processes that innervate the esophageal tissue. (Aii) Diagram summarizing the morphology of the B2 motoneuron. (Aiii) Suppression of B2 spikes activity eliminates gut contractile activity. The left (BL2) and right (BR2) neurons were both hyperpolarized to suppress firing, and this resulted in a decrease in muscle tonus (descending arrow), contraction rate, and amplitude. Release of the B2 neurons from hyperpolarization resulted in the restoration of normal gut contractile activity. Aiv Activation of a burst of spikes inB2 initiates rhythmic gut contractions in a preparation where the esophagus showed no intrinsic contractile activity. (B) ACh is co-localized with the myomodulin and SCP peptides in B2 (see text), and when it is applied for 2 min to the isolated esophagus, it mimics the effects of B2 stimulation by increasing the tonus and frequency of spontaneous contractions. (C) Pharmacological actions of *Lymnaea* myomodulins on the isolated esophagus. (Ci) Dose-dependent increases in the tonus and frequency of oesophageal contractions and tonus when PMSMLRLamide is applied for 2 min. (Cii) Application of a myomodulin mixture (PMSMLRlamide, SLSMLRlamide, SMSMLRlamide, GLQMLRlamide, pQIPMLRLamide) in a concentration ratio reflecting the number of copies in the myomodulin precursor protein including the B2 motoneuron produces similar increases in tonus and frequency to the peptides when applied alone except that that the responses to the mixture are greater (Ciii). D Both types of SCP (SCP$_A$ and SDC$_B$) induce tonic contractions of the isolated esophagus again mimicking the effects of B2 stimulation. The pharmacological effects are delayed in the time of onset and take longer to reach their peak compared with AChand the co-localized myomodulin peptides. (Di) Saline alone application is accompanied by sporadic oesophageal contractions. Dose-dependent increases in tonic contractions of the esophagus with SCP$_A$ (Dii) and SCP$_B$ (Diii). The records in D are from the same preparation.

eggs or more, are deposited on the substrate, and tiny snails in adult form eventually emerge.

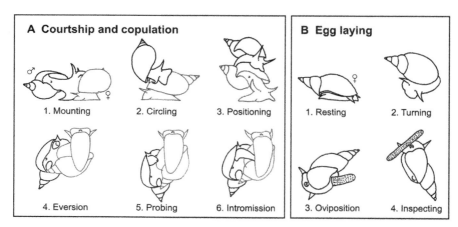

FIGURE 7.7   Male and female reproductive behavior. A. Drawings showing the different stages of male courtship and copulatory behavior. (A1) Courtship starts with the male (in black) on the shell of the female (in grey). For some 5 min, the male then performs counter-clockwise circling (A2), first towards the shell tip and then towards its lateral margin. It takes about 17 min from positioning (A3) to intromission, the copulation stage (A4), during which eversion and probing (A5) of the preputium occurs (see text). Once intromission is reached, sperm insemination takes place over a period of about 35 min. (B) Drawings showing the different stages of egg-laying behavior. It begins with a resting phase (B1). On average about 40 min in duration, when the animal remains stationary. The shell is held still and pulled forward over the tentacles. Rasping movements are absent. During the turning phase (B2), (about 60 min in duration) locomotion is resumed, the shell is turned backwards and forwards through a 90° angle and the substrate 'cleaned' by continuous rasping movements of the radula. Oviposition (B3) takes about 10 min during which rasping continues, shell turning stops and the snail ceases to move. During inspection (Figure 7.B4.) the snail crawls along with the egg mass for up to 30 min without rasping or turning (Reprinted with permission from Ref. [45]. © 2010 Koene.)

7.7.1   MALE COURTSHIP AND COPULATORY BEHAVIOR AND THE ROLE OF NEUROPEPTIDES

7.7.1.1   BEHAVIOR

During courtship (Figure 7.7A1–5.), the snail playing the male role climbs on the shell of the prospective female (mounting), moves over the shell in a counterclockwise direction (circling) until it reaches the area of the female gonophore (positioning). The preputium (muscular structure that surrounds the penis) is then partially everted through the male pore (Figure 7.7A4.). This is followed by probing for the female pore by the preputium (Figure 7.7A5.). Insertion of this organ during copulation is followed by eversion by the penis into the female pore and intromission (Figure 7.7A6.) (Positioning to intromission takes about 17 min) followed by insemination Each of the 5 stages of courtship prior to intromission is variable in duration but the 6th intromission stage is more constant and lasts for about 36 min. The whole mating behavior may last for several hours (male courtship and copulation reviewed in [46]).

7.7.1.2   PREPUTIUM EVERSION

Most neural and hormonal information on mating concerns the motor control of the preputium. Five groups of neurons on the right-hand side of the CNS are likely to be involved in the control of this organ and other parts of the male system because they project along the penis nerve (PN) that exits from the right cerebral ganglion (Figure 7.2). The role of one of these five groups of neurons located in the ventral part of the right anterior lobe (rAL) (Figure 7.2) of the cerebral ganglion has been investigated in detail [47]. Recording from these rAL cells *in vivo* using fine-wire electrodes shows that the cells are normally silent but increase their spiking during preputium eversion and throughout intromission [47]. Artificial electrical stimulation of the rAL neurons causes eversion of the preputium in all the animals tested [47]. Eversion of the preputium involves relaxation of preputial retraction muscle bands and the circular muscles surrounding the male gonophore. These muscles are innervated by nerve fibers in the pedal nerve from rAL neurons that contain the peptide APGWamide (Figure 7.2). Significantly, injection of the APGW peptide into intact snails caused reliable eversion of the preputium, so one of the roles of the APGW peptide is to relax the preputial muscle bands to cause preputial eversion [47]. However, the rAL neurons that contain APGW also co-express other peptides. Lys-conopressin is often co-localized with APGW in the rAL neurons, but conopressin does not interact with APGW on the preputium or have any effect on the relaxation or contraction when applied alone [47]. In other rAL neurons, APGW is co-expressed with LyNPY [47], and in this type of AL lobe neuron, both peptides act together as synergistic co-transmitters to relax retractor muscles. There are a number of other peptides present in the penial complex, e.g., myomodulins [48] and LIPs [6] that have the ability to relax the preputial retractor muscles and these, like LyNPY, could also be involved in preputial eversion, although the details of this are unknown.

7.7.1.3   PREPUTIAL PROBING MOVEMENTS

Contractions of longitudinal muscles inside and near the body contribute to the subsequent fine probing movements of the preputium that are the next phase of the mating behavior after eversion (Figure 7.7A). These movements are required for the preputium to find and insert into the female gonophore. During probing, the fully everted preputium makes movements under the lips of the recipient shell (Figure 7.7A5.) in search of the female gonopore. This requires a sensory mechanism to achieve accurate positioning. Candidates that could serve this function are neurons that lay on the distal tip of the preputium. Several of these neurons are Lys-conopressin-containing sensory neurons with dendrites extending through the preputial epithelium [49]. They send their axons into the PN [50].

7.7.1.4   INTROMISSION OF THE PENIS AND INSEMINATION

The end of these courtship behaviors and the start of the copulation phase are characterized by the intromission of the penis into the female gonopore (Figure 7.7A6.). Eversion of the penis resulting from alternate contractions of longitudinal and circular muscular layers of the penis. Once intromission is achieved sperm, and seminal fluid are transferred into the vaginal duct of the recipient to complete the mating sequence [51]. Presumably, this is also controlled by the release of neuropeptides, but the details of this are unknown.

7.7.1.5   RETRACTION OF THE PENIS AFTER COPULATION

Once ejaculation has finished the penis and preputium are retracted. Whilst retraction of the penis is most likely due to the relaxation of the preputial muscles and a decrease in local hydrostatic pressure, retraction of the much larger preputium also requires contraction of preputium muscle fibers because when these muscles are cut snails cannot retract the organ [52]. How this is controlled is unknown.

7.7.1.6   PEPTIDE CONTROL OF THE VAS DEFERENS

The vas deferens, which transports semen to the penis by peristaltic contractions, is another target for peptide neurohormones of the male reproductive system. The anterior lobe peptide Lys-conopressin increases this rhythmic activity whilst APGW acts antagonistically to modulate the pattern of peristaltic movements [53]. There is a pacemaker region in the vas deferens that seems essential for peristalsis [50]. Electrical stimulation of a branch of the PN that supplies the pacemaker region increases the frequency of the peristaltic contractions. Members of another large family of peptides present in the pineal complex, the F(X) RIamides (Table 7.1), inhibit spontaneous vas deferens contractions [9] and they could also be involved in suppressing male reproductive activity.

7.7.2   FEMALE HORMONAL CONTROL

7.7.2.1   DORSAL BODY HORMONE: MATURATION OF THE FEMALE SYSTEM

The paired dorsal bodies (DBs) of *Lymnaea* (locations in Figure 7.2) are endocrine gonadotropic organs that play a vital role in female reproduction. The development of the female reproductive system is severely retarded by the removal of the DBs, resulting in no egg production in juveniles and low egg production in adults [54]. Female function recovers following re-implantation of the DBs, providing evidence that they secrete a hormone the DBH (Dorso Body Hormone) that targets peripheral organs. Later experiments showed that DBH stimulates the follicle cells in the ovotestis, resulting in the increased growth of oocytes (Dogterom et al., 1993) . DBH also stimulates the development and growth of the female accessory organs. For example, DBH stimulates growth and synthesis in the galactogen-producing albumen glands [54]. The molecular structure of DBH in *Lymnaea* remains to be elucidated.

7.7.2.2   *EGG-LAYING BEHAVIOR AND ITS HORMONAL CONTROL*

Egg-laying consists of a sequence of behavioral events composed of various postures and action patterns (Figure 7.7B) triggered in the intact animal by exposure to a clean water stimulus. It begins with a rest period (Figure 7.7B1.), when the animal ceases to locomote, then a turning phase characterized by counterclockwise shell movements (Figure 7.7B2.) and high frequency rasping to clean the substrate, followed by oviposition (Figure 7.7B3.) and a final phase called inspection when the snail moves along the length of the egg-mass, brushing it with lips and tentacles (Figure 7.7B4.). Resting and turning last for about an hour each, oviposition 10 min and inspection about 2 min [55].

Egg-laying behavior in *Lymnaea* is a prime example of a complex behavior programmed by the central release of multiple types of peptides encoded by a small family of genes. The CDCs express three different CDCH (caudodorsal cell hormone) genes, CDCH I-III, that encode related but diverse peptide hormones [56]. CDCH-1 and CDCH-2 both encode eleven different peptides. So far the structure of 9 of the CDCH-1 peptides have been confirmed by peptide sequencing or MALDI-TOF MS. These are the CDCH peptide, the ε peptide, the δ peptide, α, β1 and β3CDC peptides, the CTP (C-terminal peptide), calfluxin (CaFl), and the αCDCP peptide [57, 58].

The best understood of these peptides is CDCH, the ovulation hormone (amino acid sequence in Table 7.1). Release of this hormone into the blood is triggered by synchronous sustained electrical firing in several hundred CDCs located in the cerebral ganglia of *Lymnaea* (Figure 7.2). When CDCH is injected into animals, it evokes ovulation, egg mass formation, and oviposition [59].

CDCH and other peptides encoded on the CDCH gene have a variety of roles in controlling the neural circuitry associated with the different phases of egg-laying behavior [60]. For example, pedal ganglion motoneurons involved in turning behavior are excited by the β3 CDC peptide but are inhibited by the ovulation hormone to suppress turning during oviposition. Injection of β3 and αCDC peptides into intact snails increase the rate of rasping movements of the radula that occur during the turning phase of the natural behavior. The rasping movements clean the substrate to allow the subsequent deposition of the egg mass. This is a distinct behavior from the normal role of rasping in food ingestion and is accompanied by changes in the firing pattern of feeding motoneurons and the modulatory CGCs that are 're-programmed' for their role in egg-laying [61, 62]. The CDCH and the αCDC peptide together also play an interesting role as auto-transmitters in the induction of the long-lasting spike discharge of the CDC cells that accompanies egg-laying. Combined application of these two peptides triggers sustained firing of the cells. Generation of after-discharges by electrical stimulation is prevented by the application of antibodies to the CDCH or α CDC peptide showing they are necessary for the maintenance of firing in the CDCs [63]. Thus the release of these two specific peptides encoded on the CDCH1 gene provides a self-sustaining mechanism for maintaining firing of the CDC neurons.

7.7.2.3 STIMULATION OF THE ALBUMEN GLAND BY CALFLUXIN (CAFL)

CaFl is a 29 amino acid peptide hormone (amino acid sequence in Table 7.1) [64] that stimulates the female accessory albumen gland to produce perivitelline fluid that is deposited around the freshly ovulated eggs during the egg-laying process. It acts by stimulating the influx of calcium into the mitochondria of the glandular cells [65]. It is thus another example of a gonadotropic hormone that is encoded on the CDCH1 gene.

7.7.2.4 MALE INHIBITION OF FEMALE SEXUAL FUNCTION

The hermaphroditic lifestyle of *Lymnaea* requires that the female and male functions should not be performed simultaneously. This is thought to be a competitive inhibitory process, the direction of the inhibition effect depending on whether the snail is performing the male or female role. Most information on mechanisms comes from male inhibition of female reproduction. There is good evidence that APGW plays a key role in female suppression. Application of APGW hyperpolarizes the CDCs (Croll et al., 1991) and prevents their after-discharge. There is evidence for a co-ordinating interneuron being responsible for the inhibitory effects of APGW. This is 'ring neuron' (RN) a single APGW-containing interneuron located in the right cerebral ganglion (Figure 7.2). Electrical stimulation of the RN by intracellular current injection shows that it has the ability to inhibit CDC after-discharge [66] to mimic the effect of APGW application to the *in vitro* preparation. Excitatory synaptic inputs to the RN from the male system occur in response to electrical stimulation of the PN [67]. This activation of the male pathway mediates the male inhibition of CDC network due to strong inhibitory synaptic connections that the RN has with the ventral CDCs [66]. Other peptides such as lys-conopressin and FMRFamide also inhibit CDC discharges and could also be involved in female suppression [68, 69].

Also interesting in the context of female reproductive suppression, is the recently discovered peptide, ovipostatin [45]. This large peptide (amino acid sequence in Table 7.1) is added to the seminal fluid prior to the transfer of sperm to the snail acting as the female partner. Injecting ovipostatin intravaginally into snails reduces the number of egg masses produced and the average number of eggs compared with controls. This appears to be a mechanism for the inhibition of female egg-laying that acts in parallel with insemination by the male partner.

7.8   ION AND WATER REGULATION: HORMONAL CONTROL OF ION TRANSPORT AND BLOOD PRESSURE

7.8.1   REGULATION OF BLOOD OSMOLARITY BY THE YELLOW CELL PEPTIDE

Lymnaea is a freshwater snail that efficiently maintains its body fluids at a higher osmolarity than the surrounding environment by the uptake of ions from the outside medium and the excretion of dilute urine from the kidney. An important part of the mechanism that underlies the maintenance of this hyper-osmolarity is the active uptake of Na^+ ions from the outside medium to the hemolymph by an integumental Na^+ pump. Na^+ is by far the most important cation in the blood [70] and is, therefore, one of the most significant contributors to hemolymph osmolarity. The pump is stimulated by the 77 amino acid peptide; SIS (sodium stimulating hormone) whose primary structure (Table 7.1) was determined by peptide sequencing [71] and confirmed by cloning of the single cDNA that encodes the peptide prohormone [72]. This peptide is proposed to be part of a negative feedback system for the control of blood osmolarity. The role of SIS in stimulating Na^+ uptake was initially demonstrated in the intact snail by injecting the extracts of SIS-containing ganglia into the hemolymph [73]. Specific effects on skin transport were later demonstrated in elegant experiments using an *in vitro* 'stripped' skin bioassay. SIS increased the size of the electrical potential difference and a short circuit current across the skin and ouabain application blocked these currents indicating that 'active' Na^+ ion transport was being stimulated (De Witt et al., 1988).

Immunocytochemistry and *in situ* hybridization showed that the SIS peptide is expressed in CNS peptidergic neurons called the YCs [74]. There are about 25 YC cell bodies scattered in the visceral, left parietal, right parietal ganglia (Figure 7.2) with a few others located in the proximal visceral nerves [75]. Most YCs release SIS hormonally into the blood from fine neuritic branches that penetrate the vascular connective tissue surrounding the central ganglia [76]. Thus the majority of the SIS peptide is released from these neurohaemal areas into the head sinus where it acts upon the sodium pumps in the epidermal cells of the head. However, there is a special identified group of YCs in the visceral ganglia (Figure 7.2) [75] whose axons project along the distal processes of the intestinal nerve to innervate a number of peripheral organs that include the pericardium of the heart and kidney ducts such as the reno-pericardial canal (Figure 7.4B) and the ureter [74]. This indicates that peripheral targeting of the SIS peptide is involved in stimulating renal uptake of Na^+ ions from the pro-urine that is produced by ultra-filtration of the blood across the auricular walls to the pericardial cavity (Figure 7.4B). This conclusion was supported by earlier electron microscopic studies that showed the peripheral release of peptides from YC terminals in these organs was increased when snails were subjected to a hypo-osmotic environment [77]. Thus the SIS peptide acts to maintain blood Na^+ concentration by stimulating Na^+ uptake across the skin but also by reducing the loss of Na^+ ions in the urine by peripheral reabsorption of Na^+ in the cardiac-renal system.

Other targeting of YC nerve fibers to the muscle fibers of peripheral organs and blood vessels, suggests that the SIS peptide might also be involved in controlling blood volume/pressure as well as hemolymph ion concentration. However, when snails were emersed in demonized water, it was shown that whilst a drop in blood sodium concentration preceded the loss of neurosecretory material from the cell bodies of YCs, a drop in blood volume only occurred after 4 days of exposure to deionized water [70]. This delayed change suggests that the primary role of the SIS hormone is to control blood sodium concentration rather than blood volume, although more work is required to prove this.

7.8.2 REGULATION OF BLOOD VOLUME/PRESSURE BY THE LYC PEPTIDES

The LYCs are another type of neuroendocrine cell that plays an important general role in body fluid regulation [78]. The 60 LYCs form a coherent group in the ventral lobe of the right parietal ganglia (ca.40, Figure 7.2) with a smaller number (ca. 20) in scattered clusters in the visceral ganglion.

Electrophysiological studies showed that the LYCs fire spontaneous bursts of spikes [79, 80]. This type of electrical activity has been shown to enhance neuropeptide release in other molluscan neuroendocrine systems (e.g., see Ref. [81]). Recordings from LYC axons with fine wire cuff electrodes in intact animals show that increases in bursting activity in the LYCs is correlated with a number of complex behavioral patterns, e.g., egg-laying, copulation, and eating (Ter Maat in Boer et al., 1994). These behaviors share a common feature that they involve movements of large volumes of fluids within the animal that require careful control of body shape and blood pressure, suggesting that the LYCs are involved with this type of regulation.

The identity of the LYC peptides is known from the extensive molecular and proteomic analysis. The cDNA that encodes the LYCP (light yellow cell protein) precursor [82] predicted that three peptides named LYCPI, LYCPII, and LYCPIII are generated from the precursor prohormone by cleavage of conventional flanking dibasic sites. However, MALDI-TOF MS) of single isolated LYCs revealed variant peptides, LYCP1' and LYPII' (amino acid sequences in Table 7.1) derived from the trimming of single amino acid residue from the N-terminus of LYCP1 and LYCP2 [43]. The variants were much more abundant than the intact peptides, indicating that LYCPI and LYCPII serve as intermediates in the peptide sequencing process. The MALDI-TOF MS findings were confirmed by amino acid sequencing and electrospray ionizing mass spectrometry of purified peptides. The 45 amino acid sequence LYCPIII (SLAQMYVGNHHFNENDLTSTRGGSRRWS-NRKHQSRIYTGAQLSEA) predicted by cDNA analysis was confirmed by similar peptide sequencing methods carried out on purified extracts of LYCs and parietal nerves, so unlike the other two LYCPs, there is no post-translational processing of LYCPIII.

A variety of immunocytochemical and *in situ* studies (e.g., [78], Li et al., 1994a), using probes derived from molecular analysis, showed that all three LYC peptides are present in the cell bodies of the LYCs and nerve processes penetrating the connective tissue surrounding the central connectives and proximal nerves. Immunostaining of nerve fibers using a specific antibody raised to LYCPII peptides showed that fibers were found in the epithelial walls of the anterior aorta indicating release into the head blood sinus. LYC nerve fibers also innervate vascular muscle fibers (electron microscopy, [78]) in the central ganglion connective tissue sheath suggesting that the control of blood pressure may occur via direct control of centrally-located blood vessels. Recovering blood pressure/volume after blood loss is an important part of whole-body body fluid regulation and it is significant that ultrastructural observations of the LYCs show considerable changes in neurosecretory activity during the rehydration phase after the bleeding of animals ([74], unpublished). Although most of the LYC release sites are centrally located, immunostaining of whole animal serial sections revealed that a minority of LYC axons innervate muscle fibers of the ureter papilla (Boer et al., 2004). This morphology underlies the suggested functional involvement of LYC peptides in the regulation of renal processes. This hypothesis is supported by observations in another mollusk the terrestrial snail, *Helix aspersa* [83]. It was shown in *Helix* that that blood vessels in the region of the secondary ureter, which are involved in hypotonic water absorption and resorption [83], are heavily innervated by axons containing αLYCPII immunoreactive material (reported in [78]). These results indicate the presence of a peripheral system in *Lymnaea* that complements the more extensive central LYCP release system. These results from a variety of studies provide evidence that the LYCPs are involved in whole-body fluid control, but further physiological investigations are required to determine the detailed mechanisms.

7.9   GROWTH AND METABOLISM: HORMONAL ROLE OF INSULIN-RELATED PEPTIDES

7.9.1   *STIMULATION OF GROWTH BY THE LGCS*

Extensive endocrinological experiments using the classical methods of extirpation and transplantation show that the LGCs are involved in the hormonal control of body growth [84]. For example, soft body growth in developing snails due to stimulation of cell multiplication requires the presence of the LGCs. This is also the case with the various processes underlying shell growth, such as the formation of the periostracum and the incorporation of calcium and bicarbonate into the shell edge [85]. LGCs affect various aspects of metabolism. For example, factors released from the LGCs keep glycogens stores low and stimulate the activity of the ornithine decarboxylase enzyme that shows high levels of activity in growing animal tissues.

7.9.2   EXPRESSION OF MIPS IN THE LGCS

About 75 LGC neurons lie in each cerebral ganglion and their axons project to the ipsilateral medial lip nerves that are the sites of the release of peptides into the blood (Figure 7.2). The LGCs exert their growth functions by producing and releasing molluscan insulin-related peptides (MIPs) that are encoded by a family of 5 related genes (MIP genes) [3]. cDNA cloning and peptide characterization demonstrate that the LGCs express five structurally diverse peptides, MIP I-II, V-VII (amino acid sequence of MIP1 shown in Table 7.1). Each MIP prohormone consists of A, B, and C domains separated by dibasic cleavage sites. Processing results in mature insulin-related peptides (A and B chains connected by disulfide bridges) and C-chains. The MIPs are structurally diverse although they share the cysteine backbone with vertebrate insulins. The structure of *Lymnaea* C peptide is more conserved. Although the population of LGCs expresses all five of the functional MIP genes (MIP IV and VI are pseudogenes) individual LGCs appear to contain only subsets of the gene products, suggesting that there may be functional differentiation of the MIP peptides. This is supported by experiments showing that food deprivation decreases the expression of the MIP genes II (transcripts completely lost) and III in the LGCs, whereas the expression of transcripts from the other genes was unaffected [86]. Furthermore, the length of the MIP III transcript is reduced in starved animals, indicating a change in mRNA stability and/or translational efficiency.

7.9.3   MIPS RECEPTOR

Differential functions for the MIP peptides might be thought to require different receptors but Southern Blot analysis and extensive screening of cDNA and genomic clones led to the discovery of only a single MIP receptor gene suggesting that the five different MIPs may exert their function by binding to the same receptor. There may be other types of MIP receptors, however, such as those related to tyrosine kinase receptors that have structural relationships to insulin-related receptors [3]. Although the regulation of growth by the LGCs appears to require the MIPs derived from all three functional MIP genes, there is an ectopic population of neurons in the buccal ganglia (Figure 7.5) that only express the MIP VII gene [3]. This suggests that the MIPs encoded on this gene have an additional role in the control of feeding.

7.9.4 INHIBITION OF GROWTH BY LATERAL LOBE NEURONS OF THE CEREBRAL GANGLION

There is a second endocrine center involved in growth regulation that is located in the lateral lobes (LL) of the cerebral ganglia (Figure 7.2). Cauterization of the LLs results in 'giant growth,' whereas re-implantation of the cerebral ganglia with the LL restores normal growth [87]. This shows that the LGCs are under inhibitory control by the LL [4]. A single large neuron, the Canopy Cell (CC), is present in each of the LL (Figure 7.2) and although it located in the LL it appears to be an ectopic LGC, expressing the MIP peptides, rather than being responsible for having inhibitory effects of the LL. More recent evidence indicates that another type of peptide-expressing neuron performs this LL inhibitory function. The LFRFamide precursor mRNA [88] is expressed in three neurons in the LL (Figure 7.2) and all five of the LFRFamide peptides have been extracted from the LL and identified by mass spectrometry [8]. When anyone the LFRF peptides (Table 7.1), e.g., GGSLFRFamide is applied to the LGCs *in vitro, the* firing of the LGCs is inhibited. A rapid hyperpolarizing response is recorded on the LGCs due to the opening of K^+ channels.

7.9.5 PHYSIOLOGICAL FACTORS STIMULATING THE LGCS

It is not known for certain what physiological factors stimulate the release of MIP peptides to cause growth, but *in vitro* electrophysiological experiments show that the LGCs respond to a variety of chemical messengers. These messengers include glucose (present in the blood), monoamines such as dopamine and several different types of peptides.

LGCs in isolated ganglion preparations are usually silent and show strong spike adaptation when brief supra-threshold depolarizing current pulses are applied [89], so they need to be switched to a more excitable state to play a role in growth regulation. Application of D-Glucose *in vitro* at physiological concentrates (0.1–10mM) excites the LGCs providing evidence that they act as a physiological stimulus for MIPs release [90]. The presence of glucose in the medium causes a slow depolarization of the LGCs, leading to sustained firing. The glucose response depends on the presence of extracellular sodium, and it is blocked by phlorizin, suggesting that the effects of glucose depend on the activation of an electrogenic Na^+-coupled glucose transporter.

Application of dopamine causes a delayed, slow depolarization of the LGCs and an increase in excitability but an early inhibitory effect that reduces excitability makes the role of dopamine in the physiological control of LGC release difficult to interpret [91].

There is evidence suggesting that the peptide lys-conopressin could be part of the mechanism for stimulating LGC MIP release. The receptor for lys-conopressin is expressed in the LGCs and application of this peptide drives spiking activity in isolated LGCs (Van Kesteren et al., 1995) by increasing their excitability.

Another peptide, schistosomin, is present in the LGCs (immunocytochemistry, [92]) and application of the peptide *in vitro* caused an increase in LGC excitability (autoexcitation?) and the induction of firing by slow depolarization [93]. Although most data on schistosomin function derives from work on parasitized snails (see section 7.10), schistisomin is present in the LGCs at low levels even when the animals are not infected with parasites so presumably it plays a role in controlling growth in non-infected animals, but the details of this are unknown.

7.10 THE ROLE OF NEUROHORMONES IN SETTING THE BALANCE BETWEEN GROWTH AND REPRODUCTION IN PARASITIZED SNAILS

Infection of *Lymnaea* by the avian schistosome parasite, *Trichobilharzi an acellata* induces abnormal giant growth and cessation of reproductive activity. Egg laying and growth both demand high levels of metabolites and energy, and this makes the two processes potentially antagonistic. Parasitism favors growth over reproduction. The increase in snail growth benefits the stage of the parasite by increasing the space for the developing intermediate stages of the parasite [94]. Reproduction is suppressed by parasitic infection to release 'energy' for growth.

The main agent involved in mediating the effects of parasitism is the peptide hormone, schistosomin. This was extracted from the CNS and sequenced to reveal a single-chain structure of 79 amino acids (Table 7.1) with eight cysteine residues giving rise to four intramolecular disulfide bridges that fold the peptide into a globular structure [92]. Immunocytochemistry revealed that schistosomin was located in the LGCs but also in the CCs, co-localized with the MIP peptide system in both cell types.

In vitro experiments showed that the release of schistosomin from the CNS is due to stimulation by an unknown factor produced by the cercariailarvae, an intermediate parasitic stage that develops inside the snail [95]. Schisto-somin released into the hemolymph has humoral effects at both the central and peripheral levels. Schistosomin inhibits the biological effects that the three female gonadotrophic hormones CDCH, CaFl, and DBH have on their target organs in the periphery. Detailed evidence for CaFl using a receptor-binding assay on a membrane fraction form its target organ the albumen showed that schistosomin blocks the binding of CaFl to its receptor, perhaps by modifying a receptor-G-protein complex [92, 96]. It is suggested that the peripheral inhibitory effects of CDH and DBH are also due to interactions at the receptor level. Schistosomin also acts centrally to block the release of CDCH, CaFl, and other peptides encoded on the CDCH gene by inhibiting the spiking after-discharges in the cerebral CDCs [93]. A slow depolarizing wave underlies the after-discharge and applying hemolymph from infected snails reduces the amplitude of this depolarization and prevents the long-lasting firing of the CDCs and consequent release of CDCH, required for egg-laying in the intact animal (see section 7.2.2). At the same time, schistosomin increases the excitability of the LGCs, leading to the increased release of MIPs and thus an increase in the growth rate. The opposing effect of the peptide on the neuronal systems that control reproduction and growth suggest that the peptide plays a key role in the coordination of the two processes [93].

Other types of neuropeptide genes also are up-regulated in the CNS of *Lymnaea* following infection by the parasite *Trichobilharzia ocellata* [88]. Three transcripts, LFRFamide, FMRFamide, and pedal peptide (PFDSISGSH-GLSGFA), are up-regulated throughout parasitic infection [88]. Others are up-regulated only during the late stage of parasitization (conopressin and LyNPY). Of particular interest are the LFRFamide peptides. It was previously described (section 9.4) that the LFRFamides inhibit LGC electrical activity. However, LFRF peptides such as GGSLFRFamide also hyperpolarize the CDCs producing the same slow inhibitory effect as the LGCs. Again, this is mediated by activation of a K^+ channel that hyperpolarizes the CDCs to block firing. Thus LFRFamide peptides inhibit the release of both MIPS and CDCH hormones. So unlike schistosomin, both growth and egg-laying behavior are suppressed.

The suppression of growth and reproduction, both high energy-consuming processes by parasitization, would conserve energy for parasite development, particularly if snails were starved due to a shortage of food. However, whether starvation leads to the selective activation of LRLFamide hormone release has not so far been investigated.

7.11   THE PEPTIDE HORMONE GRANULARIN ACTS IN *LYMNAEA* DEFENSIVE RESPONSES TO PARASITISM

A comprehensive multidisciplinary study has investigated the role of the *Lymnaea* peptide hormone granularin in defensive responses to parasitization [97]. The infection of *Lymnaea* by miracidia of the parasitic schistosome, *Trichobilharzia ocelli,* and the subsequent development of sporocysts leads to a phagocytic response by blood hemocytes to these foreign invaders as part of a mechanism to prevent harm to the host. We will review evidence that granular in facilitates this phagocytic immune response. The granular cells that are located in the connective tissue sheath of the CNS ganglia and nerves play an important role in the internal defensive system of *Lymnaea*. It has been known for some time that secretary activity is induced in these cells upon injection of foreign particles (Sminia and Van der Knapp, 1987) . The granular cells are indirect contact with the circulatory system and release hormonal factors into the hemolymph. One of these secretary molecules is granularin [97]. Granularin was identified as a parasite-induced peptide in the CNS by MALDI-TOF MS screening techniques. Purified peptide was sequenced by Edman degradation to yield a 35 amino acid sequence. This was found to be a part of a full-length 62 amino acid sequence (Table 7.1) that was predicted by the sequencing of a cDNA clone that encoded the whole of the granularin pro-peptide.

An important result from the molecular analysis was that granularin gene expression was up-regulated by parasitization. Transcript levels were measured by reversed Northern blotting in the CNS of parasitized snails at 1.5 h, 5 h and 6 weeks after infection with maracidia and compared with non-parasitized snails. Granularin mRNA was significantly elevated compared to controls at 5h after infection, and this was sustained until six weeks post-infection.

Another significant finding, using complementary molecular mRNA probes for *in situ* hybridization, was that the granularin gene is expressed in the granular cells of the connective tissue. No expression was detected in neurons.

To show that granularin is a defensive peptide it was necessary to provide evidence that granularin facilitated phagocytosis of foreign bodies by blood hemocytes [97]. Hemocytes were allowed to adhere to glass slides. Fine particle suspension (zymosan derived from yeast cell walls) was added to the wells containing the hemocytes, and the number of particles taken up by the hemocytes countered as a measure of phagocytosis. It was observed that zymosan particles that were pre-incubated with granularin were taken up more efficiently than untreated particles. These findings suggest that granulin-binding sites on hemocytes play a role in phagocytosis. It appears that granularin, on release into the blood, first binds to foreign substances (zymosan and presumably parasitic intermediate stages), enabling a subsequent binding to hemocytes and consequent phagocytosis.

These results, taken together, suggest that granularin acts as a classical 'opsonin,' a serum factor that binds to a foreign substance and renders the substance more susceptible to phagocytosis by the hemocytes. It has been shown in other systems (e.g., white blood cells) that the Von Willebrand factor type C molecular Domaine encoded on the granularin gene of *Lymnaea*, is involved in the formation and binding of large protein complexes in other systems (e.g., blood clotting). Thus it may well be possible (Smit et al., 1984) that granulin oligomerizes around foreign particles such as sporocysts, and that the resultant oligomers are recognized by hemocytes to promote phagocytosis. The detailed mechanism for this proposed recognition process is unknown in *Lymnaea*, and so this molecular model for granularin function is still speculative.

7.12 CONCLUSIONS: *LYMNAEA* NEUROENDOCRINOLOGY NOW AND IN THE FUTURE

Lymnaea are extremely tractable organisms for the study of neurohormonal mechanisms. The neurohormones are almost all peptides with the exception of DBH that may be a steroid [98]. The peptides are located in neurons of the CNS that are identifiable either as single large neurons or as clusters of neurons with similar molecular content (Figure 7.2). The advantage of identifiable neurons is that molecular, histological (including electron microscopy) and electrophysiological/pharmacological analysis can be carried out on neurons with known function. MALDI-TOF MS analysis allows the pattern of peptide expression to be determined in dissected single cells (e.g., Li et al., 1984; [5, 27]. Accumulating the data from molecular mapping studies on the central ganglia of Lymnaea indicate that about 2,500 neurons are peptidergic representing 12.5% of the estimated total CNS population of ~20,000 neurons [99]. However, this is likely to be an underestimate of the total number of peptidergic neurons in *Lymnaea* as many of the neurons in the CNS or PNS have not been analyzed. When considering the control of heavily chemically modulated muscular organs like the heart, penis, or gut, it is likely that all the controlling neurons are peptidergic. A very large number of neuropeptides are involved in controlling the muscular movements of these organs. For example, a peptidergic analysis of the peptides involved in copulatory behavior indicates that the penis complex contains 44 different peptides arising from 10 different genes [9]. The single gut modulatory motoneuron, B2, expresses seven different peptides. Four other peptides are also known to occur in the esophagus, so there are a total of 11 pre-oesophageal peptides encoded on four different genes [5, 7, 100]. Having groups of identified neurons in the CNS also has been a major advantage for endocrinological experiments using extirpation and transplantation techniques. These types of experiments have been extremely important in analyzing the function of neurohormones in reproduction and growth. The presence of identified groups of neurons has also facilitated the study of neural circuits that underlie the neurohormonal control of whole-body behavior. This has been particularly significant in the study of egg-laying, where the role of the CDC neurons has been elucidated in detail (reviewed in [45]).

Although the molecular studies of neurohormone genes and gene products have been extremely successful in *Lymnaea*, some of the physiological mechanisms related to peptide function are less well understood. For instance, a lot is known about the peptide effects on a heartbeat, but little is understood about how they regulate cardiac function in intact animals. Another example arises from work on the LYCs. Peptides produced by the LYCs are thought to be involved in blood pressure and volume, but the detailed mechanisms are not understood. The levels of LYC hormone in the blood need to be measured in various behaviors and the physiological effects of injections of the LYC hormone into intact animals ascertained.

The lack of genomic information from *Lymnaea* is currently a drawback in the molecular analysis of neurohormonal mechanisms; for example, it is needed to allow the development of gene-editing techniques to knock-out specific genes [2]. The recent publication of a whole-genome analysis of another pulmonate pond snail, *Biomphalaria* [101] will hopefully provide the stimulus for a similar project in *Lymnaea*. An earlier study based on a cDNA analysis of the *Lymnaea* CNS (Feng et al., 2009) produced a large collection of transcriptome data that encoded a number or proteins of functional interest. A more general point is that genomic and transcriptome studies have the advantage over the more conventional cloning of single genes and cDNAs for comparative and evolutionary studies (e.g., [102]).

KEYWORDS

- **caudodorsal cell hormone**
- **caudodorsal cells**
- **FMRFamide-related peptides**
- **median lip nerves**
- **molluscan insulin-related peptides**
- **sodium influx-stimulating peptide**

REFERENCES

1. Benjamin, P. R., (2008). Lymnaea. *Scholarpedia, 3*(1), 4124.
2. Kemenes, G., & Benjamin, P. R., (2009). *Lymnaea. Curr. Biol., 19,* 9–11.
3. Smit, A. B., Van Kesteren, R. E., Li, K. W., Van Minnen, J., Spijker, S., Van Heerikhuizen, H., & Geraerts, W. P. M., (1998). Towards understanding the role of insulin in the brain: Lessons from insulin-related signaling systems in the invertebrate brain. *Progr. Neurobiol., 54,* 35–54.
4. Geraerts, W. P. M., Smit, A. B., Li, K. W., Vreugdenhil, E., & Van Heerikhuizen, H., (1991). Neuropeptide gene families that control reproductive behavior and growth in Molluscs. In: Osborne, N. N., (ed.), *Current Aspects of the Neurosciences* (Vol. 3, pp. 244–304). McMillan Press, London.
5. Perry, S. J., Dobbins, A. C., Schofield, M. G., Piper, M. R., & Benjamin, P. R., (1999). Small cardioactive peptide gene: Structure expression and mass spectrophotometric analysis reveals a complex pattern of co-transmitters in a snail feeding neuron. *Eur. J. Neurosci., 11,* 655–662.
6. Smit, A. B., Van Kesteren, R. E., Spijker, S., Van Minnen, J., Golen, F. A., Jiménez, C. R., & Li, K. W., (2003). Peptidergic modulation of male sexual behavior in *Lymnaea stagnalis*: Structural and functional characterization of -FVamide neuropeptides. *J. Neurochem., 87,* 1245–1254.
7. Kellett, E., Perry, S. J., Santama, N., Worster, B. M., Benjamin, P. R., & Burke, J. F., (1996). Myomodulin gene of *Lymnaea*: Structure, expression and analysis of neuropeptides. *J. Neurosci., 16,* 4949–4957.
8. Hoek, R. M., Li, K. W., Van Minnen, J., Lodder, J. C., De Jong-Brink, M., Smit, A. B., & Van Kesteren, R. E., (2005). LFRFamides: a novel family of parasitization-induced –RFamide neuropeptides that inhibit the activity of neuroendocrine cells in *Lymnaea stagnalis. J. Neurochem., 92,* 1073–1080.
9. Filali, El, Z., Van Minnen, J., Liu, W. K., Smit, A. B., & Li, K. W., (2006). Peptidomics analysis of neuropeptides involved in copulatory behavior of the mollusc *Lymnaea stagnalis. J. Proteome Res., 5,* 1611–1617.
10. Benjamin, P. R., & Burke, J. F., (1994). Alternative mRNA slicing of the FMRFamide gene and its role in neuropeptidergic signaling in a defined network. *Bioessays, 16,* 335–342.

11. Santama, N., & Benjamin, P. R., (2000). Gene expression and function of FMRFamide-related neuropeptides in the snail *Lymnaea. Microsc. Res. Tech., 49,* 547–556.

12. Bright, K., Kellett, E., Saunders, S. E., Brierley, M., Burke, J. F., & Benjamin, P. R., (1993). Mutually exclusive expression of alternatively spliced FMRFamide transcripts in identified neuronal systems of the snail *Lymnaea. J. Neurosci., 13,* 2719–2729.

13. Smit, A. B., Spijker, S., Nagle, G. T., Knock, S. L., Kurosky, A., & Geraerts, W. P. M., (1992). Molluscan putative prohormone convertases: structural diversity in the central nervous system of *Lymnaea stagnalis. FEBS Lett., 312,* 213–218.

14. Smit, A. B., Spijker, S., Nagle, G. T., Knock, S. L., Kurosky, A., & Geraerts, W. P. M., (1994). Structural characterization of a *Lymnaea* putative endoprotease related to human furin. *FEBS Lett., 343,* 27–31.

15. Spijker, S., Smit, A. B., Martens, G. J. M., & Geraerts, W. P. M., (1997). Identification of a molluscan homologue of the neuroendocrine polypeptide 7B2. *J. Biol. Chem., 272,* 4116–4120.

16. Spijker, S., Smit, A. B., Eipper, B. A., Malik, A., Mains, R. E., & Geraerts, W. P. M., (1999). A molluscan peptide α-amidating enzyme precursor that generates five distinct enzymes. *FASEB, J., 13,* 735–748.

17. Spijker, S., Sharp-Baker, H. E., Geraerts, W. P. M., Van Minnen, J., & Smit, A. B., (2004). Stimulus-dependent regulation and cellular expression of genes encoding neuropeptides, prohormone convertases, α-amidating enzyme and 7B2 in identified *Lymnaea* neurons. *J. Neurochem., 90,* 287–297.

18. Roovers, E., Vincent, M. E., Van Kesteren, E., Geraerts, W. P. M., Planta, R. J., Vreugdenhil, E., & Van Heerikhuizen, H., (1995). Characterization of a putative molluscan-related receptor. *Gene, 162,* 181–188.

19. Tensen, C. P., Cox, K. J. A., Smit, A. B., Van Der Shors, R. C., Meyerhof, W., Richter, D., et al., (1998). The *Lymnaea* cardioexcitatory peptide (LyCEP) receptor: A G-protein-coupled receptor for a novel member of the RFamide neuropeptide family. *J. Neurosci., 18,* 9812–9821.

20. Perry, S. J., Straub, V. A., Schofield, M. G., Burke, J. F., & Benjamin, P. R., (2001). Neuronal expression of an FMRFamide-gated Na$^+$ channel and its modulation by acid pH. *J. Neurosci., 21,* 5559–5567.

21. Saunders, S. E., Burke, J. F., & Benjamin, P. R., (2000). Multimeric CREB binding sites in the promotor regions of a family of G-protein-coupled receptors treated to the vertebrate galanin and nociceptive/orphanin-FQ receptor families. *Eur. J. Neurosci., 12,* 2345–2353.

22. Van Kesteren, R. E., Tensen, C. P., Smit, A. B., Van Minnen, J., Van Soest, P. F., Kits, K. S., Meyerhof, W., Richter, D., Heerikhuizen, H., Vreugdenhil, E., & Geraerts, W. P. M., (1995b). A novel G protein-coupled receptor mediating both vasopressin and oxytocin-like functions of lys-conopressin in *Lymnaea stagnalis. Neuron, 15,* 897–908.

23. Van Kesteren, R. E., & Geraerts, W. P. M., (1998). Molecular evolution of ligand-binding specificity in the vasopressin/oxytocin receptor family. *Ann. N. Y. Acad. Sci., 839,* 25–24.

24. Buckett, K. J., Peters, M., Dockray, G. J., Van Minnen, J., & Benjamin, P. R., (1990b). Regulation of heart beat in *Lymnaea* by motoneurons containing FMRFamide-like peptides. *J. Neurophysiol., 63,* 1462–1435.

25. Buckett, K. J., Peters, M., & Benjamin, P. R., (1990c). Excitation and inhibition of the heart of the snail, *Lymnaea*, by non-FMRFamidergic motoneurons. *J. Neurophysiol., 63,* 1436–1447.

26. Worster, B. M., (1996). *Maldi-Tof-MS and Neuropeptide Signaling in the Nervous System*. D. Phil thesis, University of Sussex.

27. Worster, B. M., Yeoman, M. S., & Benjamin, P. R., (1998). Matrix-assisted laser desorption/ionization time of flight mass spectrometric analysis of the pattern of peptide expression in single neurons resulting from alternative mRNA slicing of the FMRFamide gene. *Eur. J. Neurosci., 10*, 3498–3507.

28. Buckett, K. J., Dockray, G. J., Osborne, N. N., & Benjamin, P. R., (1990a). Phamacology of the myogenic heart of the pond snail *Lymnaea stagnalis. J. Neurophysiol., 63*, 1413–1425.

29. Benjamin, P. R. Buckett, K. J., & Peters, M., (1988). Neurones containing FMRFamde-like peptides in the model invertebrate system *Lymnaea*. In: Salanki, J., & Rozsa, K. S., (eds.), *Neurobiology of Invertebrate Transmitters, Modulators and Receptors* (pp. 247–259). Pergamon Press, Oxford.

30. Dobbins, A. C., (1998). *Co-Localization and Release of Multiple Peptide Neurotransmitters in Lymnaea stagnalis*. D. Phil thesis, University of Sussex.

31. Benjamin, P. R., & Kemenes, I., (2013). *Scholarpedia, 8*(7), 11520.

32. Willoughby, D., Yeoman, M. S., & Benjamin, P. R., (1999a). Inositol-1, 4, 5-triphosphate and inositol-1, 3, 4, 5-tetrakisphosphate are second messenger targets for cardioactive neuropeptides encoded on the FMRFamide gene. *J. Exp. Biol., 202*, 2581–2593.

33. Willoughby, D., Yeoman, M. S., & Benjamin, P. R., (1999b). Cyclic AMP is involved in cardioregulation by multiple neuropeptides encoded on the FMRFamide gene. *J. Exp. Biol., 202*, 2595–2607.

34. Brezden, B. L., Benjamin, P. R., & Gardner, D. R., (1991). The peptide FMRFamide activates a bivalent cation-conducting channel in heart muscle cells of the snail *Lymnaea stagnalis. J. Physiol., 443*, 727–738.

35. Brezden, B. L., Yeoman, M. S., Gardner, D. R., & Benjamin, P. R., (1999). FMRFamide-activated Ca^{2+} channels in *Lymnaea* heart muscle cells are modulated by 'SEEPLY,' a neuropepeptide encoded on the same gene. *J. Neurophysiol., 81*, 1818–1826.

36. Benjamin, P. R., & Winlow, W., (1981). The distribution of three wide-acting synaptic inputs to identified neurons in the isolated brain of *Lymnaea stagnalis*(L.). *Comp. Biochem. Physiol., 70A*, 293–307.

37. Jeménez, C. R., Li, K. W., Smit, A. B., Van Minnen, J., Janse, C., Van Veelen, P., Dreisewerd, K., Zeng, J., Van Der Greef, J., Hillenkamp, F., Karas, M., & Geraerts, W. P. M., (1998). Direct neuropeptide profiling of single neurons and target tissue by Matrix-assisted laser desorption ionization mass spectrometry. *Biochemistry, 37*, 2070–2076.

38. Jiménez, C. R., Spijker, S., De Schipper, S., Lodder, J. C., Janse, C. K., Geraerts, W. P. M., Van Minnen, J., Syed, N. I., Burlingame, A. L., Smit, A. B., & Li, K. W., (2006). Peptidomics of a single identified neuron reveals diversity of multiple neuropeptides with convergent actions on cellular excitability. *J. Neurosci., 26*, 518–529.

39. Yeoman, M. S., Brezden, B. L., & Benjamin, P. R., (1999). LVA and HVA Ca^{2+} currents in the ventricular muscle cells of the *Lymnaea* heart. *J. Neurophysiol., 82*, 2428–2440.

40. Wiens, B. L., & Brownell, P. H., (1995). Neurotransmitter regulation of the heart in the nudibranch *Archidorismontereyensis. J. Neurophysiol., 74*, 1639–1651.

41. Staddon, J., (1996). *Peptidergic Co-Transmission in the Pond Snail*. D-Phil Thesis of the University of Sussex.

42. Perry, S. J., Straub, V. A., Kemenes, G., Santama, N., Worster, B. M., Burke, J. F., & Benjamin, P. R., (1998). Neural control of gut motility by myomodulin peptides and acetylcholine in the snail *Lymnaea. J. Neurophysiol., 79*, 2460–2474.

43. Li, K. W., Hoek, R. M., Smith, F., Jeménez, C. R., Van Der Schors, R. C., Van Veelen, P. A., Chen, S., Van Der Greef, J., Parish, D. C., Benjamin, P. R., & Geraerts, W. P. M., (1984a). Direct peptide profiling by mass spectrometry of single identified neurons reveals a complexneuropeptide-processing pattern. *J. Biol. Chem., 269*, 30288–30992.

44. Park, Ji-Ho, Straub, V., & O'Shea, M., (1988). Anterograde signaling by nitric oxide: Characterization and *in vitro* reconstitution of SN identified nitrergic synapses. *J. Neurosci., 18*, 5463–5476.

45. Koene, J. M., Sloot, W., Montagne-Wajer, K., Cummins, S. F., Degnan, B. M., Smith, J. S., Nagle, G. T., & Ter Maat, A., (2010). Male accessory gland protein reduces egg laying in a simultaneous hermaphrodite. *Plos One, 5*(4), e10117.

46. De Boer, P. A. C. M., Jansen, R. F., & Ter Maat, A., (1996). Copulation in the hermaphroditic snail *Lymnaea stagnalis. Invert. Reprod. Dev., 30*, 167–176.

47. De Boer, P. A. C. M., Ter Maat, A., Pieneman, A. W., Croll, R. P., Kurokawa, M., & Jansen, R. F., (1997). Functional role of peptidergic anterior lobe neurons in the male sexual behaviour of the snail *Lymnaea stagnalis. J. Neurophysiol., 78*, 2823–2833.

48. Van Golen, F. A., Li, K. W., Chen, S., & Geraerts, W. P. M., (1996). Various isoforms of myomodulin identified form the male copulatory organ of *Lymnaea* show overlapping yet distinct modulatory effects on the penis muscle. *J. Neurochem., 66*, 321–329.

49. Zijlstra, U., (1972). Distribution and ultrastructure of epidermal sensory cells in the freshwater snails *Lymnaea stagnalis* and *Biomphilaria pfeifferi. Neth. J. Zool., 22*, 283–298.

50. De Lange, R. P. J., Joosse, J., & Van Minnen, J., (1998a). Multi-Messenger innervation of the male sexual system of *Lymnaea stagnalis. J. Comp. Neurol., 390*, 564–577.

51. Koene, J. M., (2010). Neuro-endocrine control of reproduction in hermaphroditic freshwater snails: Mechanisms and evolution. *Front. Beh. Neurosci., 4, article 167*. doi: 10.3389/fnbeh.2010.00167.

52. De Boer. P. A. C. M., Jansen, R. F., Ter Maat, A., Van Stralen, N. M., & Koene, J. M., (2010). The distinction between retractor and protractor muscles of the freshwater snail's male organ has no physiological basis. *J. Exp. Biol., 213*, 40–44.

53. Van Golen, F. A., Li, K. W., De Lange, R. P. G., Van Kesteren, R. E., Van Der Schors, R. C., & Geraerts, W. P. M., (1995). Co-localized neuropeptides conopressin and Ala-Pro-Gly-TRP-NH$_2$have an antagonistic effects on the vas deferens of *Lymnaea. Neuroscience, 4*, 1275–1287.

54. Geraerts, W. P. M., & Joosse, J., (1975). Control of vitellogenesis and of growth of female sex organs by the dorsal body hormone (DBH) in the hermaphroditic freshwater snail *Lymnaea stagnalis. Gen Comp. Endocrinol., 27*, 450–464.

55. Ter Maat, A., Van Duivenboden, Y. A., & Jansen, R. E., (1987). Copulation and egg laying in the pond snail. In: Boer, H. H., Geraerts, W. P. M., & Joosse, J., (eds.), *Neurobiology: Molluscan Models* (pp. 255–261). North Holland Publishing Company, Amsterdam.

56. Li, K. W., Geraerts, W. P. M., Van Loenhout, & Joosse, J., (1992). Biosynthesis and axonal transport of multiple molluscan insulin-related peptides by the neuroendocrine light green cells of *Lymnaea stagnalis. Gen. Comp. Endocrin., 87*, 79–86.

57. Li, K. W., Jimenez, C. R., Van Veelen, P. A., & Gerearts, W. P. M., (1994). Processing and targeting of a molluscan egg-laying peptide prohormone as revealed by mass spectrometric peptide fingerprinting and peptide sequencing. *Endocrinology, 134*, 1812–1819.

58. Jiménez, C. R., Ter Maat, A., Pieneman, A., Burlingame, A. L., Smit, A. B., & Li, K. W., (2004). Spatio-temporal dynamics of the egg-laying-induced peptides during an

egg-laying cycle: Semi-quantitative matrix-assisted laser desorption/ionization mass spectrometry approach. *J. Neurochem., 89*, 865–878.

59. Ter Maat, A., Pieneman, A. W., Goldschmeding, J. T., Smelik, W. F. E., & Ferguson, G. P., (1989). Spontaneous and induced egg laying behaviour of the pond snail, *Lymnaea stagnalis. J. Comp. Physiol. A, 164*, 673–683.

60. Hermann, P. M., De Lange, R. P. J., Pieneman, A. W., Ter Maat, A., & Jansen, R. F., (1997). Role of neuropeptides encoded on the CDCH-1 gene in the organization of egg-laying behavior in the pind snail, *Lymnaea stagnalis. J. Neurophysiol., 78*, 2859–2869.

61. Jansen, R. F., Pieneman, A. W., & Ter Maat, A., (1997). Behavior-dependent activities of a central pattern generator in freely behaving *Lymnaea stagnalis. J. Neurophysiol., 78*, 3415–3427.

62. Jansen, R. F., Pieneman, A. W., & Ter Maat, A., (1999). Pattern generation in the buccal system of freely behaving *Lymnaea stagnalis. J. Neurophysiol., 82*, 3378–3391.

63. Brussaard, A. B., Schluter, N. C. M., Ebberink, K. K. S., & Ter Maat, A., (1990). Discharge induction in molluscan peptidergic cells requires a specific set of autoexcitatory neuropeptides. *Neuroscience, 39*, 479–449.

64. Dictus, W. J. A. G., & Ebberink, R. H. H., (1988). Structure of one of the neuropeptides of the egg-laying hormone prercuror of *Lymnaea. Mol. Cell. Endocrin., 60*, 23–29.

65. Dictus, W. J. A. G., De Jong-Brink, & Boer, H. H., (1987). A neuropeptide (calfluxin) is involved in the influx of calcium into the mitochondria of the albumen gland of the freshwater snail *Lymnaea stagnalis. Gen. Comp. Endocrin., 65*, 439–450.

66. Jansen, R. F., & Bos, N. P. A., (1984). An identified neuron modulating the activity of the ovulation hormone-producing caudodorsal cells of the pond snail *Lymnaea stagnalis. J. Neurobiol., 15*, 161–167.

67. Ter Maat, A., & Jansen, R. F., (1984). The egg-laying behavior of the pond snail: Electrophysiological aspects, In: Hoffman, J. A., & Porchet, M., (eds.), *Biosynthesis, Metabolism and Mode of Action of Invertebrate Hormones* (pp. 57–62). Springer Verlag, Heidelberg.

68. Brussaard, A. B., Lodder, J. C., Ter Maat, A., De Vlieger, T. A., & Kits, K., (1991). Inhibitory modulation by FMRFamde of the voltage gated- sodium current in identified neurones in Lymnaea stagnalis. *J. Physiol., 441*, 385–404.

69. Van Kesteren, R. E., Smit, A. B., De Lange, R. P., Kits, K. S., Van Golen, F. A., Van Der Schors, R. C., De Witt, N. D., Burke, J. F., & Geraerts. W. P. M., (1995a), Structural and functional evolution of the vasopressin/oxytocin superfamily: Vasopressin-related conopressin is the only member, present in *Lymnaea,* and is involved in the control of sexual behavior. *J. Neurosci., 15*, 5989–5998.

70. Soffe, S. R., Benjamin, P. R., & Slade, C. T., (1978). Effects of environmental osmolarity on blood composition and light microscope appearance of neurosecretory neurones in the snail *Lymnaea stagnalis* (L.). *Comp. Biochem. Physiol., 61A*, 577–584.

71. De Witt, N. D., Van Der Schors, R. C., Boer, H. H., & Ebberink, R. H. M., (1993). The sodium influx stimulating peptide of the pulmonate freshwater snail *Lymnaea stagnalis. Peptides, 14,* 783–789.

72. Smit, A. B., Thijsen, S. F. T., & Geraerts, W. J. M., (1993b). cDNA cloning of the sodium-influx-stimulating peptide ion the mollusc, *Lymnaea stagnalis. Eur. J. Biochem., 215*, 397–400.

73. De Witt, N. D., & Van der Schors, R. C., (1986). Neurohormonal control of Na⁺ and Cl⁻ metabolism in the pulmonate freshwater snail *Lymnaea stagnalis. Gen. Comp. Endocrin., 63*, 344–352.

74. Boer, H. H. Montagne-Wajer, C., Van Minnen, J., Ramkema, M., & De Boer, P., (1992). Functional morphology of neuroendocrine sodium influx-stimulating peptide system by *in situ* hybridization and immunocytochemistry. *Cell Tissue Res., 268,* 559–566.

75. Swindale, N. V., & Benjamin, P. R., (1976). The anatomy of neurosecretory neurones in the pond snail *Lymnaea stagnalis. Phil. Trans. Roy. Soc. B., 274,* 169–202.

76. Wendelaar, B. S. E., (1970). Ultrastructure and histochemistry of neurosecretory and neurohaemal areas in the pond snail *Lymnaea stagnalis* (L.). *Z. Zellforsch., 108,* 190–224.

77. Wendelaar-Bonga, S. E., (1972). Neuroendocrine involvement in osmoregulation in a freshwater mollusc *Lymnaea stagnalis* (L.). *Gen. Comp. Endocrinol., 3,* 308–316.

78. Boer, H. H., Montagne-Wajer, C., Smith, F. G., Parish, D. C., Ramkema, M. D., Hoek, R. M., Van Minnen, J., & Benjamin, P. R., (1994). Functional morphology of the light yellow cell and yellow cell (sodium influx-stimulating peptide) neuroendocrine systems of the pond snail, *Lymnaea stagnalis. Cell Tissue Res., 275,* 361–368.

79. Van Swichgem, H., (1979). On the endogenous bursting properties of the 'light yellow' neurosecretory cells in the freshwater snail *Lymnaea stagnalis. J. Exp. Biol., 80,* 55–67.

80. Van Swichgem, H., (1981). Electrotonic coupling within a cluster of endogenous oscillators in *Lymnaea stagnalis. Comp. Biochem. Physiol. A, 68,* 199–201.

81. Velim, F. S., Cropper, E. C., Price, D. A., Kupfermann, I., & Weiss, K. R., (1996). Release of peptide co-transmitters in *Aplysia. J. Neurosci., 16,* 105–114.

82. Smit, A. B. Hoek, R. M., & Geraerts, W. P. M., (1993a). The isolation of a cDNA encoding a neuroptide prohormone from the light yellow cells. *Cell Mol. Neurobiol., 13,* 263–270.

83. Vorhwohl, G., (1961). Zur function der exkretionorgane von *Helix Pomatia* und *Achatinaventricosa* Gould. *Z. Zellforsch , 108,* 190–224.

84. Geraerts, W. P. M., (1976a). Control of growth by the neurosecretory hormone of the light green cells in the freshwater snail *Lymnaea stagnalis. Gen. Comp. Endocrinol., 29,* 61–71.

85. Geraerts, W. P. M., Smit, A. B., Li, K. W., & Hordijk, P. L., (1992). The light green cells of *Lymnaea*: A neuroendocrine model for stimulus-induced expression of multiple peptide genes in a single cell type. *Experientia, 48,* 464–473.

86. Geraerts, W. P. M., Smit, A. B., Li, K. W., Vreugdenhil, E., & Heerikhuizen, H., (1991). Neuropeptide gene families that control reproductive behavior and growth in molluscs. In: Osborne, N. H., (ed.), *Current Aspects of the Neurosciences* (Vol. 3, pp. 255–305). Macmillan, London.

87. Geraerts, W. P. M., (1976b). The role of the lateral lobes in the control of growth and reproduction in the hermaphrodite freshwater snail *Lymnaea stagnalis. Gen. Comp. Endocrinol., 29,* 97–108.

88. Hoek, R. M., Van Kesteren, R. E., Smit, A. B., De Jong-Brink, M., & Geraerts, W. P. M., (1997). Altered gene expression in the host brain caused by a trematode parasite: neuropeptide genes are preferentially affected during parasitosis. *Proc. Acad. Sci. USA, 94,* 14072–14076.

89. Benjamin, P. R., & Rose, R. M., (1984). Electrotonic coupling and after discharges in the light green cells: A comparison with two other cerebral ganglia neurosecretory cell types in the pond snail, *Lymnaea. Comp. Biochem. Physiol., A 77,* 67–74.

90. Kits, K. S., Bobeldijk, R. C., Crest, M., & Loder, J. C., (1991). Glucose-induced excitation in molluscan central neurons producing insulin-related peptides. *Pflugers Arch., 417,* 597–604.

91. De Vlieger, T. A., Lodder, J. C., Werman, T. R., & Stoof, J. C., (1987). Changes in excitability in growth hormone producing cells induced by dopamine receptor stimulation. In: Boer, H. H., Geraerts, W. P. M., & Joosse, J., (eds.), *Neurobiology: Molluscan Models* (pp. 172–178). North-Holland Publishing, Amsterdam.

92. Hordijk, L., Schallig, H. D. F., Ebberink, R. H. M., De Jong-Brink, M., & Joosse, J., (1991). Primary structure and origin of schistosomin an anti-gonadotropic neuropeptide of the pond snail, *Lymnaea stagnalis. Biochem, J., 279*, 837–842.

93. Hordiijk, P. L., De Jong-Brink, M., Ter Maat, A., Pieneman, A. W., Lodder, J. C., & Kits, K. S., (1992). The neuropeptide schistosomin and haemolymph from parasitized snails induce similar changes in excitability in neuroendocrine cells controlling reproduction and growth. *Neurosc. Lett., 136*, 193–197.

94. Joosse, J., & Van Elk, R., (1986). *Trichobilharzia ocellata*: Physiological characterization of giant growth, glycogen depletion, and absence of reproductive activity in the intermediate snail host, *Lymnaea stagnalis. Exp. Parasitol., 62*, 1–13.

95. Schallig, H. D. F. H., Sassen, M. J. M., & De Jong-Brink, M., (1992). *In vitro* release of an anti-gonadotropic hormone, schistisomin, from the central nervous system of *Lymnaea stagnalis* with the methanolic extract of circariae of *Trichobilharzia ocellata. Parasitology, 104*, 309–314.

96. De Jong-Brink, M., Hordijk, D. P. E. J., Vergeest, H. D. F. H., Schallig, K. S., Kits, K. S., & Ter Maat, A., (1992). The anti-gonadal neuropeptide schistosomin interferes with peripheral and central neuroendocrine mechanisms involved in the regulation of reproduction and growth in the schistosome-infected snail *Lymnaea stagnalis. Progress in Brain Research, 92*, 385–396.

97. Smit, A. B., De Jong-Brink, Li, K. W., Sassen, M. J., Spijker, S., Van Elk, R., Buijs, S., Van Minnen, J., & Kesteren, R. E., (2004). Granularin, a novel molluscan opsomin comprising a single vWF type C domain is up-regulated during parasitization. *FASEB, J., 10, 1096*/fj. 03–0590fje.

98. Teunissen, Y., Geraerts, W. P. M., Van Heerikhuizen, H., Planta, R. J., & Joosse, J., (1992). Molecular cloning of a cDNA encoding a member of a novel cytochrome P450 family in the mollusc *Lymnaea stagnalis. J. Biochem., 112*, 249–252.

99. Benjamin, P. R., Elliott, C. J. H., & Ferguson, G. P., (1985). Neural network analysis in the snail brain. In: Allen, I. S., (ed.), *Model Neural Networks & Behavior* (pp. 87–108). Plenum Press, New York and London.

100. Li, K. W., Holling, T., De Witt, N. D., & Geraerts, W. P. M., (1993). Purification and characterization of a novel tetradecapeptide that modulates esophagus motility in *Lymnaea stagnalis. Biochem. Biophys. Res. Commun., 197*, 1056–1061.

101. Adema, C. M., et al., (2017). Whole genome analysis of a schistosomisis-transmitting freshwater snail. *Nat. Commun., 8*, 151451. doi: 10., 1038/ncomm15451s.

102. Moroz, L. L., et al., (2006) Neuronal transcriptome of *Aplysia*: Neuronal compartments and circuitry. *Cell, 127*, 1453–1467.

103. Benjamin, P. R., & Pilkington, J. B., (1986). The electrotonic location of low-resistance intercellular junctions between a pair of giant neurones in the snail *Lymnaea. J. Physiol., 370*, 111–126.

104. Croll, R., & Van Minnen, J., (1992). Distribution of the peptide Ala-Pro-Gly-Trp-NH$_2$ (APGW) in the nervous system and periphery of the snail *Lymnaea* as revealed by immunocytochemistry and *in situ* hybridization. *J. Comp. Neurol., 324*, 567–574.

105. De Lange, R. P. J., Van Golen, F. A., & Van Minnen, J., (1997). Diversity in cell specific co-expression of four neuropeptide genes involved in control of male copulation behavior in *Lymnaea stagnalis*. *Neuroscience, 78,* 289–299.
106. DeWitt, N. D., Slootstra, J. W., & Van Der Schors, R. C., (1988). The bioelectrical activity of the body wall of the pulmonate freshwater snail *Lymnaea stagnalis*: Effects of transmitters and the sodium influx stimulating neuropeptides. *Gen. Comp. Endocrin., 70,* 216–223.
107. Doghterom, G. E., & Van Loenhout, H., (1983). Specificity of ovulation hormone of some basommatophoran species studied by means of iso- and heterospecific injections. *Gen. Comp. Endocrin., 52,* 121–125.
108. Jansen, R. F., & Ter Maat, A., (1985). Ring neuron control of columellar motor neurons during egg-laying behavior in the pond snail. *J. Neurobiol., 16,* 1–14.
109. Jeménez, C. R. Van Veelen, P. A., Li, K. W., Wildering, W. C., Geraerts. W. P. M., Tjaden, U. R., & Van Der Greef, J., (1994). Neuropeptide expression and processing as revealed by direct matrix-assisted laser desorption ionization mass spectrometry of single neurons. *J. Neurochem., 62,* 404–407.
110. Skingsley, D. R., Bright, K., Santama, N., Van Minnen, J., Brierley, M. J., Burke, J. F., & Benjamin, P. R., (1993). A molecularly defined cardiorespiratory interneuron expressing SDPFLRFamide/GDPFLRamide in the snail *Lymnaea*: monosynaptic connections and pharmacology. *J. Neurophysiol., 69,* 915–927.
111. Sminia, T., & Van Der Knapp, W. P., (1981). The internal defense system of the freshwater snail, *Lymnaea stagnalis*. *Dev. Comp. Immunol., 5,* 87–97.
112. Smit, A. B., Vreugdenhil, E., Ebberink, R. H. M., Geraerts, W. P. M., Klootwijk, J., & Joosse, J., (1988). Growth controlling molluscan neurons produce a precursor of an insulin-related peptide. *Nature, 331,* 335–338.

CHAPTER 8

A Critical Review of Sex Steroid Hormones and the Induction Mechanisms of Imposex in Gastropod Mollusks

TOSHIHIRO HORIGUCHI[1] and YASUHIKO OHTA[2]

[1]*Ecosystem Impact Research Section, Center for Health and Environmental Risk Research, National Institute for Environmental Studies, Tsukuba, Ibaraki 305-8506, Japan, E-mail: thorigu@nies.go.jp*

[2]*Laboratory of Experimental Animals, Department of Veterinary Medicine, Faculty of Agriculture, Tottori University, Tottori, Tottori 680-8553, Japan, E-mail: yohta1022@gmail.com*

8.1 INTRODUCTION

For many years, mollusks were thought to have sex steroid hormones, as is the case in vertebrates (for example, Ref. [1]). Therefore, it was also thought that mollusks could synthesize or metabolize such steroids in their tissues (for example, Ref. [1]). Recently, however, doubts have arisen regarding whether mollusks possess vertebrate-type steroids that function as sex hormones [2–4].

In the first part of this chapter, we review basic information on the endocrinology and reproductive physiology of mollusks, with a focus on gastropods, in terms of vertebrate-type steroids. With the exception of neuropeptides, we critically review existing knowledge regarding the vertebrate-type steroids detected in mollusks, focusing on the possibility that vertebrate-type steroids may function as molluscan sex hormones. The major points we review are the presence in molluscan tissues of steroid-producing cells, enzymes to synthesize or metabolize steroids, and functional receptors for steroids. We also discuss the latest information on genes encoding vertebrate-type steroid hormone receptors in mollusks.

Later in this chapter, we review mechanisms for inducing and promoting the development of imposex (i.e., superimposition of male-type genitalia [penis and vas deferens] on female prosobranch gastropods). Six mechanisms have been proposed to explain how organotin compounds, such as tributyltin (TBT) and triphenyltin (TPhT), which originate in antifouling paints, induce the development of imposex in gastropod mollusks. These mechanisms are: (1) an increase in androgen (e.g., testosterone) levels as a result of the TBT-mediated inhibition of aromatase [5]; (2) an increase in testosterone levels owing to the inhibition of acyl CoA-steroid acyltransferase [6, 7]; (3) TBT-mediated inhibition of the excretion of androgen sulfate conjugates, with a consequent increase in androgen levels [8]; (4) TBT interference with the release of penis morphogenetic or retrogressive factors from the pedal or cerebropleural ganglia [9]; (5) an increase in the level of an alanine-proline-glycine-tryptophan amide (APGWamide) neuropeptide in response to TBT [10]; and (6) activation of the retinoid X receptor (RXR) [11]. Although the debate is ongoing, several studies support the RXR activation hypothesis [12–19]. We also describe the latest information regarding nuclear receptors other than RXR in gastropod mollusks, namely, retinoic acid receptor (RAR) and peroxisome proliferator-activated receptor (PPAR) [20, 21].

8.2   A CRITICAL REVIEW OF SEX STEROID HORMONES IN MOLLUSKS

Because sex steroid hormones such as testosterone and 17β-estradiol play physiologically important roles in the development of sex organs and the maturation of gonads (i.e., oogenesis, and spermatogenesis) in vertebrates, similar hormones have been hypothesized to regulate reproduction in invertebrates, including gastropod mollusks [1]. For example, oogenesis, and spermatogenesis were observed in the gonads of 17β-estradiol-treated females, and testosterone-treated males, respectively, of the slug *Limax marginatus* after the removal of the hermaphroditic organ, and egg-laying was induced by 17β-estradiol in female slugs. These finding simply the existence of vertebrate-type sex steroid hormones in this species [22, 23]. However, do vertebrate-type sex steroid hormones really exist in gastropod mollusks?

Many studies have reported the presence of vertebrate-type sex steroid hormones in mollusks (Table 8.1). Le Guellec et al., [30] reported the *in vitro* metabolism of androstenedione and the identification of endogenous steroids (androsterone, dehydroepiandrosterone, androstenedione, 3α-androstanediol, estrone, 17β-estradiol, and estriol) by gas chromatography with mass spectrometry (GC-MS) in *Helix aspersa*. Meanwhile, several vertebrate-type sex steroids (androsterone, estrone, 17β-estradiol, and testosterone) and synthetic estrogen (ethynylestradiol) were identified by high-resolution GC-MS in the gonads of the rock shell (*Thais clavigera*) and the ivory shell (*Babylonia japonica*) [33]. However, the synthetic estrogen ethynylestradiol detected in the gonads is presumably of environmental rather than endogenous origin—indicating the contamination of *B. japonica* habitat [33]. Therefore, the presence of the other vertebrate-type sex steroids in *T. clavigera* and *B. japonica* likely was similarly due to environmental exposure rather than *in vivo* synthesis. The detection of certain steroids, for example, by high-resolution GC-MS in mollusks only indicates the presence of such steroids in the analytical samples, which may include steroids of exogenous origin, and is consistent with possible environmental pollution. It is also possible that the detection of vertebrate-type sex steroids in mollusks may be due to contamination during the analytical process. If this is true, how do we interpret the many reports of vertebrate-type sex steroid hormones in molluscan tissue samples that interpret these data to imply that mollusks also have endogenous vertebrate-type sex steroid hormones (Table 8.1)? To judge whether mollusks have endogenous vertebrate-type sex steroid hormones, we need critical reviews regarding the presence of steroid-producing cells, enzymes to synthesize or metabolize steroids, and functional receptors for steroids in molluscan tissues [2].

In his review, Scott [3] assessed data regarding the presence, biosynthesis, and uptake of steroids in mollusks and concluded that there was no convincing evidence for the biosynthesis of vertebrate steroids by mollusks. Furthermore, Scott [3] also pointed out that the mollusk genome does not contain the genes for key enzymes that are necessary to transform cholesterol through progressive steps into vertebrate-type steroids. To our knowledge, steroid-producing cells in mollusks have not been reported. However, very strong evidence supports that mollusks are able to absorb vertebrate steroids from the environment and that they are able to store some of them (by conjugating them to fatty acids) for weeks to months [3]. In addition, we should remember that the three steroids that have been proposed as functional hormones in mollusks (i.e., progesterone, testosterone, and 17β-estradiol) are the same as those of humans. Because humans (and indeed all vertebrates) continuously excrete steroids though not only urine and feces but also through the body surface (and, in fish, the gills), it is impossible to rule out contamination as the sole reason for the presence of vertebrate steroids in mollusks (even in animals kept under supposedly clean laboratory conditions). Essentially, the presence of vertebrate steroids in mollusks cannot be taken as reliable evidence of either endogenous biosynthesis or an endocrine role [3].

Meanwhile, testosterone biotransformation has been characterized in the mud snail (*I. obsoleta*) [58]. However, given the lack of scientific verification of the presence of an AR in gastropods (see below), we likely should interpret the biological significance of the transformation of testosterone in *I. obsoleta* exposed to a relatively high dose (1.0 μM (150,000 DPM) [14C] testosterone) with caution [58]. Alternatively, this apparent biotransformation of high doses of steroids might represent a kind of metabolism of xenobiotics in mollusks.

The aromatase-like activity has been measured and reported in several gastropod species [32, 59]. However, this activity does not necessarily confirm the existence of vertebrate-type aromatase in gastropods. To our knowledge, aromatase has not yet been isolated from invertebrates. Further evidence of the presence of steroid-producing cells and of synthetic or metabolic enzymes for steroid biosynthesis are needed to clarify the existence of vertebrate-type sex steroid hormones in gastropod mollusks.

Table 8.2 provides summary information regarding the nuclear receptors that have been identified or suggested to exist in the tissues of mollusks. To our knowledge, AR has not yet been cloned from the tissues of invertebrates, including that of mollusks. Although an estrogen receptor (ER)-like cDNA has been isolated from the California sea hare (*Aplysia californica*; Gastropoda: Opisthobranchia), and even though the protein it encodes functions as a constitutively activated transcription factor, estrogen cannot bind this protein [66]. Similarly, an ER-like protein has been isolated from *T. clavigera*. Like that from *A. californica*, the *T. clavigera* protein fails to bind estrogen [60, 72] and is a constitutively activated transcription factor [72]. In the absence of direct evidence for the presence of ER and AR, their physiological roles in mollusks remain in doubt despite the detection of estrogens and androgens in their tissues. According to a study of fully sequenced invertebrate genomes, homologs of ER and AR have yet to be found in invertebrates [73]. In fact, recent studies on nuclear receptors in mollusks (*Crassostrea gigas, Biomphalaria glabrata*, and *Lottia gigantea*) have confirmed the lack of functional nuclear receptors, such as AR and ER, in bivalves and gastropods, although homologs of ER and estrogen-related receptor (ERR) were identified [74–76]. Therefore, the mollusk genome does not seem to contain genes for functioning classic nuclear steroid receptors [3, 75, 77], and thus whether gastropods inherently have verte-brate-type steroids that act as sex hormones are doubtful. The absence of a molluscan AR and the constitutive expression of the ER *in vitro* suggest that alternative pathways exist for spermatogenesis and oogenesis in mollusks [76]. Further studies are necessary to identify the critical factors in such alternative pathways for spermatogenesis and oogenesis and to clarify their functions in mollusks.

After reviewing the evidence regarding whether vertebrate sex steroids (e.g., testosterone, estradiol, and progesterone) have hormonal actions in mollusks, Scott [4] criticized the experimental designs of almost all papers on this subject (i.e., tested compounds or mixtures were only presumed to behave as steroids [or modulators of steroids] on the basis of their effects in vertebrates) and pointed out the absence of both blinding procedures (implying the possibility of operator bias) and the evaluation of results (i.e., there was no statistical analysis).

8.3 MECHANISMS FOR THE DEVELOPMENT OF IMPOSEX IN PROSOBRANCH GASTROPOD MOLLUSKS: INVOLVEMENT OF THE RETINOID X RECEPTOR (RXR)

Regarding the mechanism by which organotins, such as TBT and TPhT, induce the development of imposex in prosobranch gastropods, as we mentioned earlier, six hypotheses have been proposed: (1) an increase in androgen (e.g., testosterone) levels due to the TBT-mediated inhibition of aromatase [5]; (2) an increase in testosterone levels owing to the inhi-bition of acyl CoA-steroid acyltransferase [6, 7]; (3) TBT-mediated inhi-bition of the excretion of androgen sulfate conjugates, with a consequent increase in androgen levels [8]; (4) TBT interference with the release of penis morphogenetic or retrogressive factor from the pedal or cere-bropleural ganglia [9]; (5) an increase in the level of an APGWamide neuropeptide in response to TBT [10]; and (6) activation of RXR [11] (Table 8.3).

TABLE 8.1 Steroids Detected in the Tissues of Mollusks

Groups	Species	Tissues Analyzed	Steroids Detected	Methods in Detection	References
Gastropods	*Nucella lapillus*	Whole tissue (females)	T (ester), E2	RIA	[24]
		Whole tissue (females)	T, E2, P	RIA	[25]
		Whole tissue (females)	T	RIA	[5]
Gastropods	*Hinia reticulata*	Whole tissue (females)	T	RIA	[5]
Gastropods	*Ilyanassa obsoleta*	Gonad/viscera	T (ester), E2 (ester)	RIA	[26]
		Whole tissue	T	RIA	[27, 28]
		Whole tissue	T (ester)	RIA	[6, 27]
Gastropods	*Marisa cornuarietis*	Whole tissue	T (ester), E2 (ester)	RIA	[29]
Gastropods	*Helix aspersa*	Gonad	T, Ad, P	RIA and GC-MS	[30]
		Hemolymph	T, Ad, P	RIA and GC-MS	[30]
Gastropods	*Achatina fulica*	Hemolymph	T, E2, P, Ad, Cortisol	RIA	[31]
Gastropods	*Bolinus brandaris*	Digestive gland + gonad	T, E2	RIA	[32]
Gastropods	*Thais clavigera*	Testis and ovary	T, E2	GC-MS	[33]
		Testis	T	EIA	[34]
Gastropods	*Babylonia japonica*	Testis and ovary	T, E1, E2, And, EE2	GC-MS	[33]
Bivalves	*Ruditapes decussatus*	Whole tissue	T, E2	RIA	[35]
		Whole tissue minus digestive gland	T, E2	RIA	[36]
		Gonad	T, E2, P	RIA	[37]
Bivalves	*Tapes philippinarum*	Whole tissue	T, E2, P	RIA	[38]
Bivalves	*Paticopecten yessoensis*	Gonad	E1, E2	EC	[39]
		Gonad	E1, E2	EC	[40]
Bivalves	*Mytilus edulis*	Mantle + gonad	T, E1, E2	RIA	[41]
		Whole tissue, gonad	P	RIA and GC-MS	[42]

TABLE 8.1 *(Continued)*

Groups	Species	Tissues Analyzed	Steroids Detected	Methods in Detection	References
Bivalves		Whole tissue	T, E1, E2, P, Ad	GC-MS	[43]
		Gonad	T (ester), E2 (ester)	RIA	[44]
		Peripheral tissue	T (ester), E2 (ester)	RIA	[44]
	Mytilus galloprovincialis	Whole tissue	E2 (ester)	RIA	[45]
		Whole tissue	T (ester)	RIA	[46]
		Mantle + gonad	E2	RIA	[47]
Bivalves	*Mytilus spp.*	Digestive gland	P, Preg, T, E1, E2, Ad, DHT, DHA	GC-MS	[48]
Bivalves	*Crassostrea gigas*	Gonad	E1, E2	RIA	[40]
Bivalves	*Scrobicularia plana*	Gonad	T, E2, P	EIA	[49]
Bivalves	*Mya arenaria*	Gonad	T, E2	EIA and GC-MS	[50]
		Gonad	P	EIA and GC-MS	[51]
		Gonad	P	EIA	[52]
Bivalves	*Dreissena polymorpha*	Whole-body (excl. gills)	T (ester), E2 (ester)	RIA	[53]
		Whole tissue	T (ester), E2 (ester)	RIA	[54]
Cephalopods	*Sepia officinalis*	Hemolymph	T, Estrogen	RIA	[55]
		Tissue	T	RIA	[55]
Cephalopods	*Octopus vulgaris*	Reproductive tissue (females)	T	RIA/EIA and HPLC	[56]
		Ovary	E2	RIA	[57]

Ester, a portion of the steroid is conjugated to fatty acids (and can be converted to free steroid by base hydrolysis); Ad, androstenedione; And, androsterone; DHA, 5α-dihydro androstenedione; DHT, 5α-dihydrotestosterone; E1, estrone; E2, 17β-estradiol; EC, electron capture detector on HPLC; EE2, ethynylestradiol; EIA, enzyme immunoassay; GC-MS, gas chromatography coupled with mass spectrometry; HPLC, high-performance liquid chromatography; P, progesterone; Preg, pregnenolone; RIA, radioimmunoassay; T, testosterone.

TABLE 8.2   Nuclear Receptors Identified or Suggested to Occur in Molluscan Tissues

Groups	Species	Tissues Analyzed	Nuclear Receptors Identified/Suggested	Methods of Identification/Analysis	Characterization/Function	References
Gastropods	*Thais clavigera*	Whole male soft tissue	RXR	Cloning of protein similar to vertebrate RXR (TcRXR), ligand binding assay	Development of penis and vas deferens in females (imposex), TcRXR binds 9cRA, TBT, and TPhT.	[11]
		Testis/ovary, digestive gland, ctenidium (gill), penis/penis forming area, head ganglia	RXR	Gene expression, injection and flow-through exposure experiments	Development of penis and vas deferens in males and females (imposex)	[13–15]
		Penis	RXR-1, -2	Cloning of *T. clavigera* RXR subtypes/isoforms (TcRXR-1 and -2), reporter gene assay	Development of penis and vas deferens in females (imposex), TcRXR-1 and -2 bind 9cRA, TBT, and TPhT. The transcriptional activity of TcRXR-2 was significantly lower than that of TcRXR-1.	[18]
			RXR-1, -2	Reporter gene assay with TcRXR-1 and -2	Transactivated by 9cRA, TBT, TPhT, TPrT, and TcHT.	[19]
		Whole soft tissue	RAR	Cloning of protein similar to vertebrate and invertebrate RAR, reporter gene assay, one- and two-hybrid assay	No ligand-dependent transactivation by ATRA, possibility to form a heterodimer with TcRXR isoforms	[20]

TABLE 8.2 (Continued)

Groups	Species	Tissues Analyzed	Nuclear Receptors Identified/Suggested	Methods of Identification/Analysis	Characterization/Function	References
		Head ganglia, testis/ovary, muscle	ER	Cloning of protein similar to vertebrate ER, ligand binding assay	No binding to E2	[60]
Gastropods	*Nucella lapillus*	Whole soft tissue	RXRa, b	Cloning of orthologue of RXR (*NlRXR*) similar to vertebrate and invertebrate RXR, ligand binding assay	Development of penis and vas deferens in females (imposex), *NlRXR* binds to 9cRA.	[12]
		Penis	RXRa, b	Cloning of *N. lapillus* RXR isoforms, reporter gene assay	Transactivated by 9cRA, TBT, TPhT, TPrT, and TcHT.	[19]
		Digestive gland, gonads, gill, penis-forming area (penis) and cerebral ganglia	PPAR	Application of rosiglitazone, a well-known vertebrate PPAR gamma (γ) ligand, to dogwhelks (*Nucella lapillus*), microarray analysis	Possible additional involvement in the development of imposex	[21]
		Whole soft tissue	ER	Cloning of protein similar to vertebrate ER, semi-static exposure experiment	Authors suggested it does not bind E2 due to its structural similarity to *Thais clavigera* ER.	[61]
Gastropods	*Babylonia japonica*	Penis	RXR-1, -2	Cloning of *B. japonica* RXR isoforms, reporter gene assay	Transactivated by 9cRA, TBT, TPhT, TPrT, and TcHT.	[19]
Gastropods	*Ilyanassa obsoleta*	Gonad-viscera complex	RXRa, b	Cloning of protein similar to vertebrate and invertebrate RXR, gene expression	Involvement in the development of imposex, gonadal maturation/recrudescence	[17]

TABLE 8.2 *(Continued)*

Groups	Species	Tissues Analyzed	Nuclear Receptors Identified/Suggested	Methods of Identification/Analysis	Characterization/Function	References
		Gonad-viscera complex	ER	Cloning of protein similar to vertebrate ER, gene expression	Changes (up to 5-fold) in the reproductive season, no gene similar to vertebrate AR	[26]
Gastropods	*Marisa cornuarietis*	Whole soft tissue	ER	Cloning of protein similar to vertebrate ER, ligand binding assay	No binding to E2	[62]
Gastropods	*Potamo-pyrgus antipodarum*	Whole soft tissue	ER	Cloning of protein similar to vertebrate ER, semi-static exposure experiment	No data shown on binding to E2	[63]
Gastropods	*Biomphalar-ia glabrata*	Whole soft tissue	RXR	Cloning of protein similar to vertebrate RXR, reporter gene assay	BgRXR binds to direct repeat elements as a homo- or heterodimer, BgRXR binds 9cRA and is transactivated by 9cRA.	[64]
Gastropods	*Lottia gigantea*		RAR	Analysis with public genomic and EST databases	Orthologs of enzymes for retinoic acid synthesis, as well as RAR sequences	[65]
Gastropods	*Aplysia californica*	Adult neural tissue and ovotestis	ER	Cloning of protein similar to vertebrate ER, binding assay, reporter gene assay	No binding to E2	[66]
Bivalves	*Chlamys farreri*	Testis/ovary	RXR	Cloning of protein similar to vertebrate and invertebrate RXR (CfRXRs), gene expression	CfRXRs are involved in germ cell differentiation in both sexes.	[67]

TABLE 8.2 *(Continued)*

Groups	Species	Tissues Analyzed	Nuclear Receptors Identified/Suggested	Methods of Identification/Analysis	Characterization/Function	References
Bivalves	*Mytilus edulis*	Gonad	ER	Cloning of protein similar to vertebrate ER, semi-static exposure experiment, gene expression	Not affected by exposure to E2 for ten days	[68]
Bivalves	*Crassostrea gigas*	Gonad	ER	Cloning of protein similar to vertebrate ER, reporter gene assay, gene expression	No binding to E2	[69]
Bivalves	*Placopecten magellanicus*	Gonad, digestive gland, adductor muscle	ER	Cloning of protein similar to vertebrate ER, amino acid sequence, Northern blotting	Contamination? (close to sequence of rainbow trout ER)	[70]
Cephalopods	*Octopus vulgaris*	Ovary and oviduct	ER	Cloning of protein similar to vertebrate ER, reporter gene assay, ligand binding assay	No binding to E2, Octopus ER LBD was insensitive to other steroid hormones (other estrogens, progesterone, androgens, corticosteroids, and estrogenic xenobiotics) as well.	[71]

9cRA, 9-*cis* retinoic acid; ATRA, all-*trans* retinoic acid; E2, 17β-estradiol; AR, androgen receptor; ER, estrogen receptor; PPAR, peroxisome proliferator-activated receptor; RAR, retinoic acid receptor; RXR, retinoid X receptor; TBT, tributyltin; TcHT, tricyclohexyltin; TPhT, triphenyltin; TPrT, tripropyltin.

TABLE 8.3 Hypothesized Mechanisms by Which Organotins, Such as TBT and TPhT, Induce the Development of Imposex in Prosobranch Gastropods

Sex Hormones Suggested	Hypotheses	References
Steroids	Inhibition of aromatase	[5]
Steroids	Inhibition of acyl CoA-steroid acyltransferase	[7]
Steroids	Inhibition of the excretion of androgen sulfate conjugates	[8]
Neuropeptides	Release of penis morphogenetic/retrogressive factor from the pedal/cerebropleural ganglia	[9]
Neuropeptides	Increase in the level of an alanine-proline-glycine-tryptophan amide (APGWamide) neuropeptide	[10]
Neuropeptides?	Activation of the retinoid X receptor (RXR)	[11]

TBT–tributyltin; TPhT–triphenyltin.

Experimental evidence is weak for all but the sixth hypothesis, which involves the activation of RXR [11]. Although it is doubtful that gastropods inherently have vertebrate-type steroids that function as sex hormones, as we mentioned previously in this chapter, the correlation between the increase in testosterone titer over time and penis growth in females is insufficient to support the aromatase inhibition hypothesis [5, 25]. Regarding Hypotheses 1, 2, and 3, Spooner et al., [25] reported that testosterone levels were significantly elevated in TBT-exposed dog whelks (*Nucella lapillus*), on days 28 and 42 when compared with controls, although penis length in female *N. lapillus* started to increase on day 14. In another study, a combination of the aromatase inhibitor fadrozole (5 µg/g wet wt) and testosterone (0.1 µg/g wet wt) had little effect on the induction or promotion of imposex in *T. clavigera*, as indicated by penis growth and the incidence of imposex [72]. Consequently, there seems to be substantial uncertainty about the mechanism by which organotins induce imposex in gastropods, assuming that vertebrate-type steroid hormones are involved (Table 8.3). Meanwhile, the results given in support of the "inhibition of testosterone excretion" hypothesis [8] (Table 8.3) may reflect a phenomenon that is at least transient or associated with acutely toxic TBT concentrations [78]. Conversely, regarding Hypothesis 2 (an increase in testosterone levels owing to the inhibition of acyl CoA-steroid acyltransferase), we should interpret the biological significance of the transformation of testosterone in gastropods with caution, as mentioned earlier.

Whether the TBT concentrations in the tissues of gastropods collected in the field at sites with low levels of TBT contamination actually inhibit aromatase-like activity is unknown. In addition, evidence regarding the relationship between the decrease in levels of aromatase-like activity and the advance of imposex symptoms in the gastropod *Bolinus brandaris* is inconsistent and inconclusive [32]. Santos et al., [24] suggested the involvement of AR, in addition to aromatase inhibition, in the development of imposex in *N. lapillus*. However, gastropods may lack inherent AR [3, 73–77], as mentioned earlier.

Several neuropeptides released from gastropod visceral ganglia, cerebral ganglia, or prostate glands (e.g., *A. californica* and *Lymnaea stagnalis*) act as ovulation, egg-laying, or egg-releasing hormones [79, 80]. Féral and Le Gall [9] suggested that TBT-induced imposex in the European sting winkle (*Ocenebra erinacea*) might be related to the release of neural morphogenetic controlling factors (Table 8.3). Their study examined *in vitro* tissue cultures derived from a presumed penis-forming area on the immature slipper limpet (*Crepidula fornicata*) and the isolated nervous systems of male or female *O. erinacea* in the presence or absence of TBT (0.2 µg/L) [9] (Table 8.3). The accumulation of TBT or TPhT in the central nervous systems of the giant abalone (*Haliotis gigantea*) [81], *N. lapillus* [82], and *T. clavigera* [83] indicate the potential for the toxic effects of TBT and TPhT on neuroendocrine systems. Oberdörster and McClellan-Green [10] reported that APGWamide, a neuropeptide released from the cerebral ganglia of gastropods such as the large pond snail (*Lymnaea stagnalis*), markedly induced the development of imposex in female *I. obsoleta* (Table 8.3). However, the effect of APGWamide on the induction of imposex or promotion of its development appears weak in light of experimental data regarding penis growth and the incidence of imposex [10, 84]: the incidence of imposex was higher, and penis growth occurred for much longer in gastropods exposed to TBT or TPhT in the laboratory [2].

Therefore, except Hypothesis 6 (activation of RXR), none of the hypothesized mechanisms for the induction of imposex in gastropod mollusks can be supported fully at present.

Nishikawa et al., [11] proposed a mechanism of action for TBT or TPhT on the development of imposex in gastropods that was completely different from the other hypotheses already proposed (Tables 8.2 and 8.3). Nishikawa et al., [11] showed that a single injection of 9-*cis* retinoic acid (9cRA), the natural ligand of human retinoid X receptors (hRXRs), into female *T. clavigera*, induced the development of imposex, together with organotin compounds (both TBT and TPhT) bound to hRXRs with high affinity. The cloning of an RXR homolog from *T. clavigera* revealed that the ligand-binding domain of the *T. clavigera* RXR was very similar to that of the vertebrate RXR and bound to both 9cRA and organotins [11]. Horiguchi et al., [14] treated female *T. clavigera* with a single injection of 9cRA at one of three different concentrations (0.1, 1, or 5 μg/g wet wt), or with TBT or TPhT at 1 μg/g wet wt (as positive controls), or with fetal bovine serum (as a negative control) to confirm the effectiveness of 9cRA in inducing the development of imposex in *T. clavigera*. 9cRA induced imposex in a dose-dependent manner; imposex incidence was significantly higher in the *T. clavigera* that received 1 μg ($p < 0.05$) or 5 μg ($p < 0.001$) 9cRA than in the controls [14]. After 1 month, the *T. clavigera* treated with 5 μg 9cRA exhibited substantial growth of a penis-like structure. The length of the structure differed between the 0.1-μg and 5-μg 9cRA treatment groups ($p < 0.05$) but not between the groups given 1 or 5 μg 9cRA ($p > 0.05$) [14]. Compared with the control, the vas deferens sequence (VDS) index was increased significantly in the 1-μg ($p < 0.05$) and 5-μg ($p < 0.001$) 9cRA groups [14]. Light-microscopic histology revealed that the penis-like structures behind the right tentacle in female *T. clavigera* treated with 5 μg 9cRA were essentially the same as the penises and vasa deferentia of normal males and of TBT-treated or TPhT-treated imposex females [14].

Horiguchi et al., [13] investigated RXR gene expression and measured RXR protein content in various tissues of wild male and female *T. clavigera*, to further elucidate the role of RXR in the development of organotin-induced imposex in gastropod mollusks (Table 8.2). By using quantitative real-time polymerase chain reaction analysis, Western blotting, and immunohisto-chemistry with a commercial antibody against human RXR α, they revealed that RXR gene expression was significantly higher in the penises of males ($p < 0.01$) and imposex females ($p < 0.05$) than in the penis-forming areas of normal females. Western blotting demonstrated that the antibody detected *T. clavigera* RXR and showed that the male penis had the highest RXR protein content among the analyzed tissues of males and morphologically normal females [13]. Moreover, immunohistochemical staining revealed nuclear localization of RXR protein in the epithelial and smooth muscle cells of the vas deferens and in the interstitial or connective tissues and epidermis of the penis in males and in imposex females [13]. In addition, the same results were obtained by using an antibody specific for *T. clavigera* RXR [16]. According to the results of these studies, RXR could be involved in the organotin-mediated induction of male-type genitalia (penis and vas deferens) in female *T. clavigera* [13].

To further examine the role of RXR in the development of imposex in gastropods, Horiguchi et al., [15] investigated the time course of expression of the RXR gene in various tissues (ctenidium, ovary or testis, digestive gland, penis-forming area or penis, and head ganglia) of female and male *T. clavigera* exposed to TPhT for 3 months in a flow-through exposure system. TPhT clearly accumulated in the tissues of exposed animals, whereas no TPhT accumulation occurred in the control groups [15]. In females, 3-month exposure to TPhT resulted in the development of imposex, and penis lengths in imposex-exhibiting females were significantly longer in small females (shell height <20 mm) than in large females (shell height ≥20 mm) [15]. RXR gene expression in the ovary, penis-forming area or penis, and head ganglia of females after 3 months of exposure was significantly higher than in control females; RXR gene expression was highest in the penis-forming area or penis [15]. Moreover, RXR gene expression in the penis-forming area or penis of each female exposed to TPhT seemed to be associated with penis length [15]. In males, the ratio of penis length to shell height was significantly larger in the exposed groups than in the controls. Although RXR gene expression in males after 3 months of exposure was not significantly higher than in control males in any tissue, RXR gene expression was greatest in the penises of exposed males [15]. These results further suggest that, in addition to other, potential physiologic functions, RXR plays an important role in the development of male genitalia (i.e., penis, and vas deferens) in gastropod mollusks (Table 8.2).

Oehlmann et al., [85] reviewed endocrine disruption in prosobranch gastropods and discussed its supporting evidence and ecological relevance; they also described the results of their laboratory experiments using 9cRA. Injection of 9cRA into female *N. lapillus* failed to initiate the development of imposex during the experimental period (56 days), even in the highest dose (2.5 µg/g wet wt.) group. In contrast, Castro et al., [12] demonstrated that injections of 9cRA at 1 µg/g wet wt. in female *N. lapillus* induced the development of imposex to the same degree as did TBT (1 µg/g) (Table 8.2). Given the contradictory results obtained by Oehlmann et al., [85] and Castro et al., [12], the experimental methodologies used by each scientific group should be evaluated carefully. For example, snails might need to be injected under low-light conditions, because 9cRA is easily photodegraded.

Although Castro et al., [12] reported that imposex in *N. lapillus* might be mediated by RXR, their experimental data showed that the level of expression of the RXR gene was highest in the gonads, unlike in *T. clavigera*, where expression was greatest in the penis-forming area or penis [13]. Castro et al., [12] discussed a mechanism for the induction of imposex in gastropods by organotins according to a scenario that integrated the inter-action between three cascades (retinoic, neuroendocrine, and steroid); however, the physiological roles of AR and ER in mollusks remain in doubt because an AR has neither been cloned nor a functional ER isolated from the tissues of gastropods.

A recent study has provided further evidence of the involvement of RXR in the development of imposex through the cloning of RXR from *I. obsoleta* [17] (Table 8.2). In this context, Sternberg et al., [17] reported the synchro-nized expression of RXR mRNA with reproductive tract (i.e., a combina-tion of gonadal and digestive gland tissues) recrudescence in *I. obsoleta*, and Sternberg et al., [86] reviewed the mechanism of imposex induction in prosobranch gastropods, focusing on the environmental–endocrine control of reproductive maturation in gastropods. However, Sternberg et al., [17] did not measure RXR gene expression in the presumptive penis-forming area or penis of females. Instead, they measured it in the "the gonad–viscera complex"—probably a combination of gonadal and digestive gland tissues—despite the fact that the development of imposex essentially or primarily involves the differentiation and growth of male genitalia (e.g., the penis and vas deferens). In addition, Sternberg et al., [17] also discussed the synchro-nized expression of RXR mRNA with recrudescence of the reproductive tract (primarily the gonad, but their samples might also have included the digestive gland) in *I. obsoleta.* Therefore, their discussion seems to have confused the development of imposex (principally the differentiation and growth of the penis and vas deferens) with events of the male reproductive cycle, such as gonadal maturation and regression.

Urushitani et al., [18] reported two isoforms of RXR cDNA, RXR isoform 1 (*TcRXR-1*) and RXR isoform 2 (*TcRXR-2*), in *T. clavigera* (Table 8.2). The deduced amino acid sequences of *TcRXR-1* and *TcRXR-2* are highly homologous with those of other gastropods [18]. These TcRXR isoforms displayed 9cRA-dependent activation of transcription in a reporter gene assay using COS-1 cells [18]. The transcriptional activity of TcRXR-2, the encoded protein of which has five additional amino acids in the T-box of the C domain, was significantly lower than that of TcRXR-1 [18]. The mean EC_{50} values of 9cRA were 1.1×10^{-7} M (95% confidence interval: 7.7×10^{-8}–1.6×10^{-7} M) for human RXR α, 6.4×10^{-8} M (5.3–7.9×10^{-8} M) for TcRXR-1, and 1.2×10^{-7} M (6.5×10^{-8}–2.2×10^{-7} M) for TcRXR-2 [18].

In addition, Urushitani et al., [18] analyzed the induction of TcRXR-1 transcriptional activity, which was induced by 10^{-8} M 9CRA, 10^{-8} M TBTCl, and 10^{-7} M TPhTCl but was unchanged by DHA [18]. In a subsequent study, they noted significant ligand-dependent transactivation of the *T. clavigera*

RXR protein by9cRA and five of the 16 organotins they tested (TBT, tetra-butyltin (TeBT), tripropyltin (TPrT), tricyclohexyltin (TcHT), and TPhT) (19) (Table 8.2); these same five organotins also induced significant transcriptional activity in the *N. lapillus* and *B. japonica* RXR isoforms [19] (Table 8.2). However, the effect of TeBT could be due to its decomposition product, TBT [19]. In contrast, the remaining four organotins (i.e., TBT, TPrT, TcHT, and TPhT) have been reported to promote the development of imposex in female *T. clavigera* within 1 month after a single injection at 1 μg/g wet wt [87, 88]. Thus, the results of *in vitro* assays substantially correspond to those of *in vivo* experimental studies in regard to the induction or promotion of the development of imposex through the interaction between various ligands, namely, 9cRA and four organotins (TBT, TPrT, TcHT, and TPhT), and RXR in *T. clavigera* [19, 87, 88].

It has been reported that 9cRA is a specific ligand of human RXR. However, 9cRA has not been detected in mammalian tissues [89, 90], and its concentration is lower than those of other retinols in fiddler crabs and mollusks [91, 92]. DHA is a natural ligand in mammals, although the activation of RXRs requires higher concentrations of DHA than 9cRA [93]. Urushitani et al., [18] used a concentration of DHA (10^{-6} M) that was lower than that required for the induction of RXR transcriptional activity in the mollusk *B. glabrata* [64], and they were unable to measure the transcriptional activity of TcRXR-1 induced by 10^{-4} M DHA because of toxicity. Similar results have been obtained from reporter assays using the daphnid RXR, for which DHA was not a ligand [94].

Decreases in TcRXR-1 transcriptional activity were observed when excess*TcRXR-2* fused expression vector was added to a co-transfection assay [18]. Overexpression of TcRXR-2 yielded lower transcriptional activity than did overexpression of TcRXR-1: the transcriptional activity of TcRXR-1 significantly declined in the presence of 0.1 or 0.5 μg *TcRXR-2*-fused expression vector [18]. These results suggest that this difference has a functional basis in the regulation of the molluscan endocrine system [18]. In mature males of *N. lapillus*, *T. clavigera*, and *I. obsoleta*, penis length varies seasonally in field locations that are lightly contaminated by organotins [95–97]. In addition, the expression of RXR mRNA changed seasonally in males of *T. clavigera* obtained from lightly contaminated sites, although the quantities of *TcRXR-1* and *TcRXR-2*transcripts were not measured separately [96]. Seasonal changes in the RXR mRNA levels of *I. obsoleta* have been reported also [17]. Furthermore, male penis length in *T. clavigera* decreased in the laboratory in autumn (regularly, after the spawning season) [15]. These findings suggest that the interaction between TcRXR-1 and TcRXR-2might contribute to seasonal changes in the penis length of *T. clavigera* males. Specifically, it appears that—together with possible changes in endogenous retinoids or natural TcRXR ligands—fluctuations in the expression of *TcRXR-2* contribute to seasonal differences in male *T. clavigera* penis length.

Meanwhile, a recent study revealed that Chinese scallop (*Chlamys farreri*) RXRs might be involved in germ cell differentiation in both males and females [67] (Table 8.2).

Together, these findings suggest that retinoic acids could play an important role in the development of male genitalia and their components and that RXR isoforms might underlie a novel mechanism regulating genes in mollusks. In fact, orthologs of enzymes for retinoic acid synthesis, as well as RAR sequences, occur in the owl limpet (*L. gigantea*) [65] (Table 8.2). In vertebrates, retinoic acids and their derivatives are involved in cell proliferation and differentiation, organ homeostasis, and tissue and organ regeneration (see reviews: [98–103]). All-*trans* retinoic acid (ATRA) and 9cRA have been detected in *L. stagnalis* [91]. Urushitani et al., [20] isolated a retinoic acid receptor (RAR)-like cDNA (*TcRAR*) in *T. clavigera* as a candidate partner of RXR, but using ATRA failed to reveal any ligand-dependent transactivation by the TcRAR protein (Table 8.2). Urushitani et al., [20] examined the transcriptional activity of the TcRAR-ligand-binding domain fused with the GAL4-DNA binding domain by using retinoic acids, retinol, and organotins, and again saw no noteworthy transcriptional induction by these chemicals. Use of a mammalian two-hybrid assay to assess the interaction of the TcRAR protein with the TcRXR isoforms suggested that TcRAR might form a heterodimer with the RXR isoforms [20] (Table 8.2). The transcriptional activity of domain-swapped TcRAR chimeric proteins (the A/B domain of TcRAR combined with the D–F domain of human RARα) was found to be ATRA-dependent [20]. These results suggest that although TcRAR is not activated by retinoic acids, it can heterodimerize with TcRXR isoforms.

Although 9cRA is a natural ligand for RXRs in vertebrates [104–107], whether the same is true for RXRs in *T. clavigera* or other gastropods is unclear because 9cRA is difficult to detect *in vivo* [108]. The natural ligand for gastropod RXR may be a compound other than 9cRA and must be identified to facilitate further analysis of the mechanism of imposex induction by organotins. In this regard, Dmetrichuk et al., [91] recently detected ATRA and 9cRA in the central nervous systems of adults of the pulmonate gastropod *L.*

stagnalis by using high-performance liquid chromatography-mass spectrometry. Because ATRA and 9cRA were detected in the tissues of *L. stagnalis*, this species likely also has metabolic enzymes for synthesizing or transforming RAs. Whether gastropod mollusks can synthesize 9cRA, ATRA, or both from β carotene must also be determined in terms of the enzymes involved in the synthesis and metabolism of RAs (e.g., Raldh2, Cyp26).

In addition to identifying the natural ligand for gastropod RXR, we must also examine the binding and activation properties of the ligand with regard to RXR and determine whether RXR forms homodimers, homotetramers, or heterodimers with other nuclear receptors. The activation of RXR–PPAR heterodimers by organotin compounds promotes adipocyte differentiation [109–111], and the binding and activation properties of various organotins with RXR–PPAR heterodimers have been analyzed by le Maire et al., [112]. Although gastropod imposex can be induced by very low concentrations (~1 ng/L) of either TBT or TPhT [87, 113–116], the mechanism through which nanomolar levels of TBT or TPhT activate gastropod RXR remains to be clarified. In this regard, Pascoal et al., [21] suggested the additional involvement of putative PPAR pathways (Table 8.2), given that the application of rosiglitazone, a well-known vertebrate PPAR γ ligand, to dog whelks (*N. lapillus*) induced imposex in the absence of TBT [21]. In contrast, TBT-induced imposex likely is linked to the induction of many genes and has a complex phenotype.

Overall, these findings suggest that RXR is involved in the induction of male-type genitalia (penis and vas deferens) in normal male and organotin-exposed female gastropods.

We should also note that there are several steps in the development of imposex induced by certain organotin compounds, such as TBT and TPhT, in gastropod mollusks. The initial stage of imposex development is the differentiation and growth (i.e., cell proliferation and the morphological formation of a structure similar to a curved penis) of male-type genitalia (i.e., penis, and vas deferens). This process can lead to ovarian spermatogenesis at the severely affected stage, through a consequence of oviduct blockage due to the proliferation of epidermal tissues surrounding the vas deferens [81, 88, 115, 117–126]. We consider that a complete and accurate explanation of the mechanism of action of TBT or TPhT in the development of imposex in gastropods will account for each of the characteristics mentioned earlier [88].

The physiological regulatory system of reproduction appears to differ between gastropod mollusks and vertebrates. Further studies involving histological, immunohistochemical, biochemical, and molecular biological techniques are needed to elucidate the basic endocrinology and complete mechanism of action of TBT or TPhT in the development of imposex in prosobranch gastropods. Achieving this goal may involve the identification of a natural ligand and target gene(s) of the gastropod RXR, as well as the determination of when and how the differentiation and proliferation of penis and vas deferens stem cells in female snails are initiated and promoted, thus leading to epidermal differentiation and proliferation and the development of these organs. Morphogenetic factors might be involved in the formation of the curved penis and vas deferens. In addition, other factors, such as various neuropeptides induced in the head ganglia due to exposure to organotins, might be associated with the RXR gene-mediated development of imposex, if these factors are induced downstream of the RXR cascade [127].

8.4 CONCLUSIONS

We have reviewed basic information on the endocrinology and reproductive physiology of mollusks, with an emphasis on gastropods, and have focused on the possibility that vertebrate-type steroids may function as molluscan sex hormones. In addition, we weighed the evidence for the presence of steroid-producing cells, enzymes to synthesize or metabolize steroids, and functional receptors for steroids in the tissues of mollusks. From this discussion, we conclude that mollusks are unlikely to inherently possess vertebrate-type steroids as sex hormones.

Regarding the mechanism by which organotins, such as TBT and TPhT, induce the development of imposex in prosobranch gastropods, six hypotheses have been proposed (Table 8.3). In support of Hypothesis 6, recent findings suggest that RXR is involved in the induction of male-type genitalia (penis and vas deferens) in normal male and organotin-exposed female gastropods.

The physiological regulatory system of reproduction likely differs between mollusks and vertebrates. Further studies involving histological, immunohistochemical, biochemical, and molecular biological techniques are needed to elucidate the basic endocrinology of mollusks and the complete mechanism of action of TBT or TPhT in the development of imposex in prosobranch gastropods.

KEYWORDS

- alanine-proline-glycine-tryptophan amide (APGWamide)
- aromatase
- enzymes to synthesize steroids
- functional receptors for steroids
- retinoid X receptor (RXR)
- sex hormones
- steroid-producing cells
- tributyltin (TBT)
- triphenyltin (TPhT)
- vertebrate-type steroids

REFERENCES

1. LeBlanc, G. A., Campbell, P. M., Den Besten, P., Brown, R. P., Chang, E. S., Coats, J. R., et al., (1999). The endocrinology of invertebrates. In: De Fur, P. L., Crane, M., Ingersoll, C., & Tattersfield, L., (eds.), *Endocrine Disruption in Invertebrates: Endocrinology, Testing, and Assessment* (pp. 23–106). Pensacola, Florida: SETAC Press.
2. Horiguchi, T., (2006). Masculinization of females caused by organotin compounds in gastropod mollusks, focusing on the mode of action of tributyltin and triphenyltin on the development of imposex. *Environ. Sci., 13*(2), 77–87.
3. Scott, A. P., (2012). Do mollusks use vertebrate sex steroids as reproductive hormones? Part I: Critical appraisal of the evidence for the presence, biosynthesis and uptake of steroids. *Steroids, 77*, 1450–1468.
4. Scott, A. P., (2013). Do mollusks use vertebrate sex steroids as reproductive hormones? I. I., Critical review of the evidence that steroids have biological effects. *Steroids, 78*, 268–281.
5. Bettin, C., Oehlmann, J., & Stroben, E., (1996). TBT-induced imposex in marine neogastropods is mediated by an increasing androgen level. *Helgoländer Meeresunter-suchungen, 50*, 299–317.
6. Gooding, M. P., Wilson, V. S., Folmar, L. C., Marcovich, D. T., & LeBlanc, G. A., (2003). The biocide tributyltin reduces the accumulation of testosterone as fatty acids esters in the mud snail (*Ilyanassa obsoleta*). *Environ Health Persp., 111*, 426–430.
7. Sternberg, R. M., & LeBlanc, G. A., (2006). Kinetic characterization of the inhibition of acyl coenzyme A: steroid acyltransferases by tributyltin in the eastern mud snail (*Ilyanassa obsoleta*). *Aquat. Toxicol. 78*, 233–242.

8. Ronis, M. J. J., & Mason, A. Z., (1996). The metabolism of testosterone by the periwinkle (*Littorina littorea*) *in vitro* and *in vivo*: Effects of tributyltin. *Marine Environmental Research, 42*, 161–166.

9. Féral, C., & Le Gall, S., (1983). The influence of a pollutant factor (TBT) on the neurosecretory mechanism responsible for the occurrence of a penis in the females of *Ocenebra erinacea*. In: Lever, J., & Boer, H. H., (eds.), *Molluscan Neuro-Endocrinology* (pp. 173–175). Amsterdam: North Holland Publishing.

10. Oberdörster, E., & McClellan-Green, P., (2000). The neuropeptide APGWamide induces *imposex* in the mud snail *Ilyanassa obsoleta*. *Peptides, 21*, 1323–1330.

11. Nishikawa, J., Mamiya, S., Kanayama, T., Nishikawa, T., Shiraishi, F., & Horiguchi, T., (2004). Involvement of the retinoid X receptor in the development of imposex caused by organotins in gastropods. *Environmental Science and Technology, 38*, 6271–6276.

12. Castro, L. F., C., Lima, D., Machado, A., Melo, C., Hiromori, Y., Nishikawa, J., Nakanishi, T., Reis-Henriques, M. A., & Santos, M. M., (2007b). Imposex induction is mediated through the Retinoid X Receptor signaling pathway in the neogastropod *Nucella lapillus*. *Aquatic Toxicology, 85*, 57–66.

13. Horiguchi, T., Nishikawa, T., Ohta, Y., Shiraishi, H., & Morita, M., (2007). Retinoid X receptor gene expression and protein content in tissues of the rock shell *Thais clavigera*. *Aquatic Toxicology, 84*, 379–388.

14. Horiguchi, T., Ohta, Y., Nishikawa, T., Shiraishi, F., Shiraishi, H., & Morita, M., (2008a). Exposure to 9-cis retinoic acid induces penis and vas deferens development in the female rock shell, *Thais clavigera*. *Cell Biology and Toxicology, 24*, 553–562.

15. Horiguchi, T., Nishikawa, T., Ohta, Y., Shiraishi, H., & Morita, M., (2010a). Time course of expression of the retinoid X receptor gene and induction of imposex in the rock shell, *Thais clavigera*, exposed to triphenyltin chloride. *Analytical and Bioanalytical Chemistry, 396*, 597–607.

16. Horiguchi, T., Urushitani, H., Ohta, Y., Iguchi, T., & Shiraishi, H., (2010b). Establishment of a polyclonal antibody against the retinoid X receptor of the rock shell *Thais clavigera* and its application to rock shell tissues for imposex research. *Ecotoxicology, 19*, 571–576.

17. Sternberg, R. M., Hotchkiss, A. K., & LeBlanc, G. A., (2008b). Synchronized expression of retinoid X receptor mRNA with reproductive tract recrudescence in an imposex-susceptible mollusc. *Environ. Sci. Technol., 42*, 1345–1351.

18. Urushitani, H., Katsu, Y., Ohta, Y., Shiraishi, H., Iguchi, T., & Horiguchi, T., (2011). Cloning and characterization of retinoid X receptor (RXR) isoforms in the rock shell, *Thais clavigera*. *Aquatic Toxicology, 103*, 101–111.

19. Urushitani, H., Katsu, Y., Kagechika, H., Sousa, A. C., A., Barroso, C. M., Ohta, Y., Shiraishi, H., Iguchi, T., & Horiguchi, T., (2018). Characterization and comparison of transcriptional activities of the retinoid X receptors by various organotin compounds in three prosobranch gastropods, *Thais clavigera, Nucalla lapillus* and *Babylonia japonica*. *Aquatic Toxicology, 199*, 103–115.

20. Urushitani, H., Katsu, Y., Ohta, Y., Shiraishi, H., Iguchi, T., & Horiguchi, T., (2013). Cloning and characterization of the retinoic acid receptor-like protein in the rock shell, *Thais clavigera*. *Aquatic Toxicology, 142, 143*, 403–413.

21. Pascoal, S., Carvalho, G., Vasieva, O., Hughes, R., Cossins, A., Fang, Y. X., Ashelford, K., Olohan, L., Barroso, C., Mendo, S., & Creer, S., (2013). Transcriptomics and *in vivo* tests reveal novel mechanisms underlying endocrine disruption in an ecological sentinel, *Nucella lapillus*. *Molecular Ecology, 22*, 1589–1608.

22. Takeda, N., (1979). Induction of egg-laying by steroid hormones in slugs. *Comp. Biochem. Physiol., 62A*, 273–278.
23. Takeda, N., (1983). Endocrine regulation of reproduction in the snail, *Euhadra peliomphala*. In: Lever, J., & Boer, H. H., (eds.), *Molluscan Neuro-Endocrinology* (pp. 106–111). Amsterdam: North Holland Publishing.
24. Santos, M. M., Castro, L. F., C., Vieira, M. N., Micael, J., Morabito, R., Massanisso, P., & Reis-Henriques, M. A., (2005). New insights into the mechanism of imposex induction in the dogwhelk *Nucella lapillus*. *Comp. Biochem. Physiol. C., 141*, 101–109.
25. Spooner, N., Gibbs, P. E., Bryan, G. W., & Goad LJ., (1991). The effect of tributyltin upon steroid titers in the female dogwhelk, *Nucella lapillus*, and the development of imposex. *Mar. Environ. Res., 32*, 37–49.
26. Sternberg, R. M., Hotchkiss, A. K., & LeBlanc, G. A., (2008a). The contribution of steroidal androgens and estrogens to reproductive maturation of the eastern mud snail *Ilyanassa obsoleta*. *Gen. Comp. Endocr., 156*, 15–26.
27. LeBlanc, G. A., Gooding, M. P., & Sternberg, R. M., (2005). Testosterone-fatty acid esterification: A unique target for the endocrine toxicity of tributyltin to gastropods. *Integr. Comp. Biol., 45*, 81–87.
28. Gooding, M. P., & LeBlanc, G. A., (2004). Seasonal variation in the regulation of testosterone levels in the eastern mud snail, *Ilyanassa obsoleta*. *Invertebr. Biol., 123*, 237–243.
29. Janer, G., Lyssimachou, A., Bachmann, J., Oehlmann, J., Schulte-Oehlmann, U., & Porte, C., (2006). Sexual dimorphism in esterified steroid levels in the gastropod *Marisa cornuarietis*: The effect of xenoandrogenic compounds. *Steroids, 71*, 435–444.
30. Le Guellec, D., Thiard, M. C., Remy-Martin, J. P., Deray, A., Gomot, L., & Adessi, G. L., (1987). *In vitro* metabolism of androstenedione and identification of endogenous steroids in *Helix aspersa*. *Gen. Comp. Endocr., 66*, 425–433.
31. Bose, R., Majumdar, C., & Bhattacharya, S., (1997). Steroids in *Achatina fulica* (Bowdich): steroid profile in haemolymph and *in vitro* release of steroids from endogenous precursors by ovotestis and albumen gland. *Comp. Biochem. Physiol. C., 116*, 179–182.
32. Morcillo, Y., & Porte, C., (1999). Evidence of endocrine disruption in the imposex-affected gastropod *Bolinus brandaris*. *Environ. Res. A., 81*, 349–354.
33. Lu, M., Horiguchi, T., Shiraishi, H., Shibata, Y., Abo, M., Okubo, A., & Yamazaki, S., (2001). Identification and quantitation of steroid hormones in marine gastropods by GC/M.S. *Bunseki Kagaku, 50*, 247–255 [in Japanese].
34. Lu, M., Horiguchi, T., Shiraishi, H., Shibata, Y., Abo, M., Okubo, A., & Yamazaki, S., (2002). Determination of testosterone in an individual shell of *Thais clavigera* by ELISA. *Bunseki Kagaku, 51*, 21–27 [in Japanese].
35. Morcillo, Y., & Porte, C., (2000). Evidence of endocrine disruption in clams – *Ruditapes decussata*–transplanted to a tributyltin-polluted environment. *Environ. Pollut., 107*, 47–52.
36. Morcillo, Y., Ronis, M. J., & Porte, C., (1998). Effects of tributyltin on the phase I testosterone metabolism and steroid titres of the clam *Ruditapes decussata*. *Aquat. Toxicol., 42*, 1–13.
37. Ketata, I., Guermazi, F., Rebai, T., & Hamza-Chaffai, A., (2007). Variation of steroid concentrations during the reproductive cycle of the clam *Ruditapes decussatus*: A one year study in the gulf of Gabès area. *Comp. Biochem. Physiol. A., 147*, 424–431.

38. Negrato, E., Marin, M. G., Bertotto, D., Matozzo, V., Poltronieri, C., & Simontacchi, C., (2008). Sex steroids in *Tapes philippinarum* (Adams and Reeve., 1850) during gametogenic cycle: preliminary results. *Fresen. Environ. Bull., 17*, 1466–1470.

39. Osada, M., Tawarayama, H., & Mori, K., (2004). Estrogen synthesis in relation to gonadal development of Japanese scallop, *Patinopecten yessoensis*: Gonadal profile and immunolocalization of P450 aromatase and estrogen. *Comp. Biochem. Physiol. B., 139*, 123–128.

40. Matsumoto, T., Osada, M., Osawa, Y., & Mori, K., (1997). Gonadal estrogen profile and immunohistochemical localization of steroidogenic enzymes in the oyster and scallop during sexual maturation. *Comp. Biochem. Physiol. B., 118*, 811–817.

41. De Longcamp, D., Lubet, P., & Drosdowsky, M., (1974). The *in vitro* biosynthesis of steroids by the gonad of the mussel (*Mytilus edulis*). *Gen. Comp. Endocr., 22*, 116–127.

42. Reis-Henriques, M. A., & Coimbra, J., (1990). Variations in the levels of progesterone in *Mytilus edulis* during the annual reproductive cycle. *Comp. Biochem. Physiol. A., 95*, 343–348.

43. Reis-Henriques, M. A., Le Guellec, D., Remy-Martin, J. P., & Adessi, G. L., (1990). Studies of endogenous steroids from the marine mollusc *Mytilus edulis, L.* by gas chromatography and mass spectrometry. *Comp. Biochem. Physiol. B., 95*, 303–309.

44. Lavado, R., Janer, G., & Porte, C., (2006). Steroid levels and steroid metabolism in the mussel *Mytilus edulis*: The modulating effect of dispersed crude oil and alkylphenols. *Aquat. Toxicol., 78S*, 565–572.

45. Fernandes, D., Navarro, J. C., Riva, C., Bordonali, S., & Porte, C., (2010). Does exposure to testosterone significantly alter endogenous metabolism in the marine mussel *Mytilus galloprovincialis. Aquat. Toxicol., 100*, 313–320.

46. Janer, G., Lavado, R., Thibaut, R., & Porte, C., (2005). Effects of 17b-estradiol exposure in the mussel *Mytilus galloprovincialis*: a possible regulating role for steroid acyltransferases. *Aquat. Toxicol., 75*, 32–42.

47. Kaloyianni, M., Stamatiou, R., & Dailianis, S., (2005). Zinc and 17b-estradiol induce modification in Na^+/H^+ exchanger and pyruvate kinase activity through protein kinase C in isolated mantle/gonad cells of *Mytilus galloprovincialis. Comp. Biochem. Physiol. C., 141*, 257–266.

48. Dévier, M. H., Labadie, P., Togola, A., & Budzinski, H., (2010). Simple methodology coupling microwave-assisted extraction to SPE/GC/MS for the analysis of natural steroids in biological tissues: application to the monitoring of endogenous steroids in marine mussels *Mytilus* sp. *Anal Chim. Acta., 657*, 28–35.

49. Mouneyrac, C., Linot, S., Amiard, J. C., Amiard-Triquet, C., Métais, I., Durou, C., Minierc, C., & Pellerin, J., (2008). Biological indices, energy reserves, steroid hormones and sexual maturity in the infaunal bivalve *Scrobicularia plana* from three sites differing by their level of contamination. *Gen. Comp. Endocr., 157*, 133–141.

50. Gauthier-Clerc, S., Pellerin, J., & Amiard, J. C., (2006). Estradiol-17b and testosterone concentrations in male and female *Mya arenaria* (Mollusca bivalvia) during the reproductive cycle. *Gen. Comp. Endocr., 145*, 133–139.

51. Siah, A., Pellerin, J., Benosman, A., Gagné J. P., & Amiard, J. C., (2002). Seasonal gonad progesterone pattern in the soft-shell clam *Mya arenaria. Comp. Biochem. Physiol. A., 132*, 499–511.

52. Siah, A., Pellerin, J., Amiard, J. C., Pelletier, E., & Viglino, L., (2003). Delayed gametogenesis and progesterone levels in soft-shell clams (*Mya arenaria*) in relation to in situ contamination to organotins and heavy metals in the St. Lawrence River (Canada). *Comp. Biochem. Physiol. C., 135*, 145–156.

53. Lazzara, R., Blázquez, M., Porte, C., & Barata, C., (2012). Low environmental levels of fluoxetine induce spawning and changes in endogenous levels in the zebra mussel *Dreissena polymorpha*. *Aquat. Toxicol., 106, 107*, 123–130.

54. Riva, C., Porte, C., Binelli, A., & Provini, A., (2010). Evaluation of 4-nonylphenol *in vivo* exposure in *Dreissena polymorpha*: Bioaccumulation, steroid levels and oxidative stress. *Comp. Biochem. Physiol. C., 152*, 175–181.

55. Carreau, S., & Drosdowsky, M., (1977). The *in vitro* biosynthesis of steroids by the gonad of the cuttlefish (*Sepia officinalis*). *Gen. Comp. Endocr., 33*, 554–565.

56. D'Aniello, A., Di Cosmo, A., Di Cristo, C., Assisi, L., Botte, V., & Di Fiore, M. M., (1996). Occurrence of sex steroid hormones and their binding proteins in *Octopus vulgaris* Lam. *Biochem. Biophys. Res. Commun., 227*, 782–788.

57. Di Cosmo, A., Di Cristo, C., & Paolucci, M., (2001). Sex steroid hormone fluctuations and morphological changes of the reproductive system of the female of *Octopus vulgaris* throughout the annual cycle. *J. Exp. Zool., 289*, 33–47.

58. Gooding, M. P., & LeBlanc, G. A., (2001). Biotransformation and disposition of testosterone in the eastern mud snail *Ilyanassa obsoleta*. *Gen. Comp. Endocrinol., 122*, 172–180.

59. Santos, M. M., Ten Hallers-Tjabbes, C. C., Vieira, N., Boon, J. P., & Porte, C., (2002). Cytochrome P450 differences in normal and imposex-affected female whelk *Buccinum undatum* from the open North Sea. *Mar. Environ. Res., 54*, 661–665.

60. Kajiwara, M., Kuraku, S., Kurokawa, T., Kato, K., Toda, S., Hirose, H., et al., (2006). Tissue preferential expression of estrogen receptor gene in the marine snail, *Thais clavigera*. *Gen Comp Endocrinol., 148*, 315–326.

61. Castro, L. F. C., Melo, C., Guillot, R., Mendes, I., Queirós, S., Lima, D., Reis-Henriques, M. A., & Santos, M. M., (2007a). The estrogen receptor of the gastropod *Nucella lapillus*: Modulation following exposure to an estrogenic effluent? *Aquat. Toxicol., 84*, 465–468.

62. Bannister, R., Beresford, N., May, D., Routledge, E. J., Jobling, S., & Rand-Weaver, M., (2007). Novel estrogen receptor-related transcripts in *Marisa cornuarietis*: A freshwater snail with reported sensitivity to estrogenic chemicals. *Environ. Sci. Technol., 41*, 2643–2650.

63. Stange, D., Sieratowicz, A., Horres, R., & Oehlmann, J., (2012). Freshwater mud snail (*Potamopyrgus antipodarum*) estrogen receptor: Identification and expression analysis under exposure to (xeno-)hormones. *Ecotoxicology and Environmental Safety, 75*, 94–101.

64. Bouton, D., Escriva, H., De Mendonca, R. L., Glineur, C., Bertin, B., Noël, C., Robinson-Rechavi, M., De Groot, A., Cornette, J., Laudet, V., & Pierce, R. J., (2005). A conserved retinoid X receptor (RXR) from the mollusk *Biomphalaria glabrata* transactivates transcription in the presence of retinoids. *J. Mol. Endocrinol., 34*, 567–582.

65. Albalat, R., & Cañestro, C., (2009). Identification of Aldh1a, Cyp26 and RAR orthologs in protostomes pushes back the retinoic acid genetic machinery in evolutionary time to the bilaterian ancestor. *Chem. Biol. Interact., 178*, 188–196.

66. Thornton, J. W., Need, E., & Crews, D., (2003). Resurrecting the ancestral steroid receptor: ancient origin of estrogen signaling. *Science, 301*, 1714–1717.

67. Lv, J., Feng, L., Bao, Z., Guo, H., Zhang, Y., Jiao, W., Zhang, L., Wang, S., He, Y., & Hu, X., (2013). Molecular characterization of RXR (Retinoid X Receptor) gene isoforms from the bivalve species *Chlamys farreri*. *PLoS One, 8*, e74290. doi: 10.1371/journal.pone.0074290.

68. Puinean, A. M., Labadie, P., Hill, E. M., Osada, M., Kishida, M., Nakao, R., Novillo, A., Callard, I. P., & Rotchell, J. M., (2006). Laboratory exposure to 17b-estradiol fails to induce vitellogenin and estrogen receptor gene expression in the marine invertebrate *Mytilus edulis*. *Aquat. Toxicol., 79*, 376–383.

69. Matsumoto, T., Nakamura, A. M., Mori, K., Akiyama, I., Hirose, H., & Takahashi, Y., (2007). Oyster estrogen receptor: cDNA cloning and immunolocalization. *Gen. Comp. Endocr., 151*, 195–201.

70. Croll, R. P., & Wang, C., (2007). Possible roles of sex steroids in the control of reproduction in bivalve molluscs. *Aquaculture, 272*, 76–86.

71. Keay, J., Bridgham, J. T., & Thornton, J. W., (2006). The *Octopus vulgaris* estrogen receptor is a constitutive transcriptional activator: Evolutionary and functional implications. *Endocrinology, 147*, 3861–3869.

72. Iguchi, T., Katsu, Y., Horiguchi, T., Watanabe, H., Blumberg, B., & Ohta, Y., (2007). Endocrine disrupting organotin compounds are potent inducers of imposex in gastropods and adipogenesis in vertebrates. *Mol. Cell Toxicol., 3*, 1–10.

73. Escriva, H., Safi, R., Hanni, C., Langlois, M. C., Saumitou-Laprade, P., Stehelin, D., Capron, A., Pierce, R., & Laudet, V., (1997). Ligand binding was acquired during evolution of nuclear receptors. *Proc. Natl. Acad. Sci. USA, 94*, 6803–6808.

74. Vogeler, S., Galloway, T. S., Lyons, B. P., & Bean, T. P., (2014). The nuclear receptor gene family in the Pacific oyster, *Crassostrea gigas*, contains a novel subfamily group. *BMC Genomics, 15*, 369 doi: 10–1186/1471–2164–15–369. The electronic version of this article is the complete one and can be found online at: http://www.biomedcentral. com/1471–2164/15/369 (Accessed on 23 August 2019).

75. Nordberg, H., Cantor, M., Dusheyko, S., Hua, S., Poliakov, A., Shabalov, I., Smirnova, T., Grigoriev, I. V., & Dubchak, I., (2014). The genome portal of the Department of Energy Joint Genome Institute: 2014 updates. *Nucleic Acids Res., 42*(1), 26–31.

76. Kaur, S., Jobling, S., Jones, C. S., Noble, L. R., Routledge, E. J., & Lockyer, A. E., (2015). The nuclear receptors of *Biomphalaria glabrata* and *Lottia gigantea*: Implications for developing new model organisms. *PLoS One, 10*(4), e0121259. Published online, 2015-April-7. doi: 10.1371/journal.pone.0121259.

77. Simakov, O., Marletaz, F., Cho, S. J., et al., (2013). Insights into bilaterian evolution from three spiralian genomes. *Nature, 493*, 526–531.

78. Matthiessen, P., & Gibbs, P. E., (1998). Critical appraisal of the evidence for tributyltin-mediated endocrine disruption in mollusks. *Environ. Toxicol. Chem., 17*, 37–43.

79. Chiu, A. Y., Hunkapiller, M. W., Heller, E., Stuart, D. K., Hood, L. E., & Strumwasser, F., (1979). Purification and primary structure of neuropeptide egg-laying hormone of *Aplysia californica*. *Proc. Natl. Acad. Sci. USA, 76*, 6656–6660.

80. Ebberink, R. H. M., Loenhout, H., Van Geraerts, W. P. M., & Joosse, J., (1985). Purification and amino acid sequence of the ovulation neurohormone of *Lymnaea stagnalis*. *Proc. Natl. Acad. Sci. USA, 82*, 7767–7771.

81. Horiguchi, T., Kojima, M., Kaya, M., Matsuo, T., Shiraishi, H., Morita, M., & Adachi, Y., (2002). Tributyltin and triphenyltin induce spermatogenesis in ovary of female abalone, *Haliotis gigantea*. *Mar. Environ. Res., 54*, 679–684.

82. Bryan, G. W., Bright, D. A., Hummerstone, L. G., & Burt, G. R., (1993). Uptake, tissue distribution and metabolism of ^{14}C-labelled tributyltin (TBT) in the dog-whelk, *Nucella lapillus*. *J. Mar. Biol. Assoc. UK, 73*, 889–912.

83. Horiguchi, T., Lee, J. H., Park, J. C., Cho, H. S., Shiraishi, H., & Morita, M., (2012). Specific accumulation of organotin compounds in tissues of the rock shell, *Thais clavigera*. *Mar. Environ. Res., 76*, 56–62.

84. Oberdörster, E., & McClellan-Green, P., (2002). Mechanism of imposex induction in mud snail, *Ilyanassa obsoleta*: TBT as a neurotoxin and aromatase inhibitor. *Mar. Environ. Res., 54*, 715–718.

85. Oehlmann, J., Di Benedetto, P., Tillmann, M., Duft, M., Oetken, M., & Schulte-Oehlmann, U., (2007). Endocrine disruption in prosobranch molluscs: Evidence and ecological relevance. *Ecotoxicology, 16*, 29–43.

86. Sternberg, R. M., Gooding, M. P., Hotchkiss, A. K., & LeBlanc, G. A., (2010). Environmental-endocrine control of reproductive maturation in gastropods: Implications for the mechanism of tributyltin-induced imposex in prosobranchs. *Ecotoxicology, 19*, 4–23.

87. Horiguchi, T., Shiraishi, H., Shimizu, M., & Morita, M., (1997). Effects of triphenyltin chloride and five other organotin compounds on the development of imposex in the rock shell, *Thais clavigera. Environ. Pollut., 95*, 85–91.

88. Horiguchi, T., (2000). Molluscs. In: Kawai, S., & Koyama, J., (eds.), *Problems of Endocrine Disruptors in Fisheries Environment* (pp. 54–72). Tokyo: Koseisha-Koseikaku [in Japanese].

89. Ulven, S. M., Gundersen, T. E., Sakhi, A. K., Glover, J. C., & Blomhoff, R., (2001). Quantitative axial profiles of retinoic acid in the embryonic mouse spinal cord: 9-*cis* retinoic acid only detected after all-trans-retinoic acid levels are super-elevated experimentally. *Dev. Dyn., 222*, 341–353.

90. Werner, E. A., & DeLuca, H. F., (2001). Metabolism of a physiological amount of all-trans-retinol in the vitamin A-deficient rat. *Arch. Biochem. Biophys., 393*, 262–270.

91. Dmetrichuk, J. M., Carlone, R. L., Jones, T. R. B., Vesprini, N. D., & Spencer, G. E., (2008). Detection of endogenous retinoids in the molluscan CNS and characterization of the trophic and tropic actions of 9-cis retinoic acid on isolated neurons. *J. Neurosci., 28*, 13014–13024.

92. Hopkins, P. M., Durica, D., & Washington, T., (2008). RXR isoforms and endogenous retinoids in the fiddler crab, *Uca pugilator. Comp. Biochem. Physiol. A Mol. Integr. Physiol., 151*, 602–614.

93. De Urquiza, A. M., Liu, S., Sjöberg, M., Zetterström, R. H., Griffiths, W., Sjövall, J., & Perlmann, T., (2000). Docosahexaenoic acid, a ligand for the retinoid X receptor in mouse brain. *Science, 290*, 2140–2144.

94. Wang, Y. H., & LeBlanc, G. A., (2009). Interactions of methyl farnesoate and related compounds with a crustacean retinoid X receptor. *Mol. Cell Endocrinol., 309*, 109–116.

95. Galante-Oliveira, S., Oliveira, I., Pacheco, M., & Barroso, C. M., (2010). *Hydrobia ulvae* imposex levels at Ria de Aveiro (NW Portugal) between 1998 and 2007: A counter-current bioindicator? *J. Environ. Monit., 12*, 500–507.

96. Horiguchi, T., Nishikawa, T., Ohta, Y., & Shiraishi, H., (2008b). Monthly changes of RXR gene expression and sexual characteristics in male rock shells (*Thais clavigera*) in Hiraiso, Japan. *SETAC Europe 18th Annual Meeting, Warsaw SETAC Europe Office, Brussels*, p. 150.

97. Oberdörster, E., Romano, J., & McClellan-Green, P., (2005). The neuropeptide APGWamide as a penis morphogenic factor (PMF) in gastropod mollusks. *Integr. Comp. Biol., 45*, 28–32.

98. Albalat, R., (2009). The retinoic acid machinery in invertebrates: Ancestral elements and vertebrate innovations. *Mol. Cell. Endocrinol., 313*, 23–35.

99. Chambon, P., (1996). A decade of molecular biology of retinoic acid receptors. *FASEB, J., 10*, 940–954.

100. De Luca, L. M., (1991). Retinoids and their receptors in differentiation, embryogenesis, and neoplasia. *FASEB, J., 5*, 2924–2933.

101. Kastner, P., & Chan, S., (2001). Function of RARalpha during the maturation of neutrophils. *Oncogene, 20*, 7178–7185.

102. Kastner, P., Mark, M., & Chambon, P., (1995). Nonsteroid nuclear receptors: What are genetic studies telling us about their role in real life? *Cell, 83*, 859–869.

103. Mark, M., Ghyselinck, N. B., & Chambon, P., (2009). Function of retinoic acid receptors during embryonic development. *Nucl. Recept. Signal., 7*, e002.

104. Heyman, R. A., Mangelsdorf, D. J., Dyck, J. A., Stein, R. B., Eichele, G., Evans, R. M., & Thaller, C., (1992). 9-cis retinoic acid is a high affinity ligand for the retinoid X receptor. *Cell, 68*, 397–406.

105. Mangelsdorf, D. J., & Evans, R. M., (1995). The RXR heterodimers and orphan receptors. *Cell, 83*, 841–850.

106. Mangelsdorf, D. J., Borgmeyer, U., Heyman, R. A., Zhou, J. Y., Ong, E. S., Oro, A. E., Kakizuka, A., & Evans, R. M., (1992). Characterization of three RXR genes that mediate the action of 9-cis retinoic acid. *Genes Dev., 6*, 329–344.

107. Levin, A. A., Sturzenbecker, L. J., Kazmer, S., Bosakowski, T., Huselton, C., Allenby, G., Speck, J., Kratzeisen, C., Rosenberger, M., & Lovey, A., (1992). 9-*cis* retinoic acid stereoisomer binds and activates the nuclear receptor RXR alpha. *Nature, 355*, 359–361.

108. Horton, C., & Maden, M., (1995). Endogenous distribution of retinoids during normal development and teratogenesis in the mouse embryo. *Dev. Dyn., 202*, 312–323.

109. Grün, F., & Blumberg, B., (2006). Environmental obesogens: Organotins and endocrine disruption via nuclear receptor signaling. *Endocrinology, 147*, S50–S55.

110. Grün, F., Watanabe, H., Zamanian, Z., Maeda, L., Arima, K., Cubacha, R., Gardiner, D. M., Kanno, J., Iguchi, T., & Blumberg, B., (2006). Endocrine-disrupting organotin compounds are potent inducers of adipogenesis in vertebrates. *Mol. Endocrinol., 20*, 2141–2155.

111. Kanayama, T., Kobayashi, N., Mamiya, S., Nakanishi, T., & Nishikawa, J., (2005). Organotin compounds promote adipocyte differentiation as agonists of the peroxisome proliferators-activated receptor γ/retinoid X receptor pathway. *Mol. Pharmacol., 67*, 766–774.

112. Le Maire, A., Grimaldi, M., Roecklin, D., Dagnino, S., Vivat-Hannah, V., Balaguer, P., & Bourguet, W., (2009). Activation of RXR-PPAR heterodimers by organotin environmental endocrine disruptors. *EMBO Rep., 10*, 367–373.

113. Bryan, G. W., Gibbs, P. E., & Burt, G. R., (1988). A comparison of the effectiveness of tri-*n*-butyltin chloride and five other organotin compounds in promoting the development of imposex in the dog-whelk, *Nucella lapillus*. *J. Mar. Biol. Assoc. U.K., 68*, 733–744.

114. Gibbs, P. E., Bryan, G. W., Pascoe, P. L., & Burt, G. R., (1987). The use of the dog-whelk, *Nucella lapillus*, as an indicator of tributyltin (TBT) contamination. *J. Mar. Biol. Assoc. U.K., 67*, 507–523.

115. Horiguchi, T., Shiraishi, H., Shimizu, M., et al., (1994). Imposex and organotin compounds in *Thais clavigera and, T. bonni* in Japan. *J. Mar. Biol. Assoc. U.K., 74*, 651–669.

116. Horiguchi, T., Shiraishi, H., Shimizu, M., Yamazaki, S., & Morita, M., (1995). Imposex in Japanese gastropods (Neogastropoda and Mesogastropoda): Effects of tributyltin and triphenyltin from antifouling paints. *Mar. Pollut. Bull., 31*, 402–405.

117. Gibbs, P. E., & Bryan, G. W., (1986). Reproductive failure in populations of the dog-whelk, *Nucella lapillus*, caused by imposex induced by tributyltin from antifouling paints. *J. Mar. Biol. Assoc. U.K., 66*, 767–777.

118. Gibbs, P. E., Pascoe, P. L., & Burt, G. R., (1988). Sex change in the female dog-whelk, *Nucella lapillus*, induced by tributyltin from antifouling paints. *J. Mar. Biol. Assoc. U.K., 68*, 715–731.

119. Gibbs, P. E., Bryan, G. W., Pascoe, P. L., & Burt, G. R., (1990). Reproductive abnormalities in female *Ocenebra erinacea* (Gastropoda) resulting from tributyltin-induced imposex. *J. Mar. Biol. Assoc. U.K., 70*, 639–656.

120. Gibbs, P. E., Spencer, B. E., & Pascoe, P. L., (1991). The American oyster drill, *Urosalpinx cinerea* (Gastropoda): Evidence of decline in an imposex affected population (R. Blackwater, Essex). *J. Mar. Biol. Assoc. U.K., 71*, 827–838.

121. Horiguchi, T., & Shimizu, M., (1992). Effects on aquatic organisms, mainly on molluscs. In: Satomi, Y., & Shimizu, M., (eds.), *Organotin Pollution and its Effects on Aquatic Organisms* (pp. 99–135). Tokyo: Koseisha-Koseikaku [in Japanese].

122. Horiguchi, T., Takiguchi, N., Cho, H. S., Kojima, M., Kaya, M., Shiraishi, H., Morita, M., Hirose, H., & Shimizu, M., (2000). Ovo-testis and disturbed reproductive cycle in the giant abalone, *Haliotis madaka*: Possible linkage with organotin contamination in a site of population decline. *Mar. Environ. Res., 50*, 223–229.

123. Horiguchi, T., Kojima, M., Takiguchi, N., Kaya, M., Shiraishi, H., & Morita, M., (2005). Continuing observation of disturbed reproductive cycle and ovarian spermatogenesis in the giant abalone, *Haliotis madaka* from an organotin-contaminated site of Japan. *Mar. Pollut. Bull., 51*, 817–822.

124. Horiguchi, T., Kojima, M., Hamada, F., Kajikawa, A., Shiraishi, H., Morita, M., & Shimizu, M., (2006). Impact of tributyltin and triphenyltin on ivory shell (*Babylonia japonica*) populations. *Environ Health Perspect., 114 Supplement* (1), 13–19.

125. Oehlmann, J., Fioroni, P., Stroben, E., & Markert, B., (1996). Tributyltin (TBT) effects on *Ocinebrina aciculata* (Gastropoda: Muricidae): Imposex development, sterilization, sex change and population decline. *Sci. Total Environ., 188*, 205–223.

126. Schulte-Oehlmann, U., Oehlmann, J., Fioroni, P., & Bauer, B., (1997). Imposex and reproductive failure in *Hydrobia ulvae* (Gastropoda: Prosobranchia). *Mar. Biol., 128*, 257–266.

127. Morishita, F., Minakata, H., Takeshige, K., Furukawa, Y., Takata, T., Matsushima, O., Mukai, S. T., Saleuddin, A. S., M., & Horiguchi, T., (2006). Novel excitatory neuropeptides isolated from a prosobranch gastropod, *Thais clavigera*: The molluscan counterpart of the annelidan GGNG peptides. *Peptides, 27*, 483–492.

CHAPTER 9

Hormones May Shape Sexual Behavior in Cephalopods

ANNA DI COSMO,[1] MARINA PAOLUCCI,[2] and VALERIA MASELLI[1]

[1]*Department of Biology, University of Naples Federico II, Italy*

[2]*Department of Sciences and Technologies, University of Sannio, Italy*

9.1 INTRODUCTION

Cephalopod is a mollusk class that separated from their ancestors, probably about 200 million years ago [1]. Cephalopods include chambered nautiluses, with external shells and anatomy that has remained virtually unchanged for 450 million years, and soft-bodied coleoids: Decapod and Octopoda [1, 2].

Coleoids are agile predators, equipped with highly efficient and flexible arms capable of a wide range of movements with no skeletal support, allowing them to face and find solutions to different environmental challenges [3–7]. These abilities are essential in exploring a new environment, problem-solving, and play-like [5, 8, 9]. They are good swimmers and are able to move rapidly by jet propulsion when threatened, and they are also able to mask themselves producing a large number of body patterns, dramatically change color with breathtaking rapidity producing chromatically matched camouflage [10] as they escape ejecting ink [11]. Not surprisingly, the coleoids' sophisticated behaviors are mediated by highly developed sensory organs and highly developed centralized brain.

Modern cephalopods colonized all oceans of the world except for the Black and Caspian seas, spreading from the surface waters down into the deep sea, and occupying a wide range of ecological niches [12].

Cephalopods' behavior and neurobiology have been extensively studied on a handful of the more than 300 species of cephalopods worldwide, focusing researches on few animal models as *Loligo vulgaris*, *Sepia officinalis*, and *Octopus vulgaris* [13].

They have been known for well-developed eyes, complex visual behavior and excellent color discrimination even with a single photoreceptor type [9, 14–17], for highly developed vestibular system, a 'lateral line analogue,' for a primitive 'hearing' system [18–21] and a discrete olfactory organ [22–28].

They possess chemoreceptors in the epidermis [29, 30] including numerous isolated sensory neurons all over the body surface [31–37] and mostly in the hundreds of suckers of octopods as well as, but in less concentration, on squid and cuttlefish suckers, lips, and mouth [38–41].

Their central nervous system (CNS), although functionally comparable to the vertebrate brain, possesses properties of invertebrate ganglia organization.

They have the largest nervous systems of any invertebrate, and the brain is much bigger in relationship to their bodies than those of fish. The doughnut-shaped brain is formed of numerous lobes, arranged around the esophagus and enclosed in a cartilaginous cranium. On either side of the brain there are two large optic lobes twice as big as the central mass [42]. In particular, in *O. vulgaris*, the peripheral nervous system of the arms and body is composed by about 5×10^7 neurons distributed along each arm, of which 5×10^5 are motor neurons [43]. The nervous system of the arms can function autonomously to some extent, as it can generate coordinated stereotypical movements [44–47]. Giant axons of cephalopods are important for the rapid activation of the mantle muscle during the jet-propelled escape, achieving extremely fast conductance of action potentials [48]. The brain also processes an immense quantity of sensory information gathered by several million tactile and chemical sensory cells in the skin and in hundreds of suckers along the arms [49, 50].

The complex nervous system allows cephalopods to display discriminative [51], observational [52], associative learning [53], and imprinting [54]. These learning abilities are associated with long-term memory [53, 55–57] and spatial memory [58].

Most cephalopods grow and reproduce quickly, mate once and die; most of them live no more than 18 months. The duration of the embryonic development is related to the temperature and the size of the eggs [6, 7].

Classically, the Cephalopods' reproductive behavior has never been described from an integrative neuroethological perspective, rather, more attention has been given to the description of different behaviors without considering the circuits and the molecules underlying them.

Here we propose a new vision considering proximate causes of reproductive behavior, focusing on *O. vulgaris*, highlighting the circuits and signal molecules that control the behavior in evaluating mates, stimulate or deter copulation, and promote fertilization, through sperm-egg chemical signaling.

9.2 CEPHALOPODS' SOCIAL AND SEXUAL BEHAVIOR

The generalization that cephalopods are solitary outside of the reproductive period would be true, with small exceptions and some variation [59–61].

Outside of the reproductive period, cuttlefish may be solitary, although field data [62] are fragmentary, and reproductive social tactics may have been selective forces for shaping cuttlefish intelligence [63]. A size-based dominance hierarchy is true for older, but not for newly hatched cuttlefish [64], however, cuttlefish react at the same way versus familiar or unfamiliar conspecifics [65]. To avoid predation and competition, many octopuses select shelters where they remain most of the time, particularly during daylight [11].

The behavior of cephalopods can reveal basic adaptations to the necessities of life. The first necessity is adapting to the physical, chemical, and biological environment around them, including predators; the second necessity is the provision of food; a third necessity is a relationship with conspecifics [65], that can bring problems for confinement itself, such as cannibalism [66]. A fourth necessity is the provision of opportunities for reproduction, from the stage of mating through guarding of eggs in octopuses to the proper environment for planktonic paralarvae. In fact, sexual behavior is an important aspect of cephalopods social behavior. A semelparous life history, parental care of eggs only in the octopuses (and a few squids) and no overlap of generations restrict the opportunity for social behavior.

9.3 SQUID

Squid is different from Octopus, in that they prefer to swim with conspecifics [67], in approximately parallel orientation and within a body length or two in captivity [68]. Because the adults do not live after egg laying and the young are planktonic, there is little likelihood of kin recognition [9].

Maturation in females produces a huge ovary that can, in fact, exceed the total body weight of males [69, 70]. As the ovary develops, a concomitant and dramatic tissue breakdown in the females is observed. This is particularly evident in the mantle and tentacles. The process of tissue breakdown with maturity is more dramatic in females than males (except for the testis). Both male and female individuals of *M. ingens* share the marked regression of the gonads. In the case of females, this is due to egg-release, while in males it represents the cessation of sperm production [70].

Males attract females swimming rapidly in very large circles [71, 72]. The females join in this swim in circles which initially can seem very strange and out of control, but after a while squids start to swim in pairs. These are the couples that have linked up for mating to take place. The male changes color, which is what attracts the females to them. Once the female shows interest, the mating takes place [73]. The male can be very aggressive during the courtship. During the mating process, the sperm is placed using a dedicated arm, inside the female's mantle cavity or around her mouth, where the eggs are waiting [9, 74, 75].

Thousands of eggs can be produced at a time by one female, up to 70,000, which she stores in her ovary. She distributes them in hidden areas of the water, including under rocks or in various holes and crevices. The squid females do not wait around for them to hatch, they leave after laying, making difficult for them to be safe from predators.

After four to eight weeks, baby squids hatch and they fend for themselves from the very start. They already know how to swim. Yet, a large part of the offspring is killed during the first few days of life. For many species of squid, the parents die soon after reproduction, due to the very short life span of these creatures.

9.4 CUTTLEFISHES

Even cuttlefishes "live fast and die young." They have a short lifespan, exceptionally high metabolic rate, an expressive behavior, and active sex life. There are about 100 species of cuttlefish worldwide, although *S. officinalis,* the European cuttlefish, is the most studied. They possess eight arms and two prehensile tentacles. Males impress the females with their display of colors using their bioluminescence. Male fighting is a rule, due to the unfavorable male:female sex ratio of 2:1. Cuttlefish are gonochoristic [76, 77–79] and reproduce sexually through head-to-head mating: the male and the female line up head to head and the male transfers spermatophores by means of the hectocotylus. The sperm competition is a central feature of their mating systems [9, 80] because the male is able to wash away any sperm from a previous male mating if she hasn't fertilized the eggs yet [81, 82]. The male deposits spermatophores on the ventral portion of the female's buccal membrane, where the paired seminal receptacle lies, but many sperm masses (spermatangia) ejected from the spermatophores are retained around the female's buccal area without being accommodated in the seminal receptacle. All ovulated eggs are most likely to be fertilized around the buccal area after having been transferred into the female's arms through her funnel [83]. Before ejaculation, males repeatedly flush strong jets of water through their funnel towards the female's buccal area [80, 84]. The female stores the eggs in a sac/pouch just below the mouth. After the transfer, the female goes to "lay" the eggs and is able to choose which sperm she wants to fertilize the eggs with. Cuttlefish reproduce in the spring, and each female lays several hundred of eggs over a period of a few days. After mating, the process of senescence is extraordinarily fast, and adults often die within days. Shortly after the female lays eggs, she dies, implying that reproduction only occurs once during the life cycle.

The ink-filled eggs are attached to seaweed in bunches, since the black pigment protects the eggs, which are rarely subject to predation [85–87]. Dead eggs are removed, as soon as there may be signs of fungi or algae infection [88–90].

The newly hatched cuttlefish resembles miniature adults and complete their life cycle in a year [9, 91].

Embryos visually and chemically experience their environment before they come into the world because egg capsule is translucent, allowing the embryos to discriminate the prey, imprint the species they see, and preferentially select them when they start hunting [54]. At hatching, juveniles quickly learn which prey to catch and which to avoid. Hatching occurs two months after the eggs are laid, depending on water temperature and the offspring hatched as miniature adults (6-9 mm), grows rapidly. Cattlefish reach the sexual maturity after 18–22 months, and their life spans between 12–24 months, depending on environmental conditions [9].

9.5 OCTOPUSES

Octopuses are mollusks that have completely lost their shell, although one species, the Argonaut, create a paper-thin replica of the nautilus shell from one of its arms, and uses it as the receptacle for its eggs [92]. Octopus is considered a solitary animal, although the high density of some species has been recently reported (i.e., the larger Pacific striped *O. gregarious*, has been seen in groups of up to 40 off the Pacific coasts of Nicaragua and Panama) [60, 61, 93]. Octopuses generally mate once and die soon after that. Many species live a little over a year, rarely living more than three years. During mating, the male approaches the female, who fends him off for a while, but then accepts him. He sits next to her or mounts her, inserting the hectocotylus in her mantle cavity to pass the spermatophores, so they may copulate for several hours. The same pair often repeats mating over a period of a week or so, but both sexes mate promiscuously. Mating often occurs when the females are immature and only females ready to lay eggs consistently fend off the males. The females, which can mature later in life, can receive and store sperm for the majority of their lives so that fertilization and egg-laying are temporally independent of mating [39, 94]. Recently it was discovered that the southern blue-ringed octopus (*Hapalochlaena maculosa*) uses chemosensory cues to discriminate the sex, and the identity of conspecifics, and this might influence their mate choice [95]. Most Octopus species produce eggs only once, with death occurring soon afterward [39, 94]. The immediate causes of this apparently universal mortality are not clearly understood, but the sequence of physiological changes brought about by the optic gland hormone seems to be irreversible [96]. The male places the spermatophores in the oviducts and the fertilization takes place in the oviductal glands as the mature eggs pass through them on their way out of the oviducts. Two secretions of the oviducal glands, together with the mucus, are used to stick the egg stalks together in strings and attach these to a substrate. In fact, pregnant females search for a sheltered place in shallow water where they can lay and brood the eggs without disturbance. The eggs are always attached to a substrate, as rocky shores, or sheltered place and females often protect them with shells, stones, and other solid objects that they gather. Estimates of individual fecundity vary widely among species, and there is a inverse relationship between egg size and fecundity [97]. Compared with other marine invertebrates, the eggs of cephalopods are large, well-protected, and produced in relatively low numbers (ranging from a few dozen to some hundreds of thousands; [97]), depending on the size of the animal as well as the species' spawning pattern. The total number of eggs laid by a female varies from 100,000 to 500,000 [40, 98].

Egg masses are attached to the rocks (in most octopuses, loliginid squid, and cuttlefish) or released into the water column in fragile gelatinous masses (in most squid families) [9].

During egg-laying and subsequent brooding, the female rarely leaves the egg mass, and she usually does not feed during the entire period of spawning and brooding (about 4–5 months). The female takes care of the eggs, cleaning them with the arm tips and directing jets of water from the funnel through the strings, pushing away predators. As a rule, females die shortly after the hatching of the last embryos after losing one-third of their pre-spawning weight [9].

9.6 PROXIMATE CAUSES UNDERLYING REPRODUCTIVE BEHAVIOR IN COLEOIDS

The studies on the mechanisms underlying the reproductive behavior in Coleoids remain scarce and not-contextualized to the relative behavior. Here, we propose a new integrative neuroethological approach in considering the proximate mechanisms that shape the reproductive behavior.

Proximate questions concern with how a behavior evolved and what are the mechanisms underlying it. The proximate causes that determine the reproductive behavior are mediated by "sensory-nervous system/neuroendocrine-effectors" [99]. Undoubtedly, in this context, the hormones are the major proximate factors mediating reproduction [100] and behavior [101, 102]. Recently, many researchers have claimed a deeper knowledge of the mechanisms underlying the reproductive behavior." The mechanisms underlying these responses and subsequent copulatory access to females by males remain unknown" [95].

9.6.1 PROXIMATE MECHANISMS UNDERLYING SENSORY AND CENTRAL CONTROL OF REPRODUCTION IN OCTOPUS VULGARIS

All organisms require coordinating contact that involves chemical messengers. The two systems responsible for this coordination are the nervous and endocrine system; both coordinate various mechanisms and physiological processes such as reproduction. In the classical concept of input-output relations different stimuli such as temperature, photoperiod, partners availability, and food, are perceived by sensory organs and specific areas of the nervous system that receive and integrate them in order to generate internal responses that affect endocrine glands, inducing gonad growth and maturation, egg-laying, and sexual behavior. A classic example of this complex neuroendocrine control is represented by the hypothalamus-pituitary-gonad axis of vertebrates. Among invertebrates, it is generally accepted that endocrine glands appear in mollusks, but their endocrine system differs through the classes in term of increasing complexity. Among Cephalopods, Coleoids show the highest level of complexity comparable to that one of vertebrates. Studies addressing this topic are focused on *O. vulgaris* and *O. maya*. On the other hand, few studies have been carried out on cuttlefish and squid [103–105].

Well and Well's pioneering studies [106] on *O. vulgaris* demonstrated the crucial inhibitory role of the light on the gonadal maturation trough the optical gland activity, and proposed the "CNS – optic gland – gonad" as the axis that controls the reproduction.

The hypothesis of the existence of a neuroendocrine axis in Octopus was based on the milestone experiments of Boycott and Young [107] and Well and Well's [106] of surgical resection of the optic trait, followed by enlargement of the optic gland and subsequent gonadal maturation. Similarly, the surgical lesions of the subpedunculate lobe resulted in an enlargement of the optic glands and subsequent hypertrophy of the gonads. In fact, the optic glands are innervated by the subpedunculate lobe at the back of the supra-esophageal brain through the optic gland nerve. Both lesions and removal of the subpedunculate lobe are followed by degeneration of at least some of the nerves which enter the gland from a bundle running along the optic stalk [106, 108]. Similar results were obtained, cutting the optic nerves or removing the optic lobes.

The optic glands are neuro-endocrine organs that lie on the optic traits close to the olfactory lobe (OL) widely described as a complex of nervous-tissue and endocrine glands [107] in *O. vulgaris* [108–112], *Sepia* [113], *O. aldrovandi* [114]. The glands are responsible for the production of the optic gland hormone, to date unknown [106, 107].

First inferences of proximate causes in the control of reproduction were inspired to the neuroendocrine role of the vena cava, localized in the ventral part of the posterior subesophageal mass of the cephalopods, particularly in the ventral median vasomotor lobe [115–118]. The vena cava releases two neuropeptides members of the vasopressin/oxytocin superfamily: cephalotocin and octopressin [22, 119–122], the latter evokes contractions of the smooth muscles of the oviduct [112].

Later, Le Gall and co-authors [123] studying the control of reproduction in *S. officinalis*, found the neuropeptide Phe-Met-Arg-Phe-NH2 (FMRFamide) in the optic gland nerve, and hypothesized that this peptide could be the inhibiting factor in the control of the optic gland activity. It was also demonstrated that in *O. vulgaris*, the optic glands are innervated by FMRFamide (type 4) immunoreactive fibers, originating from neurons in both the subpedunculate and the OLs [124]. FMRFamide seems to have the axo-axonic connection with gonadotropin GnRH fibers that reach the optic gland in the optic trait, so when FMRFamide is released, the result is the inhibition of GnRH release. On the contrary, when the FMRFamide inhibition is removed, the GnRH stimulates, in an excitatory way, the optic gland, affecting gland maturation. Moreover, the cGnRH-i isoform was observed in neurons of the posterior olfactory lobule, close to the optic gland, that sends fibers to the chief cells of the optic gland [124, 125].

The OL lies on the optic trait close to the optic gland. It was first described by Young [112], in *O. vulgaris*, as anatomically organized into three lobules: anterior, middle, and posterior. Such lobules are interconnected to each other, and they also receive fibers, through the olfactory nerve, from the olfactory organ, that perceives odors from the environment [24]. Moreover, the OL is neuroanatomically connected to the dorsal and basal optic lobes and to the subpedunculate areas [126]. The OL integrates information from the nervous lobes involved in the control of chemical perception, vision, motor program, and reproduction [9, 22, 41, 55, 56, 102, 112, 119, 124]. The findings of several neuropeptides, neurotransmitters, and hormones in the OL, provide new insights to the Wells and Wells' model. The OL appears to be at the crossroad of different sensory pathways playing an integrative function that only recently has been highlighted [24, 25, 102].

Recently, in the olfactory epithelium of the olfactory organ have been localized FMRFamide, APGWamide, olfactory marker protein (OMP) and NPY immunoreactivities in both Olfactory Receptor Neurons (ORNs) and fibers, GnRH immunoreactivity, resulted confined only to the fibers arising from the OLs [22, 24, 25, 124], that reach the epithelium giving rise to a neural net. This suggests a modulatory role of the OL on the activity of the ORNs and different and integrative regulation of *O. vulgaris* reproductive behavior involving the OL.

The OL plays its role in the regulation of the optic gland activity via the multipeptidergic control, exerted by GnRH, FMRFamide, APGWamide [22, 119, 124, 127], NPY, and Galanin [128].

FMRFamide is widely distributed in all three lobules of the OL [119, 124, 127], whereas GnRH and APGWa are mainly localized in the posterior lobule neurons and all peptides send their fibers to the optic glands [119, 124, 127]. It is likely that the OL influences the optic gland activity by releasing a positive factor (gonadotropin-releasing hormone, GnRH) to counterbalance the inhibitory effect of the tetrapeptide FMRFamide.

Other peptides and hormones found in the OL neurons have a key role in the energy allocation choice between metabolism and reproduction: NPY [128], Galanin [128], estrogen receptor (ER, [56]) and Corticotropin-releasing Factor (CRF) [128].

Given that the OL is also the site on which converges olfactory information coming from the olfactory organ, which in turn is under the same peptidergic control of OL. Therefore, the olfactory organ may be modulated resulting more sensitive to food odors during the energy storage period, with a consequential switch off for reproduction. Conversely, when the octopus female is in the reproductive period, the OL could modulate the olfactory organ to be more sensitive to sex odors resulting in a switch on for reproduction. A similar modulation of olfaction could be responsible for the food avoidance of the female during maternal care with a switch off for food intake [24, 25].

NPY and Galanin neuropeptides [128, 129] may be responsible for the amount of stored energy of the animal. On the other hand, it is known that in octopus, an optic gland factor inhibits muscle protein synthesis [130], inducing the release of free amino acids in the blood [131]. This physiological mechanism induces fast growth of the ovary *versus* weight loss that results in the egg yolk production by female cephalopods.

Other chemical messengers contribute to this picture, such as the neurosteroids that have a crucial role in the control of reproduction. In *O. vulgaris*, the stimulation with estradiol increases octGnRH and the estrogen receptor transcripts (octER), both expressed in the OL [56, 132]. These results link the expression of GnRH to the presence of estrogens.

The presence of these molecules in the circuit (CNS, Subpeduncolate area – OL, Olfactory organ – Optic gland – Gonads), supports the key role played by the OL and the olfactory organ, modulating each other in the balancing energy storage and reproduction.

9.6.2 PROXIMATE MECHANISMS UNDERLYING PERIPHERAL CONTROL OF REPRODUCTION IN O. VULGARIS

The reproductive system is controlled at the peripheral nervous level by the fusiform ganglia, which directly innervate the male and the female gonads. The neuropeptides FMRFamide and GnRH innervating the circular muscle layers of the reproductive ducts are involved in the motor coordination, egg-laying, and spermatophore transport as well [119, 125].

At the peripheral level, the presence of the steroidogenic enzyme, 3β-hydroxysteroid dehydrogenase (3β-HSD; [133]), in the tubular and interstitial cells of the testis establishes that the gonad is a steroidogenic organ. The localization of both progesterone and 17β-estradiol receptors in the nuclei of the follicle cells in the ovary of *O. vulgaris* supports not only the steroidogenic nature of the ovary, but supports the interplay between the sex steroids and their receptors in the regulation of egg yolk synthesis [125, 134]. In fact, progesterone induces vitellogenin synthesis in previtellogenic ovaries of *O. vulgaris* and is responsible for the follicle cell proliferation [135].

The overlapping fluctuations of the morphological changes in the oviduct and oviducal gland along with the progesterone and estradiol levels suggest that both sex steroids work in synergy to regulate the maturation of the female reproductive system in *O. vulgaris* [136].

The stimulation of spermatozoa collected from the spermatophores in the octopus male reproductive traits with progesterone or Ca ionophore induces an acrosome-like reaction. In addition, progesterone may play a role in the pre-activation of spermatozoa stored in the female tract [137]. When mature eggs fall in the fertilization chamber, they release a chemo-attractant factor, the sperm-attracting peptide (Octo-SAP), that mobilizes the spermatozoa to reach and fertilize the eggs in the presence of stored Ca2+ mobilization and membrane protein phosphorylation [138]. In other coleoids, such as *S. officinalis*, many sperm attractant peptides have been characterized, and they are released during egg laying to facilitate fertilization [139].

O. vulgaris may be considered an excellent model in which we can correlate, in an integrative way, proximate causes with the relative reproductive behavior.

9.7 CRITICAL ANALYSIS OF REPORTED FINDING

The peripheral control of reproduction in Octopus seems to be dominated by the so-called vertebrate sex steroids. The existence of an integrated network of chemical signals, among which the vertebrate-like sex steroids, responsible for the control of the reproductive functions in cephalopods is a sought after hypothesis. However, the presence and the role played by vertebrate–like sex steroids in mollusks have always been a much-disputed topic since the pioneering studies of Steidle [140] and later Hagerman et al., [141].

Probably the greatest impulse provided to this research field comes from studies of ecotoxicology, for the obvious implication due to the negative role played by endocrine disruptors released into the environment and capable of irreversibly altering in animal's crucial functions such as reproduction [142]. In many cases, the presence of steroids has been associated with different periods of the reproductive cycle, reinforcing the belief that steroids come from an endogenous production [136, 143, 144].

Numerous studies report the presence of vertebrate-like steroids in mollusks. Highly specific methods, such as gas chromatography-mass spectrometry identified the presence of steroids, although their exact quantitation is not always possible due to the high detection limits of the method [145, 146]. Very popular is the steroid identification and quantitation by the radio immune assay (RIA), based on the use of antibodies with sometimes very high levels of cross-reactivity and low level of sensitivity, which makes this method quite unreliable for unique identification of a certain steroid [147].

However, it remains one of the most employed detection methods for steroid identification in both vertebrates and invertebrates. In general, the levels of androgens and progestins are in the order of nanograms (0.1–10 ng/g tissue) and at least one order of magnitude higher than estrogens. Usually, the levels are higher in reproductive tissues, suggesting that the gonads may be a steroidogenic tissue or a target tissue of steroid action.

Even though the presence of vertebrate-like sex steroids has been unequivocally demonstrated, the following two basic critical issues cannot be ignored: where the steroids come from and how they act.

9.8 NUCLEAR RECEPTORS (NR)

Sex steroids need intracellular receptors to carry on their functions [148]. The NR superfamily is the product of an ongoing process of genomic diversification within and among species. NR has a conserved domain structure with very high homology in their DNA binding domain (DBD) and ligand binding domain (LBD) domains [149]. BLAST searches indicate that there are not statistically significant similarities to any other gene family. This indicates that all NRs evolved by descent from a single ancestral NR, through a series of gene duplications [150, 151]. The analysis of a large set of nuclear receptor sequences indicates that there are several evolutionary classes of NRs. According to the method of phylogenetic inference and the number of sequences analyzed, different classes of NRs have been identified [152].

Phylogenetic analysis reveals that the NR superfamily had already established well before the lineages of protostomes and deuterostomes split from each other at least 670 million years ago, and the NR gene duplication pattern during the evolution of metazoans likely arose from two waves of gene duplications [153, 154]. NRs are found in sponges and simple multicellular organisms [155]. They had probably an important role in the transition from unicellular to multicellular organisms [156].

The phylogenetic distribution of steroid receptors (SRs) in the animal kingdom, along with the estradiol presence in almost all animals, suggest that estrogenically active chemicals may be evolutionarily conserved signals [157, 158]. It also suggests the possibility that all animals are sensitive to estrogens, whether endogenous or environmental. This explains the vulnerability to xenobiotics in both vertebrates and invertebrates [159].

Among SRs, the ERs regulate an enormity of processes starting in early life and continuing through sexual reproduction, development, and end of life. The wide repertoire of ER actions is mediated mostly through ligand-activated transcription factors and many DNA response elements in most tissues and organs. Their versatility, however, comes with the drawback of promiscuous interactions with structurally diverse exogenous chemicals with potential for a wide range of adverse health outcomes [160].

But when did SRs evolve? Using both phylogenetic methods and experimental methods, the ancient SR has been reconstructed in the laboratory [161]. Evidence indicates that the ancient SR was an ERs with specific binding capacity. By means of gene duplications, different forms of ERs emerged with an affinity for steroids, other than estrogens, probably steroid pathways intermediates, such as progestins and androgens [162]. This means that steroids were present before a specific receptor evolved.

Early studies suggest that the steroid-binding capacity arose in a primitive deuterostome. Later on, however, the cloning of an ER in the mollusk Aplysia, cast some doubts about the SR presence only in deuterostomes. Moreover an ER has been cloned also in other mollusks. The same authors sustain that the ER is only constitutively active and therefore, does not respond to estrogens. However, we demonstrated beyond doubt that the ER in Octopus is responding to estradiol and shows both fast and classic effects. Up to date, the evolution of SRs in metazoan is still an open question. However, the ancient origin of estrogen signaling, even a Progesterone receptor (PR) in rotifers, suggests the possibility that some ecdysozoans have retained the sex SRs during the evolution.

9.9 ENDOGENOUS VS. EXOGENOUS STEROIDS

While the evolution of SRs has gained further confirmations by comparison of genomes of metazoans, the evolution of their ligands is still a matter of debate. The absence in the genome of the mollusks of steroid biosynthesis enzymes has challenged the hypothesis of the presence of an endogenous synthesis of steroids in mollusks, as well as in other invertebrates, and prompted the view that such steroids would not come from internal biochemical pathways, but instead would be absorbed from the environment and stored within the tissues even for long periods of time, bound to fatty acids [163].

Using the parsimony analysis, Markov, and his collaborators [164] started unraveling the mystery of the origin and evolution of steroid ligands. Although their phylogenetic analysis did not include mollusks, the results favor the independent elaboration of different steroid synthesis pathways within each phylum in metazoan. An example of convergent evolution for complex molecules. This may explain the apparent contradiction between the genomic and biochemical data provided by the numerous studies on mollusks. We may hypothesize that biochemically similar sex steroids could arise from different biochemical pathways, but still bind to SRs and elicit a physiological response compatible with a role in reproduction.

In the future, it will be necessary to identify steroids with sensitive methods such as GC-MS and to verify that these molecules are endogenously synthesized by a precursor, identifying potential enzymes capable of catalyzing the different phases of the synthesis pathway. Finally, the physiological effect is still open and should be addressed in future studies.

9.10   FUTURE PERSPECTIVES

Future perspectives could be focused on an integrative approach to resolve these major challenges:

- Has the olfactory organ any regulatory action on the OL during reproduction?
- What is the nature of the factor released by the optic gland?
- What is the role of egg laying-like peptide (ELH) in reproductive behavior?
- What is the physiological mechanism underlying anorexic mother behavior?
- How do chemical cues affect communication in cephalopod communities?
- Steroids identification with sensitive methods

However, next-generation sequencing technology will help elucidate what specific genes and regulatory sequences support reproductive behavior, especially in Octopus, whose behavior, such as maternal care of eggs and maternal death are described in general terms.

The analysis of differential gene expression in different organs involved in reproduction at both central and peripheral levels (Subpeduncolate lobe, OL, optic gland, white bodies, gonads, olfactory organ) in different life cycle stages and reproduction in the male and female could help to further elucidate the several genes related with reproductive functions, linking behavior and gene expression, deepening remote and proximate causes of behavior (Figure 9.1).

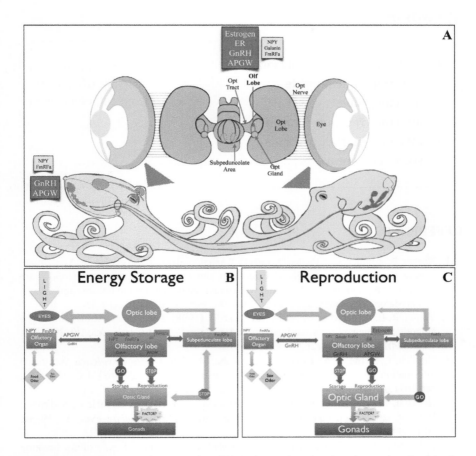

FIGURE 9.1 *Octopus vulgaris* mating interaction and molecules players involved in the control of reproduction. The male (blue) transfers spermatophores by the hectocotylus into the female (pink) mantle cavity near the opening of the oviduct. The olfactory organ (green) is connected through the olfactory nerve to the olfactory lobe (Olf lobe), that lies on the optic trait (Opt Tract) close to the optic gland. During the reproduction, GnRH, APGWaminde, ER, and NMDA stimulate reproduction inhibiting sensitiveness to food odors.

KEYWORDS

- **corticotropin-releasing factor**
- **egg laying-like peptide**
- **olfactory lobe**
- **olfactory receptor neurons**
- **olfactory marker proteins**
- **steroid receptors**

REFERENCES

1. Kroger, B., Vinther, J., & Fuchs, D., (2011). Cephalopod origin and evolution: A congruent picture emerging from fossils, development and molecules: Extant cephalopods are younger than previously realized and were under major selection to become agile, shell-less predators. *Bioessays, 33*, 602–613. 10.1002/bies.201100001.
2. Young, R. E., Vecchione, M., & Donovan, D. T., (1998). The evolution of coleoid cephalopods and their present biodiversity and ecology. *S. Afr. J. Mar. Sci., 20*, 393–420. 10.2989/025776198784126287.
3. Gutnick, T., Byrne, R. A., Hochner, B., & Kuba, M., (2011). *Octopus vulgaris* uses visual information to determine the location of its arm. *Curr. Biol., 21*, 460–462. 10.1016/j.cub.2011.01.052.
4. Hochner, B., (2013). How nervous systems evolve in relation to their embodiment: What we can learn from octopuses and other molluscs. *Brain Behav. Evol., 82*, 19–30. 10.1159/000353419.
5. Kuba, M., Meisel, D. V., Byrne, R. A., Griebel, U., & Mather, J. A., (2003). Looking at play in *Octopus vulgaris. Berliner Pala Ontologische Abhandlungen, 3*, 163–169.
6. Mather, J. A., (1991). Navigation by spatial memory and use of visual landmarks in octopuses. *J. Comp. Physiol. A Sens. Neural. Behav. Physiol., 168*, 491–497.
7. Mather, J. A., (2008). Cephalopod consciousness: Behavioral evidence. *Consciousness and Cognition, 17*, 37–48. 10.1016/j.concog.2006.11.006.
8. Bertapelle, C., Polese, G., & Di Cosmo, A., (2017). Enriched environment increases PCNA and PARP1 levels in *Octopus vulgaris* central nervous system: First evidence of adult neurogenesis in lophotrochozoa. *J. Exp. Zool. B Mol. Dev. Evol., 10*. 1002/jez.b., 22735.
9. Hanlon, R. T., & Messenger, J. B., (1996). *Cephalopod Behavior*. Cambridge University Press.
10. Hanlon, R., Chiao, C., Mäthger, L., Barbosa, A., Buresch, K., & Chubb, C., (2009). Cephalopod dynamic camouflage: Bridging the continuum between background matching and disruptive coloration. *Philos. Trans R. Soc. Lond. B. Biol. Sci., 364*, 429–437.
11. Mather, J. A., (1988). Daytime activity of juvenile *Octopus vulgaris* in Bermuda. *Malacologia, 29*, 69–76.

12. Doubleday, Z. A., Prowse, T. A. A., Arkhipkin, A., Pierce, G. J., Semmens, J., Steer, M., Leporati, S. C., Lourenço, S., Quetglas, A., Sauer, W., & Gillanders, B. M., (2016). Global proliferation of cephalopods. *Curr. Biol., 26*, R406–R407. 10.1016/j.cub.2016.04.002.

13. Osorio, D., (2014). Cephalopod behavior: Skin flicks. *Curr. Biol., 24*, 684–685. 10.1016/j.cub.2014.06.066.

14. Grable, M. M., Shashar, N., Gilles, N. L., Chiao, C. C., & Hanlon, R. T., (2002). Cuttlefish body patterns as a behavioral assay to determine polarization perception. *Biol. Bull., 203*, 232–234. 10.2307/1543414.

15. Stubbs, A. L., & Stubbs, C. W., (2016). Spectral discrimination in color blind animals via chromatic aberration and pupil shape. *Proc. Natl. Acad. Sci. USA, 113*, 8206–8211. 10.1073/pnas.1524578113.

16. Yoshida, M. A., Ogura, A., Ikeo, K., Shigeno, S., Moritaki, T., Winters, G. C., Kohn, A. B., & Moroz, L. L., (2015). Molecular evidence for convergence and parallelism in evolution of complex brains of cephalopod molluscs: Insights from visual systems. *Integr. Comp. Biol., 55*, 1070–1083. 10.1093/icb/icv049.

17. Zylinski, S., Osorio, D., & Shohet, A. J., (2009). Perception of edges and visual texture in the camouflage of the common cuttlefish, *Sepia officinalis. Philos. Trans. R. Soc. Lond. B. Biol. Sci., 364*, 439–448.10.1098/rstb.2008.0264.

18. Bleckmann, H., Budelmann, B. U., & Bullock, T. H., (1991). Peripheral and central nervous responses evoked by small water movements in a cephalopod. *Journal of Comparative Physiology. A, Sensory, Neural, and Behavioral Physiology, 168*, 247–257.

19. Budelmann, B. U., & Williamson, R., (1994). Directional sensitivity of hair cell afferents in the *Octopus* statocyst. *J. Exp. Biol., 187*, 245–259.

20. King, A. J., Adamo, S. A., & Hanlon, R. T., (2003). Squid egg mops provide sensory cues for increased agonistic behavior between male squid. *Anim. Behav., 66*, 49–58. 10.1006/anbe.2003.2197.

21. Williamson, R., & Chrachri, A., (2007). A model biological neural network: The cephalopod vestibular system. *Philos. Trans. R. Soc. Lond. B. Biol. Sci., 362*, 473–481. 10.1098/rstb.2006.1975.

22. Di Cosmo, A., & Polese, G., (2014). Cephalopods meet neuroecology: The role of chemoreception in *Octopus vulgaris* reproductive behavior. In: Di Cosmo, A., & Winlow, W., (eds.), *Neuroecology and Neuroethology in Molluscs—the Interface Between Behavior and Environment* (pp. 117–132). NOVA Science Publisher, Hauppauge, NY.

23. Di Cosmo, A., & Polese, G., (2017). *Cephalopod Olfaction.* Oxford University Press.

24. Polese, G., Bertapelle, C., & Di Cosmo, A., (2015). Role of olfaction in *Octopus vulgaris* reproduction. *Gen. Comp. Endocrinol., 210*, 55–62. 10.1016/j.ygcen.2014.10.006.

25. Polese, G., Bertapelle, C., & Di Cosmo, A., (2016). Olfactory organ of *Octopus vulgaris*: morphology, plasticity, turnover and sensory characterization. *Biology Open, 5*, 611–619. 10.1242/bio.017764.

26. Wildenburg, G., (1990). Ultrastructure of the olfactory organ of *Sepia officinalis* at the end of embryonic development. *Actes. Prem. Symp. Int. Sur la Seiche. Caen., 1989*, 294–295.

27. Wildenburg, G., (1995). The structure of the so-called olfactory organ of octopods after hatching indicates its chemoreceptive function. *Abstr. 12h Int. Malac, Vigo*, pp. 91–93.

28. Woodhams, P. L., & Messenger, J. B., (1974). A note on the ultrastructure of the octopus olfactory organ. *Cell Tissue Res., 152*, 253–258.

29. Budelmann, B. U., (1996). Active marine predators: The sensory world of cephalopods. *Mar. Freshw. Behav. Physiol., 27*, 59–75. 10.1080/10236249609378955.

30. Packard, A., (1972). Cephalopods and fish: The limits of convergence. *Biological Reviews, 47*, 241–307. 10.1111/j.1469–185X.1972.tb00975.x.
31. Baratte, S., & Bonnaud, L., (2009). Evidence of early nervous differentiation and early catecholaminergic sensory system during *Sepia officinalis* embryogenesis. *J. Comp. Neurol., 517*, 539–549. 10.1002/cne.22174.
32. Boletzky, S., (1989). Recent studies on spawning, embryonic development, and hatching in the Cephalopoda. *Adv. Mar. Biol., 25*, 85–115. 10.1016/S0065–2881(08)60188–1.
33. Fioroni, P., (1990). Our recent knowledge of the development of the cuttlefish (*Sepia officinalis*). *Zool. Anz., 224*, 1–25.
34. Graziadei, P., (1964). Receptors in the Sucker of the Cuttle Fish. *Nature, 203*, 384–387.
35. Graziadei, P., (1965). Sensory receptor cells and related neurons in cephalopods. *Cold Spring Harb. Symp. Quant. Biol., 30*, 45–57.
36. Mackie, G. O., (2008). Immunostaining of peripheral nerves and other tissues in whole mount preparations from hatchling cephalopods. *Tissue Cell, 40*, 21–29. 10.1016/j.tice.2007.08.005.
37. Sundermann-Meister, V. G., (1978). A new type of ciliated cells in the epidermis of late embryonic stages and juveniles of *Loligo vulgaris* (Mollusca, Cephalopoda). *Zool. Jb. Anat. Bd., 99*, 493–499.
38. Anraku, K., Archdale, M. V., Hatanaka, K., & Marui, T., (2005). Chemical stimuli and feeding behavior in octopus, *Octopus vulgaris*. *Phuket Mar. Biol. Center Bull., 66*, 221–227.
39. Boyle, P. R., (1983). Ventilation rate and arousal in the octopus. *J. Exp. Mar. Biol. Ecol., 69*, 129–136. 10.1016/0022–0981(83)90062-X.
40. Wells, M. J., (1964). Detour experiments with octopuses. *J. Exp. Biol., 41*, 621–642.
41. Wells, M. J., (1978). *Octopus: Physiology and Behavior of an Advanced Invertebrate.* Chapman and Hall.
42. Young, J. Z., (1977). Brain, behavior and evolution of cephalopods. *Symp. Zool. Soc. Land., 38*, 377–434.
43. Young, J. Z., (1965). The Croonian Lecture, 1965 – The organization of a memory system. *Proceedings of the Royal Society of London. Series, B. Biological Sciences, 163*, 285–320. 10.1098/rspb.1965.0071.
44. Altman, J. S., (1971). Control of accept and rect reflexes in the octopus. *Nature, 229*, 204–206.
45. Hochner, B., (2012). An embodied view of octopus neurobiology. *Curr. Biol., 22*, 887–892. 10.1016/j.cub.2012.09.001.
46. Levy, G., Flash, T., & Hochner, B., (2015). Arm coordination in octopus crawling involves unique motor control strategies. *Curr. Biol., 25*, 1195–1200. 10.1016/j.cub.2015.02.064.
47. Sumbre, G., Gutfreund, Y., Fiorito, G., Flash, T., & Hochner, B., (2001). Control of octopus arm extension by a peripheral motor program. *Science, 293*, 1845–1848. 10.1126/science.1060976.
48. Giuditta, A., Dettbarn, W. D., & Brzin, M., (1968). Protein synthesis in the isolated giant axon of the squid. *Proc. Natl. Acad. Sci. USA, 59*, 1284–1287.
49. Graziadei, P. P. C., (1971). The nervous system of the arms. In: Young, J. Z., (ed.), *The Anatomy of the Nervous System of Octopus Vulgaris.* Clarendon Press, Oxford.
50. Yekutieli, Y., Sagiv-Zohar, R., Aharonov, R., Engel, Y., Hochner, B., & Flash, T., (2005). Dynamic model of the octopus arm. I. Biomechanics of the octopus reaching movement. *J. Neurophysiol., 94*, 1443–1458. 10.1152/jn.00684.2004.

51. Cole, P. D., & Adamo, S. A., (2005). Cuttlefish (*Sepia officinalis*: Cephalopoda) hunting behavior and associative learning. *Anim. Cogn., 8*, 27–30. 10.1007/s10071–004–0228–9.

52. Suboski, M. D., Muir, D., & Hall, D., (1993). Social learning in invertebrates. *Science, 259*, 1628.

53. Agin, V., Chichery, R., Dickel, L., & Chichery, M. P., (2006). The "prawn-in-the-tube" procedure in the cuttlefish: Habituation or passive avoidance learning? *Learn Mem., 13*, 97–101.

54. Darmaillacq, A. S., Chichery, R., & Dickel, L., (2006). Food imprinting, new evidence from the cuttlefish *Sepia officinalis. Biol. Lett., 2*, 345–347.

55. De Lisa, E., De Maio, A., Moroz, L. L., Moccia, F., Mennella, M. R., & Di Cosmo, A., (2012a). Characterization of novel cytoplasmic PARP in the brain of *Octopus vulgaris. Biol. Bull., 222*, 176–181. 10.1086/BBLv222n3p176.

56. De Lisa, E., Paolucci, M., & Di Cosmo, A., (2012b). Conservative nature of oestradiol signaling pathways in the brain lobes of *Octopus vulgaris* involved in reproduction, learning and motor coordination. *J. Neuroendocrinol., 24*, 275–284. 10.1111/j.1365–2826.2011.02240.x.

57. Sanders, G. D., (1975). *Invertebrate Learning: Cephalopods.* Plenum Press.

58. Alves, C., Boal, J. G., & Dickel, L., (2007). Short-distance navigation in cephalopods: A review and synthesis. *Cognitive Processing, 9*, 239. 10.1007/s10339–007–0192–9.

59. Boal, J. G., (2006). Social recognition: a top down view of cephalopod behavior. *Vie Milieu, 56*, 69–79.

60. Godfrey-Smith, P., (2016). *Other Minds: The Octopus, the Sea, and the Deep Origins of Consciousness.* Farrar, Straus and Giroux.

61. Scheel, D., Chancellor, S., Hing, M., Lawrence, M., Linquist, S., & Godfrey-Smith, P., (2017). A second site occupied by *Octopus tetricus* at high densities, with notes on their ecology and behavior. *Mar. Freshw. Behav. Physiol., 50*, 285–291. 10.1080/10236244.2017.1369851.

62. Corner, B. D., & Moore, H. T., (1980). Field observations on the reproductive behavior of *Sepia latimanus. Micronesia, 16*, 235–260.

63. Brown, E. R., & Piscopo, S., (2013). Synaptic plasticity in cephalopods, more than just learning and memory? *Invertebr. Neurosci., 13*, 35–44. 10.1007/s10158–013–0150–4.

64. Warnke, K., (1994). Some aspects of social interaction during feeding in *Sepia officinalis* (Mollusca: Cephalopoda) hatched and reared in the laboratory. *Vie Milieu, 44*, 125–131.

65. Boal, J. G., (1996). A review of simultaneous visual discrimination as a method of training octopuses. *Biol. Rev. Camb. Philos. Soc., 71*, 157–190.

66. Ibáñez, C. M., & Keyl, F., (2010). Cannibalism in cephalopods. *Rev. Fish Biol. Fish, 20*, 123–136.

67. Hurley, A. C., (1978). School structure of the squid *Loligo opalescens. Fish Bull., 76*, 433–442.

68. Mather, J. A., & O'Dor, R. K., (1984). Spatial organization of schools of the Squid Illex illecebrosus. *Mar. Behav. Physiol., 10*, 259–271. 10.1080/10236248409378623.

69. Jackson, G. D., & Mladenov, P. V., (1994). Teminal spawning in the deepwater squid *Moroteuthis ingens* (Cephalopoda: Onychoteuthidae). *Journal of Zoology, 234*, 189–201.

70. Jackson, G. D., Semmens, J. M., Phillips, K. L., & Jackson, C. H., (2004). Reproduction in the deepwater squid Moroteuthis ingens, what does it cost? *Mar. Biol., 145*, 905–916. 10.1007/s00227–004–1375-x.

71. Perez, J. A. A., Aguiar, D. C., & Oliviera, U. C., (2002). Biology and population dynamics of the long-finned squid *Loligo plei* (Cephalopoda: Loliginidae) in southern Brazilian waters. *Fishery Research, 58*, 267–279.

72. Rodrigues, A. R., & Gasalla, M. A., (2008). Spatial and temporal patterns in size and maturation of *Loligo plei* and *Loligo sanpaulensis* (Cephalopoda: Loliginidae) in southeastern Brazilian waters, between 23°S and 27°S. *Sci. Mar., 72*, 631–647.

73. Hanlon, R. T., (1998). Mating systems and sexual selection in the squid Loligo: How might commercial fishing on spawning grounds affect them? *CalCOFI Reports, 39*, 92–100.

74. Drew, G. A., (1911). Sexual activities of the squid *Loligo pealii* (Les). I. Copulation. Egg-laying and fertilization. *J. Morphol., 22*, 327–359.

75. Sauer, W. H. H., Roberts, M. J., & Lipnski, M. R., (1997). Choreography of the squid's "nuptial dance." *Biological Bulletin, 192*, 203–207.

76. Hall, R., (2002). Cenozoic geological and plate tectonic evolution of SE Asia and the SW Pacific: Computer-based reconstructions, model and animations. *J. Asian Earth Sci., 20*, 353–431.

77. Hanlon, R. T., Smale, M. J., & Sauer, W. H. H., (2002). The mating system of the squid *Loligo vulgaris* reynaudii (Cephalopoda, Mollusca) off South Africa: Fighting, guarding, sneaking, mating and egg laying behavior. *Bull. Mar. Sci., 71*, 331–345.

78. Jantzen, T. M., & Havenhand, J. N., (2003a). Reproductive behavior in the squid Sepioteuthis australis from South Australia: Ethogram of reproductive body patterns. *Biol. Bull., 204*, 290–304. 10.2307/1543600.

79. Jantzen, T. M., & Havenhand, J. N., (2003b). Reproductive behavior in the squid Sepioteuthis australis from South Australia: Interactions on the spawning grounds. *Biol. Bull., 204*, 305–317. 10.2307/1543601.

80. Hanlon, R. T., Forsythe, J. W., & Joneschild, D. E., (1999). Crypsis, conspicuousness, mimicry and polyphenism as antipredator defences of foraging octopuses on Indo-Pacific coral reefs, with a method of quantifying crypsis from video tapes. *Biol. J. Linn. Soc., 66*, 1–22. 10.1111/j.1095–8312.1999.tb01914.x.

81. Wada, T., Takegaki, T., Mori, T., & Natsukari, Y., (2005). Sperm displacement behavior of the cuttlefish *Sepia esculenta* (Cephalopoda: Sepiidae). *J. Ethol., 23*, 85–92. 10.1007/s10164–005–0146–6.

82. Wada, T., Takegaki, T., Mori, T., & Natsukari, Y., (2010). Sperm removal, ejaculation and their behavioral interaction in male cuttlefish in response to female mating history. *Anim. Behav., 79*, 613–619. https://doi.org/10.1016/j.anbehav.2009.12.004.

83. Naud, M. J., Shaw, P. W., Hanlon, R. T., & Havenhand, J. N., (2005). Evidence for biased use of sperm sources in wild female giant cuttlefish (Sepia apama). *Proceedings Biological Sciences 272*, 1047–1051. 10.1098/rspb.004.3031.

84. Hall, K., & Hanlon, R., (2002). Principal features of the mating system of a large spawning aggregation of the giant Australian cuttlefish *Sepia apama* (Mollusca: Cephalopoda). *Mar. Biol., 140*, 533–545. 10.1007/s00227–001–0718–0.

85. Boletzky, S., (1998). *Cephalopod Eggs and Egg Masses*.

86. Langmead, B., Trapnell, C., Pop, M., & Salzberg, S. L., (2009). Ultrafast and memory-efficient alignment of short DNA sequences to the human genome. *Genome Biol., 10, R25*. 10.1186/gb-2009–10-3-r25.

87. Okamoto, K., Yasumuro, H., Mori, A., & Ikeda, Y., (2017). Unique arm-flapping behavior of the pharaoh cuttlefish, *Sepia pharaonis*: Putative mimicry of a hermit crab. *J. Ethol., 35*, 307–311. 10., 1007/s10164–017–0519–7.

88. Anderson, R. C., & Wood, J. B., (2001). Enrichment for giant Pacific octopuses: Happy as a clam? *Journal of Applied Animal Welfare Science, 4*, 157–168.
89. Batham, E. J., (1957). Care of eggs by *Octopus maorum. Transactions of the Royal Society of New Zealand, 84*, 629–638.
90. Gabe, S. H., (1975). Reproduction in the giant octopus of the North Pacific, *Octopus dofleini martini. Veliger, 18*, 146–150.
91. Guibé, M., Poirel, N., Houdé, O., & Dickel, L., (2012). Food imprinting and visual generalization in embryos and newly hatched cuttlefish, *Sepia officinalis. Anim. Behav., 84*, 213–217. https://doi.org/10.1016/j.anbehav.2012.04.035.
92. Mather, J. A., (1985). Behavioural interactions and activity of captive Eledone moschata: Laboratory investigations of a 'social' octopus. *Anim. Behav., 33*, 1138–1144. https://doi.org/10.1016/S0003–3472(85)80173–1 (Accessed on 23 August 2019).
93. Caldwell, R. L., Ross, R., Rodaniche, A., & Huffard, C. L., (2015). Behavior and body patterns of the larger pacific striped octopus. *PLoS ONE, 10*, e0134152. 10.1371/journal.pone.0134152.
94. Boyle, P. R., (1987). *Cephalopod Life Cycles.* Comparative reviews. Academic Press, London, England.
95. Morse, P., Zenger, K. R., McCormick, M. I., Meekan, M. G., & Huffard, C. L., (2017). Chemical cues correlate with agonistic behavior and female mate choice in the southern blue-ringed octopus, *Hapalochlaena maculosa* (Hoyle, 1883) (Cephalopoda: Octopodidae). *J. Molluscan Stud., 83*, 79–87. 10.1093/mollus/eyw045.
96. Wodinsky, J., (1977). Hormonal inhibition of feeding and death in Octopus: Control by optic gland secretion. *Science, 198*, 948–951.
97. Boyle, P. R., & Rodhouse, P. G., (2005). *Cephalopods: Ecology and Fisheries.* Blackwell, Oxford, England.
98. Papini, M. R., & Bitterman, M. E., (1991). Appetitive conditioning in *Octopus cyanea. J. Comp. Psychol., 105*, 107–114.
99. Pfaff, D. W., Ogawa, S., Kia, H. K., Vasudevan, N., Krebs, C., Frohlich, J., & Kow, L. M., (2002). Genetic mechanisms in neural and hormonal controls over female reproductive behaviors. In: Pfaff, D. W., Arnold, A., Etgen, A., Fahrbach, S., & Rubin, R., (eds.), *Hormones, Brain and Behavior* (pp. 441–460). Academic Press, San Diego.
100. Chedrese, P. J., (2009). *Reproductive Endocrinology: A Molecular Approach.* Business Media, LLC, New York.
101. Adkins-Regan, E., (2005). *Hormones and Animal Social Behavior.* Princeton University Press.
102. Di Cosmo, A., & Polese, G., (2016). Neuroendocrine-immune systems response to environmental stressors in the cephalopod *Octopus vulgaris. Frontiers in Physiology, 7*, 434. 10.3389/fphys.2016., 00434.
103. Endress, M., Zatylny-Gaudin, C., Corre, E., Le Corguille, G., Benoist, L., Leprince, J., Lefranc, B., Bernay, B., Leduc, A., Rangama, J., Lafont, A. G., Bondon, A., & Henry, J., (2018). Crustacean cardioactive peptides: Expression, localization, structure, and a possible involvement in regulation of egg-laying in the cuttlefish *Sepia officinalis. Gen. Comp. Endocrinol., 260*, 67–79. 10.1016/j.ygcen.2017.12.009.
104. Lu, Z. M., Liu, W., Liu, L. Q., Wang, T. M., Shi, H. L., Ping, H. L., Chi, C. F., Yang, J. W., & Wu, C. W., (2016). Cloning, characterization, and expression profile of estrogen receptor in common chinese cuttlefish, *Sepiella japonica. Journal of experimental zoology. Part, A., Ecological Genetics and Physiology, 325*, 181–193. 10.1002/jez.2011.

105. Onitsuka, C., Yamaguchi, A., Kanamaru, H., Oikawa, S., Takeda, T., & Matsuyama, M., (2009). Molecular cloning and expression analysis of a GnRH-like dodecapeptide in the swordtip squid, *Loligo edulis. Zoolog. Sci., 26*, 203–208. 10.2108/zsj.26. 203.

106. Wells, M. J., & Wells, J., (1959). Hormonal control of sexual maturity in octopus. *J. Exp. Biol., 36*, 1–33.

107. Boycott, B. B., & Young, J. Z., (1956). Reaction to shape in *Octopus vulgatis* Lamarck. *Proceedings of the Zoological Society of London, 126*, 491–547. 10.1111/j.1096-3642.1956. tb00451.x.

108. Froesch, D., (1974). The subpedunculate lobe of the octopus brain: evidence for dual function. *Brain Res., 75*, 277–285.

109. Arrieche, D., (1999). Ultrastructure of the optic gland of the squid Sepiotheutis sepioidea (Cephalopoda: Loliginidae). *Rev. Biol. Trop., 47*, 831–842.

110. Bjorkman, N., (1963). On the ultrastructure of the optic gland in *Octopus. Journal of Ultrastructure Research, 8*, 195.

111. Nishioka, R. S., Bern, H. A., & Golding, D. W., (1970). Innervation of the cephalopod optic gland. In: Scharrer, W. B. B., (ed.), *Aspects of Neuroendocrinology* (pp. 47–54). Springer-Verlag, Berlin, Germany.

112. Young, J. Z., (1971). *The Anatomy of the Nervous System of Octopus Vulgaris.* Oxford University Press, New York.

113. Owen, R., (1832). Memoir on the pearly nautilus (Nautilus pompilius Linn.) with illustrations of its external form and internal structure. *Council of Royal College Surgeons*, London, England.

114. Delle Chiaie, S., (1828). Memorie sulla storia e notomia degli animali senza vertebre del Regno di Napoli. Società Tipografica , Napoli, Italy.

115. Alexandrowicz, J. S., (1964). The neurosecretory system of the vena cava in Cephalopoda, I. Eledone cirrosa. *Journal of the Marine Biology Association of the United Kingdom, 44*, 111–132.

116. Alexandrowicz, J. S., (1965). The neurosecretory system of the vena cava in Cephalopoda, I. I., *Sepia officinalis* and *Octopus vulgaris. Journal of the Marine Biology Association of the United Kingdom, 45*, 209–228.

117. Bonichon, A., (1967). Contribution a l'etude de la neurosecretion et de l'endocrinologie chez les Cephalopodes. I. *Octopus vulgaris. Vie Milieu, 18*, 227–263.

118. Martin, R., (1968). Fine structure of the neurosecretory system of the vena cava in Octopus. *Brain Res., 8*, 201–205.

119. Di Cosmo, A., & Polese, G., (2013). Molluscan bioactive peptide. In: Kastin, A. J., (ed.), *Handbook of Biologically Active Peptides* (pp. 276–286). Academic Press/Elsevier, Amsterdam.

120. Henry, J., Cornet, V., Bernay, B., & Zatylny-Gaudin, C., (2013). Identification and expression of two oxytocin/vasopressin-related peptides in the cuttlefish *Sepia officinalis. Peptides, 46*, 159–166. 10.1016/j.peptides.2013.05.004.

121. Reich, G., (1992). A new peptide of the oxytocin/vasopressin family isolated from nerves of the cephalopod *Octopus vulgaris. Neurosci. Lett., 134*, 191–194.

122. Takuwa-Kuroda, K., Iwakoshi-Ukena, E., Kanda, A., & Minakata, H., (2003). Octopus, which owns the most advanced brain in invertebrates, has two members of vasopressin/oxytocin superfamily as in vertebrates. *Regul. Pept., 115*, 139–149.

123. Le Gall, S., Feral, C., Van Minnen, J., & Marchand, C. R., (1988). Evidence for peptidergic innervation of the endocrine optic gland in *Sepia* by neurons showing FMRFamide-like immunoreactivity. *Brain Res., 462*, 83–88.

124. Di Cosmo, A., & Di Cristo, C., (1998). Neuropeptidergic control of the optic gland of *Octopus vulgaris*: FMRF- amide and GnRH immunoreactivity. *J. Comp. Neurol., 398*, 1–12. 10.1002/(SICI)1096–9861(19980817)398:1&lt, 1:AID-CNE1&gt, 3.0.CO, 2–5.

125. Di Cristo, C., Paolucci, M., Iglesias, J., Sanchez, J., & Di Cosmo, A., (2002). Presence of two neuropeptides in the fusiform ganglion and reproductive ducts of *Octopus vulgaris*: FMRFamide and gonadotropin-releasing hormone (GnRH). *J. Exp. Zool., 292*, 267–276.

126. Messenger, J. B., (1967). The peduncle lobe: A visuo-motor centre in octopus. *Proceedings of the Royal Society of London, Series, B., Biological Sciences, 167*, 225–251.

127. Di Cristo, C., Minnen, J. V., & Di Cosmo, A., (2005). The presence of APGWamide in *Octopus vulgaris*: A possible role in the reproductive behavior. *Peptides, 26*, 53–62. 10.1016/j.peptides.2004.07.019.

128. Suzuki, H., Yamamoto, T., Nakagawa, M., & Uemura, H., (2002). Neuropeptide Y-immunoreactive neuronal system and colocalization with FMRFamide in the optic lobe and peduncle complex of the octopus (*Octopus vulgaris*). *Cell Tissue Res., 307*, 255–264. 10.1007/s00441–001–0492–9.

129. Suzuki, H., Yamamoto, T., Inenaga, M., & Uemura, H., (2000). Galanin-immunoreactive neuronal system and colocalization with serotonin in the optic lobe and peduncle complex of the octopus (*Octopus vulgaris*). *Brain Res., 865*, 168–176.

130. O'Dor, R. K., & Wells, M. J., (1978). Reproduction versus somatic growth: Hormonal control in *Octopus vulgaris*. *J. Exp. Biol., 77*, 15–31.

131. Bertolucci, E., Conti, M., Di Cosmo, A., Maiorino, M., Mettivier, G., Montesi, M. C., Paolella, G., Pecorella, T., Russo, P., & Scognamiglio, R., (2002). Real time β-imaging with silicon hybrid pixel detectors: Kinetic measurements with C-14 amino acids and P-32 nucleotides. *IEEE Transactions on Nuclear Science, 49*I, 2213–2217. 10.1109/T. N., S2002803811.

132. Keay, J., Bridgham, J. T., & Thornton, J. W., (2006). The *Octopus vulgaris* estrogen receptor is a constitutive transcriptional activator: Evolutionary and functional implications. *Endocrinology, 147*, 3861–3869. 10.1210/en.2006–0363.

133. D'Aniello, A., Di Cosmo, A., Di Cristo, C., Annunziato, L., Petrucelli, L., & Fisher, G., (1996). Involvement of D-aspartic acid in the synthesis of testosterone in rat testes. *Life Sci., 59*, 97–104. 10.1016/0024–3205(96)00266–4.

134. Di Cosmo, A., Paolucci, M., Di Cristo, C., Botte, V., & Ciarcia, G., (1998). Progesterone receptor in the reproductive system of the female of *Octopus vulgaris*: Characterization and immunolocalization. *Mol. Reprod. Dev., 50*, 451–460. 10.1002/(SICI)1098–2795(199808)50:4&lt, 451:AID-MRD9&gt, 3.0.CO, 2-H.

135. Di Cristo, C., Paolucci, M., & Di Cosmo, A., (2008). Progesterone affects vitellogenesis in *Octopus vulgaris*. *Open Zoology Journal, 1*, 29–36.

136. Di Cosmo, A., Di Cristo, C., & Paolucci, M., (2001). Sex steroid hormone fluctuations and morphological changes of the reproductive system of the female of *Octopus vulgaris* throughout the annual cycle. *J. Exp. Zool., 289*, 33–47. 10.1002/1097–010X(20010101 /31)289:1&lt, 33:AID-JEZ4&gt, 3.0.CO, 2-A.

137. Tosti, E., di Cosmo, A., Cuomo, A., Di Cristo, C., & Gragnaniello, G., (2001). Progesterone induces activation in *Octopus vulgaris* spermatozoa. *Mol. Reprod. Dev., 59*, 97–105. 10.1002/mrd., 1011.

138. De Lisa, E., Carella, F., De Vico, G., & Di Cosmo, A., (2013). The gonadotropin releasing hormone (GnRH)-like molecule in prosobranch *Patella caerulea*: Potential

biomarker of endocrine-disrupting compounds in marine environments. *Zoolog. Sci.,* *30*, 135–140. 10.2108/zsj.30.135.

139. Zatylny, C., Marvin, L., Gagnon, J., & Henry, J., (2002). Fertilization in *Sepia officinalis*: The first mollusk sperm-attracting peptide. *Biochem. Biophys. Res. Commun., 296*, 1186–1193.

140. Steidle, H., (1930). Uber die verbreitung des weiblichen sexualhormons . *Arch. Exp. Path. Pharm., 157*, 89.

141. Hangerman, D. D., Wellington, F. M., & Viillee, C. A., (1957). Estrogens in marine invertebrates. *The Biological Bulletin, 112*, 180–183. 10.2307/1539196.

142. Janer, G., & Porte, C., (2007). Sex steroids and potential mechanisms of non-genomic endocrine disruption in invertebrates. *Ecotoxicology, 16*, 145–160. 10.1007/s10646–006–0110–4.

143. Croll, R. P., & Wang, C., (2007). Possible roles of sex steroids in the control of reproduction in bivalve molluscs. *Aquaculture, 272*, 76–86. https://doi.org/10.1016/j. aquaculture.2007.06.031.

144. Fernandes, D., Loi, B., & Porte, C., (2011). Biosynthesis and metabolism of steroids in molluscs. *J. Steroid Biochem. Mol. Biol., 127*, 189–195. 10.1016/j. jsbmb.2010. 12. 009.

145. Stanczyk, F. Z., & Clarke, N. J., (2010). Advantages and challenges of mass spectrometry assays for steroid hormones. *J. Steroid Biochem. Mol. Biol., 121*, 491–495. 10.1016/j. jsbmb.2010.05.001.

146. Zullo, L., Sumbre, G., Agnisola, C., Flash, T., & Hochner, B., (2009). Nonsomatotopic organization of the higher motor centers in octopus. *Curr. Biol., 19*, 1632–1636. 10.1016/j.cub.2009.07.067.

147. Taieb, J., Benattar, C., Birr, A. S., & Lindenbaum, A., (2002). Limitations of steroid determination by direct immunoassay. *Clin. Chem., 48*, 583–585.

148. Clark, J. H., & Peck, E. J., (2012). *Female Sex Steroids: Receptors and Function.*

149. Huang, P., Chandra, V., & Rastinejad, F., (2010). Structural overview of the nuclear receptor superfamily: Insights into physiology and therapeutics. *Annu. Rev. Physiol., 72*, 247–272. 10.1146/annurev-physiol-021909–135917.

150. Laudet, V., (1997). Evolution of the nuclear receptor superfamily: Early diversification from an ancestral orphan receptor. *J. Mol. Endocrinol., 19*, 207–226.

151. Maglich, J. M., Sluder, A., Guan, X., Shi, Y., McKee, D. D., Carrick, K., Kamdar, K., Willson, T. M., & Moore, J. T., (2001). Comparison of complete nuclear receptor sets from the human, Caenorhabditis elegans and Drosophila genomes. *Genome Biol., 2*, RESEARCH0029.

152. Moore, L. B., Maglich, J. M., McKee, D. D., Wisely, B., Willson, T. M., Kliewer, S. A., Lambert, M. H., & Moore, J. T., (2002). Pregnane X receptor (PXR), constitutive androstane receptor (CAR), and benzoate X receptor (BXR) define three pharmacologically distinct classes of nuclear receptors. *Mol. Endocrinol., 16*, 977–986. 10.1210/ mend.16.5.0828.

153. Escriva, H., Safi, R., Hänni, C., Langlois, M. C., Saumitou-Laprade, P., Stehelin, D., Capron, A., Pierce, R., & Laudet, V., (1997). Ligand binding was acquired during evolution of nuclear receptors. *Proc. Natl. Acad. Sci. USA, 94*, 6803–6808.

154. Saez, P. J., Lange, S., Pérez-Acle, T., & Owen, G., (2010). Nuclear receptor genes: Evolution. In: Sons, J. W., (ed.), *Encyclopedia of Life Sciences*. Chichester.

155. Fang, H., Oates, M. E., Pethica, R. B., Greenwood, J. M., Sardar, A. J., Rackham, O. J., Donoghue, P. C., Stamatakis, A., De Lima Morais, D. A., & Gough, J., (2013). A daily-updated tree of (sequenced) life as a reference for genome research. *Sci. Rep., 3*, 10.1038/srep02015.

156. Csaba, G., (2000). Hormonal imprinting: Its role during the evolution and development of hormones and receptors. *Cell Biol. Int., 24*, 407–414. 10.1006/cbir.2000.0507.
157. Markov, G. V., (2011). *Evolution of Steroid Signaling in Metazoans*. Agricultural sciences. Ecole normale supérieure de lyon-ENS LYON.
158. Tarrant, A. M., Reitzel, A. M., Blomquist, C. H., Haller, F., Tokarz, J., & Adamski, J., (2009). Steroid metabolism in cnidarians: Insights from Nematostella vectensis. *Mol. Cell Endocrinol., 301*, 27–36. 10.1016/j.mce.2008.09.037.
159. Adeel, M., Song, X., Wang, Y., Francis, D., & Yang, Y., (2017). Environmental impact of estrogens on human, animal and plant life: A critical review. *Environ. Int., 99*, 107–119. 10.1016/j.envint.2016.12.010.
160. Markov, G. V., & Laudet, V., (2011). Origin and evolution of the ligand-binding ability of nuclear receptors. *Mol. Cell Endocrinol., 334*, 21–30. 10.1016/j.mce.2010.10.017.
161. Thornton, J. W., Need, E., & Crews, D., (2003). Resurrecting the ancestral steroid receptor: ancient origin of estrogen signaling. *Science, 301*, 1714–1717. 10.1126/science.1086185.
162. Thornton, J. W., (2001). Evolution of vertebrate steroid receptors from an ancestral estrogen receptor by ligand exploitation and serial genome expansions. *Proc. Natl. Acad. Sci. USA, 98*, 5671–5676. 10.1073/pnas.091553298.
163. Scott, A. P., (2018). Is there any value in measuring vertebrate steroids in invertebrates? *Gen Comp Endocrinol*. 10.1016/j.ygcen.2018.04.005.
164. Markov, G. V., Tavares, R., Dauphin-Villemant, C., Demeneix, B. A., Baker, M. E., & Laudet, V., (2009). Independent elaboration of steroid hormone signaling pathways in metazoans. *Proc. Natl. Acad. Sci. USA, 106*, 11913–11918. 10.1073/pnas.0812138106.
165. Di Cosmo, A., Di Cristo, C., & Paolucci, M., (2002). A Estradiol-17β receptor in the reproductive system of the female of *Octopus vulgaris*: Characterization and immunolocalization. *Mol. Reprod. Dev., 61*, 367–375. 10.1002/mrd.10014.

CHAPTER 10

Physiological Functions of Gastropod Peptides and Neurotransmitters

SPENCER T. MUKAI[1] and FUMIHIRO MORISHITA[2]

[1]Department of Multidisciplinary Studies, Glendon College, 2275 Bayview Avenue, Toronto, ON, Canada M4N 3M6. E-mail: smukai@yorku.ca

[2]Department of Biological Science, Graduate School of Science, Hiroshima University, 1-3-1 Kagamiyama, Higashi-Hiroshima 739-8526, Hiroshima, Japan. E-mail: fumi425@hiroshima-u.ac.jp

10.1 INTRODUCTION

10.1.1 WHY STUDY GASTROPOD PEPTIDES AND NEUROTRANSMITTERS?

Among the Mollusca, the Gastropoda is the largest and most diverse class, containing an estimated 40,000–150,000 species which constitute about 85% of all described molluscs (Haszprunar and Wanninger, 2012; Ponder and Lindberg, 2008). Besides being the most numerous molluscan group, there are several important reasons why regulatory peptides and neurotransmitters are the subject of intensive research in this group. From a scientific standpoint, many gastropod molluscs have large, identifiable neurons, and their anatomically simpler nervous systems are highly amenable for the study of peptide structure and function. For example, the study of egg-laying-inducing neuropeptides in *Aplysia californica* and *Lymnaea stagnalis* has uncovered important cellular mechanisms involved in neuropeptide biosynthesis, processing, transport and release (for excellent reviews see Arch and Berry, 1989; Nagle et al., 1989a; Sossin et al., 1990a,b). Other studies in bivalves and in gastropods have identified novel peptides involved in shell repair and biomineralization (Mann et al., 2012; Werner et al., 2013; Zhao et al., 2012a).

Gastropod molluscs are of economic importance to humans as well. In many regions of the world, abalone, conch, escargot, periwinkles, and whelks are considered delicacies. They are also and provide income to local human populations that harvest them. Studies focused on peptides and neurotransmitters governing growth and reproduction may have direct applications to molluscan aquaculture and food security. In addition, many novel chemicals which have potential biomedical or commercial value have been discovered in gastropods (Cimino and Gavagnin, 2006; Benkendorff, 2010). For example, numerous peptide venoms from cone snails that modulate ion channels have been isolated and identified as sources of novel medicinal or research chemicals (Dang et al., 2015; Vetter and Lewis, 2012).

In various regions of the world, terrestrial slugs and snails are major agricultural and horticultural pests. Grazing by slugs is known to damage or to reduce the yield of crops such as canola, soybeans, cereals, and legumes, whereas terrestrial snails such as *Helix aspersa* are a major pest in citrus orchards (Barker, 2002). Some freshwater pulmonates are intermediate hosts of trematode parasites which infect numerous mammalian species, including humans, and thus are of direct significance to human health. Freshwater pulmonates such as *Biomphalaria glabrata*, *Bulinus truncatus*, and *Oncomelania hupensis* are intermediate hosts of *Schistosoma* spp., whereas several species of freshwater snails can serve as hosts for other parasitic flukes such as *Fasciola hepatica* and *Clonorchis sinensis* (Roberts et al., 2013). Therefore, the study of peptides and neurotransmitters in these snails may lead to the development of strategies aimed at halting the transmission of diseases such as schistosomiasis and fasciolosis.

Ecologically, terrestrial and aquatic gastropods are an important part of numerous ecosystems as they have large influences over both animal and plant communities (Russell-Hunter, 1983). Because of their importance in both terrestrial and aquatic ecosystems and their sensitivity to pollutants, perturbation of chemical signaling molecules of the gastropod neuroendocrine system has been used as a bioindicator of environmental health (Lagadic et al., 2007; Oehlmann et al., 2007). Finally, many molluscan species are experiencing global decline so a better understanding of the peptides and neurotransmitters regulating physiological processes is important for the conservation and maintenance of biodiversity (Harley, 2011; Lydeard et al., 2004).

10.1.2 PEPTIDES

Animal cells can receive information from the external and internal environment in the form of light–dark cues, nutrients, temperature, water, and ion concentrations, and then process this information to coordinate cellular activities in the whole organism. The communication of information between animal cells is an essential activity that maintains homeostasis. The two major systems that maintain homeostasis in animals are the nervous and endocrine systems. The cells that belong to these two systems communicate with one another is through chemical signaling molecules such as peptides and neurotransmitters.

Peptides are the largest and most diverse group of extracellular signaling molecules, and probably represent an evolutionarily ancient form of chemical communication. For example, in the Cnidaria, peptides are utilized as the main form of chemical communication during development (Grimmelikhuijzen et al., 1996). In invertebrates, many peptide signaling molecules are synthesized and secreted from neurons where they can act on target cells as neurotransmitters, neuromodulators, or neurohormones (Altstein and Nässel, 2010; Kiss, 2011; Thorndyke and Goldsworthy, 1988). Peptides are generally grouped together based on the structural similarities of their primary amino acid sequences (e.g., RFamide-like peptides, insulin-like peptides, myomodulin-like peptides). However, peptides with structural similarities can often have different functions in the same animal. In addition, peptide nomenclature can sometimes be confusing as the peptide's name may not reflect its main physiological function as many peptides are assigned a name based on the functional assay that was originally used to characterize the peptide. In some cases, peptides may even have the same acronym (e.g., MIP-*m*olluscan *i*nsulin-related *p*eptide or *Mytilus* *i*nhibitory *p*eptide).

10.1.3   METHODS USED TO IDENTIFY AND ISOLATE MOLLUSCAN PEPTIDES

Since the publication of the Physiology series of "The Mollusca" (Volumes 4 and 5) more than 30 years ago, there have been enormous advances in peptide biochemistry which has enabled the isolation and identification of an unprecedented number of novel peptides from molluscs. In general, there are two different ways to isolate and identify peptides—the first is a "function first or forward approach", and the second is a "peptide first or reverse approach". The function first approach seeks to purify a peptide based on its physiological activity in a bioassay. Historically, the bioassay is the oldest method used to identify biologically active peptides. It is generally very reliable since it detects activity based on physiological function. Bioassays can be performed on whole animals (in vivo bioassay) such as the induction of egg-laying in *A. californica* by crude abdominal ganglion extracts

(Kupfermann, 1967) or in vitro using isolated cells, tissues, or whole organs in culture.

The in vitro bioassays provide greater sensitivity and have high throughput compared to in vivo assays; however, they do not reflect the physiological complexity of the entire animal. An example of an in vitro bioassay that is highly successful in isolating novel myotropic peptides is the effect of peptide-containing tissue extracts on pieces of muscle attached to a sensitive force transducer (Kuroki et al., 1990; Muneoka and Kobayashi, 1992). Typically, cell or tissue extracts are fractionated by chromatography (e.g., high-performance liquid chromatography, HPLC) and then the separated fractions are tested in the bioassay for physiological activity. Additional rounds of chromatography coupled with further bioassays are performed until a chromatographically pure peptide fraction is obtained for sequence analysis. The first molluscan neuropeptide isolated in such a fashion was the cardioexcitatory peptide FMRFamide (Price and Greenberg, 1977).

The peptide first or reverse approach typically seeks to identify the primary sequences of peptides in a cell or tissue extract, and then determine what their functions are. A commonly used approach is to identify molluscan peptides employs radioimmunoassay (RIAs) or enzyme-linked immuno-sorbent assays directed against the amidated C-terminus of peptides, since α-amidation is a common chemical modification of secreted peptides. These immunological detection methods are usually combined with chromato-graphic separation of cell or tissue extracts and have the ability to identify groups of related peptides in different molluscan species or from different animal pkyla (Greenberg and Price, 1992; Nassel, 1999).

During the 1980s, molecular methods such as the cDNA cloning of precursor proteins enabled the identification of genes encoding the deduced amino acid sequence of peptide precursors (Nambu et al., 1983; Scheller et al., 1982; Taussig and Scheller, 1986). For example, the first insulin-like peptide (ILP) in invertebrates was isolated from the neuroendocrine light green cells (LGCs) of the pond snail *L. stagnalis* using a subtractive hybrid-ization approach (Smit et al., 1988). Furthermore, since the vast majority of peptides bind to cell surface G-protein-coupled receptors (GPCRs), another frequently used strategy involves the cloning of orphan GPCRs, then expressing these receptors in a heterologous expression system. Peptide extracts or purified peptides are then screened for their ability to bind and activate the expressed receptors. A novel cardioactive peptide has been isolated from *L. stagnalis* using this approach (Tensen et al., 1998).

More recently, HPLC or capillary electrophoresis coupled with sensitive techniques such as mass spectrometry allows for the identification of peptides from complex tissue mixtures or from cell releasates in vitro (Croushore et al., 2012; Nemes et al., 2011). Moreover, during the last decade, technical improvements to massively parallel or next-generation sequencing techniques (e.g., Illumina, Roche 454, PacBio) have allowed the entire transcriptome of the cell or tissue to be rapidly sequenced and then in silico predictions of function can be inferred (Puthanveettil et al., 2012; Senatore et al., 2015). Sequences can be matched against a reference genome, if one is available, or in the case of less studied species, de novo transcriptome assembly is undertaken. In addition, these next generation sequencing methods allow for the quantitation of transcript levels in tissues, and even single cells, under different experimental conditions allowing one to infer function (Chu et al., 2014). Antibodies may also be generated against specific peptide fragments derived from proteomic or RNA sequencing studies, and then localized in cells by immunohistochemistry or used in functional assays (De Haes et al., 2014).

This chapter deals with the physiological functions of selected peptides in gastropod molluscs, as well as the contribution of some classic neurotransmitters such as serotonin and dopamine as it relates to selected physiological processes in gastropods. Since it is nearly impossible to describe all the peptides and neurotransmitters regulating the diverse physiological processes in gastropods, we will focus our attention to those peptides and neurotransmitters where there is a substantial body of evidence pertaining to specific physiological functions, namely reproduction, growth, cardiac function, osmoregulation, and ciliary activity.

10.2 REGULATION OF REPRODUCTION

Reproduction is a process that is essential for the continuation of a species and contributes to genetic diversity in sexually reproducing organisms. The Gastropoda have radiated into varied and diverse habitats which include trees, deserts, lakes, rivers, estuaries, marine intertidal zones, pelagic, and benthic habitats. There is a staggering array of reproductive strategies used by gastropod molluscs, and the organization of the gastropod reproductive system reflects this considerable diversification (Tompa et al., 1984). In gastropods, the Archaeogastropods and Neogastropods are gonochorists. Fertilization occurs externally in Archaeogastropods, whereas in Neogastropods it is internal. By contrast, the pulmonates and opisthobranchs are hermaphroditic, and fertilization occurs internally after the exchange of gametes with conspecifics. Synchronized gamete release is essential for species with external fertilization, hence the maturation of the gonads as well as the accessory sex organs must be synchronized to environmental cues (e.g., photoperiod or temperature), which trigger gamete release. On the other hand, courtship and mating behaviors contribute to successful copulation with conspecifics and are generally associated with species where fertilization is internal.

The neuroendocrine control of egg-laying has been studied extensively in two species of gastropods, the sea hare *A. californica* and the pond snail *L. stagnalis*. In these two gastropods, and likely in others, egg-laying involves a series well defined physiological processes including ovulation, packaging the fertilized eggs into the egg string or egg mass, and finally oviposition. In addition, specific stereotyped behaviors have been documented during oviposition, which are controlled by neuropeptides (Geraerts et al., 1988). The peptidergic control of egg-laying behavior in *A. californica* will be used as an example, since this the cell and molecular biology of the neuroendocrine system is well understood.

10.2.1   NEUROPEPTIDES REGULATING FEMALE REPRODUCTION IN APLYSIA

Egg laying in *A. californica* and *L. stagnalis* consists of a series of stereotyped behaviors including ovulation, the packaging of fertilized eggs into the egg string/mass, and oviposition. The egg-laying hormone (ELH) and ELH-related peptides such as califin in *Aplysia*, as well as the caudodorsal cell hormone (CDCH) in *Lymnaea* are well studied peptide hormones controlling egg-laying. In particular, there is a long history of investigation on the regulatory actions of ELH and its related peptides on reproduction in *A. californica* that began almost 50 years ago and continues to this day.

10.2.2   THE BAG CELLS AND ATRIAL GLAND

In *Aplysia*, two distinct secretory centers are known to regulate egg-laying behavior. The first center is the bag-cell neurons that are located around proximal region of the pleuro-abdominal connectives on rostral side of the abdominal ganglion (see Fig. 10.1). The second center is the atrial gland, an accessory sex gland that is located on the large hermaphroditic duct (Hadfield and Switzer-Dunlap, 1984). The bag cells are clusters of neurosecretory cells consisting of around 400 cells in the left and right portions of the abdominal ganglion (Coggeshall et al., 1966). Axon terminals of the bag cells are found in the connective tissue sheath covering the pleuro-abdominal connectives, as well as in the space between the surface of the abdominal ganglion and the ganglionic sheath. Neurosecretory material enters the systemic circulation via the caudal artery (Chiu and Strumwasser, 1981; Kaczmarek et al., 1979).

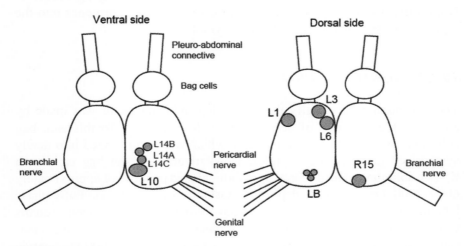

FIGURE 10.1 Ventral and dorsal view of the abdominal ganglion of *Aplysia* showing the location of the neurosecretory bag cells, bursting neuron R15, neurons L1, L3, L6, L10, neurons L14A/B/C, neuronal cluster LB and LC.

The function of the bag cells in *A. californica* was first demonstrated by Irving Kupfermann in the late 1960s and early 1970s. Injection of a crude homogenate of the bag cells into sexually mature *A. californica* induced egg-laying (Kupfermann, 1967, 1970). Brief electrical stimulation to the abdominal ganglia induces depolarization of the bag cells, initiating the synchronized discharge of action potentials, the afterdischarge, which continues for more than 20 min (Kupfermann, 1970). The afterdischarge of the bag cells can also be recorded in vivo with extracellular electrodes placed in the vicinity of the bag cells (Pinsker and Dudek, 1977; Dudek et al., 1979). Thus, the afterdischarge of the bag cells is an important event for the initiation of egg-laying.

It has also been recognized that the freshly deposited egg strings of *A. californica* release factors that promote egg-laying in conspecifics (Nagle et al., 1985). The source of these egg-laying stimulatory factors was determined to be an accessory sex organ, the atrial gland, an exocrine organ located on the reproductive tract. Several peptides with egg-laying activity have been isolated from the atrial gland such as califin and Peptide-A (Nagle et al., 1985; Rothman et al., 1986). Califin and Peptide-A are packaged into the secretory granules of the glandular cells of the atrial gland, and trigger egg-laying in *A. californica* when injected into mature animals. However, the involvement of the atrial gland in the regulation of egg-laying behaviors is not fully understood since the gland secretes peptide hormones into the reproductive tract but not into the hemolymph.

10.2.3   IDENTIFICATION OF ELH

The first attempt to identify the egg-laying inducing factor was made by analyzing the secretion of peptides from the bag cells. To this end, bag cells were incubated with [^{3}H]-leucine which was incorporated into newly synthesized proteins and peptides. Depolarization of the bag cells with K$^+$-rich solution induced secretion of radiolabeled peptides of approximately 6000–7000 Da in a Ca^{2+}-dependent fashion (Arch, 1972a). Pulse-chase labeling with [^{3}H]-leucine revealed that most newly synthesized proteins were initially present as larger protein around 25,000 Da, and then the quantity of radioactively labeled proteins increased in a 6000–7000-Da peptide (Arch, 1972b). Moreover, the egg-laying activity of the peptide extract of the bag cells coeluted with the 6000–7000 Da material from gel filtration chromatography (Arch et al., 1976a). These results strongly suggested that the 6000–7000-Da peptide was the ELH of *Aplysia*.

The bag cells project their nerve endings into the space between the surface of the abdominal ganglion and the overlying connective tissue sheath. Since this extraganglionic space is connected with the systemic circulation via the caudal artery, it is likely that ELH released in this space is delivered to various parts of the abdominal ganglion. To determine if the bag cells release ELH peptide during the afterdischarge, the bag cells were pulse-labeled with radioactive amino acid which was followed by electrical stimulation and collection of releasate. It was clearly demonstrated that the radioactive label and egg-laying activity coeluted in a peak of 6000–7000 Da (Mayeri et al., 1985). Subsequent purification of ELH from bag cells extracts was accomplished using HPLC and bioassay of the fractions possessing egg-lying inducing activity. A peptide of 36 amino acids was identified after automated Edman degradation (Chiu et al., 1979).

10.2.4 PHYSIOLOGICAL FUNCTIONS OF ELH

The physiological actions of ELH are diverse. Application of ELH-containing extracts to isolated pieces of ovotestis in vitro induces the release of oocytes into the bathing media (Dudek and Tobe, 1978; Dudek et al., 1980; Rothman et al., 1983b). ELH appears to have a direct action on the ovotestis by inducing ovulation. Another important action of ELH is the modification of the neuronal activities in various neurons in the central ganglia (see Table 10.1). Perfusion of ELH into the isolated abdominal ganglion from the caudal artery initiates discharge in the silent motoneuron, the LC cell, which regulates muscle tone in in the gill, pericardial region, and vasculature (Mayeri et al., 1979a,b). ELH also increases the frequency of discharge in the bursting pacemaker neuron R15. Neuron R15 is itself a peptidergic neuron that releases R15α peptide as well as other peptides which are thought to be involved in cardiovascular function and osmoregulation (Weiss et al., 1989). Immunohistochemical localization of R15α peptide-containing nerve processes revealed that R15 peptides are distributed in the in the auricle, pericardium and arteries (Alevizos et al. 1991a,b) as well as in the large hermaphroditic duct (Alevizos et al., 1991d). Moreover, R15α peptide is myoactive on those tissues (Alevizos et al., 1991d). Thus, R15 affects transport of the egg-strings by regulating the contraction of the muscular tissues of the large hermaphroditic duct. The left lower quadrant cells in the abdominal ganglion also mediate the regulation of the arterial system by the bag cells (Ligman and Brownell, 1985). The increase in cardiovascular activity during egg-laying may help control locomotor activity by altering the hydrostatic pressure and redistributing hemolymph in the body.

The left upper quadrant (LUQ) cells express two neuropeptide precursors, LUQ-1 that encodes bradykinin-related peptides and L5–67 that encodes ACEP-1-related peptide. LUQ-1 peptides are found in the nerve processes innervating the ovotestis, whereas L5–67 peptide is found in axonal processes projecting to the heart. Although the physiological actions of bradykinin- and ACEP-1-like peptides are not fully understood in *Aplysia*, it is likely that these peptides are involved in the regulation of egg-laying behaviors because bag cell discharge negatively regulates the discharge of the LUQ cells. These diverse actions of ELH on neural activity are fundamental to the complex, integrated behaviors that regulate egg-laying.

TABLE 10.1 The Actions of ELH on the Neurons and Neural Clusters in the Abdominal Ganglion of *Aplysia* (Source: Mayeri, E. *Fed. Proc.* **1979,** *38,* 2103–2108).

	Neurons	Functions	Response to ELH	End response
Neurons	L3/L6	Bursting pacemaker neuron	Prolonged inhibition	
	R1/L1	Mechanoreceptor neuron	Transient excitation	Unknown
	L10	Cardiovascular motoneuron	Transient inhibition followed by prolonged excitation	Increase in cardiac output
	L14A/B/C	Ink-gland motoneurons	Slow inhibition	Suppression of ink release
	R15			Salt and water balance
Neuronal clusters	LB/LC	Motoneurons on gill, siphon, and vasculature	Prolonged excitation	Regulation of muscle tone

10.2.5　ELH-ASSOCIATED PEPTIDES IN APLYSIA

Although purified ELH mimics various actions induced by bag cell after-discharge or by the application of the bag cell extract to the abdominal ganglion, it does not fully reproduce specific effects on other neurons such as the inhibition of the left-upper quadrant (LUQ) cells. Therefore, the bag cells release other signaling molecules together with ELH during the after-discharge. One group of peptides that was identified was α-, β-, and γ-bag cell peptides (BCPs). The β-BCP and γ-BCP share the common Arg-Leu-Arg-Phe (RLRF) sequence with the α-BCP (see Table 10.2).

The specific functions for some of the BCPs have been reported. For instance, the α-bag cell peptide (α-BCP) is known to inhibit neuronal activity of the LUQ cells (Rothman et al., 1983a; Sigvardt et al., 1986). Another important function of the BCPs is auto-excitation of the bag cells (Brown et al., 1989). All the BCPs depolarized membrane potential in the neurites of the bag cells with the following rank potency order, α-BCP > β-BCP > γ-BCP. In addition, the BCPs modulate intracellular cAMP levels in the bag cells by changing the activity of adenylate cyclase in the cells (Redman and Berry, 1992). Interestingly, the effects on intracellular cAMP levels is temperature-dependent, namely, α- and γ-BCP reduced cAMP at 15°C, whereas at 30°C, it increased cAMP (Loechner et al., 1990). β-BCP, as well as the RLRF peptide (the common structure shared by the three BCPs), increased intracellular cAMP in the bag cells at both low and high temperatures (Berry et al., 1994). Since the inhibitory actions of the α- and γ-BCPs on cAMP are GTP-dependent and pertussis toxin-sensitive, this indicated the involvement of the inhibitory GTP-binding protein (Gi) (Redman and Berry, 1993). Although it is unclear whether the increase in cAMP is mediated by the stimulatory GTP-binding protein (Gs), temperature-dependent switching between Gi and Gs is a suggested mechanism in the bag cells. These temperature-dependent modulations of bag cell activity may be involved in the initiation of egg-laying behavior which is associated with changes in seasonal temperature.

All the BCPs also induce depolarization of the cell membrane in cultured bag cells. The depolarization response induced by the BCPs can be desensitized by repeated BCP application. During desensitization there is an augmentation of K^+-current via the delayed voltage-gated potassium channel (Loechner and Kaczmarek, 1990, 1994). Interestingly, the cultured bag cells that had been previously desensitized with α-BCP responded to β-BCP by depolarization, suggesting that the two peptides bind to different bag cell autoreceptors from one another (Loechner and Kaczmarek, 1994).

Immunohistochemistry using anti-ELH and anti-α-BCP antibodies demonstrate the localization of ectopic bag cells in the cerebral and right pleural ganglia, respectively (Painter et al., 1989; Chiu and Strumwasser, 1984). The electrical properties of the α-BCP-immunopositive cells in the central ganglia resemble that of the bag cells in the abdominal ganglion (Brown et al., 1989). The α-BCP-containing cells send their processes to the neuropile, but not to the connective tissues surrounding the ganglia (Brown et al., 1989). In fact, burst discharge of the α-BCP-containing cells initiates the burst discharge in the bag cells with a 10-s delay, and vice versa, suggesting a functional coupling between the α-BCP-containing cells with the bag cells (Brown and Mayeri, 1989). Furthermore, retrograde labeling (Shoppe et al., 1991), as well as the detection of α-BCP by mass spectrometry (Hatcher and Sweedler, 2008), supports this functional connection. Since sensory inputs are integrated in the cerebral and pleural ganglia, it is likely that sensory stimuli including temperature, photoperiod, and chemical cues (e.g., pheromones) may trigger burst discharge of the α-BCP-containing cells in the head ganglia, which then initiates afterdischarge in the bag cells.

TABLE 10.2 Structures of Egg-laying Hormone and Related Peptides Found on the ELH-related Peptide Precursors in Molluscs.

Source animal	Peptide name	Structure
Aplysia	ELH	ISINQDLKAITDMLLTEQIRERQRYLADLRQRLLEKa
	Califin (large)	ISINQDLKAITDMLLTEQIQARRRCLDALRQRLLDLa
Lymnaea	CDCH-1	LSITNDLRAIADSYLYDQNKLRERQEENLRRRFLELa
	CDCH-II	SITNDLRAIADSYLYDQHKLREQQEENLRRRFY-ELSLRPYPDNLa
Peptides found on the ELH and CDCH precursors		
Aplysia	Peptide A	IFVPNRAVKLSSDGNYPFDLSKEDG AQPYFMTPRLRFYPI
	α-BCP	APRLRFYSL
	β-BCP	RLRFH
	γ-BCP	RLRFSD
Lymnaea	α-CDCP	EPRLRFHDV
	β_1-CDCP	RLRFH
	β_2-CDCP	RLRAS
	β_3-CDCP	RLRFN
Aplysia	δ-BCP	DQDEGNFRRFPTNAVSMSADENSPFDL SNEDGAVYQRDL
Lymnaea	Calfluxin	RVDS---ADESNDDGFD
Aplysia	Califin (small)	DSDVSLFNGDLLPNGRCS
	Acidic peptide	SSGVSLLTSNKDEEQRELLKAISNLLD

The *Aplysia* atrial gland is known to produce the peptide califin and Peptide A. Califin consists of a large subunit (36 amino acids) and small subunit (18 amino acids) joined by a disulfide bond. The N-terminal 19 amino acids of califin's large subunit are identical to that of ELH. Peptide A is composed of 34 amino acids and shares a common C-terminal Arg-Leu-Arg-Phe sequence to that of α-BCP. Both califin and Peptide A induce egg-laying when injected to sexually mature *A. californica* and activate putative receptors for ELH and α-BCP, respectively.

It is clearly established that ELH and ELH-associated peptides govern various aspects of female reproduction from ovulation to oviposition in *Aplysia*. However, it is unlikely that these peptides induce gonadal maturation in the animal since injection of ELH and BCPs into animals outside the breeding season does not induce gonadal maturation, nor egg-laying.

Information on the extracellular signaling molecules that induce gonadal maturation in *Aplysia* is largely unknown (reviewed by Hadfield and Switzer-Dunlap, 1984).

10.2.6 INTRACELLULAR MECHANISMS CONTROLLING AFTERDISCHARGE IN THE BAG CELLS

The bag cell afterdischarge is triggered by a brief electrical stimulation and is essential for the prolonged release of ELH. What is happening in the bag cell to initiate the afterdischarge? Electrical stimulation depolarizes the bag cell membrane and induces the opening of voltage-gated Ca^{2+}-channels resulting in Ca^{2+} influx across the cell membrane. The local increase in Ca^{2+} in the vicinity of cell membrane causes the opening of Ca^{2+}-sensitive, nonselective cation channels. The elevation of internal Ca^{2+} induces sustained depolarization of membrane potential, which results in the prolonged discharge of the bag cells.

Although the afterdischarge is essential for triggering ELH release, the time course of ELH release is somewhat different from that of discharge (Loechner et al., 1990). The frequency of discharge is maximal within a few minutes, then it decreases gradually and terminates in 15–20 min. By contrast, ELH release is maximal after 15–20 min, then decreases to basal levels as the discharge is terminated. In fact, extracellular Ca^{2+} influx via the plasma membrane is not necessary for ELH-release (Wayne and Frumovitz, 1995). For instance, depletion of extracellular Ca^{2+} by the Ca^{2+} chelator BAPTA, reduced the duration of afterdischarge, while it had little effect on the ELH-secretion (Wayne et al., 1998). Any treatments that increase intracellular Ca^{2+} levels including Ca^{2+} mobilization from intracellular stores or Ca^{2+}-influx by sustained depolarization of membrane potential induces ELH-release without afterdischarge (Wayne et al., 2004). Recent evidence suggests the importance of Ca^{2+} buffering by intracellular Ca^{2+} stores and mitochondria for ELH release (Groten et al., 2013; Hickey et al., 2010, 2013).

The molecular mechanisms that induce the afterdischarge and prolonged release of ELH from the bag cell are complex and are not yet fully understood. However, studies by the Greengard and Kaczmarek labs demonstrated that several protein kinases mediated these phenomena. For instance, the fact that some calmodulin (CaM) inhibitors such as TFP and W-7 inhibit the afterdischarge, and ELH-secretion induced by electrical stimulation of the bag cells (DeRiemer et al., 1985) suggests the involvement of CaM-dependent enzymes. One of the enzymes is CaM-dependent kinase (CaM-kinase). In the

CaM-free soluble fraction prepared from the bag cells, phosphorylation of a 51-kDa protein is augmented in the presence of CaM and Ca^{2+} (DeRiemer et al., 1984). Although the entity of the 51-kDa protein is unclear, phosphorylation by CaM-kinase seems to modulate the activity of nonselective cation channels because a CaM-kinase inhibitor attenuated the spike-broadening of action potentials observed in the afterdischarge (Hung and Magoski, 2007).

The involvement of PKC in the afterdischarge is also suggested by the fact that action potential during the afterdischarge from the cultured bag cell is augmented by a PKC activator, TPA, whereas it is reduced by a PKC inhibitor, H-7 (Conn et al., 1989). Using various preparations such as whole-cell or excised patch clamp, it was demonstrated that PKC-dependent protein phosphorylation increases the open probability of a nonselective cation channel, which promotes Ca^{2+} influx (Gardam and Magoski, 2009; Tam et al., 2011). Single channel-containing patch clamping suggests a direct interaction between the PKC and the nonselective cation channel (Magoski and Kaczmarek, 2005; Magoski et al., 2002; Wilson et al., 1998).

PKC augments Ca^{2+} influx to the bag cells in different ways. In addition to the Ca^{2+} channel in the plasma membrane, the bag cells have vesicular pools of distinct Ca^{2+} channels (Strong et al., 1987; Wayne et al., 1999; White and Kaczmarek, 1997). The Ca^{2+} channel in the plasma membrane, classified as Apl Ca_v1, has a smaller Ca^{2+}-conductance and constitutively locates in the plasma membrane. By contrast, the Ca^{2+} channel in the vesicular pool, classified as Apl Ca_v2, has a larger Ca^{2+}-conductance and translocates to the plasma membrane, when the cells are excited. The discharge in the bag cells promotes translocation of Apl Ca_v2-containing vesicle through the activation of PKC. Apparently, it results in the increase of Ca^{2+} influx because the number of the functional Ca^{2+} channels is increased in the plasma membrane (Zhang et al., 2008b). Actin filaments appear to mediate channel-containing vesicle translocation and insertion to the plasma membrane because actin inhibitors disrupt vesicle fusion to the membrane, as well as the increase in the Ca^{2+} influx (Groten et al., 2013).

Conversely, PKC is also involved in the desensitization and refractory period of the bag cells. It is known that an injection of inositol trisphosphate (IP_3) into the cultured bag cell augmented the outward K^+-current. Since Ca^{2+} imaging with Fura-2 demonstrated that injected IP_3 mobilized intracellular Ca^{2+}, it is likely that K^+ efflux through the Ca^{2+}-sensitive K^+ channel is augmented. PKC-dependent phosphorylation appears to positively regulate the K^+ efflux, since a PKC activator augmented the Ca^{2+}-sensitive K^+-current, which results in the hyperpolarization of membrane potential and attenuation of the action potential (Zhang et al., 2002). These results suggest that activation of PKC promotes afterdischarge by increasing Ca^{2+} influx, whereas it attenuates the discharge by increasing K^+ efflux. Thus, PKC mediates both of the initiation and cessation of egg-laying.

The aforementioned results demonstrate that PKC plays important roles for the induction of the afterdischarge by elevating intracellular Ca^{2+} levels. How is PKC is activated in *Aplysia* bag cells? The hydrophilic IP_3 that mobilized Ca^{2+} from intracellular Ca^{2+} stores and hydrophobic diacylglycerol are important factors for the activation of PKC. These substances are generated by degradation of phosphatidyl inositol phosphate (PIP_2) catalyzed by phospholipase C (PLC). PLC is activated by interaction between ligand-bound GPCR and GTP-binding protein (Gq).

When the bag cells are labeled with [^{3}H]-inositol, incorporation of [^{3}H]-inositol into both of the soluble and membrane fractions of the cell are increased following the afterdischarge (Fink and Kaczmarek, 1988). Incorporation of [^{3}H]-inositol into the soluble fraction indicates IP_3 synthesis, while that into the membrane fraction reflects de novo synthesis of PIP_2 for replenishment. FMRFamide is one of the endogenous ligands that activate PLC in the bag cells, since the peptide modulates IP_3 production in various molluscan tissues (Falconer et al., 1993; Willoughby et al., 1999b). In fact, FMRFamide is known to terminate the afterdischarge in the bag cells (Fisher et al., 1993).

During the afterdischarge, a transient increase in intracellular cAMP levels also occurs (Kaczmarek et al., 1984). The cAMP concentration reached a peak around 2 min after the initiation of discharge, and returned to basal levels in 8 min. In cultured bag cells, the membrane-permeable cAMP analog 8-benzylthio-cAMP increased the inward Ca^{2+}-current through the attenuation of the delayed and transient K^+-currents. In the whole-mount cell patch on the cultured bag cell, forskolin, an adenylate cyclase activator, and theophylline, a phosphodiesterase inhibitor, showed similar effects (Strong and Kaczmarek, 1986). As mentioned above, BCPs released from the bag cells areknown to elevate cAMP level in the cell by activating adenylate cyclase.

To investigate the PKA-dependent phosphorylation during the afterdischarge, a crude membrane preparation of the bag cells was incubated with γ-^{32}P-ATP and analyzed by SDS-PAGE and autoradiography. Phosphorylated proteins with molecular weight at 33 kDa (BC-I) and 21 kDa (BC-II) were detected and markedly augmented by cAMP, suggesting phosphorylation by PKA (Jennings et al., 1982). Of the two proteins, BC-II seemed to be the bag cell specific protein, because it was not found in other ganglia.

When isolated bag cells were labeled with $Na_2H[^{32}P]O_4$ and then analyzed by SDS-PAGE/autoradiography at different times after the initiation of afterdischarge, radiolabeled BC-I and BC-II was detected at 20 min (but not at 2 min) following the afterdischarge. Such a delayed PKA-dependent phosphorylation during afterdischarge, together with the cAMP-dependent increase in K^+-current in the bag cell suggests that cAMP mediates termination of prolonged discharge of the bag cells. Unfortunately, the structure and function of the radiolabeled proteins are not known.

Using an excised inside-out patch clamp of the bag cells, Magoski's lab postulated a model in which the cation channel makes a complex with PKC and protein phosphatase (Magoski, 2004). In this hypothesis, PKC-dependent phosphorylation of the channel increases the open-probability of the cation channel in the initial phase of afterdischarge, while the protein phosphatase turns the channel back to the resting state by removing phosphate from the channel. On the other hand, in the late phase of the discharge, PKC can be replaced with PKA, and then PKA-dependent phosphorylation of the channel reduces the open-probability, which results in the termination of the discharge.

10.2.7   THE ELH PRECURSOR

The structure of the ELH gene was identified in the early 1980s. Messenger RNAs were prepared from the bag cells, abdominal ganglia, and digestive gland, and then cDNAs were synthesized by reverse-transcription. Using the cDNA as a probe, a genomic DNA-library of *A. californica* was screened and clones that exclusively hybridized with cDNA from the bag cells were isolated (Scheller et al., 1982). By analyzing the nucleotide sequences of the positive DNA fragments, the amino acid sequence of the ELH precursor protein was determined (Scheller et al., 1983). The precursor protein for ELH consisted of 271 amino acids, including N-terminal signal peptides. The ELH sequence was found at the C-terminal region of the precursor, which is followed by a C-terminal acidic peptide (AP). At the N-terminal region, α-, β-, δ-, and γ-BCPs were aligned in tandem. Thus, egg-laying associated peptides in the bag cell were encoded on the same precursor protein.

There are several variants of precursors encoding for ELH-related peptides such as Peptide-A and califin (Scheller et al., 1983). For instance, one precursor cDNA has a nucleotide deletion which results in the deletion of 83 amino acids in the N-terminal region of the ELH-precursor. As the result, this variant does not have β-, γ-, and δ-BCPs found in the ELH-precursor but encodes Peptide-A. The Peptide-A precursor also includes the ELH-related peptide, califin, at its C-terminal region. The nucleotide sequence of another variant encoding the Peptide-B precursor is similar to that of the Peptide-A precursor. However, the encoded precursor protein is much shorter than that of the Peptide-A precursor because a deletion of a nucleotide in the Peptide-B precursor causes a frame shift, resulting in a stop codon in the middle of the precursor protein. As a result, Peptide B does not include the califin-coding region. The ELH precursor is expressed in the bag cells, whereas the Peptide-A and Peptide-B precursors are expressed in the atrial gland (Nagle et al., 1986; Scheller et al., 1982).

The *Aplysia* ELH-gene consists of three exons (Mahon et al., 1985a). The short exons, exon I and II, are connected by a splicing consensus nucleotides (GT), while exon II and III are connected by an intron-spanning 5.3-kb sequence. The entire length of the open-reading frame for all the ELH precursors is located within the confines of exon III. Thus, alternative splicing produces mRNA variants with different 5'-UTRs.

Besides *A. californica*, the ELH precursor is conserved among the other *Aplysia* species including, *A. parvula, A. punctate, A. brasiliana, A. dacty-lomela, A. oculifera, A. juliana*, and *A. vaccaria* (Li et al., 1999; Nambu and Scheller, 1986). By contrast, Peptide-A and -B precursors are not found in these marine snails. It is likely that the Peptide-A and -B precursor genes evolved after the divergence of *A. californica* and other *Aplysia* species from their common ancestor (Nambu and Scheller, 1986).

10.2.8   PROCESSING OF THE ELH PRECURSOR AND RELEASE OF ELH

As with other precursor proteins for short neuropeptides, the ELH-precursor protein is subject to posttranslational modifications (PTMs) before the bioactive mature peptide is produced. The processing and cleavage of the polypeptide chain at appropriate sites is fundamental to the PTM of ELH-precursor (Arch et al., 1976b). Processing enzymes, such as furin (Nagle et al., 1993), prohormone convertase PC2 (Ouimet et al., 1993), as well as an amidating enzyme (Fan et al., 2000), have been cloned in *Aplysia* and were demonstrated to be expressed in the bag cells (Chun et al., 1994; Nagle et al., 1993; Ouimet et al., 1993). In fact, ELH and BCPs have consensus sites for cleavage by processing enzymes such as prohormone convertase 2 and furin (Sossin et al., 1990a,b). Cleavage of the ELH-precursor by furin generates the N-terminal fragment including BCPs and C-terminal fragments including ELH. Using a radiolabeled ELH precursor, localization of the ELH-containing fragment and the BCP-containing fragment in the bag cells were detected by electron microscope autoradiography (Sossin et al., 1990a,b). In the soma of the bag cell, the ELH-containing fragment was packaged into small, immature dense-core vesicles (DCVs) as it passed though the trans-Golgi network. These vesicles are then subjected to the PTM pathway and transported toward the axon terminal in the ganglionic space. By contrast, the BCP-containing fragment was packaged into a larger immature DCV and transported elsewhere in the cell. Therefore, the axon terminals located on the abdominal ganglion mainly release ELH, rather than BCPs as the bag cells are undergoing discharge (Fisher et al., 1988; Jung and Scheller, 1991).

Interestingly, the axon terminals on the pleuro-abdominal connectives contained two classes of vesicles, namely, one containing the BCP fragment and the other containing the ELH fragment (Sossin et al., 1990a,b). Thus, both the ELH and BCPs are released from nerve endings on the pleuro-abdominal connectives. A similar distribution of BCP- or ELH-containing vesicles is also observed by immunoelectron microscopy using antibodies that recognize the N-terminal and C-terminal fragments, respectively. These results suggest that the bag cells release different combinations of peptides in the pleuro-abdominal connectives and in the abdominal ganglion. The ELH-containing vesicles are released into the peripheral circulation, whereas the BCPs are secreted onto other bag cells or nearby neurons and have neurotransmitter-like effects. The differential sorting and release of ELH and the BCPs is important in coordinating the stereotyped set of behaviors involved in egg-laying.

The axon terminals of the bag cells in the pleuro-abdominal connective make synaptic contacts and receive neural input that triggers the afterdischarge. When the axon terminals are depolarized, co-release of BCPs with ELH, augments the excitability in the axon terminals of other nearby bag cells, and induces prolonged discharge of action potentials. This auto-excitatory action of BCPs, together with the electrical coupling of the bag cells, triggers the synchronized afterdischarge in the cells. The afterdischarge then propagates to the axon terminals located on the surface of the abdominal ganglion and initiates massive release of ELH into the ganglionic space (Jung and Scheller, 1991).

Synthesis and processing of ELH-precursor is modified by the neuronal activities in the bag cell. The involvement of depolarization-sensitive Ca^{2+} influx in neurites of the bag cells is suggested, because high K^+-solution augmented the incorporation of ^3H-leucine in the ELH-precursor. Incorporation of radiolabeled leucine was not observed in bag cells without neurites, and it was sensitive to Ca^{2+}/Mg^{2+} concentration. Moreover, elevation of intracellular cAMP in the bag cells also augmented radiolabeling of the ELH-precursor.

On the other hand, Azhderian and Kaczmarek (1990) reported that the afterdischarge of the bag cells reduced the amount of ELH-precursor. Since the reduction of the precursor was observed in the presence of the translation inhibitor, anisomycin, it was suggested that increased precursor processing occurred rather than decreased de novo synthesis (Bruehl and Berry, 1985). Precursor processing was augmented via elevation of bag cell cAMP levels as was previously shown. During the afterdischarge, it is likely that the bag cells initiate the processing of preexisting ELH-precursors, and then promote de novo synthesis of the ELH precursor to replenish the depleted precursor pool.

As is the case with other eukaryotic cells, protein synthesis in the bag cells is regulated by a cap-dependent initiation cascade. During this process, phosphorylated eukaryotic initiation factor 4E (eif-4E) and the 40S subunit of ribosome bind to the 5'-cap end of mRNA, and then scans on the mRNA to find the methionine initiation site. The 60S ribosomal subunit then binds to the complex at the initiation methionine site and polypeptide synthesis begins. However, in *Aplysia* bag cells Lee and Wayne (1998) reported that de-phosphorylation of eif-4E is dominant during the afterdischarge, which results in a reduction of protein synthesis. Dyer et al. (2003) suggested the involvement of a novel cap-independent internal ribosome entry system in ELH-precursor synthesis as a mechanism by which the bag cells promote the synthesis of ELH-precursor during the afterdischarge.

10.2.9 LOCAL SYNTHESIS OF ELH

It is generally accepted that the neuronal soma is the site at which all protein synthesis that is essential to maintain neural activity occurs. However, the axon terminal also has ability to synthesize proteins. For instance, in situ hybridization for ovulation hormone precursor mRNAs demonstrated the localization of some neuropeptide mRNAs in the axon hillock and in the axons of the caudodorsal cells of *L. stagnalis* (Van Minnen, 1994), suggesting the transport of mRNA toward the nerve endings. In fact, in *L. stagnalis* poly-somes can be found in the nerve endings by in situ hybridization (Spencer et al., 2000). Furthermore, injection of ovulation hormone precursor mRNA into the isolated neurites resulted in the de novo synthesis of ELH (Van Minnen et al., 1997). In *Aplysia* bag cells, reverse transcription-PCR using mRNA isolated from the axon terminal of the bag cells amplified cDNA encoding ELH precursor (Lee et al., 2002). It was demonstrated that the afterdischarge of the bag cell induces local synthesis and processing of ELH-precursor in the neurites, which contributes to the massive release of ELH from the bag cells in vivo (Lee and Wayne, 2004).

These results demonstrate that machinery for protein synthesis is present in the nerve endings. However, the mechanism that enables these axonal proteins to be released into the extracellular space is still unknown. Electron microscopy demonstrated that polysomes for local protein synthesis are located in the axoplasm of cultured neurons of *Lymnaea*, but not associated with membranous structures, such as the Golgi apparatus, mitochondria, and small vesicles located in the axon terminals (Spencer et al., 2000). Therefore, it is assumed that synthesized polypeptides remain in the axoplasm. A plausible explanation may be the existence of a bacterial Sec translocase-like protein (Du Plessis, 2011) that is functional in the axon terminal and translocates axoplasmic proteins to the extracellular space.

10.2.10   REPRODUCTIVE PEPTIDES IN OTHER MOLLUSCS

In the freshwater snail *L. stagnalis*, there are functionally equivalent neurosecretory cells to the bag cells *of A. californica*, called the CDCs (see Fig. 10.2). The CDCs are located in in two clusters of 50–100 cells on the dorsal side of the left and right cerebral ganglia (Wendelaar-Bonga, 1970, 1971a). The CDCs project their axons to the neurohemal area, the cerebral commissure, and release their secretory products in this region (Schmidt and Roubos, 1987, 1989). A similar anatomical arrangement of neurosecretory cells involved in egg-laying is present in the planorbid snail *Helisoma duryi* (Khan et al., 1990; Saleuddin et al., 1990), as well as in other gastropods (Ram et al., 1998). Ablation of the CDCs in adult *L. stagnalis* results in a cessation of egg-laying which could be restored by injection of a crude extract of the cerebral commissure (Geraerts and Bohlken, 1976). The ovulation-inducing hormone synthesized and secreted by the CDCs was referred to as CDCH, also referred to as ovulation hormone. Like the *Aplysia* bag cells, the *Lymnaea* CDC somata are located near the surface of ganglia and are large (~90 μm in diameter) neurosecretory cells (Boer, 1965; Joosse, 1964). This makes the CDCs an experimentally tractable system for studying neuropeptide synthesis and secretion using electrophysiological, molecular and proteomic techniques.

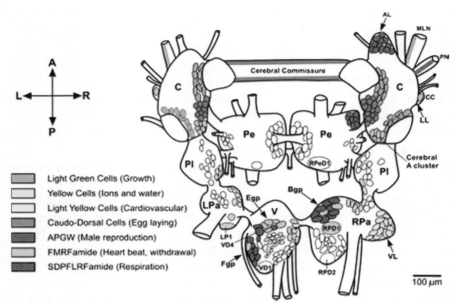

FIGURE 10.2 Map of peptidergic neurons in the CNS of *Lymnaea*. The cerebral commissure is the neurohemal organ of the CDCs. The median lip nerves (MLN) are the neurohemal organ of the LGCs. The anterior lobe (AL) APGW neurons project along the right penis nerve (PN) and innervate the penis complex. Note that the cluster of Yellow Cells that lie on the right side of the visceral ganglion are a special group of cells whose axons project along the distal processes of the intestinal nerve to innervate the pericardium and the reno-pericardial canal (see text). The cerebral A cluster cells are an example of whole-body withdrawal response motoneurons. Abbreviations: C, cerebral ganglion; CC, Canopy Cell; B/E gp, B/E group; LL, lateral lobe; LP1, left parietal 1; LPa, left parietal ganglion; Pe, pedal ganglion; Pl, pleural ganglion; RPa, right parietal ganglion; RPD1/2, right parietal dorsal 1/2; RPeD1, right pedal dorsal 1; V, visceral ganglion; VD1, visceral dorsal 1; VD4, visceral dorsal 4; VL ventral lobe. (Source: Benjamin, P. R., Kemenes, I. Lymnaea Neuropeptide Genes. *Scholarpedia*, 8(7), 11520). (Used with permission.)

Various environmental factors are known to stimulate the activity of the CDCs including, adequate nutrition (Ter Maat et al., 1982), long-day photoperiods (Dogterom et al., 1983), and a clean, oxygenated water stimulus (Ter Maat et al., 1983). Once the CDCs are induced to fire, they undergo a bout of electrical bursting activity which is followed by the secretion of CDC neurosecretory products, and finally egg-laying (De Vlieger et al., 1980; Ter Maat, 1992). Thus, the pattern of electrical activity and secretion of CDC neuropeptides is similar to the electrical events seen in the *Aplysia* bag cells that lead to ELH secretion. The CDCH is a 36 amino acid peptide and shares 44% sequence homology to *Aplsyia* ELH (Ebberink et al., 1985). Based on biochemical and molecular cloning studies (Geraerts et al., 1988; Li et al., 1992; Vreugdenhil et al., 1988), the CDCs are known to produce neuropeptides that are part of as small multigene family. Two genes, CDCH-I and CDCH-II, encode preprohormones from which ovulation hormone or CDCH (now referred to as CDCH-I) and CDCH-II are derived from, respectively. CDCH-II has been biochemically purified from cerebral commissure extracts and is a neuropeptide of 44 amino acids and is nearly identical to the N-terminus of CDCH-I except for a truncated amino acid but it differs from CDCH-I by having a C-terminal extension of nine amino acids (Li et al., 1992). The function of CDCH-II has not been demonstrated but it is believed to play a role in egg-laying behavior as it is colocalized with CDCH-I in the same neurosecretory granules (Van Heumen and Roubos, 1991). In addition to CDCH-I, nine other peptides are encoded on the CDCH-I preprohormone gene (Hermann et al., 1997; Li et al., 1994a–c; Vreugdenhil et al., 1988), including calfluxin, a peptide that stimulates calcium influx into the gland (AG) (Dictus and Ebberink, 1988; Dictus et al., 1987), and four types of CDC peptides (α-, β-, γ-, and δ-CDCPs). All the CDCPs are auto-excitatory on the CDCs; however, only specific combinations of CDCPs together with CDCH are capable of inducing characteristic CDC discharges in vitro (Brussaard et al., 1990). Furthermore, injection of β3-CDCP and α-CDCP into intact *L. stagnalis* stimulated electrical activity in the right pedal motoneurons which are involved in coordinating the stereotypical turning behavior during oviposition (Hermann et al., 1997). These physiological studies suggest that the N-terminal peptides on the CDCH-I precursor are involved in controlling neuronal activity in the CNS, whereas the peptides at the C-terminal region (e.g., CDCH-I) act as true hormones as they are transported in the hemolymph to distant organs in the periphery. Analysis of the spatiotemporal dynamics of peptide secretion by the CDCs using mass spectrometry showed a large depletion of CDC-derived peptides from the cerebral commissure during an electrical discharge as predicted by electrophysiological and cell biology experiments (Jiménez et al., 2004). In addition, two novel peptides of unknown function that were not encoded on the CDCH-I precursor were identified, suggesting that CDCs produce other peptides which may be involved in coordinating ovulation or egg-laying behaviors.

The activity of the two major reproductive centers *in L. stagnalis*, the CDCs and dorsal bodies (DBs) is known to modulated by infection with the schistosome *Trichobilharzia ocellata*, a parasite that causes swimmer's itch. Parasitic platyhelminths are well known to modulate the immunological and metabolic activities of their hosts, an adaptive trait that is thought to shift host resources toward parasite multiplication. It is thought that a factor secreted from *T. ocellata* causes the release of a peptide known as schistosomin from the CNS of *Lymnaea*. Schistosomin is then secreted into the hemolymph where it acts as an antigonadotropin to shift resources away from host reproduction (Schallig et al., 1992). Infection of *L. stagnalis* by *T. ocellata* inhibits reproduction by acting as an antagonist to calfluxin's action on the AG (De Jong-Brink et al., 1988) and blocks the action of the putative dorsal body hormone (DBH), which controls vitellogenesis (Hordijk et al., 1991a). Furthermore, application of hemolymph from parasitized snails to the CDCs inhibited their electrical activity suggesting that schistosomin has central effects on the CDCs as well its peripheral actions on the female reproductive system (De Jong Brink et al., 1992). Schistosomin is a 79-amino-acid peptide containing eight cysteine residues which are predicted to form four disulfide bonds (Hordijk et al., 1991b). Antibodies raised against schistosomin stain the neurosecretory LGCs, which are known to produce a hormone that regulates growth in *L. stagnalis*. In *Biomphalaria glabrata*, a schistosomin-like peptide was also identified which showed 62–64% amino acid sequence identity to *L. stagnalis* schistosomin (Zhang et al., 2009). However, based on mRNA and protein expression studies, schistosomin peptide appears to be most abundant during embryonic development and in juvenile snails. In addition, schistosomin expression does not change between parasitized versus non-parasitized snails, suggesting that in *B. glabarata*, schistosomin does not play a direct role in parasite-mediated shifts in metabolic activities in the adult snail (Zhang et al., 2009).

In *L. stagnalis*, other CNS neuropeptides that can be altered by infection with *T. ocellata* have been identified. For example, *Lymnaea* neuropeptide Y (LyNPY) is upregulated during parasitic infection with *T. ocellata*, and experimentally elevating in vivo NPY concentrations in snails suppressed egg-laying and inhibited growth (De Jong Brink et al., 1999). Restoring basal levels of NPY using implanted peptide-containing capsules restored both reproductive activity and growth. In addition, double labeling immunohistochemistry for LyNPY and CDCH demonstrated that NPY-positive axons are closely associated with CDCH-positive axons. Collectively, these data suggest that *in L. stagnalis*, infection with *T. occellata* upregulates NPY titers in the snail acting to suppress both reproduction and growth. Parasitic infection also upregulates FMRFamide and LFRFamide-related peptides which is thought to suppress host metabolism and allow the parasite to exploit the energy resources of the snail (Hoek et al., 1997, 2005).

ELH-like and CDCH-like peptides have been reported in other gastropods as well as some bivalves. In whelks, injection of CNS extract induces the laying of egg capsules (each capsule contains about 50 eggs) in *Busycon canaliculatus* and *B. carica* (Ram, 1977; Ram et al., 1982). The egg-laying-inducing activity of the brain extracts is abolished by protease treatment and the putative egg-laying peptide in CNS extract coelutes with *Aplysia* ELH using gel-filtration chromatography. This suggests that the nervous system of *B. canaliculatus* and *B. carica* contains a peptide that stimulates egg-laying and has a similar molecular mass as *Aplysia* ELH. However, the sequence of *Busycon* ELH has not been determined.

The distribution of ELH-related peptides or α-CDCP-related peptide has been detected in the CNS of several prosobranch gastropods by immuno-histochemistry. For example, in *B. canaliculatum*, immunoreactive neurons are widespread in in the CNS, and in particular, there are clusters of cells near the medial margins of the cerebral ganglia that stain strongly with both anti-α-CDCP and anti-ELH (Ram et al., 1998). In the abalone, *Haliotis rubra*, the gene encoding for an abalone ELH has been cloned (Wang et al., 1998). Antibodies raised against authentic abalone ELH (aELH) and used in immunohistochemistry stain neurosecretory cells in the brain, neurites in the neuropil, and the connective tissue sheath of the cerebral, pleuropedal, and visceral ganglia in *H. asinia* (Saitongdee et al., 2005). Interestingly, follicular and granular cells in the ovary also stain positive for aELH as well as the cytoplasm of mature oocytes. The authors suggest that the ovary may be a local source of ELH in *H. asinia*. Also, injection of aELH into juveniles caused early sex differentiation, whereas injection of aELH into ripe adults of *H. asinia* induced spawning in both sexes, suggesting an ELH-like peptide is involved in the maturation of the reproductive system and egg-laying (Nuurai et al., 2010).

Recently, by analyzing genome databases or by transcriptome analysis, several ELH-related peptide precursors have been identified in the owl limpet, *Lottia gigantea* (Veenstra, 2010) and two bivalves, *Crassostrea gigas* (Stewart et al., 2014) and *Pinctada fucata* (Matsumoto et al., 2013). The structures of ELH-related peptides encoded on the precursor proteins share similarities with *Aplysia* ELH. However, the overall structures of the precursors are somewhat different from the *Aplysia* ELH precursor. For example, two distinct ELH-related peptides are found on the same precursor in *L. gigantea*, while two distinct ELH-related peptides are found on precursor protein in *C. gigas*. None of the precursors contained BCP-related peptides; therefore, the interaction between ELH peptides and BCPs that coordinate egg-laying behavior in *Aplysia* is unlikely to occur in in the life history of *L. gigantea* and in bivalves. The elucidation of the physiological functions of ELH- and CDCH-related peptides in gastropods other than *Aplysia* and *Lymnaea* is an important issue that needs further research.

10.2.11   GONADOTROPIN-RELEASING HORMONE

Gonadotropin-releasing hormone (GnRH) is 10-mer peptide that plays central roles in regulation of reproductive activities in vertebrates. In mammals, GnRH is synthesized by neurosecretory cells in the hypothalamus and secreted in the capillaries of the median eminence of the pituitary gland. The secreted GnRH is transported to the anterior pituitary and promotes the synthesis and secretion of the gonadotropins, follicle-stimulating hormone, and luteinizing hormone in the parenchymal cells (Iversen et al., 2000; Schally, 1978). Structurally related peptides to GnRH have been found in invertebrates, including molluscs.

10.2.12   STRUCTURES OF GnRH AND GnRH-RELATED PEPTIDES

The structural sequence of the GnRH was first determined in mammals (Baba et al., 1971; Matsuo et al., 1971). Since then, more than 20 kinds of GnRHs have been found in chordates (Okubo et al., 2008; Tsai, 2006). In addition, GnRH-related peptides have been identified in invertebrates such as molluscs and annelids (Sun et al., 2012a,b). There are clear differences in the basic structures of vertebrate GnRHs and invertebrate GnRH-related peptides. For instance, all of vertebrate GnRHs share the N-terminal pyroglutamine and C-terminal Pro-Gly-NH$_2$. By contrast, the invertebrate GnRHs have insertions of two amino acids between the N-terminal pyroglutamine and penultimate histidine, and the C-terminal glycine residue found in vertebrate GnRHs is missing. As the result, invertebrate GnRHs are 11 mer peptides. An exception is octopus GnRH which consists of 12 amino acids since it has a C-terminal glycine like the vertebrate GnRHs.

10.2.13   LOCALIZATION AND FUNCTION OF MOLLUSCAN GnRH-RELATED PEPTIDES

Localization of GnRH-like peptide in the central and peripheral nervous systems was demonstrated by immunohistochemistry in the chiton (Amano et al., 2010a), oyster and mussel, abalone (Amano et al., 2010b; Nuurai et al., 2014), freshwater snail (Goldberg et al., 1993; Young et al., 1999), sea hare (Zhang et al., 2000), and octopus (Di Cosmo and Di Cristo, 1998). In most cases, GnRH-like peptides were found in the cerebral, pleural, and pedal ganglia of the central nervous system, while numerous GnRH-positive nerve processes were detected in the peripheral nervous system. However, some of these studies were conducted with antibodies that recognize vertebrate-type GnRH or by using vertebrate bioassays which tested the actions of vertebrate-type GnRH. Considering the differences in the structures of GnRHs between vertebrates and invertebrates, careful consideration may be necessary for the interpretation of the results.

The first report suggesting the existence of GnRH-like peptide in gastropods appeared in 1980. Application of mammalian GnRH to the abdominal ganglion of *A. californica* augmented the bursting activity of the right-upper quadrant (RUQ) neurons (Seaman et al., 1980). The initial investigations on the distribution of GnRH-like peptides in nervous and reproductive tissues in *Aplysia* were conducted by immunohistochemistry or RIA with different antivertebrate GnRH antibodies. Although it is practical and convenient to use neuropeptide antibodies to explore structurally related peptides in different animals, careful interpretation of the data are necessary. For instance, some antibodies detect the presence of GnRH-like peptides in hemolymph (Zhang et al., 2000), whereas others detect it in the central nervous system (Tsai et al., 2003). Moreover, subsequent molecular cloning studies revealed that the structure of *Aplysia* GnRH-related peptide is somewhat different from vertebrate GnRHs at both of the N-terminal and C-terminal regions (Zhang et al., 2008a). Thus, some anti-GnRH antibodies may have recognized potentially different substances in *Aplysia* tissues. Subsequent studies using immunohistochemistry with an antibody against authentic *Aplysia* GnRH-related peptide, as well as in situ hybridization, demonstrated that neurons in the cerebral and pedal ganglia contain GnRH-related peptides (Jung et al., 2014; Sun and Tsai, 2011; Zhang et al., 2008a). By contrast, no GnRH-like immunoreactivity was detected in the reproductive organs, although the RT-PCR detected the expression of precursor GnRH mRNA.

Functional studies on the *Aplysia* GnRH suggest that the peptide does not directly play a role reproduction. Since GnRH-containing nerve endings are not found in reproductive organs such as ovotestis, it is assumed that GnRH release from the neurons in CNS may have hormonal action on the reproductive organs. However, repetitive injection of GnRH into sexually immature *Aplysia* failed to induce gonadal maturation or egg-laying behavior (Tsai et al., 2010). Moreover, in vitro application of GnRH on the abdominal ganglion reduces the number of bag cells exhibiting afterdischarges. These results suggest that GnRH acts as a negative regulator of reproduction in *Aplysia* as opposed to its stimulatory role in vertebrates.

Interestingly, *Aplysia* individuals that received an injection of GnRH showed some unique behaviors such as frequent parapodia opening and reduced detachment from rigid substrates (Tsai et al., 2010). In this context, it is noteworthy that GnRH neurons in the pedal ganglion send their axons through the P1, P6, and P9 nerves. These nerve bundles which emanate from the pedal ganglion project onto the foot and parapodial region. Therefore, it is likely that GnRH neurons in the pedal ganglion mediate neural regulation of the parapodia and foot muscles.

In pulmonates, the presence of a GnRH-like peptide was first reported in *Helisoma trivolvis* by immunostaining and RIA with an anti-GnRH antibody, as well as its gonadotropin-releasing activity on fish pituitary cells (Goldberg et al., 1993). Young et al. (1999) found that neurons immunopositive to antimammalian GnRH antibody were diffusely distributed in all the circumesophageal ganglia. Immunopositive nerve fibers were also found in the reproductive organs such as penial complex, vas deferens, oviduct and ovotestis, as well as in the neuropile in the ganglia. A similar localization of GnRH-related peptide occurs in *L. stagnalis*. Although detailed physiological investigations regarding the function of GnRH in pulmonates is lacking, the localization of GnRH in the central and peripheral nervous systems suggest that GnRH may be involved in controlling male reproductive behavior by affecting the muscles of the penial complex and vas deferens.

In prosobranch gastropods, Amano et al. (2010a,b) found GnRH-like peptide-containing neurons in the cerebral ganglion of *Haliotis discus hannai* whereas, Nuurai et al. (2014) found GnRH-immunopositive neurons in both of the cerebral and pleuro-pedal ganglia of another abalone, *Haliotis asinina*. Since immunopositive nerve fibers were not found on the mature gonad, the authors speculate that *Haliotis* GnRH has hormonal action on this tissue. Nevertheless, GnRH may mediate the initial phase of oocyte maturation because oocytes in the early stage of maturation in the ovary of *H. asinine* are immunopositive to anti-lamprey GnRH antibody (Nuurai et al., 2010). Although it was reported that repetitive injections of salmon GnRH analogue induced maturation of the gonad in Hawaiian limpet, *Cellana* (Hua et al., 2013), the functional significance of GnRH in prosobranchs reproduction remains unclear.

10.2.14 PHYLOGENETIC RELEVANCE OF GnRH AND GnRH-RELATED PEPTIDES

The structures of GnRH-related peptides in protostomes, including molluscs, show resemblance to those of insect corazonin or adipokinetic hormone (AKH). Corazonin is a cardioactive peptide that regulates heartbeat in insects (Veenstra, 1989), while AKH is a peptide hormone that regulates energy metabolism during insect flight (Stone et al., 1976). The relationship between corazonin/AKH and GnRH has also been suggested in *Drosophila* where the endogenous ligand to the GnRH-like receptor is AKH (Park et al., 2002).

Different analyses based on the similarity in amino acid sequences of receptors for corazonin, AKH, and GnRH place the molluscan GnRH receptors with insect AKH receptors or corazonin receptors, but not with vertebrate GnRH receptors (Hauser et al. 1998; Hauser and Grimmelikhuijzen, 2014; Roch et al., 2011). GnRH-like peptides appear to be widespread in metazoans as GnRH-like immunoreactivity is present in the diffuse nerve net of a sea pansy, suggesting that GnRH-related peptides have evolved in animals with the first primitive nervous systems (Anctil, 2000). With respect to GnRH receptor evolution, the ancestral GnRH receptor is thought to have emerged before the divergence of deuterostomes and protostomes. As the two animal groups diverged, the ancestral GnRH receptor also split into the AKH/corazonin receptor trait in protostomes and the GnRH receptor trait in the deuterostomes (Hauser and Grimmelikhuijzen, 2014). As the divergence of the Mollusca and Arthropoda occurred, the AKH/corazonin receptor then diverged again into the corazonin receptor trait and the AKH receptor trait. Peptide ligands for the receptors also followed a similar pathway of molecular evolution, where the ancestral GnRH diverged into corazonin, AKH, and GnRH. In molluscs, the endogenous ligand for the corazonin receptor is a GnRH-related peptide (Hauser and Grimmelikhuijzen, 2014).

10.2.15 CONTROL OF FEMALE ACCESSORY SEX ORGANS

Gastropod molluscs possess morphologically elaborate reproductive systems with unique accessory sex glands. The neuroendocrine regulation of the secretory activities of various accessory sex glands has been studied mainly in pulmonates and opisthobranchs. We will focus our attention to the AG as it is the best studied accessory sex gland and it is known to be regulated by peptides and neurotransmitters. The AG is an exocrine gland that secretes a viscous perivitelline fluid that coats each fertilized egg as it proceeds along the reproductive tract and is packed into an egg mass/string. It is the main source of nutrition to the developing embryos in the egg mass (Duncan, 1975; Geraerts and Joosse, 1984). In addition, several novel proteins have been isolated from gastropod AGs, including antibacterial proteins (Kamiya et al., 1986), neurotoxins (Dreon et al., 2013), and peptide pheromones (Cummins and Bowie, 2012).

The effect of neuroendocrine regulators of the pulmonate AG has been investigated using biochemical techniques or bioassays that measure the two main classes of substances synthesized by the AG, a large molecular weight polysaccharide known as galactogen, and proteins (for reviews see Cummins et al., 2006; Dreon et al., 2006; Geraerts and Joosse, 1984; Runham, 1983). The neuroendocrine control of the pulmonate AG has historically been studied using organ culture techniques that measure the incorporation of radiolabeled monosaccharide precursors into large molecular weight polysaccharides (Goudsmit and Ashwell, 1965; Wijdenes et al., 1983). Putative factors from the brain and endocrine dorsal bodies are known to stimulate the synthesis of AG polysaccharides, however, some of these factors have not been fully purified and characterized (Goudsmit, 1975; Miksys and Saleuddin, 1985, 1988; Wijdenes et al., 1983). In *L. stagnalis*, a novel 14-residue peptide named calfluxin has been chemically identified as a peptide that is processed from the CDCH precursor protein (Dictus and Ebberink, 1988; Dictus et al., 1987). Calfuxin stimulates the influx of extracellular calcium and mobilizes calcium from internal stores as demonstrated by intracellular calcium deposition in the AG of *L. stagnalis* (Dictus et al., 1988). The rise in AG intracellular calcium levels is presumably linked to secretion of macromolecules from the AG, although this has not been shown. In *H. duryi*, in vitro protein secretion by the AG can be induced with a peptide-containing extract of the CNS (Morishita et al., 1998).

In addition to neuroendocrine regulation via peptides, the AG and reproductive tract of gastropods is innervated by catecholaminergic axons (Croll, 2001; Croll et al., 1999; Hartwig et al., 1980; Kiehn et al., 2001). For instance, the AG of planorbid snails has intrinsic catecholamine-positive cell bodies as well as extrinsic catecholaminergic innervation by varicosities from the CNS (Brisson, 1983; Brisson and Collin, 1980; Kiehn et al., 2001). In *H. duryi*, a large glycoprotein, *H. duryi* AG protein (HdAGP), is the major secretory protein synthesized and secreted by the AG (Mukai et al., 2004b). The in vitro secretion of HdAGP can be induced by treatment with dopamine (Saleuddin et al., 2000). Dopamine binds to and activates a D1-like dopamine receptor on the AG membrane which then elevates intracellular cAMP

(Mukai et al., 2004a) and calcium levels (Kiehn et al., 2004), resulting in the release of HdAGP. Catecholamine localization in the AG demonstrates intense staining around the carrefour, a small sac-like structure into which secretory material from AG flows (Kiehn et al., 2001). The carrefour has sensory cells lining its lumen and it is where each fertilized egg receives an equal coating of perivitelline fluid. The neural stimulation of the AG by dopamine is thought to trigger the precise release of perivitelline fluid from the AG cells into the carrefour to ensure each egg receives a coating of perivitelline fluid (Mukai et al., 2004a).

In *B. glabrata*, dopamine content in the AG was highest when perivitelline fluid secretion was expected to occur (Boyle and Yoshino, 2002), and application of dopamine to the AG stimulates in vitro protein secretion from the AG (Santhanagopalan and Yoshino, 2000). In *B. glabrata*, serotonin also plays a role in reproduction as culturing snails in water containing serotonin increases egg production (Manger et al., 1996), and addition of serotonin to the AG in vitro stimulates protein secretion (Santhanagopalan and Yoshino, 2000).

10.2.16 REGULATION OF MALE MATING BEHAVIOR

Although pulmonates such as *L. stagnalis* are hermaphroditic, during mating one animal acts as the donor "male" and the other as the recipient "female." Prior to mating, there is a distinct courtship behavior that occurs when the male actively mounts the shell of the female snail. In *L. stagnalis*, male mating behavior involves the extrusion of the preputium and the penis which is followed by probing for the female genital opening. The male snail then assumes the copulation position on the female shell rim and inserts the penis into the female genital opening (intromission) and then the transfer of semen occurs. Lastly, retraction of the penis complex occurs which may be followed by reversal of role where the donor male now becomes the recipient female (Runham, 1983).

In addition to the extensive studies on the peptidergic control of the female reproductive system, a growing body of evidence suggests that peptides are also involved in controlling distinct aspects of male reproductive behavior. For example, in *L. stagnalis*, mating behavior is known to be regulated by the action of numerous neuropeptides that are produced in a specific region of the brain. In *L. stagnalis*, the gene for a vasopressin-like peptide, named Lys-conopressin has been cloned from the nervous system (Van Kesteren et al., 1992a,b). *L. stagnalis* preproconopressin has a similar organization to its vertebrate counterparts having a signal peptide, and a mature peptide flanked by dibasic residues at the C-terminus. Furthermore, the snail neurophysin (carrier protein of conopressin) domain shares 49% sequence identity to human vasopressin neurophysin, including all the conserved cysteine residues, suggesting that the tertiary structure of *L. stagnalis* neurophysin assumes a similar conformation to that of vertebrate neurophysins (van Kesteren et al., 1992b).

In situ hybridization and immunohistochemical analyses of conopressin expression demonstrated that conopressin is expressed in clusters of neurons in the right cerebral ganglion, a region known to control male reproductive behavior, as well as staining nerve fibers innervating the penis and vas deferens (van Kesteren et al., 1995). Application of conopressin to electrically stimulated CDCs in vitro rapidly hyperpolarized the membrane potential indicating that conopressin exerts an inhibitory effect on neurosecretory cells controlling ovulation and egg-laying behavior in *L. stagnalis*. In contrast to its inhibitory effect on female reproduction, conopressin had a dose-dependent stimulatory effect on smooth muscle contractions of the anterior vas deferens, suggesting a functional role in the regulation of male reproductive behavior (van Kesteren et al., 1995). In mammals, oxytocin/vasopressin-related peptides are well known to regulate sexual behavior (Smock et al., 2008; Veenema and Newman, 2008), and recent studies in invertebrate systems also demonstrate that these peptides have similar functions as it relates to the modulation of sexual behavior (Garrison et al., 2012; Gruber, 2013). Collectively, these studies on the behavioral effects of oxytocin/vasopressin-related peptides suggest a conserved function for oxytocin/vasopressin-related peptides in diverse animal phyla.

In *L. stagnalis*, subsequent studies revealed that conopressin was co-localized with APGWamide in a subset of neurons in the right anterior lobe of *L. stagnalis* and then are transported to the penis complex (preputium and penis with retractor muscles) and the vas deferens via the penis nerve (Van Golen et al., 1995a). These two peptides were structurally identified by HPLC and immunoassays in the penis complex and co-localized by immunohistochemistry to some axon bundles innervating the vas deferens as well. Although conopressin had no effect on the contractions of the penis retractor muscle, it had potent excitatory effects on the total number and frequency of contractions of the anterior vas deferens (Van Golen et al., 1995a). On the other hand, APGWamide had a dose-dependent inhibitory effect on the posterior vas deferens and inhibited conopressin-induced contractions of the anterior vas deferens. The coordinated actions of these two peptides are thought to facilitate the transport of semen during copulation. Although APGWamide-like immunoreactivity is widespread in the central nervous systems of *L. stagnalis*, *A. californica*, and *H. aspersa*, there appears to be a conserved anatomical region of the brain, namely clusters of peptidergic neurons in the right cerebral ganglion that control male reproductive behaviors (Croll and Van Minnen, 1992; de Lange and van Minnen, 1998; Fan et al., 1997; Griffond et al., 1992; Koene et al., 2000). In addition to APGWamide and conopressin, other peptides have been identified from neurons in the right cerebral ganglion of *L. stagnalis*. For example, FMRFamide-related peptides in are present in the penis nerve and modulate the activity of the penis retractor muscle (Van Golen et al., 1995b). The FMRFamide gene undergoes alternative splicing resulting in two transcripts coding for the tetrapeptides (FMRFamide FLRFamide) and the heptapeptides (GDPFLRFamide and SDPFLRLamide) which are expressed in neurons of the right cerebral ganglia, and in the B-group neurons of the right parietal ganglion and a few neurons in the right pleural ganglion (Van Golen et al., 1995b). The tetrapeptides induce rapid contractions of the penis retractor muscle whereas the the heptapeptides cause relaxation (Van Golen et al., 1995b). Therefore, the tetrapeptides have distinct effects from the heptatpeptides and are likely involved in the modulation of the penis retractor muscle during the distinct phases of male mating behavior. In the terrestrial snail *H. aspersa*, parts of the reproductive system such as the hermaphroditic duct (ovotestis duct), seminal vesicle and fertilization pouch-spermathecal complex are innervated by branches of the intestinal nerve (Geoffroy et al., 2005). FMRFamide causes the relaxation of the muscles of the hermaphroditic duct and is thought to be involved in the regulation of ejaculation which is important for the proper transfer of sperm to the mating partner.

Four myomodulin-related peptides have also been identified from a cluster of neurons in the right cerebral ganglion of *L. stagnalis*, and each of these myomodulin isoforms have distinct but overlapping functions on the penis retractor muscle (Van Golen et al., 1996). It is clear from the physiological biochemical studies on male mating behavior in *Lymnaea* that there are multiple neuropeptides with distinct functions which are released from the penis nerve. The various secreted peptides permit a range of contraction and relaxation states and thus provides a mechanism for the fine-tuning of the penis retractor muscle (and other male accessory organs) to coordinate complex mating behaviors.

10.3 REGULATION OF GROWTH AND METABOLISM

The presence of a shell is a major diagnostic feature of the Mollusca and the majority of gastropods possess a shell or a reduced/absent shell as in the terrestrial slugs and some marine gastropods. When discussing the regulation of growth in shelled gastropods, there are two distinct aspects to consider, the growth of the soft body parts, and the growth of the shell. Growth of the soft body tissues occurs through increases in the weight of organic material mainly by the synthesis of various macromolecules (Wilbur, 1964; Wilbur and Saleuddin, 1983). In gastropods, new shell growth occurs at the shell edge via secretion of the periostracum, the outermost organic layer of the shell. Calcium carbonate crystals are then deposited on the periostracum, resulting in an increase in shell growth. In general, shell growth occurs simultaneously with body growth to provide space for the proportional growth of the soft body tissues.

The shell is a product of biologically controlled mineralization or biomineralization, and it is secreted by the underlying mantle epithelium. The mantle is composed of three main regions: the inner epithelium which is in contact with the environment (e.g., water for a shelled, aquatic gastropod), underlying internal tissues such as muscle, connective tissue including nerve fibers, and the outer epithelium which secretes proteins, carbohydrates, lipids, and ions that are required for shell formation (Marin et al., 2012).

In gastropods, the anterior portion of the mantle has a fold known as the periostracal groove, which possesses specialized cells that secrete the outermost organic covering of the shell called the periostracum. This tough proteinaceous periostracal layer functions and serves as an organic matrix for shell deposition (Saleuddin and Petit, 1983). The periostracum is secreted as a fluid containing soluble tyrosine-rich precursors and fibrous proteins, which becomes hardened and insoluble in the extracellular environment via sclerotization and quinone-tanning.

The inorganic portion of the shell is mainly composed of two polymorphs of calcium carbonate, calcite, or aragonite. The calcium used in shell formation is taken up from the environment in aquatic gastropods and from dietary sources in terrestrial gastropods. The carbonate ions are supplied from the reaction of carbon dioxide and water in the animal, or via the uptake of bicarbonate from the external medium. Free calcium and carbonate are then transported by the hemolymph to calcification sites along the mantle epithelium. During the last decade, significant advances have been made toward the identification of various macromolecules involved in molluscan biomineralization, especially in commercially important gastropods and bivalves (Aguilera et al., 2014; Marie et al., 2012; O'Neill et al., 2013). Despite these advances in the identification and characterization of shell matrix proteins and the genes encoding them, we know very little about how these unique biomineralization proteins are regulated by nervous and/or endocrine mechanisms.

Classic endocrinological extirpation and re-implantation studies in *Crepidula fornicata* suggested that the cerebral ganglia produced a factor that stimulated body growth (Lubet, 1971; Lubet and Silberzahn, 1971). The putative growth-stimulating factors from the cerebral ganglia were hypothesized to stimulate growth via nerves as well as hormones. However, the putative growth-stimulating factor was never identified in *C. fornicata*. In *L. stagnalis*, paired groups of neurosecretory cells in the cerebral ganglia, the LGCs, were determined to be the source of a putative growth hormone as extirpation of the LGCs from growing juveniles resulted in reduction of body growth (Geraerts, 1976). Functionally similar neurosecretory medial cells in the slug *Deroceras reticulatum* (Widjenes and Runham, 1977), and the mediodorsal cells (MDCs) of the freshwater snail *H. duryi* were also demonstrated to be involved in the regulation of body growth (Kunigelis and Saleuddin, 1985; Saleuddin and Kunigelis, 1984).

In addition to causing cessation of growth, LGC extirpation in *L. stagnalis* results in a secondary accumulation of glycogen stores and higher glucose levels in the hemolymph (Dogterom, 1980; Geraerts, 1976). Extirpation of the LGCs also diminishes incorporation of protein into periostracum formation, whereas shell matrix formation remains unaffected (Dogterom and Jentjens, 1980). The putative growth hormone from the LGCs affects calcium transport primarily at the mantle edge as LGC removal decreased radiolabeled calcium incorporation in this region of the mantle by 30% when compared with control animals (Dogterom et al., 1979). Furthermore, LGC ablation also caused reduction of a putative calcium-binding protein in the mantle edge (Dogterom and Doderer, 1981). Based on studies in *L. stagnalis*, the growth hormone from the LGC is thought to exert its effect on shell growth by increasing the synthesis of periostracal proteins and maintaining high concentrations of calcium at the mantle edge which is required for calcification.

Initial attempts to characterize the chemical nature of the *L. stagnalis* growth hormone relied on in vivo bioassays that involved the injection of peptide-containing extracts of the median lip nerves (neurohemal area of the LGCs), then measuring ornithine decarboxylase activity or cAMP production at the mantle edge, or the incorporation of $^{45}Ca^{2+}$ at the shell edge (Dogterom and Robles, 1980; Ebberink and Joosse, 1985). These bioassays were generally unreliable because they showed extreme variability and relatively low specificity to be of use in screening crude extracts for growth hormone activity (Joosse, 1988). However, Ebberink and Joosse (1985) were able to partially purify a hydrophobic peptide of approximately 1000 Da that stimulated calcium incorporation into the shell and increased cAMP levels in the mantle edge. However, no further biochemical characterization of this 1000-Da peptide has been achieved.

Vertebrate peptides have also been reported in the LGCs of *L. stagnalis*. Immunohistochemical staining of the CNS using antibodies against various vertebrate peptides have shown the presence of somatostatin-like immuno-reactivity in the LGCs and the median lip nerves (Grimm-Jørgensen, 1983; Schot et al., 1981). However, vertebrate somatostatin did not have any growth-promoting activity in *L. stagnalis* (Grimm-Jørgensen, 1985). ILPs have also been reported in the LGCs of *L. stagnalis* and the release of these ILPs could be induced in vitro (Ebberink et al., 1987).

As molecular cloning techniques became widely available in the 1980s, they were applied to molluscan nervous systems to identify and clone neuropeptide genes regulating numerous physiological processes. Using a subtractive hybridization approach, Smit et al. (1988) searched for mRNAs specifically expressed in the LGCs of *L. stagnalis* and identified a cDNA clone that encoded the precursor of an insulin-like molecule which was named MIP, now called MIP-I. The pre-proMIP-I possessed both A and B chains and a C-peptide; therefore, it resembled the overall structure of mammalian insulin precursors. The A and B chains of MIP-I had approximately 50% and 18% sequence identity, respectively, when compared with vertebrate insulin chains but MIP-I possessed no sequence homology to vertebrate C peptide (Smit et al., 1988).

Subsequent molecular biological and peptide isolation studies by the Amsterdam researchers found that there were seven MIPs in *L. stagnalis*, named MIP-I to VII (Li et al., 1992a,b, Smit et al., 1991, 1992, 1993c, 1996). MIP IV and VI are considered pseudogenes since they are not transcribed in *L. stagnalis* (Meester et al., 1992). Sequence analyses of the MIPs revealed that their surface residues and N-termini were highly divergent compared with human insulin thus, MIPs were not expected to have similar biophysical properties in terms of receptor binding, the ability to form hexamers and other molecular interactions (Geraerts et al., 1992).

Although *L. stagnalis* MIP-I did not share high amino acid sequence identity with human insulin, the position of the cysteine residues in the A and B chains and the hydrophobic core amino acids were found to be conserved, suggesting that the MIPs have a similar overall organization of the A and

B chains and the C-peptide to vertebrate insulins (Smit et al., 1988). MIP-I and all subsequent MIPs identified have conserved cysteine residues (see Fig. 10.3) in their A- and B-chains which are predicted to form interchain disulfide bonds. The MIPs also possess an extra pair of conserved cysteine residues in the A-chain, suggesting that the MIPs form an intrachain disulfide bond. Furthermore, the sequences of the C-peptides of the MIPs are highly conserved in *L. stagnalis* suggesting that they might have important physiological functions in this snail (Smit et al., 1998). The MIPs, however, are not considered true insulin molecules because their surface residues, which are important for receptor recognition, solubility, conformation, and intermolecular interactions, are completely divergent (Geraerts et al., 1992; Roovers et al., 1995; Smit et al., 1998).

```
                    Signal peptide                           B-chain
                ────────────────────────────────     ──────────────────────
    MIP II      -MVGVRLVFTNAFVVTVLLTLLLDVVVKPAEGQ-SSCSLSSRPHPRGICG
    MIP V       -MAGVRLVFTKAFMVTVLLTLLLNIGVKPAEGQFSACSFSSRPHPRGICG
    MIP I       -MAGVRLVFTKAFMVTVLLTLLLNIGVKPAEGQFSACNINDRPHRRGVCG
    MIP III     -MASVHLTLTKAFMVTVFLTLLLNVSITRGTTQ-HTCSILSRPHPRGLCG
    MIP VII     MNASVESCLT---FTFVLVALCVGLTIG---QQVNTCTMFSRQHPRGLCG

                    B-chain
                ──────────────────────
    MIP II      SNLAGFRAFICSNQNSP---------------------------------
    MIP V       SDLADLRAFICSRRNQP---------------------------------
    MIP I       SALADLVDFACSSSNQP---------------------------------
    MIP III     STLANMVQWLCSTYTTS---------------------------------
    MIP VII     NRLARAHANLCFLLRNTYPDIFPRKRSVDNTFEKVYSIPLSVLAELDLSD

                                        C-peptides
                ──────────────────────────────────────────────────────
    MIP II      ----SMVKRDAETGWLLPETMVKRNAETDL-DDPLRNIKLSSESALTYLT
    MIP V       ----AMVKRDAETGWLLPETMVKRNAQTDL-DDPLRNIKLSSESALTYLT
    MIP I       ----AMVKR--------------NAETDL-DDPLRNIKLSSESALTYLT
    MIP III     ----SKVKR--------------QAEPDEEDDAMSKIMISKKRALSYLT
    MIP VII     DDWGAYVSKKDIPYRSETNGLSGANFESSAFDKQLELPAMKSTTSQLFRI

                                    A-chain
                ──────────────────────────────────
    MIP II      KRQR---------TTNLVCECCFNYCTPDVVRKYCY
    MIP V       KRQR---------TTNLVCECCYNVCTVDVFYEYCY
    MIP I       KRQG---------TTNIVCECCMKPCTLSELRQYCP
    MIP III     KRES---------RPSIVCECCFNQCTVQELLAYC-
    MIP VII     LKLRGSRLKREVMAEPSLVCDCCYNECSVRKLATYC-
```

FIGURE 10.3   Clustal W sequence alignment of the five functionally transcribed molluscan insulin-related peptides (MIPs) from *Lymnaea stagnalis* showing the regions of the signal peptide, A- and B-chains, and the C-peptides. Two pairs of conserved cysteine residues (bolded and underlined) in the A- and B-chains, respectively, are predicted to form interchain disulfide bonds, and a third pair of cysteine residues in the A-chain is predicted to form an intrachain disulfide bond.

The various MIP genes appear to be differentially expressed in the brain of *L. stagnalis*. Using oligonucleotide probes to specific MIP transcripts, the LGCs were shown to co-express MIP I, II, III, V, and VII (Meester et al., 1992; Smit et al., 1996). MIP VII appears to be exclusively expressed in neurons of the buccal ganglia, suggesting that it may be involved in the regulation of feeding (Meester et al., 1992). Furthermore, two types of LGCs could be distinguished with in situ hybridization, type A cells express all MIP transcripts, whereas type B cells generally do not express or weakly express MIP I (Meester et al., 1992). MIP III is strongly expressed in type B cells, whereas MIP II and V are moderately expressed in both cell types (Meester et al., 1992). The physiological significance of differential MIP gene expression is unknown but may be linked to stimuli from outside the animal as well as signals within the animal. For example, ablation of the lateral lobe, a specialized neuron on each side of the cerebral ganglia which is thought to regulate the activity of the LGCs, accelerates growth and enhances expression of MIP I, II, II, and V (Meester et al., 1992). On the other hand, snails subjected to starvation show a reduction in MIP II and V (Meester et al., 1992). Therefore, expression of the different MIP genes appears to be regulated independently, suggesting that some MIPs have overlapping function as well as potentially distinct functions.

In vertebrates, insulin binds to its receptor on target cells and stimulates specific intracellular signaling cascades linked to the activation of tyrosine kinases (Boucher et al., 2014; Du and Wei, 2014). Several insulin receptors have been cloned in gastropod molluscs, including *L. stagnalis* (Roovers et al., 1995), *A. californica* (Jonas et al., 1996), and *B. glabrata* (Lardans et al., 2001). In addition, the presence of insulin receptors in other gastropod nervous systems has been demonstrated by immunohistochemistry (Saavedra et al., 1989; Sonetti and Bianchi, 1993), radioreceptor autoradiography (Kerschbaum et al., 1993), or the production of known intracellular messengers linked to insulin receptor activation (Sossin et al., 1996a,b). However, binding of specific gastropod ILPs to their cognate receptors has not been demonstrated.

10.3.1   INSULIN-LIKE PEPTIDES AS REGULATORS OF GROWTH AND METABOLISM IN GASTROPODS

Since the discovery of ILPs in the LGCs of *L. stagnalis* (Smit et al., 1988), several labs have demonstrated the presence of ILPs in the nervous systems of gastropods. Using antibodies against mammalian insulin, the growth-controlling MDCs of the freshwater pulmonate *H. duryi* (Khan et al., 1992) and the mesocerebral neurosecretory cells of the terrestrial pulmonate *H. aspersa* were shown to contain ILPs (Gomot et al., 1992). In addition, immunohistochemical approaches using antibodies raised against specific peptide fragments deduced from molecular cloning demonstrate the presence of *L. stagnalis* pro-MIP-like immunoreactivity in the dorsal cells and other neurons of the CNS of freshwater snail *Planorbarius corneus* (Sonetti et al., 1992), whereas in the marine gastropod *A. californica*, the F and C clusters of the cerebral ganglia show intense staining for *Aplysia* insulin (AI) with an antibody raised against a portion of the predicted sequence of the C-peptide (Floyd et al., 1999).

In the freshwater pulmonate *H. duryi*, a circadian fluctuation of insulin-like immunoreactive material in the hemolymph was detected with a mammalian insulin RIA (Sevala et al., 1993a). Peaks of insulin-like material were detected during the photophase, which preceded daily increases in shell deposition rates, suggesting ILPs may be involved in shell growth in *H. duryi* (Sevala et al., 1993a). To further investigate the possible role of ILPs in the regulation of growth, Sevala et al. (1993b) demonstrated that partially purified ILPs in *H. duryi* stimulated in vitro protein synthesis in mantle tissue explants. In the related planorbid snail *B. glabrata*, application of bovine insulin to Bge cell lines increased the rate of amino acid incorporation (Lardans et al., 2001).

ILPs appear to play a role in growth and metabolism in stylommatophorans as well as basommatophorans. In *H. aspersa*, removal of the mesocerebral green cells, a group of insulin-positive neurosecretory cells, causes a cessation of growth and the accumulation of glycogen in the epithelium of the mantle edge (Gomot et al., 1992). Immunohistochemical and Western blotting studies in *Otala lactea* demonstrated the presence of ILPs in the cerebral ganglia and other parts of the CNS, as well as the gut (Abdraba and Saleuddin, 2000a,b). Both crude and partially purified ILPs from brain extracts applied to an mantle collar explants in vitro, a bioassay for growth, stimulated total protein synthesis (Abdraba and Saleuddin, 2000b). These studies in aquatic and terrestrial pulmonates suggest that ILPs stimulate cellular protein synthesis and have functional characteristics of a growth hormone.

In the opisthobranch *A. californica*, injection of extracts derived from the upper labial and anterior tentacular nerves, the neurohemal regions of ILPs in this animal, decreased glucose concentration in the hemolymph but did not affect feeding behavior (Horn et al., 1998). Furthermore, expression of AI mRNA decreased when the animals were starved, whereas injections of highly purified AI caused a reduction in hemolymph glucose concentration (Floyd et al., 1999). Therefore, ILPs appear to function as metabolic regulators of carbohydrate metabolism in *A. californica*, similar to the functions of insulin in vertebrates. The role of ILPs as a potential growth regulator in in *Aplysia* is not known. Taken together, the studies in *Aplysia*, *Lymnaea*, *Helisoma*, and *Helix* strongly suggest a conserved function of ILPs in influencing growth and metabolic activities in molluscs.

10.3.2   NOVEL FUNCTIONS OF ILPs IN GASTROPODS

ILPs in gastropods have also been reported to have other functions not directly related to carbohydrate metabolism and growth. For instance, in *L. stagnalis* neuronal extracts containing MIP were purified by HPLC, and then applied to cultured neurons in vitro. Purified MIP stimulated neurite outgrowth suggesting that ILPs may play a role in neuronal development (Kits et al., 1990). ILPs have also been reported to be important in the formation of long-term memory as it relates to conditioned taste aversion behavior in *L. stagnalis* (Mita et al., 2014). Indeed, insulin is increasingly recognized as an important signaling molecule in the brain which regulates neuronal function in both vertebrates and invertebrates (Derakhshan and Toth, 2013; Liu et al., 2014).

Recently, a novel function of ILPs has been reported in two fish-hunting cone snails, *Conus geographus* and *Conus tulipa*. The ILPs in these two snails are closely related to insulin in their fish prey. The cone snail insulin is released together with other toxins into the water and induces hypoglycemic shock in fishes, thereby facilitating prey capture (Safavi-Hemami et al., 2015). In *A. californica*, application of bovine insulin to isolated bag cells induces ELH secretion in vitro without triggering action potentials (Jonas et al., 1997). Vertebrate insulin appears to bind to and activate an insulin receptor on the bag cells resulting in the mobilization of calcium from a unique intracellular calcium store. Therefore, ILPs in *Aplysia* can communicate with the neuroendocrine system controlling egg-laying behavior and potentially influence reproductive processes as well carbohydrate metabolism.

10.3.3   REGULATION OF GROWTH HORMONE-PRODUCING CELLS

In *L. stagnalis*, the MIP-producing LGCs appear to be capable of detecting extracellular glucose in vitro as concentrations of 1 mM glucose or higher evoked long-lasting spiking activity (Kits et al., 1991). Thus, the LGCs are functionally somewhat similar to mammalian pancreatic β-cells in that they can detect changes in extracellular glucose concentrations. However, the cellular mechanism for glucose uptake by the LGCs is different than the β-cells, and the secretion of specific MIPs in response to increased glucose has not been shown. Other neuroactive agents have been demonstrated to affect the electrical excitability of the LGCs. The catecholamine dopamine (Stoof et al., 1984; De Vlieger et al., 1986), and the peptides FMRFamide and APGWamide (van Tol-Steye et al., 1997, 1999) cause a hyperpolarization of the LGCs. Dopamine, FMRFamide, and APGWamide appear to converge on a potassium channel and activate an S-like potassium conductance via different G proteins.

In freshwater pulmonates, there is a close physiological interdependence between body growth and reproduction since these two processes require considerable energy investments. For example, experimental or parasite-induced castration markedly increases shell growth compared to reproducing snails which grow at slower rates (Geraerts and Mohammed, 1981; Miksys and Saleuddin, 1987). Hemolymph from parasitized snails or application of the neuropeptide schistosomin reduced the excitability of the CDCs, whereas the same treatments had the opposite effect on LGC excitability (Hordijk et al., 1992). Furthermore, differential mRNA screening of parasitized versus nonparasitized snails showed that several genes in the CNS coding for various peptides are differentially expressed (Hoek et al., 1997). Application of LFRFamide (an FMRFamide-related peptide) to the LGCs or the CDCs inhibits their electrical activity (Hoek et al., 2005). It has also been shown that there is a close morphological association of NPY (neuropeptide Y)-positive axons and the axons from the LGCs, suggesting that a *Lymnaea* NPY-like molecule (LNPY) may regulate growth. For example, implantation of slow-release pellets containing LNPY inhibited both growth and reproduction in a dose-dependent fashion (De Jong-Brink et al., 1999). It is thought that parasites manipulate the host snail's neuroendocrine system by suppressing metabolism and reproduction and exploiting the host's energy reserves for their own use.

The LGCs in *L. stagnalis* receive input from other regions of the CNS such as the lateral lobes (Geraerts, 1976; Roubos et al., 1980) and from sensory cells located near in the epidermis near the base of the tentacle which can perceive environmental stimuli (Roubos and Wal-Divendal, 1982). The precise cellular mechanisms by which these cells regulate the LGCs are unknown.

10.4 REGULATION OF CARDIAC ACTIVITY AND CIRCULATION BY NEUROPEPTIDES

10.4.1 STRUCTURE OF THE CIRCULATORY SYSTEM IN MOLLUSCS

The basic architecture of the molluscan heart consists of a pair of auricles that collect blood from the kidney and gill, and a single ventricle that pumps the blood out to the sinus. There are flaps between the auricle and ventricle (auricloventricular valve) that prevent back flow from the ventricle to the auricle. The heart is nestled in the pericardial cavity located on the dorsal side of the body. There are well developed arteries and veins connected to the heart, although there are no direct connections between vessels (Jones, 1983). Hemolymph pumped out from the ventricle flows toward the sinus surrounding tissues and organs through the arteries and aortae. Hemolymph then leaks out to the hemocoel through lacunae on the sinus wall. Hemolymph is collected to the auricle after it passes through the gill or kidney. Thus, the molluscan circulatory system is considered semi-open.

In gastropods, it is believed that ancestral form of the heart consisted of a pair of auricles, each of them connected to the right and left gills, respectively. One ventricle is connected to the auricle and receives blood supply from both of the auricles. However, most gastropods (except for the Archaeogastropods such as limpets) lost the auricle and gill on the left side as a result of torsion. Therefore, in the hypsobranchia, including neogastropods, the heart consists of single auricle connected to the ventricle. This basic architecture of the heart is unchanged in opisthobranchs and pulmonates that have not undergone torsion.

An important function of the circulatory system in molluscs is the transport of hemolymph, which contributes to the delivery of nutrients and O_2 to the cell as well as to the removal of waste and CO_2 from the cell. Moreover, the circulatory system in molluscs is also involved in locomotion. For example, the redistribution of hemolymph is known to be important for the extension of the foot in bivalves. In *Aplysia*, the anterior aorta is the major artery supplying hemolymph to the CNS, buccal mass, the genital organs, opaline gland, and other somatic tissues at the anterior end of the animal. The hydrostatic pressure generated by the redistribution of hemolymph is the motive force for the extension of various body parts. For instance, closure of the sphincter muscle in the region of the abdominal aorta introduces hemolymph in the posterior part of the body and is associated with an increase in heartbeat, and resultant hemolymph accumulation in the head region (Sasaki et al., 2004; Skelton et al., 1992). Concomitant relaxation of body wall in the anterior region results in the extension of the head forward. Thus, the cardiovascular system is important for locomotion in soft-bodied animals. The regulation of aortic blood flow is known to be under control of classical transmitters as well as peptides.

Another function of the circulatory system is in urine formation. In some gastropods, a portion of the hemolymph in the heart spills out to the pericardial cavity through the wall of heart or crista aortae (Andrews, 1988). The pericardial cavity is connected to the kidney via the renal-pericardial pore and hemolymph in the cavity drains into the kidney. Absorption and secretion of salts, nutrients and wastes occurs in the kidney, and, finally, urine is excreted to the outside through the renal pore. Thus, in some gastropods, the heart is the proposed site of filtration.

The heartbeat of the molluscs is generated by a myogenic pacemaker and the beating amplitude and frequency are regulated by both excitatory and inhibitory neural inputs. The following sections will summarize how neuropeptides are involved in the neural regulation of cardiac activity. For more information, refer to other excellent reviews (Kodirov, 2011; Skelton et al., 1992).

10.4.2   REGULATION OF HEARTBEAT BY NEUROPEPTIDES

In *Aplysia*, heart rate changes in response to various external and internal stimuli such as feeding (Dieringer et al., 1978), food-induced arousal (Koch et al., 1984), sensitizing stimuli (Krontiris-Litowitz, 1999), and hypoxia (Pinsker et al., 1974). The neural regulation of the cardiovascular system has been well studied in this gastropod (Koester et al., 1974; Liebeswar et al., 1975; Mayeri et al., 1974; Rozsa, 1979). In *Aplysia*, for instance, heart excitatory neurons such as RB_{HE}, are serotonergic, and heart inhibitory neurons such as RB_{HI} are cholinergic. On the other hand, some neurosecretory cells that release peptides are also known to be involved in the regulation of the cardiovascular system.

In the abdominal ganglion of *Aplysia*, the R3-R13 neurons located in the RUQ region, as well as the R14 and R15 neurons of the caudal side of the ganglion, project axons toward the cardiovascular region, including the gill and efferent veins, anterior and gastroesophageal aortae, ganglionic artery dorsal, and mantle arteries, auricle and the auricloventricular valve (Rittenhouse and Price, 1986a,b; Skelton and Koester, 1992). These neurons are involved in the regulation of heartbeat and the contraction of the aorta (Rozsa et al., 1980; Sawada et al., 1981).

R3–14 neurons appear to be neurosecretory cells because of their whitish appearance, the presence of DCVs and blunted nerve-endings (Coggeshall et al., 1966). In fact, R3–14 neurons synthesize a 12-kDa-protein (Aswad et al., 1978). Nambu et al. (1983) cloned a cDNA expressed specifically in R3–14 cells, and predicted that the 12-kDa-protein is a neuropeptide precursor consisting of 108 amino acids (Nambu et al., 1983). The predicted precursor is posttranslationally cleaved into three fragments since the precursor has two dibasic processing sites. Of the three fragments, the one consisting of 43-amino acids was purified from the abdominal ganglion and its structure was confirmed (Campanelli and Scheller, 1987; Knock et al., 1989; Nagle et al., 1989a,b). Although the effective concentration used in bioassays is relatively high (over 10^{-6} M), the peptide referred to as histidine-rich basic peptide (HRBP), augments heartbeat of *A. californica* (Campanelli and Scheller, 1987). Precursor genes for LYCPs (light-yellow cell peptides) in *Lymnaea stagnalis* (Smit et al., 1993a) and HSD-1/-2 in *Helix lucorum* (Bogdanov et al., 1996) also encode HRBP-related peptides. On the other hand, the R15 neuron releases two distinct peptides, R15α1 and R15α2. R15α2 peptide-containing nerve processes are distributed in the arterial system (Skelton and Koester, 1992) and have moderate cardioexcitatory action on the *Aplysia* heart (Alevizos et al., 1991b).

Morishita et al. (1997) isolated a D-tryptophan-containing cardioexcitatory Asn-D-Trp-Phe-NH$_2$ (NdWFamide) from *Aplysia kurodai* heart (Morishita et al., 1997). Immunohistochemistry with specific anti-NdWFamide antibody revealed that some neurons in the RUQ region of the abdominal ganglion contain NdWFamide. Based on cell location and cell size, NdWFamide-containing neurons include some of the R3–14 cells, as well as other smaller white cells in the RUQ region (Morishita et al., 2003c). Moreover, the distri-bution of NdWFamide-containing nerve processes in the cardiovascular region overlaps with axons of the R3–14 neurons. Therefore, it is likely that NdWFamide mediates the physiological action of the RUQ neurons including some R3–14 neurons.

Besides the aforementioned neuropeptides, many peptides such as FMRFamide, SCP$_B$, achatin-1, and ACEP-1 are also involved in the regu-lation of heartbeat in molluscs. In the following section, we will focus on some selected neuropeptides and summarize how they are involved in the regulation of heartbeat.

10.4.3 FMRFAMIDE

FMRFamide consists of four amino acids (Phe-Met-Arg-Phe) with a C-terminal amide and it was originally purified from the ganglia of the bivalve, *Macrocallista nimbosa* (Price and Greenberg, 1977). The discovery of FMRFamide by Greenberg and his colleagues was a landmark publication that opened up new avenues in peptide research and further strengthened the concept that peptides play crucial roles in metazoan neural communication. Immunohistochemical staining with anti-FMRFamide antibody has demonstrated the distribution of FMRFamide in the central and peripheral nervous systems in numerous animal groups (Grimmelikhuijzen, 1983). In fact, various neuropeptides having a C-terminal $RF-NH_2$ structure have been isolated in other animal phyla including Cnidaria, Platyhelminthes, Arthropoda, and Chordata (Grimmelikhuijzen and Graff, 1985; Johnston et al., 1995; Schneider and Taghert, 1988; Ukena et al., 2002). It is thought that the RFamide-related peptides belong to an ancestral peptide family because the C-terminal $Arg-Phe-NH_2$ structure was conserved during the diversification and molecular evolution of neuropeptides.

FMRFamide and structurally related peptides (FMRFa-RP), FLRFamide have been identified in many molluscan groups. Moreover, longer forms of FMRFamide such as GDPFMRFamide and *Mytilus*-FFRFamide are also found in bivalves, pulmonates and cephalopods (Fujisawa et al., 1992; Lopez-Vera et al., 2008; Price et al., 1987). FMRFamide augments the amplitude and frequency of the isolated heart of *Mercenaria* (threshold = 10^{-9}–10^{-8} M) (Price and Greenberg, 1979), *L. stagnalis*, *H. aspersa* (around 10^{-7}–10^{-6} M) (Price et al., 1990), *Achatina fulica* (threshold = 10^{-10}–10^{-9} M) (Koch et al., 1993), *Rapana thomasiana* (threshold = 10^{-9}–10^{-8} M) (Fujiwara-Sakata and Kobayashi, 1992), and *Archidoris montereyensis* (around 10^{-6}–10^{-4} M) (Wiens and Brownell, 1995). FMRFamide-related peptides such as pQDP-FLRFamide and SDPFLRFamide are also cardioactive on *Helix*, *Lymnaea*, and *Achatina*. By contrast, the cardioexcitatory action of FMRFamide on *Aplysia* heart is unclear (Harris et al., 1995).

The signal transduction pathway mediating the cardioexcitatory action of FMRFamide is rather complex. Unlike small cardioactive peptide B (SCP_B) and serotonin, FMRFamide does not elevate intracellular cAMP levels in *Helix* heart (Reich et al., 1997a). However, in *L. stagnalis*, the cardioexcitatory action of FMRFamide and related peptides correlate with increases in intracellular cAMP (Willoughby et al., 1999a). In the optic lobe of a squid, *Loligo pealei*, FMRFamide activated adenylate cyclase in a GTP-dependent manner (Chin et al., 1994). Thus, one possible signaling pathway is that FMRFamide activates adenylate cyclase by activating the Gs subunit of a G-protein. Another possible pathway is via the elevation of inositol phosphates by activating PLC. FMRFamide promotes generation of inositol trisphosphate in molluscan hearts (Bayakly and Deaton, 1992; Ellis and Huddart, 2000; Willoughby et al., 1999b), which triggers Ca^{2+}-mobilization from intracellular stores. In addition, it is suggested that activation of protein kinase C is involved in the cardioexcitatory action of FMRFamide.

In both of these signaling pathways, FMRFamide binds to a 7-transmembrane spanning GPCR. FMRFamide receptors in molluscs are predicted by annotation of genome database sequences (e.g., accession number: XM_005110697 for *C. gigas*, XP_005110754.1 for *Aplysia*). However, the biochemical and physiological characterization of these receptors is lacking at this time.

In addition to the metabotropic receptor system, FMRFamide modulates ion channel activities on cardiac muscle cells. In *L. stagnalis*, FMRFamide increases the open-probability of Ca^{2+}-channels in isolated cardiac muscle cells, which gradually decreases during repetitive application of the peptide (Brezden et al., 1999). In the heart muscle cells of the squid *Loligo forbesii*, FMRFamide shows opposing actions on L-type Ca^{2+}-channels in distinct cardiac cell types(Chrachri et al., 2000). As is in *Lymnaea* heart cells, FMRFamide increased Ca^{2+}-current through an L-type Ca^{2+}-channel in the type-II heart cells. By contrast, the peptide decreased Ca^{2+}-current in the type-I heart cells through a pertussis toxin-sensitive G-protein. It can be argued that the modulatory actions of FMRFamide on molluscan heart cells are mediated by G-proteins, rather than the direct action on a channel protein.

FMRFamide is known to modulate sodium channels by binding to a channel protein. FMRFamide-gated Na^+-channels have been identified in the nervous tissues of *H. aspersa* (Lingueglia et al., 1995), *H. trivolvis* (Jeziorski et al., 2000), and *A. kurodai* (Furukawa et al., 2006) (accession number: X92113 for *H. aspersa*, AF254118 for *H. trivolvis* and AB206707 for *A. kurodai*, respectively). The FMRFamide-gated channel is an amiloride-sensitive Na^+-channel. It has two transmembrane domains and functions as tetramer (Coscoy et al., 1998), whereas the ligand-binding site is located near the first transmembrane region (Cottrell et al., 2001). This was an important discovery in the field of peptide and ion channel research as it demonstrated the presence of a ligand-gated ion channel that is activated by direct binding of a neuropeptide.

The structures of FMRFamide precursors have been cloned in several molluscan species. For instance, the precursor protein for FMRFamide in *Aplysia* consists of 552 amino acids and 28 copies of the sequence FMRFGKR repeats on the precursor (Taussig and Scheller, 1986). A single copy of FLRFGKR sequence was found in the N-terminal region of the precursor. In *Mytilus edulis*, the cloned precursor consists of 403 amino acids. All the FMRFa-RPs identified from *Mytilus* (FLRFamide and *Mytilus*-FFRFamide) are encoded on this precursor, together with 16 copies of FMRFamide (Favrel et al., 1998). Therefore, when the precursor proteins are subjected to PTMs, all the FMRFa-RPs are generated as a minor component of mature peptides. By contrast, in *Lymnaea*, tetra- and penta-FMRFa-RPs including authentic FMRFamide are encoded on the precursor distinct from that encoding longer hepta-FMRFa-RPs. The *L. stagnalis* FMRFamide precursor gene consists of five exons. All the precursors are generated by alternative splicing from the single precursor gene (Santama and Benjamin, 1995). For instance, precursor 1 which encodes the tetrapeptides (e.g., FMRFa, FLRFa, and others) is generated by the combination of exons I and II, whereas precursor II that encodes longer FMRFamide-RPs (e.g., GDPFLRFa, SDPFLRFa) is generated by exons I, III, IV, and V (see Fig. 10.4). As shown by in situ hybridization and mass spectrometry, the heart exciter neuron, Ehe, expresses precursor 1 and expresses the tetrapeptides, whereas the visceral white interneuron expresses precursor 2 (heptapeptides) in the visceral ganglion, suggesting that alternative splicing generates two mutually exclusive neuronal populations (Santama and Benjamin, 2000). Therefore, the *Lymnaea* FMRFamide-related peptides demonstrate how alternative splicing generates peptide diversity and how the cell-specific expression of these different peptides is capable of controlling cardiac function.

10.4.4   THE SMALL CARDIOACTIVE PEPTIDES

As described above, the ganglia of gastropod molluscs contain several cardioactive substances. In *H. aspersa*, three cardioexcitatory and one cardioinhibitory substance was separated by size-exclusion chromatography (Lloyd, 1982, 1978). The low molecular weight cardioexcitatory substance appeared to be serotonin because the substance coeluted with serotonin, and methysergide, a serotonin antagonist, blocked its cardioexciatory action on the *Helix* ventricle. On the other hand, the cardioinhibitory substance was acetylcholine (Ach) since it coeluted with Ach, and its physiological effect was diminished by treatment with cholinesterase. The other two cardioexcitatory substances seem to be peptides because protease treatment diminished cardioexcitatoy actions of the substances on the heart. The cardioexcitatory peptide with larger molecular weight (6000–8000 Da) was named as large cardioexcitatory peptide (LCP), while the other with low molecular weight (ca. 1500 Da) was named as small cardioexcitatory peptide (SCP).

A Alternative mRNA splicing of the FMRFamide gene

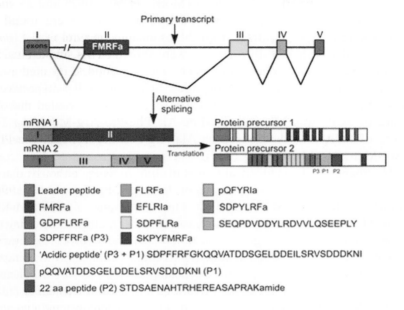

- Leader peptide
- FMRFa
- GDPFLRFa
- SDPFFRFa (P3)
- FLRFa
- EFLRIa
- SDPFLRa
- SKPYFMRFa
- pQFYRIa
- SDPYLRFa
- SEQPDVDDYLRDVVLQSEEPLY

'Acidic peptide' (P3 + P1) SDPFFRFGKQQVATDDSGELDDEILSRVSDDDKNI

pQQVATDDSGELDDELSRVSDDDKNI (P1)

22 aa peptide (P2) STDSAENAHTRHEREASAPRAKamide

B Mutually exclusive expression of exon II and exon III

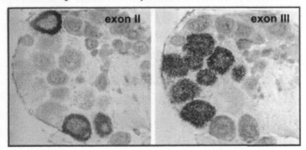

FIGURE 10.4 Alternate mRNA splicing of the FMRFamide gene. (A) Two mRNA variants are spliced from the primary transcript. Protein precursor 1 encodes 5 different peptides, including multiple copies of the tetrapeptides, FMRFamide and FLRFamide. Post-translational processing of QFYRIamide converted Q into pQ. Protein precursor 2 encodes 7 peptides including multiple copies of the heptapeptides, SDPFLRFamide and GDPFLRFamide. Post-translational cleavage of the "acidic peptide" resulted in two further peptides P3 and P1. Only peptides that were confirmed by sequencing and mass spectrometry are included in the list of peptides. (B) In situ hybridization of the alternatively spliced transcripts shows the mutually exclusive mRNA expression at the single neuron level. The same neurons can be identified in these adjacent sections of the visceral ganglion. (Source: Benjamin, P. R., Kemenes, I. Lymnaea Neuropeptide Genes. *Scholarpedia*, 8(7), 11520). (Used with permission.)

An SCP-like substance was also found in the peptide-containing extract of the central ganglia of *A. brasiliana* (Morris et al., 1982) and *L. stagnalis* (Geraerts, 1981), whereas no LCP-like substances were found in *Aplysia*. Using a peptide extract from the esophagus and crop of *A. californica*, peptides with SCP-like activity were separated by size-exclusion, ion exchange and reverse-phase column chromatography and resulted in the isolation of two distinct SCP-like peptides, SCP_A and SCP_B, respectively. Analysis by fast-atom bombardment mass spectrometer revealed that the structure of SCP_B was Met-Asn-Tyr-Leu-Ala-Phe-Pro-Arg-Met-NH$_2$. The structure of SCP_A was determined as Ala-Arg-Pro-Gly-Tyr-Leu-Ala-Phe-Pro-Arg-Met-NH$_2$ (Lloyd et al., 1987).

In *H. aspersa*, SCP_B is identical to that of *Aplysia* SCP_B, although that of *Helix* SCP_A is slightly different from that of *Aplysia* SCP_A (Price et al., 1990). Nevertheless, all of the SCPs share the same C-terminal YLAFPRM-NH$_2$ structure. SCP-related peptides are also present in other molluscs, including the freshwater pulmonates, *L. stagnalis* and *H. trivolvis*, the bivalves *Mercenaria mercenaria*, *Crassostrea virginica* and *Mytilus edulis* (Candelario-Martinez et al., 1993; Fujisawa et al., 1993), and the cephalopod, *Octopus vulgaris* (Kanda and Minikata, 2006). The structures of various molluscan SCPs are listed in Table 10.3. Molecular cloning of the SCP precursors in *Aplysia* and *Lymnaea* has shown that SCP_A and SCP_B are encoded in tandem in the N-terminal region of the same precursor (Mahon et al., 1985a; Perry et al., 1999).

TABLE 10.3 Structures of cardioactive peptides referred in this review

Peptides		Structure	Animals
FMRFa-RP		FMRFa	*Lymnaeastangalis*
		FLRFa	
		QFYRIa	
		EFLRIa	
		GDPFLRFa	
		pQDPFLRFa	
		SDPFLRFa	
SCP-RP	A	SGYLAFPRMa	*Helix pomatia*
	B	MNYLAFPRMa	
	A	SGYLAFPRMa	*Lymnaeastagnalis*
	B	pQNYLAFPRMa	
	A	SQGYLAFPRMa	*Helisomatrivolvis*
	B	SGYLAFPRMa	

TABLE 10.3   *(Continued)*

Peptides		Structure	Animals
	A	PGYLAFPRMa	*Aplysia californica*
	B	NMYLAFRPMa	
		SQPYIAFPRMa	*Littorina littorea*
		NYLAFPRMa	*Thais heamastoma*
	A	IAMSFYFPRMa	*Mercenaria mercenaria*
	B	YFAFPRQa	
		APNFLAYPRLa	*Mytilus edulis*
	A	AMSFYFPRMa	*Dinocardium robustum*
	B	YFAFPRQa	
ACEP-RP	ACEP-1	SGQSWRPQGRFa	*A. californica*
	LyCEP	TPHWRPQGRFa	*L. stagnalis*
	LUQIN	APSWRPQGRFa	*A. californica*
HRBP-RP	HRBP	pQVAQMHVWRAVNHDRN-HGTGSGRHGRFLIRN-RYRYGGGHLSDA	*A. californica*
	LYCP	SLAQMYVGNHHFNENDLT-STHGGSRRWSNRKHQSRI-YTGAQLSEA	*L. stagnalis*
	HDS-1	NTAQWLQGRQHFN-NYDHESNRRFSH-PRYNIRNNGRGVIAEA	*Helix lucorum*
R15α peptide	R15α peptide-2	MGSGGLKLHCQVHPANCP-GGLMVT	*A. californica*
	VD1/RPD2 α-peptide	DMYEGLAGRCQHH-PRNCPGFN	*L. stagnalis*
DAACP	Achatin-1	GdFGD	*Achatina fulica*
	fulicin	FdNEFVa	*A. fulica*
	NdWFa	NdWFa	*Aplysia kurodai, L. stagnalis,*
Cardioinhibitory peptide			
	Helix HCIP	VFQNQFKGIQGRFa	*Helix aspersa*
	APGWamide	APGWa	*H. aspersa*
	CARP	AMPMLRFa	*Rapanatho masiana*
	FRFamide	GSLFRFa	*A. fulica*
		SSLFRFa	*A. fulica*

ACEP: *Achatina* cardioexcitatory peptide, DAACP: D-amino acid containing peptide, HRBP: Histidine-rich basic peptide, SCP: Small cardioactive peptide.
Note that "a" represents C-terminal amide, while "d" represents D-amino acid.

SCPs increase both the frequency and amplitude of the heartbeat in *A. californica* (Cawthorpe et al., 1985) and in *H. aspersa* (Lesser and Greenberg, 1993). One of the remarkable features of the SCPs is its wide effective concentration range on the heart. For example, in the isolated and perfused preparation of the *Aplysia* heart, the threshold concentration of SCP_B was as low as 10^{-12} M whereas the maximum action was obtained at concentration above 10^{-7} M (Cawthorpe et al., 1985). Application of SCP concentrations of higher than 10^{-10} M required more than 20-min before the heartbeat returned to basal. SCPs augment intracellular levels of cAMP in the myocardial fibers of *Aplysia* (Cawthorpe et al., 1985) and in *Helix* (Reich et al., 1997a) through the activation of adenylate cyclase. In the dispersed myocardial cells of *Helix*, SCP-dependent elevation of cAMP promotes phosphorylation of a 53-kDa protein (Reich et al., 1997b), although the chemical characterization of the protein is not known.

Immunohistochemistry using anti-SCP_B antibody demonstrates SCP-staining neurons in the central and peripheral nervous systems in several gastropods including *Aplysia*, *Helix*, and *Lymnaea*. In *Aplysia*, the most prominent signals were found in the buccal ganglia, although a few neurons in the other ganglia also react with anti-SCP antibody (Lloyd et al., 1985). The B1 and B2 motoneurons that regulate the motility of buccal mass as well as many small neurons in the ganglion are immunopositive. Therefore, it is likely that in addition to cardiac regulation, SCPs are involved in the regulation of feeding behavior.

In the abdominal ganglion, SCP-containing nerve processes are abundant in the neuropil, connectives and the nerve bundles that emanate from the ganglia. However, only a few small neurons are immunoreactive to the anti-SCP antibody in this ganglion. In addition, the distribution of SCP-containing nerve processes is quite diffuse in the cardiovascular region (Lesser and Greenberg, 1993). Thus, despite of the potent action on the molluscan hearts, it is currently unknown which SCP-positive neuron regulates heartbeat. Nevertheless, the fact that SCPs are effective on the heart at low concentrations, even trace amounts of SCPs released from those nerve endings may have action on the myocardium.

10.4.5   *ACHATINA CARIDOEXCITATORY PEPTIDE-1*

Achatina cardioexcitatory peptide (ACEP)-1 is an undecapeptide having C-terminal RF-NH$_2$ structure (Ser-Gly-Gln-Ser-Trp-Arg-Pro-Gln-Gly-Arg-Phe-NH$_2$). ACEP-1 was originally isolated from the auricle of *A. fulica* using HPLC fractionation and a bioassay measuring ventricular contraction (Fujimoto et al., 1990). Although ACEP-1 was isolated from the auricle it has no effect on the auricle itself, whereas it augments the beating amplitude of the *Achatina* ventricle. Fujiwara-Sakata and Kobayashi (1994) subsequently examined the localization of ACEP-1 in *Achatina* by immunohisto-chemistry. ACEP-1-like immunoreactivity is localized to the left and right pedal nerve large neurons (L- and R-PeNLN), ventral visceral large neuron (v-VLN), and ventral right parietal large neuron (v-RPLN). Moreover, other large neurons in the visceral ganglion send ACEP-1-containing nerve processes toward the atria through the intestinal nerve. In addition, dense ramifications of ACEP-1 containing nerve endings are seen in the auricle. It is likely that ACEP-1 released from the auricle acts in a paracrine fashion on the ventricle.

ACEP-1-related peptides have also identified in *A. californica* and *L. stagnalis*. In *Aplysia*, biochemical identification of ACEP-1-related peptides (LUQIN) was preceded by the molecular cloning of its precursor. Shyamala et al. (1986) found that the L5 neuron in the LUQ region of the abdominal ganglion in *Aplysia* was immunoreactive to anti-FMRFamide antibody (Shyamala et al., 1986). When one of the cDNA clones designated as L5–67, which was derived from the mRNAs specifically synthesized in the L5 neuron was sequenced, it was found that the predicted protein included the Arg-Phe-Gly-Lys-Arg sequence in the N-terminal region. The mature form of the novel RFamide peptide (LUQIN) was subsequently purified from LUQ cells and its structure was determined as Ala-Pro-Ser-Trp-Arg-Pro-Gln-Gly-Arg-Phe-NH$_2$ (Aloyz et al., 1995). LUQIN share the same C-terminal seven amino acids with ACEP-1 and alternative splicing generates another form of the LUQIN precursor N-terminal region including the first 45 amino acids which are identical to the original L5–67 propeptide (Angers and DesGro-seillers, 1998). In situ hybridization analysis localized the expression of the LUQIN precursor in the identified neurons L2, L3, L4, and L6 in the LUQ region of the *Aplysia* abdominal ganglion (Giardino et al., 1996; Landry et al., 1992). Tissue contents of the LUQIN precursor estimated by immunoreactivity to anti-LUQIN precursor antibody detected significant immunoreactivity in the abdominal, pleural, pedal, and cerebral ganglia.

In the freshwater snail *L. stagnalis*, a novel cardioexcitatory peptide was isolated by screening for ligands that bound to some previously cloned GPCRs (Cox et al., 1997; van Kesteren et al., 1995; Tensen et al., 1998). Endogenous ligands from ganglionic extracts were screened for their ability to bind to cloned receptors that were expressed in *Xenopus* oocytes. From this heterologous screening procedure, Tensen et al. (1998) identified of a novel ACEP-1-related peptide which bound to clone GRL 106 and was named *Lymnaea* cardioexcitatory peptide (LyCEP). ACEP-1, LUQIN, and LyCEP have cardioexcitatory action on *A. fulica* and *L. stagnalis,* respectively (Tensen et al., 1998). Unlike ACEP-1, LyCEP augments beating frequency of *Lymnaea* auricle. However, ACEP-1 and related peptides appear to be multifunctional. For instance, cholinergic motoneuron B10 in buccal ganglion of *A. fulica* is colocalized with ACEP-1. ACEP-1 is thought to function as a peptide co-transmitter and regulates the radula protractor muscle. ACEP-1 also functions to increase the phasic contractions of the penis retractor muscle and buccal muscle that was previously excited by electrical stimulation. In *L. stagnalis*, LyCEP appears to have reproductive functions as well. LyCEP-immunopositive nerve endings are found in the vicinity of the CDCs and the LyCEP receptor is also expressed in some CDCs. LyCEP is known to decrease the electrical discharge of CDCs and may regulate feeding behaviors by inducing the spontaneous discharge of buccal neuron B4 (Tensen et al., 1998).

10.4.6   *CARDIOACTIVE PEPTIDES CONTAINING D-AMINO ACIDS*

Amino acids, except for glycine, have an asymmetrical carbon atom in their molecules. Accordingly, there are stereoisomers, D-type and L-type, for each amino acid. Nevertheless, peptides and proteins found in living organisms usually consist of L-amino acids. Although it is a rare phenomenon, several cardioactive peptides found in molluscs possess a D-amino acid residue in their structure. Achatin-1 (G_DFAD), fulicin (Phe-D-Asn-Glu-Phe-Val-NH$_2$) and NdWFamide are examples of such D-amino acid containing cardioactive neuropeptides in molluscs (Kamatani et al., 1991; Ohta et al., 1991; Morishita et al., 1997). A common structural feature shared by those D-amino acid-containing peptides is that the D-/L-conversion occurs on the N-terminal penultimate amino acid. Other peptides such as fulyal from *A. fulica*, FLRFamide-related peptide from *M. edulis*, and crustacean cardioactive peptide (CCAP)-related peptide from the Roman snail, *Helix pomatia*, are other examples of D-amino acid containing neuropeptides, although it is unclear if they modulate heartbeat these animals (Fujisawa et al., 1992; Minakata, 1996; Yasuda-Kamatani et al., 1997).

Achatin-1 is a tetrapeptide containing D-phenylalanine identified in the African giant snail, *A. fulica* (Kamatani et al., 1991). The same peptide was also purified from the *Achatina* aorta by a different group (Fujimoto et al.,

1991). Structurally related peptides to achatin-1 are found in *Octopus* and these peptides also have cardioexcitatory action on the isolated systemic heart (Iwakoshi et al., 2000). An achatin-1-like peptide has also been identified in *Aplysia*, however, a cardioexcitatory action of *Aplysia* achatin-1 on the heart has not been reported (Bai et al., 2013).

Achatin-1 has a direct action on the *Achatina* heart augmenting the beating amplitude and frequency (Fujimoto et al., 1990). Immunostaining using specific anti-achatin-1 antibody demonstrated dense ramifications of achatin-1-containing nerve endings in the atria. However, physiological experiments reveal that achatin-1 is only effective on the ventricle but not on the atria, indicating that ahcatin-1 probably acts in a paracrine fashion. Achatin-1 also has an indirect action on cardiac regulation. Achatin-1 augments the excitability of the heart exciter neuron, PON, in the visceral ganglion of the snail (Takeuchi et al., 1995). In situ hybridization localized the expression of achain-1 precursor in some neurons in medial region of right and left hemisphere of the pedal ganglion of *Achatina*, although morphological and chemical characterization of the neurons is still lacking (Satake et al., 1999).

Fulicin is another D-amino acid containing neuropeptide isolated from the ganglia of *A. fulica* (Ohta et al., 1991). Recently, structurally related peptides were identified in *Aplysia*. Interestingly, *Aplysia* fulicin consists of all L-amino acids. Fulicin at 10^{-6} M clearly increases beating amplitude but not the frequency of *Achatina* ventricle. Fulicin may also be involved in the regulation of reproduction, because it has a more potent effect on the penis retractor muscle and the oviduct of the snail (Fujisawa et al., 2000). Fulicin-containing neurons have been localized to the left and right parietal ganglion of *Achatina* by in situ hybridization (Satake et al., 1999).

NdWFamide is a potent cardio-excitatory peptide in *Aplysia*, *Lymnaea* and the terrestrial snail, *Euhadora congenita*. The peptide augments both of the beating rate and amplitude of the auricle and ventricle in these snails (Morishita et al., 2003b,c). Presumably, *A. californica*, *Limax valentianus*, *Tritonia*, and *L. gigantea* have NdWFamide or structurally related peptides because precursor cDNAs encoding precursor of the peptide have been identified in these molluscs (Matsuo et al., 2011).

FMRFamide and SCP_B are also cardioactive on the heart of *A. kurodai* (Harris et al., 1995; Lloyd et al., 1985). However, FMRFamide is effective at relatively higher concentrations than NdWFamide. Although SCP_B is effective on the heart at the lower concentrations than NdWFamide, this effect is not dose-dependent (Morishita et al., 2001). In this context, NdWFamide is the most dominant mediator that regulates heartbeat in *A. kurodai*. Moreover, of the three cardioactive peptides, NdWFamide alone induced contraction in the arteries of *A. kurodai* (Morishita et al., 2001). Therefore, NdWFamide may have diverse action on the cardiovascular system.

As described earlier, immunostaining with specific anti-NdWFamide antibody localized the peptide in some of the R3–14 neurons and white cells in RUQ region of the abdominal ganglion (Morishita et al., 2003c). NdWFamide containing nerve processes emanating from these neurons project onto the cardiovascular region including the branchial and kidney vein, and the pericardial cavity via the branchial nerve.

In *A. kurodai*, the abdominal aorta supplies hemolymph toward the digestive gland which is located in the tail region. The aorta has a sphincter that regulates blood flow through the aorta. Sasaki et al. (2004) reported that NdWFamide reduces the blood flow to the digestive gland by contracting the sphincter muscle. If NdWFamide induces both the contraction of sphincter and augments heartbeat at the same time, it may promote the accumulation of hemolymph to the head region, generating hydrostatic pressure that contributes to the forward extension of the head.

Unlike SCP_B and serotonin, NdWFamide has no effect on the intracellular concentration of cAMP in the ventricle of *Aplysia* (Morishita, F., unpublished observations). By whole-cell patch-clamp recording, Kanemaru et al. (2002) demonstrated that NdWFamide augmented a high voltage-activated inward Ca^{2+}-current across the plasma membrane of dispersed myocardial cells of *Aplysia*. Since the increase in Ca^{2+}-current is blocked by the nifedipine and GTP-γS, it was suggested that NdWFamide activates an L-type Ca^{2+}-channel through the mediation of certain GTP-binding proteins.

NdWFamide is highly resistant to digestion by peptidase (Morishita et al., 2003a). Arguably, the unique molecular shape of NdWFamide prevents the peptide from fitting into the binding pocket of the peptidases. Although the *Aplysia* CNS does possess deamidase-like activity that inactivates NdWFamide by removing C-terminal amide (Morishita et al., 2003a), the molecular mechanism that inactivates NdWFamide is not fully understood.

Generally, the D-configuration of amino acids is essential for the bioactivity of D-amino acid-containing peptides (Fujita et al., 1995; Kim et al., 1991; Morishita et al., 2003a). In achatin-1, structural analysis demonstrated that replacement of D-phenylalanine to L-phenylalanine greatly modifies the molecular shape of the peptide (Ishida et al., 1992; Kamatani et al., 1990). For NdWFamide, the D-configuration of tryptophan, as well as the C-terminal amide, is essential for the cardioexcitatory action on the heart because an analog peptide with L-tryptophan, or the one without C-terminal amide, markedly reduces bioactivity. Structure-modeling by analysis with nuclear magnetic resonance (Yokotani et al., 2004) predicted that the molecular shape of NdWFamide is compactly folded by two intramolecular interactions, namely, the one between the phenyl ring on the phenylalanine residue and the amide moiety on the side chain of the N-terminal aspara-gine, and the other between the indole ring on the D-tryptophan residue and the C-terminal amide on the phenylalanine. Apparently, the aforementioned analog peptides disrupted the native molecular shape of the peptide which resulted in diminished bioactivity.

Precursor cDNAs for achatin-1, fulicin and NdWFamide have been identi-fied in some species of molluscs (Morishita et al., 2012; Satake et al., 1999; Yasuda-Kamatani et al., 1995). In all the precursor cDNAs encoding the precur-sors for D-amino acid neuropeptides, the D-amino acid residue is encoded by the codon for the corresponding L-amino acid. Since transfer RNA specifically binds to L-type amino acids and carries it to the ribosome for protein synthesis, the precursor protein for the D-amino acid-containing neuropeptides is synthe-sized with all L-amino acids on ribosomes, and then the D-/L-conversion of an amino acid in neuropeptides takes place during PTM of the precursor.

Peptidyl-aminoacyl-L-/D-isomerase which catalyzes the L-/D-conversion of an amino acid has been partially characterized in the spider, platypus and leaf frog (Jilek et al., 2011; Torres et al., 2007). The L-/D-conversion of amino acids in neuropeptides by this enzyme is a slow reaction which requires long incubation times (hours) to obtain a detectable end product. Using sophis-ticated mass-spectrometry, Song and Liu (2008) detected NWFamide, the diastereomer of NdWFamide in *Aplysia* abdominal ganglia extracts. It was suggested that NWFamide is the precursor to NdWFamide which undergoes post-translational modification in *Aplysia* abdominal ganglion (Song and Liu, 2008). Characterization of a peptidyl-aminoacyl-L-/D-isomerase has not yet been reported in molluscs.

10.4.7 CARDIOINHIBITORY PEPTIDES

In molluscs, several neuropeptides are known to have inhibitory actions on the heart. A tetrapeptide, APGWamide (Ala-Pro-Gly-Trp-NH$_2$) at 10^{-8} M order terminates regular beating of isolated ventricle of *H. aspersa* (Reich et al., 1997a). When Price et al. (1990) examined the cardiac actions of several SCP-related peptides, they found that some peptides terminated the heartbeat of the snail. However, transient, but marked increases in the beating ampli-tude preceded the termination of heartbeat (Price et al., 1990). Therefore, the possibility remains that termination of heartbeat reflects the systolic arrest of the heart. Baud et al. (1998) isolated a tridecapeptide from *H. aspersa* which shares the same C-terminal tetra-peptide with ACEP-1 of *Achatina*. The authors named the peptide as *Helix* cardioinhibitory peptide (HCIP) because it reduced the beating frequency of the heart. However, HCIP also augmented the beating amplitude. Considering the fact that cardiac muscle has a longer refractory period than other muscles, prolonging heartbeat intervals via HCIP action could be a consequence of elevated heart muscle excitability.

On the other hand, Walker and colleagues examined the action of RFamide peptides on isolated perfused heart of *A. fulica*, and found that FMRFamide and FLRFamide had an excitatory action, whereas SSLFR-Famide and GSLFRFamide had an inhibitory action. This is an interesting result in view of structure-activity relationship of peptides because the difference in peptide sequence reversed its physiological action on the heart. However, since the two cardioinhibitory peptides are not endogenous in *Achatina*, it is unclear if *Achatina* has a cardioinhibityory RFamide.

Catch-relaxing peptide (CARP), a myomodulin-related peptide isolated from *Mytilus*, inhibited the heartbeat in the isolated heart of a whelk, *R. thomasiana* (Fujiwara-Sakata et al., 1992), suggesting that myomodulin-related peptide is cardioinhibitory peptide in the whelk. However, the cardioinhibitory action of myomodulin may not be universal because the peptide has cardioexcitatory effects on the heart of a nudibranch (Wiens and Brownell, 1995).

The inhibition or termination of heartbeat may play an important physiological role during hibernation or during prolonged defensive closure of the shell in molluscs. It is likely that many peptidergic signaling molecules mediate the inhibitory regulation of the molluscan heart.

10.5   REGULATION OF HYDROMINERAL BALANCE AND RENAL FUNCTION

The majority of studies involving peptides or neurotransmitters that regulate hydromineral balance in molluscs have been done on gastropods. In general, most strictly marine gastropods are isosmotic (but not isoionic) with respect to the environment, however, they do show considerable neuroendocrine control over body volume and ionic composition of hemolymph. In contrast, all freshwater pulmonates are hyperosmotic regulators whereas the terrestrial pulmonates live in environments where the availability of water may change dramatically over time.

Freshwater snails live in an environment that is hypotonic relative to their hemolymph and tissues, therefore water is continually entering the animal's body via surfaces directly exposed to the water, and major ions (Na^+, K^+, Ca^{2+}, Mg^{2+}, Cl^-, and HCO_3^-) are continually being lost to the environment. To counterbalance osmotic influx and solute losses, water entering the snail by osmosis is removed by the kidney which excretes a hypotonic urine (Potts, 1967), whereas ions are actively taken up from the external medium and via the consumption of food (Burton, 1983). The following sections will deal with these two processes and the peptides that regulate them.

In most molluscs, the formation of prourine is thought to occur by ultrafiltration of the hemolymph (Potts, 1967). The site of ultrafiltration in pulmonates is believed to be the kidney as ultrastructural studies have shown the presence of podocyte-like processes in the kidney sac as well as septate junctions between kidney epithelial cells (Khan and Saleuddin, 1979a,b; Matricon-Gondran, 1990; Newell and Skelding, 1973a,b; Wendelaar-Bonga and Boer, 1969). The neuroendocrine control of osmoregulation has been studied mainly in the freshwater pulmonates *L. stagnalis* and *H. duryi*. In the visceral and left parietal ganglia of *H. duryi*, there are two identified neuro-secretory cell types that are respond to osmotic stress, the type 2 and type 3 cells, as identified by PAF staining (Khan and Saleuddin, 1979a,b). Based on ultrastructural studies, type 2 cells show increased synthetic activity when snails are under hypoosmotic stress (distilled water). A functionally similar response to hypoosmotic stress is seen in the dark green cells (DGCs) of *L. stagnalis* which are mainly located in the pleural and parietal ganglia (Wendelaar-Bonga, 1971b, 1972).

In *H. duryi*, hypoosmotic treatment causes the extracellular spaces between kidney epithelial cells to become wider, whereas under hyperosmotic stress the spaces between cells become reduced (Khan and Saleuddin, 1979a,b). This expansion of intercellular spaces suggests increased urine production to excrete the excess water that has entered the body by osmosis. Extracts of visceral ganglia obtained from hypoosmotically stressed snails, and then applied to kidneys from snails acclimated to isosmotic medium, resulted in a widening of the extracellular spaces. These data suggest that the type 2 cells in *H. duryi* produce a diuretic factor that acts on the kidney epithelium to control paracellular transport (Khan and Saleuddin, 1979a,b). In contrast, the type 3 cells show increased synthetic activity under hyperosmotic stress when snails are placed in isosmotic medium, suggesting they produce an antidiuretic factor. Khan and Saleuddin (1981) further demonstrated that brain extracts from hyposmotic snails induced structural changes in the septate junctions of the kidney epithelial cells and suggest that a brain factor might act on the septate junctions of the kidney to alter paracellular permeability.

The identity of the diuretic and antidiuretic factor(s) in *H. duryi* are presumed to be peptides based on the presence of many electron-dense, membrane-bound vesicles which is typical of neuropeptide synthesizing cells (Khan and Saleuddin, 1979a). At the time these studies were conducted, the identity of the peptide(s) in the type 3 cells was unknown. Subsequent immunogold labeling studies, however, showed neurosecretory granules in the type 3 cell stained positive for FMRFamide-like peptides (Saleuddin et al., 1992). FMRFamide-immunoreactive axons in the intestinal nerve project to and innervate the kidney smooth muscle cells and epithelium. In addition, *H. duryi* kidneys incubated in vitro with FMRFamide showed an increase in wet weight, indicating that this peptide (or a related peptide) might have an antidiuretic function in *H. duryi* (Saleuddin et al., 1992). When the concentration of FMRFamide-related peptides was measured by RIA in two major osmoregulatory organs, the kidney and mantle, higher concentrations of FMRFamide-like peptides were found in both these tissues under hyperosmotic stress (Madrid et al., 1994).

During the embryonic development in gastropods, one of the earliest neuronal elements to form are neurons that produce FMRFamide (Croll, 2000; Croll and Voronezhskaya, 1996). In the embryos of the freshwater snails *L. stagnalis* and *H. trivolvis*, hyperosmotic conditions are thought to elevate FMRFamide levels and induce developmental irregularities such as hydropia as well as earlier synthesis of serotonin in the visceral ganglia (Chaban and Voronezhskaya, 2008). This suggests that FMRFamide may also have an osmoregulatory role in developing embryos as it does in adult snails. A role for serotonin in osmoregulation has been suggested in in *H. duryi*. The kidney of *H. duryi* is innervated by serotonergic fibers from the visceral ganglion, suggesting that serotonin may regulate some aspect of kidney function (Khan et al., 1998). Depleting serotonin levels in the mantle and kidney with selective neurotoxins resulted in a significant reduction of FMRFamide-like peptides in these tissues. Therefore, serotonin may be involved in the synthesis and/or release of FMRFamide-like peptides in osmoregulatory organs of *H. duryi*.

The role of peptides in hydromineral balance has been studied extensively in *L. stagnalis*. Removal of the cerebral ganglia in this snail resulted in a reduction of Na^+ and Cl^- concentrations in the hemolymph, suggesting the presence of a presumptive ion transport stimulating factor from the cerebral ganglia (De With and van der Schors, 1986). Injection of a crude peptide extract from the cerebral ganglia into intact snails stimulated Na^+ uptake from the external medium as measured by $^{22}Na^+$ uptake (De With and van der Schors, 1986). To aid in the purification of a putative sodium-influx stimulating peptide (SIS) in *L. stagnalis*, a bioassay that measured ion transport across the membrane using *Lymnaea* skin attached in an Ussing chamber was developed (De With et al., 1988). Brain extracts or median lip nerve extracts containing SIS peptide were capable of increasing the potential difference and short circuit current across skin epithelium. Using this SIS peptide bioassay, a 76-amino acid peptide was biochemically purified from *Lymnaea* CNS by HPLC and Edman sequencing (De With et al., 1993).

The cloned precursor cDNA of SIS peptide encodes a precursor protein which consists of 100 amino acids, including an N-terminal 24 amino acid signal peptide. In addition, six cysteine residues are thought to form three disulfide bonds which contribute to the peptide's tertiary structure (Smit et al., 1993b). Synthetic SIS peptide stimulates ion transport across the skin epithelium and is localized to specific neurons of the visceral and right parietal ganglia, and in axons of the intestinal nerve and right internal pallial nerve (De With et al., 1993). SIS peptide-positive nerve endings are abundant in the cardiovascular and renal regions of *Lymnaea*, suggesting that SIS peptide is involved in regulating cardiac and renal activities (Boer et al., 1992). Based on the pattern SIS peptide localization by immunohistochemistry and in situ hybridization, SIS peptide is believed to be synthesized and secreted by the neurosecretory Yellow Cells (YCs) in the CNS and regulates ion uptake across the integument, and perhaps in the renopericardial system as well (Boer et al., 1992; De With et al., 1993; Smit et al., 1993a).

When *L. stagnalis* is acclimated in water with low osmolarity, the YCs become electrically active and the numbers of neurosecretory granules in the neurons stained with paraldehyde fuchsin (PAF) are greatly reduced, suggesting the peptide is being released into the extracellular space (Soffe and Slade, 1979). These results collectively demonstrate that in *L. stagnalis*, hyposomotic stress causes the release of SIS peptide from the yellow cells resulting in Na^+ uptake in osmoregulatory tissues.

In addition to SIS peptide, the monoamine serotonin also increases short circuit current and electrical potential difference in *L. stagnalis* skin preparations via an intracellular cAMP signaling pathway (De With et al., 1988, 1993). Serotonin is an abundant neurotransmitter in the CNS of *L. stagnalis* (Audesirk, 1985; Croll and Chiasson, 1989; Marois and Croll, 1992) and other gastropods (Bernocchi et al., 1998; Croll, 1987, 1988, 2000; Salimova et al., 1987), and it has been implicated to play a role in osmoregulation in *H. duryi* (Khan et al., 1998). Although the mantle and renopericardial system of gastropods has been implicated in ion uptake, the role of specific transmitters regulating ion uptake in these tissue has not been demonstrated. It would be interesting to test if other known neuroactive substances in the gastropod nervous system are capable of stimulating active ion uptake from freshwater similar to the mechanism used in the gills of freshwater fishes. In this regard, the development of an in vitro model using kidney or mantle epithelial cells cultured on a solid support with distinct apical and basal sides would be of exceptional utility (Wood et al., 2002). These "surrogate" kidney or mantle models could be used as bioassays to screen for peptides and other chemical mediators regulating ion transport across osmoregulatory tissues in snails.

The role of peptides in the control of osmoregulation in marine gastropods has been studied mainly in the opisthobranch, *A. californica*. Injections of identified neuron R15 from the abdominal ganglion into the hemocoel of intact animals resulted in the rapid uptake of water in the recipients, whereas an extract of other neurons such as R2 and R14 in the abdominal ganglion failed to increase body weight (Kupfermann and Weiss, 1976). The putative osmoregulatory factor from neuron R15 was estimated to be a peptide with a molecular mass of 1500 Da (Gainer et al., 1977). R15 is a bursting pacemaker neuron whose firing is inhibited when the animal is subjected to hyposmotic stress, suggesting that R15 may produce an antidiuretic peptide (Bablanian and Treistman, 1983). The peptides synthesized by neuron R15 were purified and isolated by Weiss et al. (1989). Two neuropeptides named R15α1 (38 amino acids) and R15β (28 amino acids) were purified from R15 extracts and stimulated weight gain when injected into intact *Aplysia califorrnica*.

The response of neuron R15 to changes in seawater osmolarity appear to be mediated through the osphradium, a chemosensory organ (Skinner and Peretz, 1989). Furthermore, osmoregulatory responses to dilute media appear to decline with age. However, in the related marine snail *A. brasiliana*, which does not regulate its hemolymph osmolarity like *A. californica*, no such response to dilute media was found in the osphradium–R15 pathway (Scemes et al., 1991).

The kidney of *A. californica* is innervated by cholinergic branches of neuron L10, and a group of five peptidergic neurons (L2–L6) in the LUQ within the abdominal ganglion (Angers et al., 2000; Koester and Alevizos, 1989). Specifically, L10 and a subset of LUQ neurons, cells L2–4, 6, and L5, send branches to the renal pore which is a sphincter that controls urine efflux. Electrical input from L10 stimulates the opening of the renal pore whereas the LUQ neurons control its closing. Two transcripts have been cloned by differential screening of an abdominal ganglion cDNA library and code for the putative neuropeptide precursors, L5–67 and LUQ-1 (Shyamala et al., 1986; Wickham and DesGroseillers, 1991). L5–67 and

LUQ-1 mRNAs are expressed in cells L2–4, 6, and L5 (Landry et al., 1992). The mRNAs and peptides are present in the axons innervating the region surrounding the renal pore as well as other parts of the kidney.

In the LUQ neurons, the L5–67 transcript is processed to produce a decapeptide LUQIN which shares sequence similarity to ACEP-1 (*Achatina* cardioexcitatory peptide 1) and LyCEP (*Lymnaea* cardioexcitatory peptide). A second peptide having 89 amino acids, proline-rich mature peptide (PRMP) is also processed from the L5–67 transcript (Aloyz and DesGroseillers, 1995). The LUQ-1 mRNA directs the production of a bradykinin-like peptide which is present in L5 and some neurons of the CNS (Giardino et al., 1996; Landry et al., 1992). It would be interesting to determine the identity of the neuropeptides responsible for the closing of the renal pore and if these peptides may have other osmoregulatory functions in *A. californica*. These results suggest that neuropeptides mediate some aspects of osmoregulatory function in *A. californica*, although the precise cellular mechanisms are not fully understood.

Recent advances in next-generation DNA sequencing technology have enabled investigators to sequence and mine data from the genomes of numerous molluscs more rapidly than ever before. The next-generation sequencing techniques can also be used to obtain entire mRNA sequences transcribed in particular cells or tissues treated under different experimental conditions or exposed to different environments. Using this technique, several papers have reported changes in mRNA expression level in response to applied osmotic stress in some molluscs (Eierman and Hare, 2014; Lockwood and Somero, 2011; Zhao et al., 2012b). Although changes in neuropeptide gene expression does not in itself demonstrate function, it may allow for the identification of specific target molecule(s) and elucidation of neuroendocrine pathways involved in the regulation of molluscan osmoregulation.

10.6 REGULATION OF EMBRYONIC CILIARY ACTIVITY

10.6.1 RESPIRATORY RESPONSE TO HYPOXIA BY FRESHWATER PULMONATE EMBRYOS

The control of respiratory behavior in adult gastropods has been studied extensively in the freshwater pulmonate *L. stagnalis* where identified central pattern-generating neurons are responsible for coordinating respiratory behaviors (for reviews see Lukowiak et al., 2006, 2014). We will not address respiratory behavior in adult snails as it have been covered elsewhere in the context of operant conditioning of respiratory behavior in response to hypoxia. In this section, we will discuss the behavioral response of embryonic snails to environmental hypoxia (reduced oxygen concentration) as this response is known to serve a respiratory function and is regulated by specific neurotransmitters.

Environmental conditions such as pH, salinity and oxygen tension are generally more variable in freshwater in comparison with the marine environment. For example, oxygen concentrations in freshwater can undergo dramatic fluctuations depending on temperature, the amount of dissolved nutrients and ions, and the presence of organisms consuming dissolved oxygen in water. The embryos of gastropods are generally encased in a gelatinous egg mass which is affixed to underwater substratum and are therefore subjected to ambient environmental conditions. Unlike marine gastropods which have a free-swimming larval stage which can escape unfavorable environmental conditions, freshwater snails undergo all their larval stages within the egg mass and develop directly into juvenile snails (Morrill, 1982). The embryos of freshwater snails and likely those of other aquatic organisms that undergo encapsulated development within egg masses have evolved adaptive mechanisms which allow them to survive periods of environmental hypoxia.

Aquatic gastropods that develop in egg masses are surrounded by a viscous capsular fluid which is derived mainly from the albumen gland. Developing embryos display a characteristic ciliary-driven rotational behavior within the egg mass which is thought to enhance the mixing of diffusion-limiting processes in the capsular fluid (Hunter and Vogel, 1986; Moran and Woods, 2007). For instance, the embryos of the freshwater snail *Helisoma trvivolvis* display two characteristic rotational behaviors, a slow basal rotation which is due to constitutive ciliary beating, and periods of transient accelerations called surges. Under normoxic conditions, the ciliary-driven rotation is predicted to function similar to a laboratory "stir-bar" which enhances the mixing of oxygen in the capsular fluid. Mixing of the capsular fluid is hypothesized to maintain a constant diffusion gradient of environmental oxygen to the developing embryo (Goldberg et al., 2008). In support of this hypothesis, egg masses kept in hypoxic pond water have a faster basal rotation rate and increased frequency of rotational surges as compared to egg masses maintained under normoxia (Kuang et al., 2002).

In *H. trivolvis*, ciliary-driven rotational behavior is mediated by a pair of embryonic neurons named embryonic neurons C1 (ENC1s). The ENC1s are specialized serotonergic sensorimotor neurons that sense environmental oxygen and synapse with target pedal ciliary cells in the embryo to control rotational behavior. Since the ENC1s develop well before the formation of the CNS neurons, they are implicated to play an important role in normal embryonic development and survival as intact ENC1s are necessary for the response to environmental hypoxia (Kuang et al., 2002). Furthermore, laser treatment of the ENC1s is known to perturb their activity and cause a transient increase in those cilia that are in close anatomical association with the ENC1s (Kuang and Goldberg, 2001). Application of serotonin to egg masses causes a marked increase of embryonic rotation rate and ciliary beat frequency, whereas serotonin-induced increases in rotational behavior and cilio-excitatory response is suppressed by mianserin, a serotonin receptor antagonist (Diefenbach et al., 1991; Goldberg et al., 1994; Kuang and Goldberg, 2001; Kuang et al., 2002). Collectively, these studies suggest that the ENC1s in *H. trivolvis* respond to hypoxia by releasing serotonin at the embryonic ciliary cells which results in increased embryonic rotational behavior and enhances the mixing of oxygen in the capsular fluid.

The ciliary beat frequency is believed to be controlled by the constitutive activity of nitric oxide in embryonic ciliary cells of *H. trivolvis*. Inhibition of nitric oxide production using nitric oxide synthase inhibitors (e.g., L-NAME and 7-NI) reduced embryonic rotational rate, whereas providing NO donors (SNAP and SNP) increased embryonic rotation rate twofold (Cole et al., 2002). Furthermore, ciliary cells bathed in the presence of NOS inhibitors abolished the cilio-excitatory response to serotonin, indicating that a functional nitric oxide signaling system plays a permissive role in regulating the cilio-excitatory response to serotonin (Doran et al., 2003). In *H. trivolvis*, two serotonin receptor subtypes have been cloned which have been named HT_{1Hel} and $5HT_{7Hel}$. Activation of these serotonin receptor subtypes is thought to regulate ciliary beating via the activation of protein kinase C (PKC) and the release of calcium from a caffeine-sensitive intracellular store (Christopher et al., 1999; Doran and Goldberg, 2006; Doran et al., 2003, 2004).

In addition to playing a central role in regulating basal ciliary beating and rotational response to hypoxia, serotonin is thought to play an important protective function in response to long-term anoxia in *H. trivolvis*. Long-term anoxia induces a transient increase in embryonic rotation rate in egg masses exposed to anoxic pond water, a response similar to that seen with hypoxia treatment. Basal rates of embryonic rotation under anoxia were maintained for up to 13 h after which the embryos perished (Shartau et al., 2010). However, when embryos were cultured in anoxic pond water in the presence of serotonin, their embryonic rotation rates persisted for up to 40 h leading to markedly prolonged survival (Shartau et al., 2010). The authors suggest that anoxia or long-term hypoxia suppresses metabolic activities at the expense of development and growth since embryonic development is delayed in hypoxic embryos. Thus, ENC1 serotonin release is proposed to have a dual role in *H. trivolvis* embryos. Serotonin is proposed to control embryonic rotational behavior under normoxia and hypoxia via the activation of 5-HT_{7Hel} receptors which are present in embryonic ciliated cells. An additional function of serotonin might be to increase ATP production via increased glycogenolysis, thereby enhancing energy production to sustain rotational behavior during periods of hypoxia.

To determine if the aminergic control of embryonic rotational response to hypoxia is an evolutionarily conserved response of other freshwater snails, Goldberg et al. (2011) studied the embryos of *L. stagnalis*. In addition to having pedal and dorsolateral ciliary bands as in *H. trivolvis*, only *L. stagnalis* has a specific ciliated apical plate region during embryonic development. The ciliated apical plate is innervated by transient apical catecholaminergic neurons which contain dopamine and are primarily responsible for driving embryonic rotational behavior (Voronezhskaya et al., 1999). Both serotonin and dopamine increase the rate of embryonic rotation in vivo (Goldberg et al., 2011), however, serotonin is more potent than dopamine in stimulating ciliary beat frequency of isolated ciliary patches in vitro. Furthermore, the rotational response to hypoxia was blocked by the dopamine receptor antagonist SKF83566 but not by mianserin, suggesting that the rotational response to hypoxia in *L. stagnalis* is mediated by a dopamine D1-like receptor. Although the major neurotransmitter regulating embryonic rotational behavior in *L. stagnalis* is dopamine, the basic neural mechanism underlying the ciliary response to hypoxia is similar to that seen in *H. trvolvis*, suggesting that transient increases in ciliary beating may be an evolutionarily conserved response in freshwater snails to maintain sufficient oxygen supply to the embryos.

10.7 CONCLUSIONS

Peptides are the most diverse extracellular chemical messengers in animals. In gastropods, there are likely dozens of peptides yet to be fully characterized at a functional level in any single animal. Most gastropod peptides that have been studied in significant functional detail have been done using the sea hare *A. californica* and the pond snail *L. stagnalis*. These two molluscan models have provided a great deal of information regarding selected peptides and their functions in various aspects of reproduction, growth, cardiac function, and osmoregulation, particularly at the cellular and molecular levels. In addition, classical neurotransmitters commonly work together with peptides to exert precise control over key physiological processes. We know many pieces of interesting pieces of information about the function many molluscan peptides and neurotransmitters. However, most of this information is derived from in vitro studies that are conducted on isolated preparations. Therefore, a major challenge in the future will be to determine how these signaling molecules function in the context of whole animal physiology. As current proteomic and next-generation sequencing techniques continue to improve, more data from the proteome and transcriptome of model and nonmodel molluscs will become available. This should open the door for greater widespread use of techniques such RNA interference in molluscs, and perhaps the future implementation of methods such as CRISPR (clustered regularly interspaced short palindromic repeat) to alter gene function. Finally, better functional assays will need to be developed to study all these newly sequenced peptides and link them to in vivo physiology.

KEYWORDS

- Gastropoda
- *Aplysia*
- *Lymnaea*
- *Helisoma*
- *Biomphalaria*
- *Helix*
- *Achatina* neuropeptides
- egg-laying hormone
- schistosomin
- APGWamide
- FMRFamide
- molluscan insulin-related peptide
- sodium influx-stimulating peptide
- gonadotropin-releasing hormone
- cardioexcitatory peptides
- cardioinhibitory peptides
- bag cells
- R-15
- caudodorsal cells
- light green cells
- yellow cells
- atrial gland
- albumen gland
- neurotransmitters
- dopamine
- serotonin
- animal behavior
- reproduction
- osmoregulation
- cardiac activity
- growth
- ciliary activity
- neuropeptide synthesis
- neuropeptide processing
- neuropeptide secretion
- identifiable neuron
- neurosecretory cell
- neuroendocrine
- signal transduction
- G protein-coupled receptors
- D-amino acid peptide

REFERENCES

1. Abdraba, A. M.; Saleuddin, A. S. M. Localization and Immunological Characterization of Insulin-like Peptide(s) in the Land Snail *Otala lactea* (Mollusca: Pulmonata). *Can. J. Zool.* **2000a,** *78,* 1515–1526.
2. Abdraba, A. M.; Saleuddin, A. S. M. Protein Synthesis In Vitro by Mantle Tissue of the Land Snail *Otala lactea*: Possible Insulin-like Peptide Function. *Can. J. Zool.* **2000b,** *78,* 1527–1535.
3. Aguilera, F.; McDougall, C.; Degnan, B. Evolution of the Tyrosinase Gene Family in Bivalve Molluscs: Independent Expansion of the Mantle Gene Repertoire. *Acta Biomater.* **2014,** *10,* 3855–3865.
4. Alevizos, A.; Karagogeos, D.; Weiss, K. R.; Buck, L.; Koester, J. R15 Alpha 1 and R 15 Alpha 2 Peptides from Aplysia: Comparison of Bioactivity, Distribution, and Function of Two Peptides Generated by Alternative Splicing. *J. Neurobiol.* **1991a,** *22*(4), 405–417.
5. Alevizos, A.; Weiss, K. R.; Koester, J. Synaptic Actions of Identified Peptidergic Neuron R15 in Aplysia. I. Activation of Respiratory Pumping. *J. Neurosci.* **1991b,** *11*(5), 1263–1274.
6. Alevizos, A.; Weiss, K. R.; Koester, J. Synaptic Actions of Identified Peptidergic Neuron R15 in Aplysia. II. Contraction of Pleuroabdominal Connectives Mediated by Motoneuron L7. *J. Neurosci.* **1991c,** *11*(5), 1275–1281.
7. Alevizos, A.; Weiss, K. R.; Koester, J. Synaptic actions of identified peptidergic neuron R15 in Aplysia. III. Activation of the large hermaphroditic duct. *J. Neurosci.* **1991d,** *11*(5), 1282–1290.
8. Altstein, M.; Nässel, D. R. Neuropeptide signaling in insects. *Adv. Exp. Med. Biol.* **2010,** *692,* 155–165.
9. Aloyz, R. S.; DesGroseillers, L. Processing of the L5–67 Precursor Peptide and Characterization of LUQIN in the LUQ Neurons of *Aplysia californica. Peptides* **1995,** *16*(2), 331–338.
10. Amano, M.; Moriyama, S.; Okubo, K.; Amiya, N.; Takahashi, A.; Oka, Y. Biochemical and Immunohistochemical Analyses of a GnRH-like Peptide in the Neural Ganglia of the Pacific Abalone *Haliotis discus hannai* (Gastropoda). *Zool. Sci.* **2010a,** *27*(8), 656–661.
11. Amano, M.; Yokoyama, T.; Amiya, N.; Hotta, M.; Takakusaki, Y.; Kado, R.; Oka, Y. Biochemical and Immunohistochemical Analyses of GnRH-like Peptides in the Nerve Ganglion of the Chiton, *Acanthopleura japonica. Zool. Sci.* **2010b,** *27*(12), 924–930.
12. Anctil, M. Evidence for Gonadotropin-releasing Hormone-like Peptides in a Cnidarian Nervous System. *Gen. Comp. Endocrinol.* **2000,** *119*(3), 317–328.
13. Andrews, E. B. Excretory Systems of Molluscs. In *The Mollusca. Form and Function*; Truman, E. R., Clarke, M. R., Eds.; Academic Press, Inc.: San Diego, 1988; Vol 11.
14. Angers, A.; DesGroseillers, L. Alternative Splicing and Genomic Organization of the L5–67 Gene of *Aplysia californica. Gene* **1998,** *208*(2), 271–277.
15. Angers, A.; Zappulla, J. P.; Zollinger, M.; DesGroseillers, L. Gene Products from LUQ Neurons in the Abdominal Ganglion are Present at the Renal Pore of *Aplysia californica. Comp. Biochem. Physiol. B* **2000,** *126*(3), 435–443.
16. Arch, S. Biosynthesis of the Egg-laying Hormone (ELH) in the Bag Cell Neurons of *Aplysia californica. J. Gen. Physiol.* **1972a,** *60,* 102–119.
17. Arch, S. Polypeptide Secretion from the Isolated Parietovisceral Ganglion of *Aplysia californica. J. Gen. Physiol.* **1972b,** *59*(1), 47–59.

18. Arch, S.; Berry, R. W. Molecular and Cellular Regulation of Neuropeptide Expression: The Bag Cell Model System. *Brain Res. Rev.* **1989**, *14*, 181–201.
19. Arch, S.; Earley, P.; Smock, T. Biochemical Isolation and Physiological Identification of the Egg-laying Hormone in *Aplysia californica*. *J. Gen. Physiol.* **1976a**, *68*(2), 197–210.
20. Arch, S.; Smock, T.; Earley, P. Precursor and Product Processing in the Bag Cell Neurons of *Aplysia californica*. *J. Gen. Physiol.* **1976b**, *68*(2), 211–225.
21. Aswad, D. W. Biosynthesis and Processing of Presumed Neurosecretory Proteins in Single Identified Neurons of *Aplysia californica*. *J. Neurobiol.* **1978**, *9*(4), 267–284.
22. Azhderian, E. M.; Kaczmarek, L. K. Cyclic AMP Regulates Processing of Neuropeptide Precursor in Bag Cell Neurons of Aplysia. *J. Mol. Neurosci.* **1990**, *2*(2), 61–70.
23. Audesirk, G. Amine Containing Neurons in the Brain of *Lymnaea stagnalis*: Distribution and Effects of Precursors. *Comp. Biochem. Physiol. A* **1985**, *81*, 359–365.
24. Azhderian, E. M.; Kaczmarek, L. K. Cyclic AMP Regulates Processing of Neuropeptide Precursor in Bag Cell Neurons of *Aplysia*. *J. Mol. Neurosci.* **1990**, *2*, 61–70.
25. Baba, Y.; Matsuo, H.; Schally, A. V. Structure of the Porcine LH- and FSH-releasing Hormone. II. Confirmation of the Proposed Structure by Conventional Sequential Analyses. *Biochem. Biophys. Res. Commun.* **1971**, *44*(2), 459–463.
26. Bablanian, G. M.; Treistman, S. N. Seawater Osmolarity Influences Bursting Pacemaker Activity in Intact *Aplysia californica*. *Brain Res.* **1983**, *271*(2), 342–345.
27. Bai, L.; Livnat, I.; Romanova, E. V.; Alexeeva, V.; Yau, P. M.; Vilim, F. S.; Weiss, K. R.; Jing, J.; Sweedler, J. V. Characterization of GdFFD, a D-Amino acid-containing Neuropeptide that Functions as an Extrinsic Modulator of the Aplysia Feeding Circuit. *J. Biol. Chem.* **2013**, *288*(46), 32837–32851.
28. Barker, G. M. *Molluscs as Crop Pests*. CABI Publishing: Wallingford, Oxon, 2002.
29. Baud, C.; Darbon, P.; Li, K. W.; Marchand, C. R. Partial Characterization of a Novel Cardioinhibitory Peptide from the Brain of the Snail *Helix aspersa*. *Cell. Mol. Neurobiol.* **1998**, *18*(4), 413–424.
30. Bayakly, N. A.; Deaton, L. E. The Effects of FMRFamide, 5-Hydroxytryptamine and Phorbol Esters on the Heart of the Mussel *Geukensia demissa*. *J. Comp. Physiol. B* **1992**, *162*(5), 463–468.
31. Benjamin, P. R.; Burke, J. F. Alternative mRNA Splicing of the FMRFamide Gene and Its Role in Neuropeptidergic Signalling in a Defined Neural Network. *Bioessays* **1994**, *16*(5), 335–342.
32. Benkendorff, K. Molluscan Biological and Chemical Diversity: Secondary Metabolites and Medicinal Resources Produced by Marine Molluscs. *Biol. Rev. Camb. Philos. Soc.* **2010**, *85*, 757–775.
33. Bernocchi, G.; Vignola, C.; Scherini, E.; Necchi, D.; Pisu, M. B. Bioactive Peptides and Serotonin Immunocytochemistry in the Cerebral Ganglia of Hibernating *Helix aspersa*. *J. Exp. Zool.* **1998**, *280*(5), 354–367.
34. Berry, R. W.; Hanu, R.; Redman, R. S.; Kim, J. J. Determinants of Potency and Temperature-dependent Function in the Aplysia Bag Cell Peptides. *Peptides* **1994**, *15*(5), 855–860.
35. Boer, H. H. A Cytological and Cytochemical Study of Neurosecretory Cells in Basommatophora, with Particular Reference to *Lymnaea stagnalis* L. *Arch. Neerl. Zool.* **1965**, *16*, 313–386.
36. Boer, H. H.; Montagne-Wajer, C.; van Minnen, J.; Ramkema, M.; de Boer, P. Functional Morphology of the Neuroendocrine Sodium Influx-stimulating Peptide

System of the Pond Snail, *Lymnaea stagnalis*, Studied by In Situ Hybridization and Immunocytochemistry. *Cell Tissue Res.* **1992**, *268*(3), 559–566.

37. Bogdanov Yu. D.; Balaban, P. M.; Zakharov, I. S.; Poteryaev, D. A.; Belyavsky, A. V. Identification of Two Novel Genes Specifically Expressed in the D-Group Neurons of the Terrestrial Snail CNS. *Invert. Neurosci.* **1996**, *2*(1), 61–69.

38. Booth, D. T. Oxygen Availability and Embryonic Development in Sand Snail (*Polinices sordidus*) Egg Masses. *J. Exp. Biol.* **1995**, *198*, 241–247.

39. Boucher, J.; Kleinridders, A.; Kahn, C. R. Insulin Receptor Signaling in Normal and Insulin-resistant States. *Cold Spring Harb. Perspect. Biol.* **2014**, *6*(1), pii: a009191. DOI:10.1101/cshperspect.a009191.

40. Boyle, J. P.; Yoshino, T. P. Monoamines in the Albumen Gland, Plasma, and Central Nervous System of the Snail *Biomphalaria glabrata* During Egg-laying. *Comp. Biochem. Physiol.* A. **2002**, *132*, 411–422.

41. Brezden, B. L.; Yeoman, M. S.; Gardner, D. R.; Benjamin, P. R. FMRFamide-activated Ca^{2+} Channels in Lymnaea Heart Cells are Modulated by "SEEPLY," a Neuropeptide Encoded on the Same Gene. *J. Neurophysiol.* **1999**, *81*(4), 1818–1826.

42. Brezina, V.; Bank, B.; Cropper, E. C.; Rosen, S.; Vilim, F. S.; Kupfermann, I.; Weiss, K. R. Nine Members of the Myomodulin Family of Peptide Cotransmitters at the B16-ARC Neuromuscular Junction of Aplysia. *J. Neurophysiol.* **1995**, *74*(1), 54–72.

43. Brisson, P. Aminergic Structures in the Genital Tract of Pulmonate Gastropods and their Possible Role in the Reproductive System. In *Molluscan Neuroendocrinology*; Lever, J., Boer, H. H., Eds. North Holland Publishing Company: Amsterdam, 1983.

44. Brisson, P.; Collin, J. P. Systèmes aminerigiques des mollusques gastéropodes pulmonés IV- Paraneurones et innervation catécholaminergiques de la région du carrefour des voies génitales; étude radioautographique. *Biol. Cell.* **1980**, *38*, 211–220.

45. Brown, R. O.; Mayeri, E. Positive Feedback by Autoexcitatory Neuropeptides in Neuroendocrine Bag Cells of Aplysia. *J. Neurosci.* **1989**, *9*(4), 1443–1451.

46. Brown, R. O.; Pulst, S. M.; Mayeri, E. Neuroendocrine Bag Cells of Aplysia are Activated by Bag Cell Peptide-containing Neurons in the Pleural Ganglion. *J. Neurophysiol.* **1989**, *61*(6), 1142–1152.

47. Bruehl, C. L.; Berry, R. W. Regulation of Synthesis of the Neurosecretory egg-laying Hormone of Aplysia: Antagonistic Roles of Calcium and Cyclic Adenosine 3′:5′-monophosphate. *J. Neurosci.* **1985**, *5*(5), 1233–1238.

48. Brussaard, A. B.; Schluter, N. C.; Ebberink, R. H. M.; Kits, K. S.; Ter Maat, A. Discharge Induction in Molluscan Peptidergic Cells Requires a Specific Set of Autoexcitatory Neuropeptides. *Neuroscience* **1990**, *39*, 479–4791.

49. Burton, R. R. Ionic Regulation and Water Balance. In *The Mollusca. Physiology Part 2*; Saleuddin, A. S. M., Wilbur, K. M., Eds.; Academic Press, Inc.: New York, 1983; Vol 5.

50. Campanelli, J. T.; Scheller, R. H. Histidine-rich Basic Peptide: A Cardioactive Neuropeptide from Aplysia Neurons R3–14. *J. Neurophysiol.* **1987**, *57*(4), 1201–1209.

51. Candelario-Martinez, A.; Reed, D. M.; Prichard, S. J.; Doble, K. E.; Lee, T. D.; Lesser, W.; Price, D. A.; Greenberg, M. J. SCP-related Peptides from Bivalve Molluscs: Identification, Tissue Distribution, and Actions. *Biol. Bull.* **1993**, *185*, 428–439.

52. Cawthorpe, D. R.; Rosenberg, J.; Colmers, W. F.; Lukowiak, K.; Drummond, G. I. The Effects of Small Cardioactive Peptide B on the Isolated Heart and Gill of *Aplysia californica*. *Can. J. Physiol. Pharmacol.* **1985**, *63*(8), 918–924.

53. Chaban, A.; Voronezhskaya, E. Involvement of Transient Larval Neurons in Osmoregulation and Neurogenesis in the Freshwater Snails, *Lymnaea stagnalis* and *Helisoma trivolvis*. *Acta Biol. Hung.* **2008**, *59*, 123–126.

54. Chin, G. J.; Payza, K.; Price, D. A.; Greenberg, M. J.; Doble, K. E. Characterization and Solubilization of the FMRFamide Receptor of Squid. *Biol. Bull.* **1994,** *187*(2), 185–199.
55. Chiu, A. Y.; Hunkapiller, M. W.; Heller, E.; Stuart, D. K.; Hood, L. E.; Strumwasser, F. Purification and Primary Structure of the Neuropeptide Egg-laying Hormone of *Aplysia californica*. *Proc. Natl. Acad. Sci. U.S.A.* **1979,** *76*(12), 6656–6660.
56. Chiu, A. Y.; Strumwasser, F. An Immunohistochemical Study of the Neuropeptidergic Bag Cells of Aplysia. *J. Neurosci.* **1981,** *1*(8), 812–826.
57. Chiu, A. Y.; Strumwasser, F. Two Neuronal Populations in the Head Ganglia of *Aplysia californica* with Egg-laying Hormone-like Immunoreactivity. *Brain Res.* **1984,** *294*(1), 83–93.
58. Chrachri, A.; Odblom, M.; Williamson, R. G Protein-mediated FMRFamidergic Modulation of Calcium Influx in Dissociated Heart Muscle Cells from Squid, *Loligo forbesii. J. Physiol.* **2000,** *525*, 471–482.
59. Christopher, K. J.; Chang, J. P.; Goldberg, J. I. Stimulation of Cilia Beat Frequency by Serotonin is Mediated by a Ca^{2+} Influx in Ciliated Cells of *Helisoma trivolvis* Embryos. *J. Exp. Biol.* **1996,** *199*, 1105–1113.
60. Christopher, K. J.; Young, K. G.; Chang, J. P.; Goldberg, J. I. Involvement of Protein Kinase C in 5-HT-stimulated Ciliary Activity in *Helisoma trivolvis* Embryos. *J. Physiol.* **1999,** *515*, 511–522.
61. Chu, N. D.; Kaluziak, S. T.; Trussell, G. C.; Vollmer, S. V. Phylogenomic Analyses Reveal Latitudinal Population Structure and Polymorphisms in Heat Stress Genes in the North Atlantic snail *Nucella lapillus*. *Mol. Ecol.* **2014,** *23*, 1863–1873.
62. Chun, J. Y.; Korner, J.; Kreiner, T.; Scheller, R. H.; Axel, R. The Function and Differential Sorting of a Family of Aplysia Prohormone Processing Enzymes. *Neuron* **1994,** *12*(4), 831–844.
63. Church, P. J.; Lloyd, P. E. Expression of Diverse Neuropeptide Cotransmitters by Identified Motor Neurons in Aplysia. *J. Neurosci.* **1991,** *11*(3), 618–625.
64. Cimino, G.; Gavagnin, M. Molluscs: From Chemo-ecological Study to Biotechnological Application. Springer: Berlin, 2006.
65. Coggeshall, R. E.; Kandel, E. R.; Kupfermann, I.; Waziri, R. A Morphological and Functional Study on a Cluster of Identifiable Neurosecretory Cells in the Abdominal Ganglion of *Aplysia californica. J. Cell. Biol.* **1966,** *31*(2), 363–368.
66. Cole, A. G.; Mashkournia, A.; Parries, S. C.; Goldberg, J. I. Regulation of Early Embryonic Behavior by Nitric Oxide in the Pond Snail *Helisoma trivolvis. J. Exp. Biol.* **2002,** *205*, 3143–3152.
67. Conn, P. J.; Strong, J. A.; Azhderian, E. M.; Nairn, A. C.; Greengard, P.; Kaczmarek, L. K. Protein Kinase Inhibitors Selectively Block Phorbol Ester- or Forskolin-induced Changes in Excitability of Aplysia Neurons. *J. Neurosci.* **1989,** *9*(2), 473–479.
68. Coscoy, S.; Lingueglia, E.; Lazdunski, M.; Barbry, P. The Phe-Met-Arg-Phe-amide-activated Sodium Channel is a Tetramer. *J. Biol. Chem.* **1998,** *273*(14), 8317–8322.
69. Cottrell, G. A.; Jeziorski, M. C.; Green, K. A. Location of a Ligand Recognition Site of FMRFamide-gated Na(+) channels. *FEBS Lett.* **2001,** *489*(1), 71–74.
70. Cox, K. J.; Tensen, C. P.; Van der Schors, R. C.; Li, K. W.; van Heerikhuizen, H.; Vreugdenhil, E.; Geraerts, W. P.; Burke, J. F. Cloning, Characterization, and Expression of a G-protein-coupled Receptor from *Lymnaea stagnalis* and Identification of a Leucokinin-like Peptide, PSFHSWSamide, as Its Endogenous Ligand. *J. Neurosci.* **1997,** *17*(4), 1197–205.

71. Croll, R. P. Distribution of Monoamines in the Central Nervous System of the Nudibranch Gastropod, *Hermissenda crassicornis. Brain Res.* **1987,** *405*(2), 337–347.
72. Croll, R. P. Distribution of Monoamines within the Central Nervous System of the Juvenile Pulmonate Snail, *Achatina fulica. Brain Res.* **1988,** *460*(1), 29–49.
73. Croll, R. P. Insights into Early Molluscan Neuronal Development Through Studies of Transmitter Phenotypes in Embryonic Pond Snails. *Microsc. Res. Tech.* **2000,** *49*, 570–578.
74. Croll, R. P. Catecholamine-containing Cells in the Central Nervous System and Periphery of *Aplysia californica. J. Comp. Neurol.* **2001,** *441*, 91–105.
75. Croll, R. P.; Chiasson, B. J. Postembryonic Development of Serotoninlike Immunoreactivity in the Central Nervous System of the Snail, *Lymnaea stagnalis. J. Comp. Neurol.* **1989,** *280*(1), 122–142.
76. Croll, R. P.; Voronezhskaya, E. E. Early Elements in Gastropod Neurogenesis. *Dev. Biol.* **1996,** *173*(1), 344–347.
77. Croll, R. P.; Van Minnen, J. Distribution of the peptide Ala-Pro-Gly-Trp-NH$_2$ (APGWamide) in the Nervous System and Periphery of the Snail *Lymnaea stagnalis* as Revealed by Immunocytochemistry and In Situ Hybridization. *J. Comp. Neurol.* **1992,** *324*, 567–574.
78. Croll, R. P. Voronezhskaya, E. E. Hiripi, L. and Elekes, K. Development of Catecholaminergic Neurons in the Pond Snail, *Lymnaea stagnalis,* II. Postembryonic Development of Central and Peripheral Cells. *J. Comp. Neurol.* **1999,** *404*, 297–309.
79. Cropper, E. C.; Evans, C. G.; Hurwitz, I.; Jing, J.; Proekt, A.; Romero, A.; Rosen, S. C. Feeding Neural Networks in the Mollusc Aplysia. *Neurosignals* **2004,** *13*(1–2), 70–86.
80. Cropper, E. C.; Miller, M. W.; Tenenbaum, R.; Kolks, M. A.; Kupfermann, I.; Weiss, K. R. Structure and Action of Buccalin: A Modulatory Neuropeptide Localized to an Identified Small Cardioactive Peptide-containing Cholinergic Motor Neuron of *Aplysia californica. Proc. Natl. Acad. Sci. U.S.A.* **1988,** *85*(16), 6177–6181.
81. Cropper, E. C.; Tenenbaum, R.; Kolks, M. A.; Kupfermann, I.; Weiss, K. R. Myomodulin: A Bioactive Neuropeptide Present in an Identified Cholinergic Buccal Motor Neuron of Aplysia. *Proc. Natl. Acad. Sci. U.S.A.* **1987,** *84*(15), 5483–5486.
82. Croushore, C. A.; Supharoek, S. A.; Lee, C. Y.; Jakmunee, J.; Sweedler, J. V. Microfluidic Device for the Selective Chemical Stimulation of Neurons and Characterization of Peptide Release with Mass Spectrometry. *Anal Chem.* **2012,** *84*, 9446–9452.
83. Cummins, S. F.; Bowie, J. H. Pheromones, Attractants and other Chemical Cues of Aquatic Organisms and Amphibians. *Nat. Prod. Rep.* **2012,** *29*(6), 642–658.
84. Cummins, S. F.; Nichols, A. E.; Schein, C. H.; Nagle, G. T. Newly Identified Water-borne Protein Pheromones Interact with Attractin to Stimulate Mate Attraction in Aplysia. *Peptides* **2006,** *27*(3), 597–606.
85. Dang, V. T.; Benkendorff, K.; Green, T.; Speck, P. Marine Snails and Slugs: A Great Place to Look for Antiviral Drugs. *J. Virol.* **2015.** pii: JVI.00287-15.
86. De Haes, W.; Van Sinay, E.; Detienne, G.; Temmerman, L.; Schoofs, L.; Boonen, K. Functional Neuropeptidomics in Invertebrates. *Biochim. Biophys. Acta* **2015,** *1854*, 812–826.
87. De Jong-Brink, M.; Elsaadany, M. M.; Boer, H. H. Schistosomin, an Antagonist of Calfluxin. *Exp. Parasitol.* **1988,** *65*, 109–118.
88. De Jong-Brink, M.; Hordijk, P. L.; Vergeest, D. P. E. J.; Schallig, H. D. F. H.; Kits, K. S.; Ter Maat, A. The Antigonadotropic Neuropeptide Schistosomin Interferes with Peripheral and Central Neuroendocrine Mechanisms Involved in the Regulation of Reproduction and Growth in the Schistosome-infected Snail *Lymnaea stagnalis. Prog. Brain Res.* **1992,** *92*, 385–396.

89. De Jong-Brink, M.; Reid, C. N.; Tensen, C. P.; Ter Maat, A. Parasites Flicking the NPY Gene on the Host's Switchboard: Why NPY? *FASEB J.* **1999**, *13*(14), 1972–1984.
90. de Lange, R. P. J.; van Minnen, J. Localization of the Neuropeptide APGWamide in Gastropod Molluscs by In Situ Hybridization and Immunocytochemistry. *Gen. Comp. Endocrinol.* **1998**, *109*, 166–174.
91. DeRiemer, S. A.; Kaczmarek, L. K.; Lai, Y.; McGuinness, T. L.; Greengard, P. Calcium/ Calmodulin-dependent Protein Phosphorylation in the Nervous System of Aplysia. *J. Neurosci.* **1984**, *4*(6), 1618–1625.
92. DeRiemer, S. A.; Schweitzer, B.; Kaczmarek, L. K. Inhibitors of Calcium-dependent Enzymes Prevent the Onset of Afterdischarge in the Peptidergic Bag Cell Neurons of Aplysia. *Brain Res.* **1985**, *340*(1), 175–180.
93. Derakhshan, F.; Toth, C. Insulin and the Brain. *Curr. Diabetes Rev.* **2013**, *9*(2), 102–116.
94. De Vlieger, T. A.; Kits, K. S.; ter Maat, A.; Lodder, J. C. Morphology and Electrophysiology of the Ovulation Hormone Producing Neuro-endocrine Cells of the Freshwater Snail *Lymnaea stagnalis* (L.). *J. Exp. Biol.* **1980**, *84*, 259–271.
95. De Vlieger, T. A.; Lodder, J. C.; Stoof, J. C.; Werkman, T. R. Dopamine Receptor Stimulation Induces a Potassium Dependent Hyperpolarizing Response in Growth Hormone Producing Neuroendocrine Cells of the Gastropod Mollusc *Lymnaea stagnalis*. *Comp. Biochem. Physiol. C* **1986**, *83*(2), 429–433.
96. De With, N. D.; Boer, H. H.; Smit, A. B. Neurosecretory Yellow Cells and Hydrominaral Regulation in the Pulmonate Freshwater Snail *Lymanea stagnalis*. In *Perspectives in Comparative Endocrinology*; Davey, K. G.; Peter, R. E.; Tobe, S. S., Eds. National Research Council of Canada: Ottawa, 1994.
97. De With, N. D.; Slootstra, J. W.; van der Schors, R. C. The Bioelectrical Activity of the Body Wall of the Pulmonate Freshwater Snail *Lymnaea stagnalis*: Effects of Neurotransmitters and the Sodium Influx Stimulating Neuropeptides. *Gen. Comp. Endocrinol.* **1988**, *70*(2), 216–223.
98. De With, N. D.; van der Schors, R. C. Neurohormonal Control of Na^+ and Cl^- Metabolism in the Pulmonate Freshwater Snail *Lymnaea stagnalis*. *Gen. Comp. Endocrinol.* **1986**, *63*(3), 344–352.
99. De With, N. D.; van der Schors, R. C.; Boer, H. H.; Ebberink, R. H. M. The Sodium Influx Stimulating Peptide of the Pulmonate Freshwater Snail *Lymnaea stagnalis*. *Peptides* **1993**, *14*(4), 783–789.
100. Di Cosmo, A.; Di Cristo, C. Neuropeptidergic Control of the Optic Gland of *Octopus vulgaris*: FMRFamide and GnRH Immunoreactivity. *J. Comp. Neurol.* **1998**, *398*(1), 1–12.
101. Dictus, W. J. A. G.; deJong-Brink, M.; Boer, H. H. A Neuropeptide (Calfluxin) is Involved in the Influx of Calcium into Mitochondria of the Albumen Gland of the Freshwater Snail *Lymnaea stagnalis*. *Gen. Comp. Endocrinol.* **1987**, *65*, 439–444.
102. Dictus, W. J. A. G. Broers-Vendrig, C. M.; deJong-Brink, M. The role of IP3, PKC, and pH in the Stimulus-response Coupling of Calfluxin-stiniulated Albumen Glands of the Freshwater Snail *Lymnaea stagnalis*. *Gen. Comp. Endocrinol.* **1988**, *70*, 206–215.
103. Dictus, W. J. A. G.; Ebberink, R. H. M. Structure of One of the Neuropeptides of the Egg-laying Precursor of Lymnaea. *Mol. Cell. Endocrinol.* **1988**, *60*, 23–29.
104. Diefenbach, T. J.; Koehncke, N. K.; Goldberg, J. I. Characterization and Development of Rotational Behavior in Helisoma Embryos: Role of Endogenous Serotonin. *J. Neurobiol.* **1991**, *22*, 922–934.

105. Diefenbach, T. J.; Koss, R.; Goldberg, J. I. Early Development of an Identified Serotonergic Neuron in *Helisoma trivolvis* Embryos: Serotonin Expression, De-expression, and Uptake. *J. Neurobiol.* **1998**, *34*, 361–376.

106. Dieringer, N.; Koester, J.; Weiss, K. R. Adaptive Changes in Heart Rate of *Aplysia californica. J. Comp. Physiol. A.* **1978**, *123*, 11–21.

107. Dogterom, A. A. Effect of the Growth Hormone of the Freshwater Snail *Lymnaea stagnalis* on Biochemical Composition and Nitrogenous Wastes. *Comp. Biochem., Physiol. B* **1980**, *65*, 163–167.

108. Dogterom, G. E.; Bohlken, S.; Joosse, J. Effect of the Photoperiod on the Time Schedule of Egg Mass Production in *Lymnaea stagnalis*, as Induced by Ovulation Hormone Injections. *Gen. Comp. Endocrinol.* **1983**, *49*, 255–260.

109. Dogterom, A. A.; Doderer, A. A Hormone-dependent Calcium-binding Protein in the Mantle Edge of the Freshwater Snail *Lymnaea stagnalis. Calcif. Tissue Int.* **1981**, *33*, 505–508.

110. Dogterom, A. A.; Jentjens, T. The Effect of Growth Hormone of the Pond Snail *Lymnaea stagnalis* on Periostracum Formation. *Comp. Biochem. Physiol.* **1980**, *66*, 687–690.

111. Dogterom, A. A.; Robles, B. R. Stimulation of Ornithine Decarboxylase Activity in *Lymnaea stagnalis* after a Single Injection with Molluscan Growth Hormone. *Gen. Comp. Endocrinol.* **1980**, *40*(2), 238–240.

112. Dogterom, A. A.; van Loenhout, H.; van der Schors, R. C. The Effect of the Growth Hormone of *Lymnaea stagnalis* on Shell Calcification. *Gen. Comp. Endocrinol.* **1979**, *39*(1), 63–68.

113. Doran, S. A.; Goldberg, J. I. Roles of Ca^{2+} and Protein Kinase C in the Excitatory Response to Serotonin in Embryonic Molluscan Ciliary Cells. *Can. J. Physiol. Pharmacol.* **2006**, *84*, 635–646.

114. Doran, S. A.; Tran, C. H.; Eskicioglu, C.; Stachniak, T.; Ahn, K.-C.; Goldberg, J. I. Constitutive and Permissive Roles of Nitric Oxide Activity in Embryonic Ciliary Cells. *Am. J. Physiol.* **2006**, *285*, R348–R355.

115. Doran, S. A.; Koss, R.; Tran, C. H.; Christopher, K. J.; Gallin, W. J.; Goldberg, J. I. Effect of Serotonin on Ciliary Beating and Intracellular Calcium Concentration in Identified Populations of Embryonic Ciliary Cells. *J. Exp. Biol.* **2004**, *207*, 1415–1429.

116. Dreon, M. S.; Frassa, M. V.; Ceolín, M.; Ituarte, S.; Qiu, J. W.; Sun, J. Fernández, P. E.; Heras, H. Novel Animal Defenses against Predation: A Snail Egg Neurotoxin Combining Lectin and Pore-forming Chains that Resembles Plant Defense and Bacteria Attack Toxins. *PLoS ONE* **2013**, *8*(5), e63782. DOI:10.1371/journal.pone.0063782.

117. Dreon, M. S.; Heras, H.; Pollero, R. J. Biochemical Composition, Tissue Origin and Functional Properties of Egg Perivitellins from *Pomacea canaliculata. Biocell* **2006**, *30*(2), 359–365.

118. Du, Y.; Wei, T. Inputs and Outputs of Insulin Receptor. *Protein Cell* **2014**, *5*(3), 203–213.

119. Dudek, F. E.; Cobbs, J. S.; Pinsker, H. M. Bag Cell Electrical Activity Underlying Spontaneous Egg Laying in Freely Behaving *Aplysia brasiliana. J. Neurophysiol.* **1979**, *42*(3), 804–817.

120. Dudek, F. E.; Tobe, S. S. Bag Cell Peptide Acts Directly on Ovotestis of *Aplysia californica*: Basis for an In Vitro Bioassay. *Gen. Comp. Endocrinol.* **1978**, *36*(4), 618–627.

121. Dudek, F. E.; Weir, G.; Acosta-Urquidi, J.; Tobe, S. S. A Secretion from Neuroendocrine Bag Cells Evokes Egg Release In Vitro from Ovotestis of *Aplysia californica. Gen. Comp. Endocrinol.* **1980**, *40*(2), 241–244.

122. Duncan, C. J. In *Pulmonates. Vol. 1. Functional Anatomy and Physiology*; Fretter, V., Peake, J., Eds. Academic Press Inc.: New York, 1975.

123. Du Plessis, D. J.; Nouwen, N.; Driessen, A. J. The Sec Translocase. *Biochim. Biophys. Acta* **2011**, *1808*(3), 851–865.

124. Dyer, J. R.; Michel, S.; Lee, W.; Castellucci, V. F.; Wayne, N. L.; Sossin, W. S. An Activity-dependent Switch to Cap-independent Translation Triggered by eIF4E Dephosphorylation. *Nat. Neurosci.* **2003**, *6*(3), 219–220.

125. Ebberink, R. H. M.; Joosse, J. Molecular Properties of Various Snail Peptides from Brain and Gut. *Peptides* **1985**, *6* Suppl., 451–457.

126. Ebberink, R. H. M.; van Loenhout, H.; Geraerts, W. P. M.; Joosse, J. Purification and Amino Acid Sequence of the Ovulation Neurohormone of *Lymnaea stagnalis*. *Proc. Natl. Acad. Sci. USA.* **1985**, *82*, 7767–7771.

127. Ebberink, R. H. M.; van Loenhout, H.; van Beek, J.; de Wilde, K.; van Minnen, J. Characterization of Peptides Isolated from Growth-controlling Neuro-endocrine Cells of *Lymnaea stagnalis* with Immunoreactivity to Ant-insulin. In *Neurobiology. Molluscan Models*; Boer, H. H., Geraerts, W. P. M., Joosse, J., Eds.; North Holland: Amsterdam, 1987.

128. Eierman, L. E.; Hare, M. P. Transcriptomic Analysis of Candidate Osmoregulatory Genes in the Eastern Oyster *Crassostrea virginica*. *BMC Genomics* **2014**, *15*, 503.

129. Elliott, C. J.; Susswein, A. J. Comparative Neuroethology of Feeding Control in Molluscs. *J. Exp. Biol.* **2002**, *205*(Pt. 7), 877–896.

130. Ellis, A. M.; Huddart, H. Excitation Evoked by FMRFamide and FLRFamide in the Heart of *Buccinum undatum* and Evidence for Inositol 1,4,5-Trisphosphate as an RF-Tetrapeptide Second Messenger. *J. Comp. Physiol. B* **2000**, *170*(5–6), 351–356.

131. Evans, C. G.; Vilim, F. S.; Harish, O.; Kupfermann, I.; Weiss, K. R.; Cropper, E. C. Modulation of Radula Opener Muscles in Aplysia. *J. Neurophysiol.* **1999**, *82*(3), 1339–1351.

132. Falconer, S. W.; Carter, A. N.; Downes, C. P.; Cottrell, G. A. The Neuropeptide Phe-Met-Arg-Phe-NH2 (FMRFamide) Increases Levels of Inositol 1,4,5-Trisphosphate in the Tentacle Retractor Muscle of *Helix aspersa*. *Exp. Physiol.* **1993**, *78*(6), 757–766.

133. Fan, X.; Croll, R. P.; Wu, B.; Fang, L.; Shen, Q.; Painter, S. D.; Nagle, G. T. Molecular Cloning of a cDNA Encoding the Neuropeptides APGWamide and Cerebral Peptide 1: Localization of APGWamide-like Immunoreactivity in the Central Nervous System and Male Reproductive Organs of Aplysia. *J. Comp. Neurol.* **1997**, *387*, 53–62.

134. Fan, X.; Spijker, S.; Akalal, D. B.; Nagle, G. T. Neuropeptide Amidation: Cloning of a Bifunctional Alpha-amidating Enzyme from Aplysia. *Brain Res. Mol. Brain Res.* **2000**, *82*(1–2), 25–34.

135. Favrel, P.; Lelong, C.; Mathieu, M. Structure of the cDNA Encoding the Precursor for the Neuropeptide FMRFamide in the Bivalve Mollusc *Mytilus edulis*. *NeuroReport* **1998**, *9*(13), 2961–2965.

136. Favrel, P.; Mathieu, M. Molecular Cloning of a cDNA Encoding the Precursor of Ala-Pro-Gly-Trp Amide-related Neuropeptides from the Bivalve Mollusc *Mytilus edulis*. *Neurosci. Lett.* **1996**, *205*(3), 210–214.

137. Filla, A.; Hiripi, L.; Elekes, K. Role of Aminergic (Serotonin and Dopamine) Systems in the Embryogenesis and Different Embryonic Behaviors of the Pond Snail, *Lymnaea stagnalis*. *Comp. Biochem. Physiol.* **2009**, *149C*, 73–82.

138. Fink, L. A.; Kaczmarek, L. K. Inositol Polyphosphates Regulate Excitability. *Tr. Neurosci.* **1988,** *11*(8), 338–339.
139. Fisher, J. M.; Sossin, W.; Newcomb, R.; Scheller, R. H. Multiple Neuropeptides Derived from a Common Precursor are Differentially Packaged and Transported. *Cell* **1988,** *54*(6), 813–822.
140. Fisher, T.; Lin, C. H.; Kaczmarek, L. K. The Peptide FMRFa Terminates a Discharge in Aplysia Bag Cell Neurons by Modulating Calcium, Potassium, and Chloride Conductances. *J. Neurophysiol.* **1993,** *69*(6), 2164–2173.
141. Floyd, P. D.; Li, L.; Rubakhin, S. S.; Sweedler, J. V.; Horn, C. C.; Kupfermann, I.; Alexeeva, V. Y.; Ellis, T. A.; Dembrow, N. C.; Weiss, K. R.; Vilim, F. S. Insulin Prohormone Processing, Distribution, and Relation to Metabolism in *Aplysia californica. J. Neurosci.* **1999,** *19*, 7732–7741.
142. Fujimoto, K.; Kubota, I.; Yasuda-Kamatani, Y.; Minakata, H.; Nomoto, K.; Yoshida, M.; Harada, A.; Muneoka, Y.; Kobayashi, M. Purification of Achatin-I from the Atria of the African giant snail, *Achatina fulica,* and Its Possible Function. *Biochem. Biophys. Res. Commun.* **1991,** 177 (2), 847–853.
143. Fujimoto, K.; Ohta, N.; Yoshida, M.; Kubota, I.; Muneoka, Y.; Kobayashi, M. A Novel Cardio-excitatory Peptide Isolated from the Atria of the African Giant Snail, *Achatina fulica. Biochem. Biophys. Res. Commun.* **1990,** *167,* 777–783.
144. Fujisawa, Y.; Furukawa, Y.; Ohta, S.; Ellis, T. A.; Dembrow, N. C.; Li, L.; Floyd, P. D.; Sweedler, J. V.; Minakata, H.; Nakamaru, K.; Morishita, F.; Matsushima, O.; Weiss, K. R.; Vilim, F. S. The Aplysia Mytilus Inhibitory Peptide-related Peptides: Identification, Cloning, Processing, Distribution, and Action. *J. Neurosci.* **1999,** *19*(21), 9618–9634.
145. Fujisawa, Y.; Ikeda, T.; Nomoto, K.; Yasuda-Kamatani, Y.; Minakata, H.; Kenny, P. T.; Kubota, I.; Muneoka, Y. The FMRFamide-related Decapeptide of *Mytilus* Contains a D-Amino Acid Residue. *Comp. Biochem. Physiol. C* **1992,** *102*(1), 91–95.
146. Fujisawa, Y.; Masuda, K.; Minakata, H. Fulicin Regulates the Female Reproductive Organs of the Snail, *Achatina fulica. Peptides* **2000,** *21*(8), 1203–1208.
147. Fujita, K.; Minakata, H.; Nomoto, K.; Furukawa, Y.; Kobayashi, M. Structure–Activity Relations of Fulicin, a Peptide Containing a D-amino Acid Residue. *Peptides* **1995,** *16*(4), 565–568.
148. Fujiwara-Sakata, M.; Kobayashi, M. Neuropeptides Regulate the Cardiac Activity of a Prosobranch Mollusc, *Rapana thomasiana. Cell Tissue Res.* **1992,** *269*(2), 241–247.
149. Fujiwara-Sakata, M.; Kobayashi, M. Localization of FMRFamide- and ACEP-1-like Immunoreactivities in the Nervous System and Heart of a Pulmonate Mollusc, *Achatina fulica. Cell Tissue Res.* **1994,** *278*(3), 451–460.
150. Furukawa, Y.; Miyawaki, Y.; Abe, G. Molecular Cloning and Functional Characterization of the Aplysia FMRFamide-gated Na^+ Channel. *Pflugers Arch.* **2006,** *451*(5), 646–656.
151. Furukawa, Y.; Nakamaru, K.; Sasaki, K.; Fujisawa, Y.; Minakata, H.; Ohta, S.; Morishita, F.; Matsushima, O.; Li, L.; Alexeeva, V.; Ellis, T. A.; Dembrow, N. C.; Jing, J.; Sweedler, J. V.; Weiss, K. R.; Vilim, F. S. PRQFVamide, a Novel Pentapeptide Identified from the CNS and Gut of Aplysia. *J. Neurophysiol.* **2003,** *89*(6), 3114–3127.
152. Furukawa, Y.; Nakamaru, K.; Wakayama, H.; Fujisawa, Y.; Minakata, H.; Ohta, S.; Morishita, F.; Matsushima, O.; Li, L.; Romanova, E.; Sweedler, J. V.; Park, J. H.; Romero, A.; Cropper, E. C.; Dembrow, N. C.; Jing, J.; Weiss, K. R.; Vilim, F. S. The Enterins: A Novel Family of Neuropeptides Isolated from the Enteric Nervous System and CNS of Aplysia. *J. Neurosci.* **2001,** *21*(20), 8247–8261.

153. Gainer, H.; Loh, Y. P.; Sarne, Y. Biosynthesis of Neuronal Peptides. In *Peptides in Neurobiology*; Gainer, H., Ed. Plenum: New York, 1977.

154. Gardam, K. E.; Magoski, N. S. Regulation of Cation Channel Voltage and Ca^{2+} Dependence by Multiple Modulators. *J. Neurophysiol.* **2009**, *102*(1), 259–271.

155. Garrison, J. L.; Macosko, E. Z.; Bernstein, S.; Pokala, N.; Albrecht, D. R.; Bargmann, C. I. Oxytocin/Vasopressin-related Peptides Have an Ancient Role in Reproductive Behavior. *Science* **2012**, *338*, 540–543.

156. Geoffroy, E.; Hutcheson, R.; Chase, R. Nervous Control of Ovulation and Ejaculation in *Helix aspersa*. *J. Mollusc. Stud.* **2005**, *71*, 393–399.

157. Geraerts, W. P. M. The Role of the Lateral Lobes in the Control of Growth and Reproduction in the Hermaphrodite Freshwater Snail *Lymnaea stagnalis*. *Gen. Comp. Endocrinol.* **1976**, *29*(1), 97–108.

158. Geraerts, W. P. M. Cardioactive Peptides of the CNS of the Pulmonate Snail *Lymnaea stagnalis*. *Experientia* **1981**, *37*, 1168–1171.

159. Geraerts, W. P. M.; Bohlken, S. The Control of Ovulation in the Hermaphroditic Freshwater Snail *Lymnaea stagnalis* by the Neurohormone of the Caudodorsal Cells. *Gen. Comp. Endocrinol.* **1976**, *28*, 350–357.

160. Geraerts, W. P. M.; de With, N. D.; Tan, B. T.; van Hartingsveldt, W.; Hogenes, T. M. Studies of the Characteristics, Distribution and Physiological Role of a Large Cardioactive Peptide in *Lymnaea stagnalis*. *Comp. Biochem. Physiol. C* **1984**, *78*(2), 339–433.

161. Geraerts, W. P. M.; Joosse, J. Freshwater Snails (Basommatophora). In *The Mollusca*; Tompa, A. S., Verdonk, N. H., van den Biggelaar, J. A. M., Eds.; Academic Press, Inc.: Orlando, 1984; Vol. 7.

162. Geraerts, W. P. M.; Mohamed, A. M. Studies on the Role of the Lateral Lobes and the Ovotestis of the Pulmonate *Bulinus truncatus* in the Control of Body Growth and Reproduction. *Invert. Reprod. Dev.* **1981**, *3*, 297–308.

163. Geraerts, W. P. M.; Smit, A. B.; Li, K. W.; Hordijk, P. L. The Light Green Cells of Lymnaea: A Neuroendocrine Model System for Stimulus-induced Expression of Multiple Peptide Genes in a Single Cell Type. *Experientia* **1992**, *48*(5), 464–473.

164. Geraerts, W. P. M.; ter Maat, A.; Vreugdenhil, E. The Peptidergic Neuroendocrine Control of Egg-laing Beahvior in Aplysia and Lymnaea. In *Endocrinology of Selected Invertebrate Types*; Laufer, H., Downer, R. G. H.; Alan R. Liss, Inc., New York, 1988.

165. Giardino, N. D.; Aloyz, R. S.; Zollinger, M.; Miller, M. W.; DesGroseillers, L. L5–67 and LUQ-1 Peptide Precursors of *Aplysia californica*: Distribution and Localization of Immunoreactivity in the Central Nervous System and in Peripheral Tissues. *J. Comp. Neurol.* **1996**, *374*(2), 230–245.

166. Goldberg, J. I.; Garofalo, R.; Price, C. J.; Chang, J. P. Presence and Biological Activity of a GnRH-like Factor in the Nervous System of *Helisoma trivolvis*. *J. Comp. Neurol.* **1993**, *336*(4), 571–582.

167. Goldberg, J. I.; Koehncke, N. K.; Christopher, K. J.; Neumann, C.; Diefenbach, T. J. Pharmacological Characterization of a Serotonin Receptor Involved in an Early Embryonic Behavior of *Helisoma trivolvis*. *J. Neurobiol.* **1994**, *25*, 1545–1557.

168. Goldberg, J. I.; Doran, S. A.; Shartau, R. B.; Pon, J. R.; Ali, D. W.; Tam, R.; Kuang, S. Integrative biology of an Embryonic Respiratory Behaviour in Pond Snails: The 'Embryo Stir-Bar Hypothesis'. *J. Exp. Biol.* **2008**, *211*, 1729–1736.

169. Goldberg, J. I.; Rich, D. R.; Muruganathan, S. P.; Liu, M. B.; Pon, J. R.; Tam, R.; Diefenbach, T. J.; Kuang, S. Identification and Evolutionary Implications of

Neurotransmitter-Ciliary Interactions Underlying the Behavioral Response to Hypoxia in *Lymnaea stagnalis* Embryos. *J. Exp. Biol.* **2011**, *214*, 2660–2670.Gomot, A.; Gomot, L.; Marchand, C.; Colard, C.; Bride, J. Immunocytochemical Localization of Insulin-related Peptide(s) in the Central Nervous System of the Snail *Helix aspersa* Muller: Involvement in Growth Control. *Cell. Mol. Neurobiol.* **1992**, *12*, 21–32.

170. Goudsmit, E. M.; Ashwell, G. Enzymatic Synthesis of Galactogen in the Snail *Helix pomatia. Biochem. Biophys. Res. Commun.* **1965**, *19*, 417–422.

171. Goudsmit, E. M. Neurosecretory Stimulation of Galactogen Synthesis within the *Helix pomatia* Albumen Gland During Organ Culture. *J. Exp. Zool.* **1975**, *191*, 193–198.

172. Greenberg, M. J; Price, D. A. FMRFamide, a Cardioexcitatory Neuropeptide of Molluscs: An Agent in Search of a Mission. *Am. Zool.* **1979**, *19*, 163–174.

173. Greenberg, M. J.; Price, D. A. Relationships among the FMRFamide-like Peptides. *Prog Brain Res.* **1992**, *92*, 25–37.

174. Griffond, B.; Van Minnen, J.; Colard, C. Distribution of APGWa-immunoreactive Substances in the Central Nervous System and Reproductive Apparatus of *Helix aspersa. Zool. Sci.* **1992**, *9*, 533–539.

175. Grimm-Jørgensen, Y. Immunoreactive Somatostatin in two Pulmonate Gastropods. *Gen. Comp. Endocrinol.* **1983**, *49*(1), 108–114.

176. Grimm-Jørgensen, Y. Distribution and Physiological Roles of TRH and Somatostatin in Gastropods. In *Advances in Comparative Endocrinology*; Lofts, B., Holmes, W. N., Eds.; Hong Kong University Press: Hong Kong, 1985.

177. Grimmelikhuijzen, C. J. FMRFamide Immunoreactivity is Generally Occurring in the Nervous Systems of Coelenterates. *Histochemistry* **1983**, *78*(3), 361–381.

178. Grimmelikhuijzen, C. J.; Graff, D. Arg-Phe-amide-like Peptides in the Primitive Nervous Systems of Coelenterates. *Peptides* **1985**, *6*(Suppl. 3), 477–483.

179. Grimmelikhuijzen, C. J.; Leviev, I.; Carstensen, K. Peptides in the Nervous Systems of Cnidarians: Structure, Function, and Biosynthesis. *Int. Rev. Cytol.* **1996**, *167*, 37–89.

180. Groten, C. J.; Rebane, J. T.; Blohm, G.; Magoski, N. S. Separate Ca^{2+} Sources are Buffered by Distinct Ca^{2+} Handling Systems in Aplysia Neuroendocrine Cells. *J. Neurosci.* **2013**, *33*(15), 6476–6491.

181. Gruber, C. W. Physiology of Invertebrate Oxytocin and Vasopressin Neuropeptides. *Exp. Physiol.* **2014**, *99*, 55–61.

182. Guillemin, R. Peptides in the Brain: The New Endocrinology of the neuron. *Science* **1978**, *202*(4366), 390–402.

183. Hadfield, M. G.; Switzer-Dunlap, M. In *The Mollusca. Reproduction*; Tompa, A. S., Verdonk, H. H., van den Biggelaar, J. A. M., Eds.; Academic Press, Inc.: Orlando, 1984, Vol 7.

184. Harley, C. D. Climate Change, Keystone Predation, and Biodiversity Loss. *Science* **2011**, *334*, 1124–1127.

185. Harris, L. L.; Lesser, W.; Ono, J. K. FMRFamide is Endogenous to the Aplysia Heart. *Cell Tissue Res.* **1995**, *282*(2), 331–341.

186. Hartwig, H. G.; Brisson, P.; Lyncker, I.; Collin, J. P. Aminergicsystems in Pulmonate Gastropod Molluscs. III. Microspectrofluorometric Characterization of the Monoamines in the Reproductive System. *Cell Tissue Res.* **1980**, *210*, 223–234.

187. Haszprunar, G.; Wanninger, A. Molluscs. *Curr. Biol.* **2012**, *22*, R15.

188. Hatcher, N. G.; Sweedler, J. V. Aplysia bag cells function as a distributed neurosecretory network. *J. Neurophysiol.* **2008**, *99*(1), 333–343.

189. Hauser, F.; Grimmelikhuijzen, C. J. Evolution of the AKH/corazonin/ACP/GnRH receptor superfamily and their ligands in the Protostomia. *Gen. Comp. Endocrinol.* **2014**, *209*, 35–49.

190. Hauser, F.; Sondergaard, L.; Grimmelikhuijzen, C. J. Molecular Cloning, Genomic Organization and Developmental Regulation of a Novel Receptor from *Drosophila melanogaster* Structurally Related to Gonadotropin-releasing Hormone Receptors for Vertebrates. *Biochem. Biophys. Res. Commun.* **1998**, *249*(3), 822–828.

191. Hermann, P. M.; de Lange, R. P.; Pieneman, A. W.; ter Maat, A.; Jansen, R. F. Role of Neuropeptides Encoded on CDCH-1 Gene in the Organization of Egg-laying Behavior in the Pond Snail, *Lymnaea stagnalis*. *J. Neurophysiol.* **1997**, *78*, 2859–2869.

192. Hickey, C. M.; Geiger, J. E.; Groten, C. J.; Magoski, N. S. Mitochondrial Ca^{2+} Activates A Cation Current in Aplysia Bag Cell Neurons. *J. Neurophysiol.* **2010**, *103*(3), 1543–1556.

193. Hickey, C. M.; Groten, C. J.; Sham, L.; Carter, C. J.; Magoski, N. S. Voltage-gated Ca^{2+} Influx and Mitochondrial Ca^{2+} Initiate Secretion from Aplysia Neuroendocrine Cells. *Neuroscience* **2013**, *250*, 755–772.

194. Hoek, R. M.; Li, K. W.; van Minnen, J.; Lodder, J. C.; de Jong-Brink, M.; Smit, A. B.; van Kesteren, R. E. LFRFamides: a Novel Family of Parasitation-induced-RFamide neuropeptides that Inhibit the Activity of Neuroendocrine Cells in *Lymnaea stagnalis*. *J. Neurochem.* **2005**, *92*(5), 1073–1080.

195. Hoek, R. M.; van Kesteren, R. E.; Smit, A. B.; de Jong-Brink, M.; Geraerts, W. P. M. Altered gene expression in the Host Brain Caused by a Trematode Parasite: Neuropeptide Genes are Preferentially Affected During Parasitosis. *Proc. Natl. Acad. Sci. U.S.A.* **1997**, *94*(25), 14072–14076.

196. Hokfelt, T. Neuropeptides in Perspective: The Last Ten Years. *Neuron* **1991**, *7*(6), 867–879.

197. Hordijk, P. L.; de Jong-Brink, M.; ter Maat, A.; Pieneman, A. W.; Lodder, J. C.; Kits, K. S. The Neuropeptide Schistosomin and Haemolymph from Parasitized Snails Induce Similar Changes in Excitability in Neuroendocrine Cells Controlling Reproduction and Growth in a Freshwater Snail. *Neurosci. Lett.* **1992**, *136*(2), 193–197.

198. Hordijk, P. L.; van Loenhout, H.; Ebberink, R. H. M.; de Jong-Brink, M.; Joosse, J. Neuropeptide Schistosomin Inhibits Hormonally-induced Ovulation in the Freshwater Snail *Lymnaea stagnalis*. *J. Exp. Zool.* **1991a**, *259*, 268–271.

199. Hordijk, P. L.; Schallig, H. D. F. H.; Ebberink, R. H. M.; de Jong-Brink, M.; Joosse, J. Primary Structure and Origin of Schistosomin, an Anti-gonadotropic Neuropeptide of the Pond Snail *Lymnaea stagnalis*. *Biochem. J.* **1991b**, *279*, 837–842.

200. Horn, C. C.; Koester, J.; Kupfermann, I. Evidence that Hemolymph Glucose in *Aplysia californica* is Regulated but Does Not Affect Feeding Behavior. *Behav. Neurosci.* **1998**, *112*(5), 1258–1265.

201. Hua, N. T.; Ako, H. Maturation and Spawning Induction in Hawaiian Opihi *Cellana* spp. by Hormone GnRH. *Commun. Agric. Appl. Biol. Sci.* **2013**, *78*(4), 194–197.

202. Hung, A. Y.; Magoski, N. S. Activity-dependent Initiation of a Prolonged Depolarization in Aplysia Bag Cell Neurons: Role for a Cation Channel. *J. Neurophysiol.* **2007**, *97*(3), 2465–2479.

203. Hunter, T; Vodel, S. Spinning Embryos Enhance Diffusion through Gelatinous Egg Masses. *J. Exp. Mar. Biol. Ecol.* **1986**, *96*, 303–308.

204. Ishida, T.; In, Y.; Inoue, M.; Yasuda-Kamatani, Y.; Minakata, H.; Iwashita, T.; Nomoto, K. Effect of the D-Phe2 Residue on Molecular Conformation of an Endogenous Neuropeptide Achatin-I. Comparison of X-ray Crystal Structures of Achatin-I (H-Gly-D-Phe-Ala-Asp-OH) and Achatin-II (H-Gly-Phe-Ala-Asp-OH). *FEBS Lett.* **1992,** *307*(3), 253–256.

205. Iversen, S.; Iversen, L.; Saper, C. B. The Autonomic Nervous System and the Hypohtalamus. In *Principles of Neural Science*; Kandel, E.; Schwartz, J. H.; Jessel, T. M., Eds.; McGraw-Hill: New York, 2000.

206. Iwakoshi, E.; Hisada, M.; Minakata, H. Cardioactive Peptides Isolated from the Brain of a Japanese octopus, *Octopus minor. Peptides* **2000,** *21*(5), 623–630.

207. Jennings, K. R.; Kaczmarek, L. K.; Hewick, R. M.; Dreyer, W. J.; Strumwasser, F. Protein Phosphorylation During Afterdischarge in Peptidergic Neurons of Aplysia. *J. Neurosci.* **1982,** *2*(2), 158–168.

208. Jeziorski, M. C.; Green, K. A.; Sommerville, J.; Cottrell, G. A. Cloning and Expression of a FMRFamide-gated Na(+) Channel from *Helisoma trivolvis* and Comparison with the Native Neuronal Channel. *J. Physiol.* **2000,** *526*(Pt. 1), 13–25.

209. Jilek, A.; Mollay, C.; Lohner, K.; Kreil, G. Substrate Specificity of a Peptidyl-aminoacyl-L/D-Isomerase from Frog Skin. *Amino Acids* **2012,** *42*(5), 1757–1764.

210. Jiménez, C. R.; ter Maat, A.; Pieneman, A.; Burlingame, A. L.; Smit, A. B.; Li, K. W. Spatio-temporal Dynamics of the Egg-laying-inducing Peptides During an Egg-laying Cycle: A Semiquantitative Matrix-assisted Laser Desorption/Ionization Mass Spectrometry Approach. *J. Neurochem.* **2004,** *89*, 865–875.

211. Johnston, R. N.; Shaw, C.; Halton, D. W.; Verhaert, P.; Baguna, J. GYIRFamide: A Novel FMRFamide-related Peptide (FaRP) from the Triclad Turbellarian, *Dugesia tigrina. Biochem. Biophys. Res. Commun.* **1995,** *209*(2), 689–697.

212. Jonas, E. A.; Knox, R. J.; Smith, T. C.; Wayne, N. L.; Connor, J. A.; Kaczmarek, L. K. Regulation by Insulin of a Unique Neuronal Ca^{2+} Pool and of Neuropeptide Secretion. *Nature* **1997,** *385*, 343–346.

213. Jonas, E. A.; Knox, R. J.; Kaczmarek, L. K.; Schwartz, J. H.; Solomon, D. H. Insulin Receptor in *Aplysia* Neurons: Characterization, Molecular Cloning, and Modulation of Ion Currents. *J. Neurosci.* 16, 1645-1658.

214. Jones, H. D. Circulatory Systems of Gastropods and Bivalves. *In The Mollusca. Physiology*; Saleuddin, A. S. M., Wilbur, K. M., Eds.; Academic Press, New York, 1983, Part 2, Vol 5.

215. Joosse, J. Dorsal Bodies and Dorsal Neurosecretory Cells of the Cerebral Ganglia of *Lymnaea stagnalis* L. *Arch. Neerl. Zool.* **1964,** *15*, 1–103.

216. Joosse, J.; Geraerts, W. P. M. Endocrinology. In *The Mollusca-Physiology*; Saleuddin, A. S. M., Wilbur, K. M., Eds.; Academic Press: New York, 1983; part 1, Vol 5.

217. Joosse, J. The Hormones of Molluscs. In *Endocrinology of Selected Invertebrate Types*; Laufer, H., Downer, R. G. H., Eds.; Alan R. Liss, Inc.: New York, 1988.

218. Jung, L. H.; Kavanaugh, S. I.; Sun, B.; Tsai, P. S. Localization of a Molluscan Gonadotropin-releasing Hormone in *Aplysia californica* by In Situ Hybridization and Immunocytochemistry. *Gen. Comp. Endocrinol.* **2014,** *195*, 132–137.

219. Jung, L. J.; Scheller, R. H. Peptide Processing and Targeting in the Neuronal Secretory Pathway. *Science* **1991,** *251*(4999), 1330–1335.

220. Kaczmarek, L. K.; Finbow, M.; Revel, J. P.; Strumwasser, F. The Morphology and Coupling of Aplysia Bag Cells within the Abdominal Ganglion and In Cell Culture. *J. Neurobiol.* **1979,** *10*(6), 535–550.

221. Kaczmarek, L. K.; Strumwasser, F. A Voltage-clamp Analysis of Currents Underlying Cyclic AMP-induced Membrane Modulation in Isolated Peptidergic Neurons of Aplysia. *J. Neurophysiol.* **1984,** *52*(2), 340–349.

222. Kamatani, Y.; Minakata, H.; Iwashita, T.; Nomoto, K.; In, Y.; Doi, M.; Ishida, T. Molecular Conformation of Achatin-I, an Endogenous Neuropeptide Containing D-Amino Acid Residue. X-ray crystal structure of its neutral form. *FEBS Lett.* **1990,** *276*(1–2), 95–97.

223. Kamatani, Y.; Minakata, H.; Nomoto, K.; Kim, K. H.; Yongsiri, A.; Takeuchi, H. Isolation of Achatin-I, a Neuroactive Tetrapeptide Having a D-Phenylalanine Residue, from *Achatina ganglia*, and Its Effects on Achatina Giant Neurones. *Comp. Biochem. Physiol. C* **1991,** *98*(1), 97–103.

224. Kamiya, H.; Muramoto, K.; Yamazaki, M. Aplysianin-A, an Antibacterial and Antineoplastic Glycoprotein in the Albumen Gland of a Sea Hare, *Aplysia kurodai*. *Experientia* **1986,** *42*, 1065–1067.

225. Kanda, A.; Minakata, H. Isolation and Characterization of a Novel Small Cardioactive Peptide-related Peptide from the Brain of *Octopus vulgaris*. *Peptides* **2006,** *27*(7), 1755–1761.

226. Kanemaru, K.; Morishita, F.; Matsushima, O.; Furukawa, Y. Aplysia Cardioactive Peptide (NdWFamide) enhances the L-type Ca^{2+} Current of Aplysia Ventricular Myocytes. *Peptides* **2002,** *23*(11), 1991–1998.

227. Kerschbaum, H.; Holzinger, K.; Hermann, A. Endocrine-like Cells and Insulin-binding Sites in the Epineurium of *Helix pomatia*. *Tissue Cell* **1993,** *25*, 237–243.

228. Khan, H. R.; Ashton, M-L.; Mukai, S. T.; Saleuddin, A. S. M. The Effects of Mating on the Fine Structure of Neurosecretory Caudodorsal Cells in *Helisoma duryi* (Mollusca). *Can. J. Zool.* **1990,** *68*, 1233–1240.

229. Khan, H. R.; Griffond, B.; Saleuddin, A. S. M. Insulin-like Peptide(s) in the Central Nervous System of the Snail *Helisoma duryi*. *Brain Res.* **1992,** *580*, 111–114.

230. Khan, H.; Price, D.; Doble, K.; Greenberg, M.; Saleuddin, A. FMRFamide-related Peptides, Partial Serotonin Depletion, and Osmoregulation in *Helisoma duryi* (Mollusca: Pulmonata). *J. Comp. Neurol.* **1998,** *393*, 25–33.

231. Khan, H. R.; Saleuddin, A. S. M. Effects of Osmotic Changes and Neurosecretory Extracts on Kidney Ultrastructure in the Freshwater Pulmonate Helisoma. *Can. J. Zool.* **1979a,** *57*, 1256–1270.

232. Khan, H. R.; Saleuddin, A. S. M. Osmotic Regulation and Osmotically Induced Changes in the Neurosecretory Cells of the Pulmonate Snail *Helisoma*. *Can. J. Zool.* **1979b,** *57*, 1371–1383.

233. Khan, H. R.; Saleuddin, A. S. M. Cell Contacts in the Kidney Epithelium of *Helisoma* (Mollusca: Gastropoda): Effects of Osmotic Pressure and Brain Extracts: A Freeze-fracture Study. *J. Ultrastr. Res.* **1981,** *75*, 23–40.

234. Kiehn, L.; Lange, A. B.; Saleuddin, A. S. M. Dopaminergic Neurons in the Brain and Dopaminergic Innervation of the Albumen Gland in Mated and Virgin *Helisoma duryi* (Mollusca: Pulmonata). *BMC Physiol.* **2001,** *1*, 9.

235. Kiehn, L.; Mukai, S. T.; Saleuddin, A. S. M. The Role of Calcium on Protein Secretion of the Albumen Gland in *Helisoma duryi* (Gastropoda). *Invert. Biol.* **2004,** *123*, 304–315.

236. Kim, K. H.; Takeuchi, H.; Kamatani, Y.; Minakata, H.; Nomoto, K. Structure–Activity Relationship Studies on the Endogenous Neuroactive Tetrapeptide Achatin-I on Giant Neurons of *Achatina fulica* Ferussac. *Life Sci.* **1991,** *48*(17), PL91–96.

237. Kiss, T. Diversity and Abundance: The Basic Properties of Neuropeptide Action in Molluscs. *Gen. Comp. Endocrinol.* **2011,** *15,* 172, 10–14.
238. Kits, K. S.; Bobeldijk, R. C.; Crest, M.; Lodder, J. C. Glucose-induced Excitation in Molluscan Central Neurons Producing Insulin-related Peptides. *Pflugers Arch.* **1991,** *417*(6), 597–604.
239. Kits, K. S.; de Vries, N. J.; Ebberink, R. H. M. Molluscan Insulin-related Neuropeptide Promotes Neurite Outgrowth in Dissociated Neuronal Cell Cultures. *Neurosci. Lett.* **1990,** *109*(3), 253–258.
240. Knock, S. L.; Nagle, G. T.; Lin, C. Y.; McAdoo, D. J.; Kurosky, A. *Aplysia brasiliana* Neurons R3–R14: Primary Structure of the Myoactive Histidine-rich Basic Peptide and Its Prohormone. *Peptides* **1989,** *10*(4), 859–867.
241. Koch, G.; Chen, M. L.; Sharma, R.; Walker, R. J. The Actions of RFamide Neuroactive Peptides on the Isolated Heart of the Giant African Snail, *Achatina fulica. Comp. Biochem. Physiol. C* **1993,** *106*(2), 359–365.
242. Koch, U. T.; Koester, J.; Weiss, K. R. Neuronal Mediation of Cardiovascular Effects of Food Arousal in Aplysia. *J. Neurophysiol.* **1984,** *51*(1), 126–135.
243. Kodirov, S. A. The Neuronal Control of Cardiac Functions in Molluscs. *Comp. Biochem. Physiol. A* **2011,** *160*(2), 102–116.
244. Koene, J. M.; Jansen, R. F.; Ter Maat, A.; Chase, R. A Conserved Location for the Central Nervous System Control of Mating Behaviour in Gastropod Molluscs: Evidence from a Terrestrial Snail. *J. Exp. Biol.* **2000,** *203,* 1071–1080.
245. Koester, J.; Alevizosi, A. Innervation of the Kidney of Aplysia by LIO, the LUQ cells, and an Identified Peripheral Motoneuron. *J. Neurosci.* **1989,** *9,* 4078–4088.
246. Koester, J.; Mayeri, E.; Liebeswar, G.; Kandel, E. R. Neural Control of Circulation in Aplysia. II. Interneurons. *J. Neurophysiol.* **1974,** *37*(3), 476–496.
247. Koh, H. Y.; Vilim, F. S.; Jing, J.; Weiss, K. R. Two neuropeptides Colocalized in a Command-like Neuron Use Distinct Mechanisms to Enhance its Fast Synaptic Connection. *J. Neurophysiol.* **2003,** *90*(3), 2074–2079.
248. Koss, R.; Diefenbach, T. J.; Kuang, S.; Doran, S. A.; Goldberg, J. I. Coordinated Development of Identified Serotonergic Neurons and their Target Ciliary Cells in *Helisoma trivolvis* Embryos. *J. Comp. Neurol.* **2003,** *457,* 313–325.
249. Krontiris-Litowitz, J. Sensitizing Stimulation Causes a Long-term Increase in Heart Rate in *Aplysia californica. J. Comp. Physiol. A* **1999,** *185*(2), 181–186.
250. Kuang, S.; Goldberg, J. I. Laser Ablation Reveals Regulation of Ciliary Activity by Serotonergic Neurons in Molluscan Embryos. *J. Neurobiol.* **2001,** *47,* 1–15.
251. Kuang, S.; Doran, S. A.; Wilson, R. J.; Goss, G. G.; Goldberg, J. I. Serotonergic Sensory–Motor Neurons Mediate a Behavioral Response to Hypoxia in Pond Snail Embryos. *J. Neurobiol.* **2002,** *52,* 73–83.
252. Kunigelis, S. C.; Saleuddin. A. S. M. Studies on the In Vitro Formation of Periostracum: The Influence of the Brain. *J. Comp. Physiol. B* **1985,** *155,* 177–183.
253. Kupfermann, I. Stimulation of Egg Laying: Possible Neuroendocrine Function of Bag Cells of Abdominal Ganglion of *Aplysia californica. Nature* **1967,** *216*(5117), 814–815.
254. Kupfermann, I. Stimulation of Egg Laying by Extracts of Neuroendocrine Cells (Bag Cells) of Abdominal Ganglion of Aplysia. *J. Neurophysiol.* **1970,** *33*(6), 877–881.
255. Kupfermann, I.; Kandel, E. R. Electrophysiological Properties and Functional Interconnections of Two Symmetrical Neurosecretory Clusters (Bag Cells) in Abdominal Ganglion of Aplysia. *J. Neurophysiol.* **1970,** *33*(6), 865–876.

256. Kupfermann, I.; Weiss, K. Water Regulation by a Presumptive Hormone Contained in Identified Neurosecretory Cell R15 of Aplysia. *J. Gen. Physiol.* **1976,** *67*(1), 113–123.

257. Kuroki, Y.; Kanda, T.; Kubota, I.; Fujisawa, Y.; Ikeda, T.; Miura, A.; Minamitake, Y.; Muneoka, Y. A Molluscan Neuropeptide Related to the Crustacean Hormone, RPCH. *Biochem. Biophys. Res. Commun.* **1990,** *167*(1), 273–279.

258. Lagadic, L.; Coutellec, M. A.; Caquet, T. Endocrine Disruption in Aquatic Pulmonate Molluscs: Few Evidences, Many Challenges. *Ecotoxicology* **2007,** *16*, 45–59.

259. Landry, C.; Crine, P.; DesGroseillers, L. Differential Expression of Neuropeptide Gene mRNA within the LUQ Cells of *Aplysia californica*. *J. Neurobiol.* **1992,** *23*(1), 89–101.

260. Lardans, V.; Coppin, J.; Vicogne, J.; Aroca, E.; Delcroix, M.; Dissous, C. Characterization of an Insulin Receptor-related Receptor in *Biomphalaria glabrata* Embryonic Cells. *Biochim. Biophys. Acta* **2001,** *1510*, 321–329.

261. Lee, W.; Wayne, N. L. The Roles of Transcription and Translation in Mediating the Effect of Electrical Afterdischarge on Neurohormone Synthesis in Aplysia Bag Cell Neurons. *Endocrinology* **1998,** *139*(12), 5109–5115.

262. Lee, W.; Wayne, N. L. Secretion of Locally Synthesized Neurohormone from Neurites of Peptidergic Neurons. *J. Neurochem.* **2004,** *88*(3), 532–537.

263. Lee, W.; Jones, A. M.; Ono, J. K.; Wayne, N. L. Regional Differences in Processing of Locally Translated Prohormone in Peptidergic Neurons of *Aplysia californica*. *J. Neurochem.* **2002,** *83*(6), 1423–1430.

264. Lesser, W.; Greenberg, M. J. Cardiac Regulation by Endogenous Small Cardioactive Peptides and FMRFamide-related Peptides in the Snail *Helix aspersa*. *J. Exp. Biol.* **1993,** *178*, 205–230.

265. Li, H.; Sheppard, D. N.; Hug, M. J. Transepithelial Electrical Measurements with the Ussing Chamber. *J. Cyst. Fibros.* **2004,** *3*(Suppl. 2), 123–126.

266. Li, K. W.; Geraerts, W. P. M. Isolation and Chemical Characterization of a Novel Insulin-related Neuropeptide from the Freshwater Snail, *Lymnaea stagnalis*. *Eur. J. Biochem.* **1992,** *205*(2), 675–678.

267. Li, K. W.; Geraerts, W. P. M.; Joosse, J. Purification and Chemical Characterization of Caudodorsal Cell Hormone-II from the Egg-laying Controlling Caudodorsal Cells of *Lymnaea stagnalis*. *Peptides* **1992a,** *13*, 215–220.

268. Li, K. W.; Geraerts, W. P. M.; Ebberink, R. H. M.; Joosse, J. Purification and Sequencing of Molluscan Insulin-related Peptide II from the Neuroendocrine Light Green Cells in *Lymnaea stagnalis*. *Endocrinology* **1992b,** *130*(6), 3427–3432.

269. Li, K. W.; Geraerts, W. P. M.; van Loenhout, H.; Joosse, J. Biosynthesis and Axonal Transport of Multiple Molluscan Insulin-related Peptides by the Neuroendocrine Light Green Cells of *Lymnaea stagnalis*. *Gen. Comp. Endocrinol.* **1992c,** *87*(1), 79–86.

270. Li, K. W.; Jiménez, C. R.; Van Veelen, P. A.; Geraerts, W. P. M. Processing and Targeting of a Molluscan Egg-laying Peptide Prohormone as Revealed by Mass Spectrometric Peptide Fingerprinting and Peptide Sequencing. *Endocrinology* **1994,** *134*, 1812–1819.

271. Li, L.; Garden, R. W.; Floyd, P. D.; Moroz, T. P.; Gleeson, J. M.; Sweedler, J. V.; Pasa-Tolic, L.; Smith, R. D. Egg-laying Hormone Peptides in the Aplysiidae Family. *J. Exp. Biol.* **1999,** *202*(Pt. 21), 2961–2973.

272. Liebeswar, G.; Goldman, J. E.; Koester, J.; Mayeri, E. Neural Control of Circulation in Aplysia. III. Neurotransmitters. *J. Neurophysiol.* **1975,** *38*(4), 767–779.

273. Ligman, S. H.; Brownell, P. H. Differential Hormonal Action of the Bag Cell Neurons on the Arterial System of Aplysia. *J. Comp. Physiol. A* **1985,** *157*(1), 31–37.

274. Lingueglia, E.; Champigny, G.; Lazdunski, M.; Barbry, P. Cloning of the Amiloride-sensitive FMRFamide Peptide-gated Sodium Channel. *Nature* **1995,** *378*(6558), 730–733.

275. Liu, J.; Spéder, P.; Brand, A. H. Control of Brain Development and Homeostasis by Local and Systemic Insulin Signalling. *Diabet. Obes. Metab.* **2014,** *16*(Suppl. 1), 16–20.

276. Lloyd, P. E. Distribution and Molecular Characteristics of Cardioactive Peptides in the Snail, *Helix aspersa. J. Comp. Physiol. A* **1978,** *128*, 269–276.

277. Lloyd, P. E. Biochemical and Pharmacological Analyses of Endogenous Cardioactive Peptides in the Snail, *Helix aspersa. J. Comp. Physiol. [A]* **1980,** *138*, 265–270.

278. Lloyd, P. E. Cardioactive Neuropeptides in Gastropods. *Fed. Proc.* **1982,** *41*, 2948–2952.

279. Lloyd, P. E.; Kupfermann, I.; Weiss, K. R. Sequence of Small Cardioactive Peptide A: A Second Member of a Class of Neuropeptides in Aplysia. *Peptides* **1987,** *8*(1), 179–184.

280. Lloyd, P. E.; Mahon, A. C.; Kupfermann, I.; Cohen, J. L.; Scheller, R. H.; Weiss, K. R. Biochemical and Immunocytological Localization of Molluscan Small Cardioactive Peptides in the Nervous System of *Aplysia californica. J. Neurosci.* **1985,** *5*(7), 1851–1861.

281. Lockwood, B. L.; Somero, G. N. Transcriptomic Responses to Salinity Stress in Invasive and Native Blue Mussels (genus *Mytilus*). *Mol. Ecol.* **2011,** *20*(3), 517–529.

282. Loechner, K. J.; Azhderian, E. M.; Dreyer, R.; Kaczmarek, L. K. Progressive Potentiation of Peptide Release During a Neuronal Discharge. *J. Neurophysiol.* **1990,** *63*(4), 738–744.

283. Loechner, K. J.; Kaczmarek, L. K. Control of Potassium Currents and Cyclic AMP Levels by Autoactive Neuropeptides in Aplysia Neurons. *Brain Res.* **1990,** *532*(1–2), 1–6.

284. Loechner, K. J.; Kaczmarek, L. K. Autoactive Peptides Act at Three Distinct Receptors to Depolarize the Bag Cell Neurons of Aplysia. *J. Neurophysiol.* **1994,** *71*(1), 195–203.

285. Lopez-Vera, E.; Aguilar, M. B.; Heimer de la Cotera, E. P. FMRFamide and Related Peptides in the Phylum Mollusca. *Peptides* **2008,** *29*(2), 310–317.

286. Lubet, P. Influence des ganglions cérébroides sur la croissance de Crepidula fornicata Phil. (Mollusque: Gastéropode) effets somatotrope et gonatdotrope. *C. R. Acad. Sci. Paris* **1971,** *273*, 2309–2311.

287. Lubet, P.; Silberzahn, N. Recherches sur les effets del'ablation bilatérale des ganglions cérébroides chez la crépidule (Crepidula fornicata Phil., Mollusque: Gastéropode) effets somatotrope et gonatdotrope. *C. R. Séances Soc. Biol.* **1971,** *165*, 590–594.

288. Lukowiak, K.; Martens, K.; Orr, M.; Parvez, K.; Rosenegger, D.; Sangha, S. Modulation of Aerial Respiratory Behaviour in a Pond Snail. *Respir. Physiol. Neurobiol.* **2006,** *154*, 61–72.

289. Lukowiak, K.; Sunada, H.; Teskey, M.; Lukowiak, K.; Dalesman, S. Environmentally Relevant Stressors Alter Memory Formation in the Pond Snail Lymnaea. *J. Exp. Biol.* **2014,** *217*, 76–83.

290. Lydeard, C.; Cowie, R. H.; Ponder, W. F.; Bogan, A. E.; Bouchet, P.; Clark, S. A.; Cummings, K. S.; Frest, T. J.; Gargominy, O.; Herbert, D. G.; Hershler, R.; Perez, K. E.; Roth, B.; Seddon, M.; Strong, E. E.; Thompson, F. G. The Global Decline of Nonmarine Molluscs. *BioScience* **2004,** *54*, 321–330.

291. Madrid, K. P.; Price, D. A.; Greenberg, M. J.; Khan, H. R.; Saleuddin, A. S. M. FMRFamide-related Peptides from the Kidney of the Snail, *Helisoma trivolvis. Peptides* **1994,** *15*, 31–36.

292. Magoski, N. S. Regulation of an Aplysia Bag-cell Neuron Cation Channel by Closely Associated Protein Kinase A and A Protein Phosphatase. *J. Neurosci.* **2004,** *24*(30), 6833–6841.

293. Magoski, N. S.; Kaczmarek, L. K. Association/Dissociation of a Channel-kinase Complex Underlies State-dependent Modulation. *J. Neurosci.* **2005,** *25*(35), 8037–8047.

294. Magoski, N. S.; Wilson, G. F.; Kaczmarek, L. K. Protein Kinase Modulation of a Neuronal Cation Channel Requires Protein–Protein Interactions Mediated by an Src Homology 3 Domain. *J. Neurosci.* **2002,** *22*(1), 1–9.

295. Mahon, A. C.; Lloyd, P. E.; Weiss, K. R.; Kupfermann, I.; Scheller, R. H. The Small Cardioactive peptides A and B of Aplysia are derived from a Common Precursor Molecule. *Proc. Natl. Acad. Sci. U.S.A.* **1985b,** *82*(11), 3925–3929.

296. Mahon, A. C.; Nambu, J. R.; Taussig, R.; Shyamala, M.; Roach, A.; Scheller, R. H. Structure and Expression of the Egg-laying Hormone Gene Family in Aplysia. *J. Neurosci.* **1985a,** *5*(7), 1872–1880.

297. Manger, P.; Li, J.; Christensen, B. M.; Yoshino, T. P. Biogenic Monoamines in the Freshwater Snail, *Biomphalaria glabrata*: Influence of Infection by the Human Blood Fluke, *Schistosoma mansoni. Comp. Biochem. Physiol. A* **1996,** *114*(3), 227–234.

298. Mann, K.; Edsinger-Gonzales, E.; Mann, M. In-depth Proteomic Analysis of a Mollusc Shell: Acid-soluble and Acid-insoluble Matrix of the limpet *Lottia gigantea. Proteome Sci.* **2012,** *10*, 28–46.

299. Marie, B.; Joubert, C.; Tayalé, A.; Zanella-Cléon, I.; Belliard, C.; Piquemal, D.; Cochennec-Laureau, N.; Marin, F.; Gueguen, Y.; Montagnani, C. Different Secretory Repertoires Control the Biomineralization Processes of Prism and Nacre Deposition of the Pearl Oyster Shell. *Proc. Natl. Acad. Sci. USA.* **2012,** *109*(51), 20986–20991.

300. Marin, F.; Le Roy, N.; Marie, B. The Formation and Mineralization of Mollusk Shell. *Front. Biosci.* **2012,** *4*, 1099–10125.

301. Marois, R.; Croll, R. P. Development of Serotoninlike Immunoreactivity in the Embryonic Nervous System of the Snail *Lymnaea stagnalis. J. Comp. Neurol.* **1992,** *322*(2), 255–265.

302. Mapara, S.; Parries, S. C.; Quarrington, C. M.; Ahn, K.-C.; Gallin, W. J.; Goldberg, J. I. Identification, Molecular Structure and Expression of Two Cloned Serotonin Receptors from the Pond Snail, *Helisoma trivolvis. J. Exp. Biol.* **2008,** *211*, 900–910.

303. Matricon-Gondran, M. The Site of Ultrafiltration in the Kidney Sac of the Pulmonate Gastropod *Biomphalaria glabrata. Tissue Cell* **1990,** *22*, 911–923.

304. Matsumoto, T.; Masaoka, T.; Fujiwara, A.; Nakamura, Y.; Satoh, N.; Awaji, M. Reproduction-related Genes in the Pearl Oyster Genome. *Zool. Sci.* **2013,** *30*(10), 826–850.

305. Matsuo, H.; Baba, Y.; Nair, R. M.; Arimura, A.; Schally, A. V. Structure of the Porcine LH- and FSH-Releasing Hormone. I. The Proposed Amino Acid Sequence. *Biochem. Biophys. Res. Commun.* **1971,** *43*(6), 1334–1339.

306. Matsuo, R.; Kobayashi, S.; Morishita, F.; Ito, E. Expression of Asn-d-Trp-Phe-NH$_2$ in the Brain of the Terrestrial Slug *Limax valentianus. Comp. Biochem. Physiol. B.* **2011,** *160*(2–3), 89–93.

307. Mayeri, E.; Brownell, P.; Branton, W. D.; Simon, S. B. Multiple, Prolonged Actions of Neuroendocrine Bag Cells on Neurons in Aplysia. I. Effects on Bursting Pacemaker Neurons. *J. Neurophysiol.* **1979a,** *42*(4), 1165–1184.

308. Mayeri, E.; Brownell, P.; Branton, W. D. Multiple, Prolonged Actions of Neuroendocrine Bag Cells on Neurons in Aplysia. II. Effects on Beating Pacemaker and Silent Neurons. *J. Neurophysiol.* **1979b,** *42*(4), 1185–1197.

309. Mayeri, E.; Koester, J.; Kupfermann, I.; Liebeswar, G.; Kandel, E. R. Neural Control of Circulation in Aplysia. I. Motoneurons. *J. Neurophysiol.* **1974,** *37*(3), 458–475.

310. Mayeri, E.; Rothman, B. S.; Brownell, P. H.; Branton, W. D.; Padgett, L. Nonsynaptic Characteristics of Neurotransmission Mediated by Egg-laying Hormone in the Abdominal Ganglion of Aplysia. *J. Neurosci.* **1985,** *5*(8), 2060–2077.

311. Meester, I.; Ramkema, M. D.; van Minnen, J.; Boer, H. H. Differential Expression of Four Genes Encoding Molluscan Insulin-related Peptides in the Central Nervous System of the Pond Snail *Lymnaea stagnalis. Cell Tissue Res.* **1992,** *269*(1), 183–188.

312. Miksys, S.; Saleuddin, A. S. M. The Effect of the Brain and Dorsal Bodies of *Helisoma duryi* (Mollusca: Pulmonata) on Albumen Gland Synthetic Activity *in vitro. Gen. Comp. Endocrinol.* **1985,** *60,* 419–426.

313. Miksys, S. L.; Saleuddin, A. S. M. Effects of Castration on Growth and Reproduction of *Helisoma duryi* (Mollusca: Pulmonata). *Invert. Reprod. Develop.* **1987,** *12,* 145–159.

314. Miksys, S. L.; Saleuddin, A. S. M. Polysaccharide Synthesis Stimulating Factors from the Dorsal Bodies and Cerebral Ganglia of *Helisoma duryi* (Mollusca: Pulmonata). *Can.. J. Zool.* **1988,** *66,* 508–511.

315. Miller, M. W.; Beushausen, S.; Cropper, E. C.; Eisinger, K.; Stamm, S.; Vilim, F. S.; Vitek, A.; Zajc, A.; Kupfermann, I.; Brosius, J.; et al. The Buccalin-related Neuropeptides: Isolation and Characterization of an Aplysia cDNA Clone Encoding a Family of Peptide Cotransmitters. *J. Neurosci.* **1993a,** *13*(8), 3346–3357.

316. Miller, M. W.; Beushausen, S.; Vitek, A.; Stamm, S.; Kupfermann, I.; Brosius, J.; Weiss, K. R. The Myomodulin-related Neuropeptides: Characterization of a Gene Encoding a Family of Peptide Cotransmitters in Aplysia. *J. Neurosci.* **1993b,** *13*(8), 3358–6337.

317. Minakata, H.; Kuroki, Y.; Ikeda, T.; Fujisawa, Y.; Nomoto, K.; Kubota, I.; Muneoka, Y. Effects of the Neuropeptide APGW-amide and Related Compounds on Molluscan Muscles—GW-amide Shows Potent Modulatory Effects. *Comp. Biochem. Physiol. C* **1991,** *100*(3), 565–571.

318. Minakata, H. Peptides Containing a D-Amino Acid isolated from Molluscs. *Nippon Kagaku Kaishi* **1996,** *7,* 595–608.

319. Mita, K.; Yamagishi, M.; Fujito, Y.; Lukowiak, K.; Ito, E. An Increase in Insulin is Important for the Acquisition Conditioned Taste Aversion in Lymnaea. *Neurobiol. Learn. Mem.* **2014,** *116,* 132–138.

320. Moran, A. L.; Woods, H. A. Oxygen in Egg Masses: Interactive Effects of Temperature, Age, and Egg-Mass Morphology on Oxygen Supply to Embryos. *J. Exp. Biol.* **2007,** *210,* 722–731.

321. Morgan, P. T.; Perrins, R.; Lloyd, P. E.; Weiss, K. R. Intrinsic and Extrinsic Modulation of a Single Central Pattern Generating Circuit. *J. Neurophysiol.* **2000,** *84*(3), 1186–1193.

322. Morishita, F.; Furukawa, Y.; Matsushima, O. Molecular Cloning of Two Distinct Precursor Genes of NdWFamide, a D-Tryptophan-containing Neuropeptide of the Sea Hare, *Aplysia kurodai. Peptides* **2012,** *38*(2), 291–301.

323. Morishita, F.; Matsushima, O.; Furukawa, Y.; Minakata, H. Deamidase Inactivates a D-Amino Acid-Containing Aplysia Neuropeptide. *Peptides* **2003a,** *24*(1), 45–51.

324. Morishita, F.; Minakata, H.; Sasaki, K.; Tada, K.; Furukawa, Y.; Matsushima, O.; Mukai, S. T.; Saleuddin, A. S. M. Distribution and Function of an Aplysia Cardioexcitatory Peptide, NdWFamide, in Pulmonate Snails. *Peptides* **2003b,** *24*(10), 1533–1544.

325. Morishita, F.; Mukai, S. T.; Saleuddin, A. S. M. Release of Proteins and Polysaccharides from the Albumen Gland of the Freshwater Snail *Helisoma duryi*: Effect of cAMP and Brain Extracts. *J. Comp. Physiol. A* **1998,** *182*(6), 817–825.

326. Morishita, F.; Nakanishi, Y.; Kaku, S.; Furukawa, Y.; Ohta, S.; Hirata, T.; Ohtani, M.; Fujisawa, Y.; Muneoka, Y.; Matsushima, O. A Novel D-Amino-Acid-Containing Peptide Isolated from Aplysia Heart. *Biochem. Biophys. Res. Commun.* **1997**, *240*(2), 354–358.
327. Morishita, F.; Nakanishi, Y.; Sasaki, K.; Kanemaru, K.; Furukawa, Y.; Matsushima, O. Distribution of the Aplysia Cardioexcitatory Peptide, NdWFamide, in the Central and Peripheral Nervous Systems of Aplysia. *Cell Tissue Res.* **2003**, *312*(1), 95–111.
328. Morishita, F.; Sasaki, K.; Kanemaru, K.; Nakanishi, Y.; Matsushima, O.; Furukawa, Y. NdWFamide: A Novel Excitatory Peptide Involved in Cardiovascular Regulation of Aplysia. *Peptides* **2001**, *22*(2), 183–189.
329. Morrill, J. G. Development of the Pulmonate Gastropod. In *Developmental Biology of the Freshwater Invertebrates*; Harrison, F. W.; Cowden, R. R., Eds., Alan R. Liss, Inc.: New York, 1982.
330. Morris, H. R.; Panico, M.; Karplus, A.; Lloyd, P. E.; Riniker, B. Elucidation by FAB-MS of the Structure of a New Cardioactive Peptide from Aplysia. *Nature* **1982**, *300*(5893), 643–645.
331. Mukai, S. T.; Kiehn, L.; Saleuddin, A. S. M. Dopamine Stimulates Snail Albumen Gland Glycoprotein Secretion through the Activation of a D1-like Receptor. *J. Exp. Biol.* **2004a**, *207*, 2507–2518.
332. Mukai, S. T.; Hoque, T.; Morishita, F.; Saleuddin, A. S. M. Cloning and Characterization of a Candidate Nutritive Glycoprotein from the Albumen Gland of the Freshwater Snail, *Helisoma duryi* (Mollusca : Pulmonata). *Invert. Biol.* **2004b**, *123*, 83–92.
333. Muneoka, Y.; Kobayashi, M. Comparative Aspects of Structure and Action of Molluscan Neuropeptides. *Experientia* **1992**, *48*, 448–456.
334. Nagle, G. T.; Knock, S. L.; Painter, S. D.; Blankenship, J. E.; Fritz, R. R.; Kurosky, A. *Aplysia californica* Neurons R3–R14: Primary Structure of the Myoactive Histidine-rich Basic Peptide and Peptide I. *Peptides* **1989b**, *10*(4), 849–857.
335. Nagle, G. T.; Painter, S. D.; Blankenship, J. E. Post-translational processing in Model Neuroendocrine Systems: Precursors and Products that Coordinate Reproductive Activity in Aplysia and Lymnaea. *J. Neurosci. Res.* **1989a**, *23*(4), 359–370.
336. Nagle, G. T.; Painter, S. D.; Blankenship, J. E.; Dixon, J. D.; Kurosky, A. Evidence for the Expression of Three Genes Encoding Homologous Atrial Gland Peptides that Cause Egg Laying in Aplysia. *J. Biol. Chem.* **1986**, *261*(17), 7853–7859.
337. Nagle, G. T.; Painter, S. D.; Kelner, K. L.; Blankenship, J. E. Atrial Gland Cells Synthesize a Family of Peptides that Can Induce Egg Laying in Aplysia. *J. Comp. Physiol. B* **1985**, *156*(1), 43–55.
338. Nagle, G. T.; van Heumen, W. R.; Knock, S. L.; Garcia, A. T.; McCullough, D. A.; Kurosky, A. Occurrence of a Furin-like Prohormone Processing Enzyme in Aplysia Neuroendocrine Bag Cells. *Comp. Biochem. Physiol. B* **1993**, *105*(2), 345–348.
339. Nambu, J. R.; Scheller, R. H. Egg-laying Hormone Genes of Aplysia: Evolution of the ELH Gene Family. *J. Neurosci.* **1986**, *6*(7), 2026–2036.
340. Nambu, J. R.; Taussig, R.; Mahon, A. C.; Scheller, R. H. Gene Isolation with cDNA Probes from Identified Aplysia Neurons: Neuropeptide Modulators of Cardiovascular Physiology. *Cell* **1983**, *35*(1), 47–56.
341. Nassel, D. R. Tachykinin-related Peptides in Invertebrates: A Review. *Peptides* **1999**, *20*, 141–158.

342. Nemes, P.; Knolhoff, A. M.; Rubakhin, S. S.; Sweedler, J. V. Metabolic Differentiation of Neuronal Phenotypes by Single-cell Capillary Electrophoresis-Electrospray Ionization–Mass Spectrometry. *Anal. Chem.* **2011**, *83*, 6810–6817.

343. Newell, P. F.; Skelding, J. M. Structure and Permeability of the Septate Junction in the Kidney Sac of *Helix pomatia* L. *Z. Zellforsch. Mikrosk. Anat.* **1973a**, *147*(1), 31–39.

344. Newell, P. F.; Skelding, J. M. Studies on the Permeability of the Septate Junction in the Kidney of *Helix pomatia* L. *Malacologia* **1973b**, *14*(1–2), 89–91.

345. Nuurai, P.; Poljaroen, J.; Tinikul, Y.; Cummins, S.; Sretarugsa, P.; Hanna, P.; Wanichanon, C.; Sobhon, P. The Existence of Gonadotropin-releasing Hormone-like Peptides in the Neural Ganglia and Ovary of the Abalone, *Haliotis asinina* L. *Acta Histochem.* **2010**, *112*(6), 557–566.

346. Nuurai, P.; Primphon, J.; Seangcharoen, T.; Tinikul, Y.; Wanichanon, C.; Sobhon, P. Immunohistochemical Detection of GnRH-like Peptides in the Neural Ganglia and Testis of *Haliotis asinina*. *Microsc. Res. Tech.* **2014**, *77*(2), 110–119.

347. Oehlmann, J.; Di Benedetto, P.; Tillmann, M.; Duft, M.; Oetken, M.; Schulte-Oehlmann, U. Endocrine Disruption in Prosobranch Molluscs: Evidence and Ecological Relevance. *Ecotoxicology* **2007**, *16*, 29–43.

348. Ohta, N.; Kubota, I.; Takao, T.; Shimonishi, Y.; Yasuda-Kamatani, Y.; Minakata, H.; Nomoto, K.; Muneoka, Y.; Kobayashi, M. Fulicin, A Novel Neuropeptide Containing a D-Amino Acid Residue Isolated from the Ganglia of *Achatina fulica*. *Biochem. Biophys. Res. Commun.* **1991**, *178*(2), 486–493.

349. Okubo, K.; Nagahama, Y. Structural and Functional Evolution of Gonadotropin-releasing Hormone in Vertebrates. *Acta Physiol.* **2008**, *193*(1), 3–15.

350. O'Neill M, Gaume B, Denis F, Auzoux-Bordenave, S. Expression of Biomineralisation Genes in Tissues and Cultured Cells of the Abalone *Haliotis tuberculata*. *Cytotechnology* **2013**, *65*(5), 737–747.

351. Orekhova, I. V.; Alexeeva, V.; Church, P. J.; Weiss, K. R.; Brezina, V. Multiple Presynaptic and Postsynaptic Sites of Inhibitory Modulation by Myomodulin at ARC Neuromuscular Junctions of Aplysia. *J. Neurophysiol.* **2003**, *89*(3), 1488–1502.

352. Ouimet, T.; Mammarbachi, A.; Cloutier, T.; Seidah, N. G.; Castellucci, V. F. cDNA Structure and In Situ Localization of the *Aplysia californica* Pro-hormone Convertase PC2. *FEBS Lett.* **1993**, *330*(3), 343–346.

353. Painter, S. D.; Kalman, V. K.; Nagle, G. T.; Blankenship, J. E. Localization of Immunoreactive Alpha-bag-cell Peptide in the Central Nervous System of Aplysia. *J. Comp. Neurol.* **1989**, *287*(4), 515–530.

354. Park, Y.; Kim, Y. J.; Adams, M. E. Identification of G Protein-coupled Receptors for *Drosophila* PRXamide Peptides, CCAP, Corazonin, and AKH supports a Theory of Ligand–Receptor Coevolution. *Proc. Natl. Acad. Sci. U.S.A.* **2002**, *99*(17), 11423–11428.

355. Perrins, R.; Weiss, K. R. A Cerebral Central Pattern Generator in Aplysia and its Connections with Buccal Feeding Circuitry. *J. Neurosci.* **1996**, *16*(21), 7030–7045.

356. Perry, S. J.; Dobbins, A. C.; Schofield, M. G.; Piper, M. R.; Benjamin, P. R. Small Cardioactive Peptide Gene: Structure, Expression and Mass Spectrometric Analysis Reveals a Complex Pattern of Co-transmitters in a Snail Feeding Neuron. *Eur. J. Neurosci.* **1999**, *11*(2), 655–662.

357. Phares, G. A.; Lloyd, P. E. Immunocytological and Biochemical Localization and Biological Activity of the Newly Sequenced Cerebral Peptide 2 in Aplysia. *J. Neurosci.* **1996**, *16*(24), 7841–7852.

358. Phares, G. A.; Walent, J. H.; Niece, R. L.; Kumar, S. B.; Ericsson, L. H.; Kowalak, J. A.; Lloyd, P. E. Primary Structure of a New Neuropeptide, Cerebral Peptide 2, Purified from Cerebral Ganglia of Aplysia. *Biochemistry* **1996**, *35*(18), 5921–5927.

359. Pinsker, H. M.; Dudek, F. E. Bag Cell Control of Egg Laying in Freely Behaving Aplysia. *Science* **1977**, *197*(4302), 490–493.

360. Pinsker, H. M.; Feinstein, R.; Gooden, B. A. Bradycardial Response in *Aplysia* Exposed to Air. *Fed. Proc.* **1974**, *33*, 361.

361. Ponder, W. F.; Lindberg, D. R. *Phylogeny and Evolution of the Mollusca.* University of California Press: Berkeley, 2008.

362. Potts, W. T. W. Excretion in Molluscs. *Biol. Rev. Camb. Philos. Soc.* **1967**, *42*, 1–41.

363. Price, D. A.; Davis, N. W.; Doble, K. E.; Greenberg, M. J. The Variety and Distribution of the FMRFamide-related Peptide in Molluscs. *Zool. Sci.* **1987**, *4*, 395–410.

364. Price, D. A.; Greenberg, M. J. Structure of a Molluscan Cardioexcitatory Neuropeptide. *Science* **1977**, *197*(4304), 670–671.

365. Price, D. A.; Greenberg, M. J. Pharmacology of the Mollusca Cardioexcitatory Neuropeptide FMRFaimide. *Gen. Pharmacol.* **1979**, *11*, 237–241.

366. Price, D. A.; Lesser, W.; Lee, T. D.; Doble, K. E.; Greenberg, M. J. Seven FMRFamide-related and Two SCP-related Cardioactive Peptides from *Helix. J. Exp. Biol.* **1990**, *154*, 421–437.

367. Puthanveettil, S. V.; Antonov, I.; Kalachikov, S.; Rajasethupathy, P.; Choi, Y. B.; Kohn, A. B.; Citarella, M.; Yu F, Karl, K. A.; Kinet, M.; Morozova, I.; Russo, J. J.; Ju, J.; Moroz, L. L.; Kandel, E. R. A Strategy to Capture and Characterize the Synaptic Transcriptome. *Proc. Natl. Acad. Sci. U.S.A.* **2013**, *110*, 7464–7469.

368. Ram, J. L. Hormonal Control of Reproduction in Busycon: Laying of Egg Capsules Caused by Nervous System Extracts. *Biol. Bull.* **1977**, *152*(2), 221–232.

369. Ram, J. L.; Gallardo, C. S.; Ram, M. L.; Croll, R. P. Reproduction-associated Immunoreactive Peptides in the Nervous Systems of Prosobranch Gastropods. *Biol. Bull.* **1998**, *195*(3), 308–318.

370. Ram, J. L.; Ram, M. L.; Davis, J. P. Hormonal Control of Reproduction in Busycon: II. Laying of Egg-containing Capsules Caused by Nervous System Extracts and Further Characterization of the Substance Causing Egg Capsule Laying. *Biol. Bull.* **1982**, *162*, 360–370.

371. Redman, R. S.; Berry, R. W. ProELH-related Peptides: Influence on Bag Cell cAMP Levels. *Brain Res. Mol. Brain Res.* **1992**, *15*(3–4), 216–220.

372. Redman, R. S.; Berry, R. W. Temperature-dependent Stimulation and Inhibition of Adenylate Cyclase by Aplysia Bag Cell Peptides. *Brain Res. Mol. Brain Res.* **1993**, *17*(3–4), 245–250.

373. Reich, G.; Doble, K. E.; Price, D. A.; Greenberg, M. J. Effects of Cardioactive Peptides on Myocardial cAMP levels in the Snail *Helix aspersa. Peptides* **1997a**, *18*(3), 355–360.

374. Reich, G.; Doble, K. E.; Greenberg, M. J. Protein Phosphorylation in Snail Cardiocytes Stimulated with Molluscan Peptide SCPb. *Peptides* **1997b**, *18*(9), 1311–1314.

375. Rittenhouse, A. R.; Price, C. H. Anatomical and Electrophysiological Study of Multitransmitter Neuron R14 of Aplysia. *J. Comp. Neurol.* **1986**, *247*(4), 447–456.

376. Rittenhouse, A. R.; Price, C. H. Electrophysiological and Anatomical Identification of the Peripheral Axons and Target Tissues of Aplysia Neurons R3–14 and their Status as Multifunctional, Multimessenger Neurons. *J. Neurosci.* **1986**, *6*, 2071–2084.

377. Roberts, L.; Janovy, Jr, J.; Nadler, S. In *Foundations of Parasitology,* 9th ed. McGraw-Hill, 2013.

378. Roch, G. J.; Busby, E. R.; Sherwood, N. M. Evolution of GnRH: Diving Deeper. *Gen. Comp. Endocrinol.* **2011,** *171*(1), 1–16.

379. Rodet, F.; Lelong, C.; Dubos, M. P.; Costil, K.; Favrel, P. Molecular cloning of a Molluscan Gonadotropin-releasing Hormone Receptor Orthologue Specifically Expressed in the Gonad. *Biochim. Biophys. Acta* **2005,** *1730*(3), 187–195.

380. Rodet, F.; Lelong, C.; Dubos, M. P.; Favrel, P. Alternative Splicing of a Single Precursor mRNA Generates Two Subtypes of Gonadotropin-releasing Hormone Receptor Orthologues and Their Variants in the Bivalve Mollusc *Crassostrea gigas*. *Gene* **2008,** *414*(1–2), 1–9.

381. Roovers, E.; Vincent, M.; van Kesteren, E.; Geraerts, W.; Planta, R.; Vreugdenhil, E.; van Heerikhuizen, H. Characterization of a Putative Molluscan Insulin-related Peptide Receptor. *Gene* **1995,** *162*, 181–188.

382. Rosen, S. C.; Teyke, T.; Miller, M. W.; Weiss, K. R.; Kupfermann, I. Identification and Characterization of Cerebral-to-buccal Interneurons Implicated in the Control of Motor Programs associated with feeding in Aplysia. *J. Neurosci.* **1991,** *11*(11), 3630–3655.

383. Rothman, B. S.; Hawke, D. H.; Brown, R. O.; Lee, T. D.; Dehghan, A. A.; Shively, J. E.; Mayeri, E. Isolation and Primary Structure of the Califins, Three Biologically Active Egg-laying Hormone-like Peptides from the Atrial Gland of *Aplysia californica*. *J. Biol. Chem.* **1986,** *261*(4), 1616–1623.

384. Rothman, B. S.; Mayeri, E.; Brown, R. O.; Yuan, P. M.; Shively, J. E. Primary Structure and Neuronal Effects of Alpha-bag Cell Peptide, A Second Candidate Neurotransmitter Encoded by a Single Gene in Bag Cell Neurons of Aplysia. *Proc. Natl. Acad. Sci. U.S.A.* **1983a,** *80*(18), 5753–5757.

385. Rothman, B. S.; Weir, G.; Dudek, F. E. Egg-laying Hormone: Direct Action on the Ovotestis of Aplysia. *Gen. Comp. Endocrinol.* **1983b,** *52*(1), 134–141.

386. Roubos, E. W.; Geraerts, W. P. M.; Boerrigter, G. H.; van Kampen, G. P. Control of the Activities of the Neurosecretory Light Green and Caudo-Dorsal Cells and of the Endocrine Dorsal Bodies by the Lateral Lobes in the Freshwater Snail *Lymnaea stagnalis* (L.). *Gen. Comp. Endocrinol.* **1980,** *40*(4), 446–454.

387. Roubos, E. W.; van der Wal-Divendal, R. M. Sensory Input to Growth Stimulating Neuroendocrine Cells of *Lymnaea stagnalis*. *Cell Tissue Res.* **1982,** *227*(2), 371–386.

388. Rozsa, K. S. Analysis of the Neural Network Regulating the Cardio-renal System in the Central Nervous System of *Helix pomatia* L. *Amer. Zool.* **1979,** *19*, 117–128.

389. Rozsa, K. S.; Salanki, J.; Vero, M.; et al. Neural Network Regulating Heart Activity in *Aplysia depilans* and Its Comparison with Other Gastropod Species. *Comp. Biochem. Physiol. A* **1980,** *65*, 61–68.

390. Runham, N. W. Mollusca. Accessory Sex Glands. In *Reproductive Biology of Invertebrates*; Adiyodi, K. G., Adiyodi, R. G., Eds. John Wiley and Sons Ltd.: Chichester, 1983.

391. Russell-Hunter, W. D. *Mollusca: Ecology*; Academic Press, Orlando, 1983, Vol 6.

392. Saavedra, J. M.; Juorio, A. V.; Shigematsu, K.; Pinto, J. E. Specific Insulin Binding Sites in Snail (*Helix aspersa*) Ganglia. *Cell. Mol. Neurobiol.* **1989,** *9*(2), 273–279.

393. Safavi-Hemami, H.; Gajewiak, J.; Karanth, S.; Robinson, S. D.; Ueberheide, B.; Douglass, A. D.; Schlegel, A.; Imperial, J. S.; Watkins, M.; Bandyopadhyay, P. K.; Yandell, M.; Li, Q.; Purcell, A. W.; Norton, R. S.; Ellgaard, L.; Olivera, B. M. Specialized Insulin is Used for Chemical Warfare by Fish-hunting Cone Snails. *Proc. Natl. Acad. Sci. U.S.A.* **2015,** *112*(6), 1743–1748.

394. Saitongdee, P.; Apisawetakan, S.; Anunruang, N.; Poomthong, T.; Hanna, P.; Sobhon, P. Egg-laying-hormone Immunoreactivity in the Neural Ganglia and Ovary of *Haliotis asinina* Linnaeus. *Invert. Neurosci.* **2005**, *5*(3–4), 165–172.
395. Saleuddin, A. S. M.; Khan, H. R.; Ashton, M. L.; Griffond, B. Immunocytochemical Localization of FMRFamide in the Central Nervous System and the Kidney ff *Helisoma duryi* (Mollusca): Its Possible Antidiuretic Role. *Tissue Cell* **1992**, *24*, 179–189.
396. Saleuddin, A. S. M.; Kunigelis, S. C. Neuro-endocrine Control Mechanisms in Shell Formation. *Am. Zool.* **1984**, *24*, 911–916.
397. Saleuddin, A. S. M.; Mukai, S. T.; Khan, H. R. Hormonal Control of Reproduction and Growth in the Freshwater Snail *Helisoma duryi* (Mollusca: Pulmonata). In *Neurobiology and Endocrinology of Selected Invertebrates*; Loughton, B. G.; Saleuddin, A. S. M., Eds. Captus Press: Toronto, 1990.
398. Saleuddin, A. S. M.; Mukai, S. T.; Almeida, K.; Hatiras, G. Membrane Transduction Pathway in the Neuronal Control of Protein Secretion by the Albumen Gland in *Helisoma* (Mollusca). *Acta Biol. Hung.* **2000**, *51*(2–4), 243–253.
399. Saleuddin, A. S. M.; Petit, H. P. The Mode of Formation and the Structure of the Periostracum. In *The Mollusca. Vol. 4. Physiology Part 1*; Saleuddin, A. S. M.; Wilbur, K. M., Eds.; Academic Press, Inc.: New York, 1983.
400. Salimova, N. B.; Sakharov, D. A.; Milosevic, I.; Turpaev, T. M.; Rakic, L. Monoamine-containing Neurons in the Aplysia Brain. *Brain Res.* **1987**, *400*(2), 285–299.
401. Santama, N.; Benjamin, P. R. Gene Expression and Function of FMRFamide-related Neuropeptides in the Snail Lymnaea. *Microsc. Res. Tech.* **2000**, *49*(6), 547–556.
402. Santama, N.; Benjamin, P. R.; Burke, J. F. Alternative RNA Splicing Generates Diversity of Neuropeptide Expression in the Brain of the Snail Lymnaea: In Situ Analysis of Mutually Exclusive Transcripts of the FMRFamide Gene. *Eur. J. Neurosci.* **1995**, *7*(1), 65–76.
403. Santhanagopalan, V.; Yoshino, T. P. Monoamines and their Metabolites in the Freshwater Snail *Biomphalaria glabrata*. *Comp. Biochem. Physiol. A* **2000**, *125*, 469–478.
404. Sasaki, K.; Morishita, F.; Furukawa, Y. Peptidergic Innervation of the Vasoconstrictor Muscle of the Abdominal Aorta in *Aplysia kurodai*. *J. Exp. Biol.* **2004**, *207*(Pt. 25), 4439–4450.
405. Satake, H.; Yasuda-Kamatani, Y.; Takuwa, K.; Nomoto, K.; Minakata, H.; Nagahama, T.; Nakabayashi, K.; Matsushima, O. Characterization of a cDNA Encoding a Precursor Polypeptide of a D-Amino Acid-containing Peptide, Achatin-I and Localized Expression of the Achatin-I and Fulicin Genes. *Eur. J. Biochem.* **1999**, *261*(1), 130–136.
406. Sawada, M.; McAdoo, D. J.; Blankenship, J. E.; Price, C. H. Modulation of Arterial Muscle Contraction in Aplysia by Glycine and Neuron R14. *Brain Res.* **1981**, *207*(2), 486–490.
407. Scemes, E.; Salomão, L. C.; McNamara, J. C.; Cassola, A. C. Lack of Osmoregulation in *Aplysia brasiliana*: Correlation with Response of Neuron R15 to Osphradial Stimulation. Am. J. Physiol. **1991**, *260*(4 Pt. 2), R777–R784.
408. Schallig, H. D. F. H.; Sassen, M. J.; De Jong-Brink, M. In Vitro Release of the Anti-gonadotropic Hormone, Schistosomin, from the Central Nervous System of *Lymnaea stagnalis* is Induced with a Methanolic Extract of Cercariae of *Trichobilharzia ocellata*. *Parasitology* **1992**, *104*, 309–314.
409. Schally, A. V. Aspects of Hypothalamic Regulation of the Pituitary Gland. *Science* **1978**, *202*(4363), 18–28.

410. Scheller, R. H.; Jackson, J. F.; McAllister, L. B.; Rothman, B. S.; Mayeri, E.; Axel, R. A Single Gene Encodes Multiple Neuropeptides Mediating a Stereotyped Behavior. *Cell* **1983**, *32*(1), 7–22.

411. Scheller, R. H.; Jackson, J. F.; McAllister, L. B.; Schwartz, J. H.; Kandel, E. R.; Axel, R. A Family of Genes that Codes for ELH, a Neuropeptide Eliciting a Stereotyped Pattern of Behavior in Aplysia. *Cell* **1982**, *28*(4), 707–719.

412. Schmidt, E. D.; Roubos, E. W. Morphological Basis for Nonsynaptic Communication within the Central Nervous System by Exocytotic Release of Secretory Material from the Egg-laying Stimulating Neuroendocrine Caudodorsal Cells of *Lymnaea stagnalis*. *Neuroscience* **1987**, *20*, 247–257.

413. Schmidt, E. D.; Roubos, E. W. Quantitative Immunoelectron Microscopy and Tannic Acid Study of Dynamics of Neurohaemal and Non-synaptic Peptide Release by the Caudodorsal Cells of *Lymnaea stagnalis*. *Brain Res.* **1989**, *489*(2), 325–337.

414. Schneider, L. E.; Taghert, P. H. Isolation and Characterization of a *Drosophila* gene that Encodes Multiple Neuropeptides Related to Phe-Met-Arg-Phe-NH2 (FMRFamide). *Proc. Natl. Acad. Sci. U.S.A.* **1988**, *85*(6), 1993–1997.

415. Schot, L. P. C.; Boer, H. H.; Swaab, D. F.; Van Noorden, S. Immunocytochemical Demonstration of Peptidergic Neurons in the Central Nervous System of the Pond Snail *Lymnaea stagnalis* with Antisera Raised to Biologically Active Peptides of Vertebrates. *Cell Tissue Res.* **1981**, *216*(2), 273–291.

416. Seaman, R. L.; Lynch, M. J.; Moss, R. L. Effects of Hypothalamic Peptide Hormones on the Electrical Activity of Aplysia Neurons. *Brain Res. Bull.* **1980**, *5*(3), 233–237.

417. Senatore, A.; Edirisinghe, N.; Katz, P. S. Deep mRNA sequencing of the *Tritonia diomedea* Brain Transcriptome Provides Access to Gene Homologues for Neuronal Excitability, Synaptic Transmission and Peptidergic Signalling. *PLoS ONE* **2015**, *10*(2), e0118321.

418. Sevala, V. M.; Sevala, V. L.; Kunigelis, S. C.; Saleuddin, A. S. M. Circadian Timing of a Daily Rhythm of Hemolymph Insulin-like Peptide Titers in *Helisoma* (Mollusca). *J. Exp. Zool.* **1993a**, *266*, 221–226.

419. Sevala, V. M.; Sevala, V. L.; Saleuddin,, A. S. M. Hemolymph Insulin-like Peptides (ILP) Titers and the Influence of ILP and Mammalian Insulin on the Amino Acid Incorporation in the Mantle Collar In Vitro in *Helisoma* (Mollusca). *Biol. Bull.* **1993b**, *185*, 140–148.

420. Shartau, R. B.; Tam, R.; Patrick, S.; Goldberg, J. I. Serotonin Prolongs Survival of Encapsulated Pond Snail Embryos Exposed to Long-term Anoxia. *J. Exp. Biol.* **2010**, *213*, 1529–1535.

421. Shoppe, S. B.; McPherson, D.; Rock, M. K.; Blankenship, J. E. Functional and Morphological Evidence for the Existence of Neurites from Abdominal Ganglion Bag Cell Neurons in the Head-ring Ganglia of Aplysia. *J. Comp. Physiol. A* **1991**, *168*(5), 539–552.

422. Shyamala, M.; Fisher, J. M.; Scheller, R. H. A Neuropeptide Precursor Expressed in Aplysia Neuron L5. *DNA* **1986**, *5*(3), 203–208.

423. Sigvardt, K. A.; Rothman, B. S.; Brown, R. O.; Mayeri, E. The bag cells of Aplysia as a Multitransmitter System: Identification of Alpha Bag Cell Peptide as a Second Neurotransmitter. *J. Neurosci.* **1986**, *6*(3), 803–813.

424. Skelton, M.; Alevizos, A.; Koester, J. Control of the Cardiovascular System of Aplysia by Identified Neurons. *Experientia* **1992**, *48*(9), 809–817.

425. Skelton, M. E.; Koester, J. The Morphology, Innervation and Neural Control of the Anterior Arterial System of *Aplysia californica*. *J. Comp. Physiol. A* **1992**, *171*(2), 141–155.

426. Skinner, T. L.; Peretz, B. Age Sensitivity of Osmoregulation and of Its Neural Correlates in Aplysia. *Am. J. Physiol.* **1989**, *256*(4 Pt. 2), R989–996.

427. Smit, A. B.; Geraerts, W. P. M.; Meester, I.; van Heerikhuizen, H.; Joosse, J. Characterization of a cDNA Clone Encoding Molluscan Insulin-related Peptide II of *Lymnaea stagnalis*. *Eur. J. Biochem.* **1991**, *199*(3), 699–703.

428. Smit, A. B.; Hoek, R. M.; Geraerts, W. P. M. The Isolation of a cDNA Encoding a Neuropeptide Prohormone from the Light Yellow Cells of *Lymnaea stagnalis*. *Cell. Mol. Neurobiol.* **1993a**, *13*(3), 263–270.

429. Smit, A. B.; Spijker, S.; Van Minnen, J.; Burke, J. F.; De Winter, F.; Van Elk, R.; Geraerts, W. P. M. Expression and Characterization of Molluscan Insulin-related Peptide VII from the Mollusc *Lymnaea stagnalis*. *Neuroscience* **1996**, *70*(2), 589–596.

430. Smit, A. B.; Thijsen, S. F.; Geraerts, W. P. cDNA Cloning of the Sodium-influx-stimulating Peptide in the Mollusc, *Lymnaea stagnalis*. *Eur. J. Biochem.* **1993b**, *215*(2), 397–400.

431. Smit, A. B.; Thijsen, S. F.; Geraerts, W. P. M.; Meester, I.; van Heerikhuizen, H.; Joosse, J. Characterization of a cDNA Clone Encoding Molluscan Insulin-related Peptide V of *Lymnaea stagnalis*. *Brain Res. Mol. Brain Res.* **1992**, *14*(1–2), 7–12.

432. Smit, A. B.; van Kesteren, R. E.; Li, K. W.; Van Minnen, J.; Spijker, S.; Van Heerikhuizen, H.; Geraerts, W. P. M. Towards Understanding the Role of Insulin in the Brain: Lessons from Insulin-Related Signaling Systems in the Invertebrate Brain. *Prog. Neurobiol.* **1998**, *54*, 35–54.

433. Smit, A. B.; van Marle, A.; van Elk, R. Bogerd, J. van Heerikhuizen, H. Geraerts, W. P. M. Evolutionary Conservation of the Insulin Gene Structure in Invertebrates: Cloning of the Gene Encoding Molluscan Insulin-related Peptide III from *Lymnaea stagnalis*. *J. Mol. Endocrinol.* **1993c**, *11*(1), 103–113.

434. Smit, A.; Vreugdenhil, E.; Ebberink, R.; Geraerts, W.; Klootwijk, J.; Joosse, J. Growth-controlling Molluscan Neurons Produce the Precursor of an Insulin-related Peptide. *Nature* **1988**, *331*(6156), 535–538.

435. Smock, T.; Albeck, D.; Stark, P. A Peptidergic Basis for Sexual Behavior in Mammals. *Prog. Brain Res.* **1998**, *119*, 467–481.

436. Soffe, S. R.; Slade, C. T.; Benjamin, P. R. Environmental Osmolarity and Neurosecretory Neurones in *Lymnaea stagnalis* (L.). *Malacologia* **1979**, *18*(1–2), 583–586.

437. Song, Y.; Liu, Y. M. Quantitation of cardioexcitatory Asn-D-Trp-Phe-NH2 diastereomers in Aplysia's Central Nervous System by Nanoscale Liquid Chromatography–Tandem Mass Spectrometry. *J. Mass Spectrom.* **2008**, *43*(9), 1285–1290.

438. Sonetti, D.; van Heumen, W. R.; Roubos, E. W. Light- and Electron-microscopic Immunocytochemistry of a Molluscan Insulin-related Peptide in the Central Nervous System of *Planorbarius corneus*. *Cell Tissue Res.* **1992**, *267* (3), 473–481.

439. Sonetti, D.; Bianchi, F. Occurrence and Distribution of Insulin Receptor-like Immunoreactivity in Molluscan Brains. *Acta Biol. Hung.* **1993**, *44*(1), 77–82.

440. Sossin, W. S.; Chen, C. S.; Toker, A. Stimulation of an Insulin Receptor Activates and Down-Regulates the Ca2+-independent Protein Kinase C, Apl II, through a Wortmannin-sensitive Signaling Pathway in Aplysia. *J. Neurochem.* **1996**, *67*(1), 220–228.

441. Sossin, W. S.; Fisher, J. M.; Scheller, R. H. Sorting within the Regulated Secretory Pathway Occurs in the Trans-Golgi Network. *J. Cell Biol.* **1990a**, *110*(1), 1–12.

442. Sossin, W. S.; Sweet-Cordero, A.; Scheller, R. H. Dale's Hypothesis Revisited: Different Neuropeptides Derived from a Common Prohormone are Targeted to Different Processes. *Proc. Natl. Acad. Sci. U.S.A.* **1990b,** *87*(12), 4845–4848.

443. Spencer, G. E.; Syed, N. I.; van Kesteren, R. E.; Lukowiak, K.; Geraerts, W. P. M.; van Minnen, J. Synthesis and Functional Integration of a Neurotransmitter Receptor in Isolated Invertebrate Axons. *J. Neurobiol.* **2000,** *44*(1), 72–81.

444. Stewart, M. J.; Favrel, P.; Rotgans, B. A.; Wang, T.; Zhao, M.; Sohail, M.; O'Connor, W. A.; Elizur, A.; Henry, J.; Cummins, S. F. Neuropeptides Encoded by the Genomes of the Akoya Pearl Oyster *Pinctata fucata* and Pacific Oyster *Crassostrea gigas*: A Bioinformatic and Peptidomic Survey. *BMC Genomics* **2014,** *15*, 840.

445. Stone, J. V.; Mordue, W.; Batley, K. E.; Morris, H. R. Structure of Locust Adipokinetic Hormone, a Neurohormone that Regulates Lipid Utilisation During Flight. *Nature* **1976,** *263*(5574), 207–211.

446. Stoof, J. C.; De Vlieger, T. A.; Lodder, J. C. Opposing Roles for D-1 and D-2 Dopamine Receptors in regulating the Excitability of Growth Hormone-producing Cells in the Snail *Lymnaea stagnalis. Eur. J. Pharmacol.* **1984,** *106*(2), 431–435.

447. Strong, J. A.; Fox, A. P.; Tsien, R. W.; Kaczmarek, L. K. Stimulation of Protein Kinase C Recruits Covert Calcium Channels in Aplysia Bag Cell Neurons. *Nature* **1987,** *325*(6106), 714–717.

448. Strong, J. A.; Kaczmarek, L. K. Multiple Components of Delayed Potassium Current in Peptidergic Neurons of Aplysia: Modulation by an Activator of Adenylate Cyclase. *J. Neurosci.* **1986,** *6*(3), 814–822.

449. Sun, B.; Kavanaugh, S. I.; Tsai, P. S. Gonadotropin-releasing Hormone in Protostomes: Insights from Functional Studies on *Aplysia californica. Gen. Comp. Endocrinol.* **2012,** *176*(3), 321–326.

450. Sun, B.; Tsai, P. S. A gonadotropin-releasing Hormone-like Molecule Modulates the Activity of Diverse Central Neurons in a Gastropod Mollusk, *Aplysia californica. Front Endocrinol (Lausanne)* **2011,** *2*, 36.

451. Sun, J.; Zhang, H.; Wang, H.; Heras, H.; Dreon, M. S.; Ituarte, S.; Ravasi, T.; Qian, P. Y.; Qiu, J. W. First Proteome of the Egg Perivitelline Fluid of a Freshwater Gastropod with Aerial Oviposition. *J. Proteome. Res.* **2012,** *11*(8), 4240–4248.

452. Sweedler, J. V.; Li, L.; Rubakhin, S. S.; Alexeeva, V.; Dembrow, N. C.; Dowling, O.; Jing, J.; Weiss, K. R.; Vilim, F. S. Identification and Characterization of the Feeding Circuit-activating Peptides, a Novel Neuropeptide Family of Aplysia. *J. Neurosci.* **2002,** *22*(17), 7797–7808.

453. Takeuchi, H.; Emaduddin, M.; Araki, Y.; Zhang, W.; Han, X. Y.; Salunga, T. L.; Wong, S. M. Further Study on the Effects of achatin-I, an Achatina Endogenous Neuroexcitatory Tetrapeptide Having a D-Phenylalanine Residue, on *Achatina neurones. Acta Biol. Hung.* **1995,** *46*(2–4), 395–400.

454. Tam, A. K.; Gardam, K. E.; Lamb, S.; Kachoei, B. A.; Magoski, N. S. Role for Protein Kinase C in Controlling Aplysia Bag Cell Neuron Excitability. *Neuroscience* **2011,** *179*, 41–55.

455. Taussig, R.; Scheller, R. H. The Aplysia FMRFamide Gene Encodes Sequences Related to Mammalian Brain Peptides. *DNA* **1986,** *5*(6), 453–461.

456. Tensen, C. P.; Cox, K. J.; Smit, A. B.; van der Schors, R. C.; Meyerhof, W.; Richter, D.; Planta, R. J.; Hermann, P. M.; van Minnen, J.; Geraerts, W. P.; Knol, J. C.; Burke, J. F.; Vreugdenhil, E.; van Heerikhuizen, H. The Lymnaea Cardioexcitatory Peptide

(LyCEP) Receptor: a G-protein-coupled Receptor for a Novel Member of the RFamide Neuropeptide Family. *J. Neurosci.* **1998,** *18*(23), 9812–9821.

457. Ter Maat, A. Egg Laying in the Hermaphrodite Pond Snail *Lymnaea stagnalis. Prog. Brain Res.* **1992,** *92*, 345–360.

458. Ter Maat, A.; Lodder, J. C.; Veenstra, J.; Goldschmeding, J. T. Suppression of Egg-laying During Starvation in the Snail *Lymnaea stagnalis* by Inhibition of the Ovulation Hormone Producing Caudo-Dorsal Cells. *Brain Res.* **1982,** *239*, 535–542.

459. Thorndyke, M. C.; Goldsworthy, G. J. *Neurohormones in Invertebrates.* Society for Experimental Biology Seminar Series 33, Cambridge University Press: Cambridge, 1988.

460. Tompa, A. S.; Verdonk, N. H.; van den Biggelaar, J. A. M. *The Mollusca. Reproduction.* Academic Press Inc.: Orlando, 1984, Vol 7.

461. Torres, A. M.; Tsampazi, M.; Kennett, E. C.; Belov, K.; Geraghty, D. P.; Bansal, P. S.; Alewood, P. F.; Kuchel, P. W. Characterization and Isolation of L-to-D-Amino-Acid-Residue Isomerase from Platypus Venom. *Amino Acids* **2007,** *32*(1), 63–68.

462. Tsai, P. S. Gonadotropin-releasing Hormone in Invertebrates: Structure, Function, and Evolution. *Gen. Comp. Endocrinol.* **2006,** *148*(1), 48–53.

463. Tsai, P. S.; Maldonado, T. A.; Lunden, J. B. Localization of Gonadotropin-releasing Hormone in the Central Nervous System and a Peripheral Chemosensory Organ of *Aplysia californica. Gen. Comp. Endocrinol.* **2003,** *130*(1), 20–28.

464. Tsai, P. S.; Sun, B.; Rochester, J. R.; Wayne, N. L. Gonadotropin-Releasing Hormone-like Molecule is Not an Acute Reproductive Activator in the Gastropod, *Aplysia californica. Gen. Comp. Endocrinol.* **2010,** *166*(2), 280–288.

465. Ukena, K.; Iwakoshi, E.; Minakata, H.; Tsutsui, K. A Novel Rat Hypothalamic RFamide-related Peptide Identified by Immunoaffinity Chromatography and Mass Spectrometry. *FEBS Lett.* **2002,** *512*(1–3), 255–258.

466. Van Golen, F. A.; Li, K. W.; De Lange, R. P. J.; Van Kesteren, R. E.; Van Der Schors, R. C.; Geraerts, W. P. M. Co-localized Neuropeptides Conopressin and ALA-PRO-GLY-TRP-NH$_2$ have Antagonistic Effects on the Vas Deferens of Lymnaea. *Neuroscience* **1995a,** *69*, 1275–1287.

467. Van Golen, F. A.; Li, K. W.; de Lange, R. P. J.; Jespersen, S.; Geraerts, W. P. M. Mutually Exclusive Neuronal Expression of Peptides Encoded by the FMRFa Gene Underlies a Differential Control of Copulation in Lymnaea. *J. Biol. Chem.* **1995b,** *270*, 28487–28493.

468. Van Golen, F. A.; Li, K. W.; Chen, S.; Jiménez, C. R.; Geraerts, W. P. M. Various isoforms of Myomodulin Identified from the Male Copulatory Organ of Lymnaea Show Overlapping Yet Distinct Modulatory Effects on the Penis Muscle. *J. Neurochem.* **1996,** *66*, 321–329.

469. Van Heumen, W. R.; Roubos, E. W. Immuno-electron Microscopy of Sorting and Release of Neuropeptides in *Lymnaea stagnalis. Cell Tissue Res.* **1991,** *264*, 185–195.

470. Van Kesteren, R. E.; Smit, A. B.; Dirks, R. W.; de With, N. D.; Geraerts, W. P. M.; Joosse, J. Evolution of the Vasopressin/Oxytocin Superfamily: Characterization of a cDNA Encoding a Vasopressin-related Precursor, Preproconopressin, from the Mollusc *Lymnaea stagnalis. Proc. Natl. Acad. Sci. U.S.A.* **1992a,** *89*, 4593–4597.

471. Van Kesteren, R. E.; Smit, A. B.; de With, N. D.; van Minnen, J.; Dirks, R. W.; van der Schors, R. C.; Joosse, J. A Vasopressin-related Peptide in the Mollusc *Lymnaea stagnalis*: Peptide Structure, Prohormone Organization, Evolutionary and Functional Aspects of Lymnaea Conopressin. *Prog. Brain Res.* **1992b,** *92*, 47–57.

472. Van Kesteren, R. E.; Tensen, C. P.; Smit, A. B.; van Minnen, J.; van Soest, P. F.; Kits, K. S.; Meyerhof, W.; Richter, D.; van Heerikhuizen, H.; Vreugdenhil, E.; et al., A Novel G Protein-coupled Receptor Mediating both Vasopressin- and Oxytocin-like Functions of Lys-conopressin in *Lymnaea stagnalis*. *Neuron* **1995**, *15*(4), 897–908.

473. Van Minnen, J. Axonal Localization of Neuropeptide-encoding mRNA in Identified Neurons of the Snail *Lymnaea stagnalis*. *Cell Tissue Res.* **1994**, *276*(1), 155–1561.

474. Van Minnen, J.; Bergman. J. J.; Van Kesteren, E. R.; Smit, A. B.; Geraerts, W. P. M.; Lukowiak, K.; Hasan, S. U.; Syed, N. I. De Novo Protein Synthesis in Isolated Axons of Identified Neurons. *Neuroscience* **1997**, *80*(1):1–7.

475. van Tol-Steye, H.; Lodder, J. C.; Mansvelder, H. D.; Planta, R. J.; van Heerikhuizen, H.; Kits, K. S. Roles of G-Protein Beta Gamma, Arachidonic Acid, and Phosphorylation Inconvergent Activation of an S-like Potassium Conductance by Dopamine, Ala-Pro-Gly-Trp-NH$_2$, and Phe-Met-Arg-Phe-NH$_2$. *J. Neurosci.* **1999**, *19*(10), 3739–3751.

476. van Tol-Steye, H.; Lodder, J. C.; H. D.; Planta, R. J.; van Heerikhuizen, H.; Kits, K. S. Convergence of Multiple G-protein-coupled Receptors onto a Single Type of Potassium Channel. *Brain Res.* **1997**, *777*(1–2), 119–130.

477. Veenema, A. H.; Neumann, I. D. Central Vasopressin and Oxytocin Release: Regulation of Complex Social Behaviours. *Prog. Brain Res.* **2008**, *170*, 261–276.

478. Veenstra, J. A. Isolation and Structure of Corazonin, a Cardioactive Peptide from the American Cockroach. *FEBS Lett.* **1989**, *250*, 231–234.

479. Veenstra, J. A. Neurohormones and Neuropeptides Encoded by the Genome of *Lottia gigantea*, with Reference to Other Molluscs and Insects. *Gen. Comp. Endocrinol.* **2010**, *167*, 86–103.

480. Vetter, I.: Lewis, R. J. Therapeutic Potential of Cone Snail Venom Peptides (conopeptides). *Curr. Top. Med. Chem.* **2012**, *12*, 1546–1552.

481. Vilim, F. S. The Enterins: A Novel Family of Neuropeptides Isolated from the Enteric Nervous System and CNS of Aplysia. *J. Neurosci.* **2001**, *21*, 8247–8261.

482. Vilim, F. S.; Cropper, E. C.; Price, D. A.; Kupfermann, I.; Weiss, K. R. Release of Peptide Cotransmitters in Aplysia: Regulation and Functional Implications. *J. Neurosci.* **1996a**, *16*, 8105–8114.

483. Vilim, F. S.; Cropper, E. C.; Price, D. A.; Kupfermann, I.; Weiss, K. R. Peptide Cotransmitter Release from Motorneuron B16 in *Aplysia californica*: Costorage, Corelease, and Functional Implications. *J. Neurosci.* **2000**, *20*, 2036–2042.

484. Vilim, F. S.; Price, D. A.; Lesser, W.; Kupfermann, I.; Weiss, K. R. Costorage and Corelease of Modulatory Peptide Cotransmitters with Partially Antagonistic Actions on the Accessory Radula Closer Muscle of *Aplysia californica*. *J. Neurosci.* **1996,b**, *16*, 8092–8104.

485. Voronezhskaya, E. E.; Hiripi, L.; Elekes, K.; Croll, R. P. Development of Catecholaminergic Neurons in the Pond Snail, *Lymnaea stagnalis*: I. Embryonic Development of Dopamine-containing Neurons and Dopamine-dependent Behaviors. *J. Comp. Neurol.* **404**, 285–296.

486. Vreugdenhil, E.; Jackson, J. F.; Bouwmeester, T.; Smit, A. B.; Van Minnen, J.; Van Heerikhuizen, H.; Klootwijk, J.; Joosse, J. Isolation, Characterization, and Evolutionary Aspects of a cDNA Clone Encoding Multiple Neuropeptides Involved in the Stereotyped Egg-laying Behavior of the Freshwater Snail *Lymnaea stagnalis*. *J. Neurosci.* **1988**, *8*, 4184–4191.

487. Wang, L.; Hanna, P. J. Isolation, Cloning and Expression of Agene Encoding an Egg-laying Hormone of the Blacklip Abalone (*Haliotis rubra* Leach). *J Shellfish Res.* **1998**, *17*, 785–793.

488. Wayne, N. L.; Frumovitz, M. Calcium Influx Following Onset of Electrical Afterdischarge is not Required for Hormone Secretion from Neuroendocrine Cells of Aplysia. *Endocrinology* **1995**, *136*, 369–372.

489. Wayne, N. L.; Kim, J.; Lee, E. Prolonged Hormone Secretion from Neuroendocrine Cells of Aplysia is Independent of Extracellular Calcium. *J. Neuroendocrinol.* **1998**, *10*, 529–537.

490. Wayne, N. L.; Lee, W.; Kim, Y. J. Persistent Activation of Calcium-activated and Calcium-independent Protein Kinase C in Response to Electrical Afterdischarge from Peptidergic Neurons of Aplysia. *Brain Res.* **1999**, *834*, 211–213.

491. Wayne, N. L.; Lee, W.; Michel, S.; Dyer, J.; Sossin, W. S. Activity-dependent Regulation of Neurohormone Synthesis and Its Impact on Reproductive Behavior in Aplysia. *Biol. Reprod.* **2004**, *70*, 277–281.

492. Weiss, K. R.; Bayley, H.; Lloyd, P. E.; Tenenbaum. R.; Kolks, M. A.; Buck, L.; Cropper, E. C.; Rosen, S. C.; Kupfermann, I. Purification and Sequencing of Neuropeptides Contained in Neuron R15 of *Aplysia californica*. *Proc. Natl. Acad. Sci. U.S.A.* **1989**, *86*, 2913–2917.

493. Weiss, K. R.; Brezina, V.; Cropper, E. C.; Heierhorst, J.; Hooper, S. L.; Probst, W. C.; Rosen, S. C.; Vilim, F. S.; Kupfermann, I. Physiology and Biochemistry of Peptidergic Cotransmission in Aplysia. *J. Physiol. Paris* **1993**, *87*, 141–151.

494. Weiss, K. R.; Brezina, V.; Cropper, E. C.; Hooper, S. L.; Miller, M. W.; Probst, W. C.; Vilim, F. S. Kupfermann, I. Peptidergic Co-transmission in Aplysia: Functional Implications for Rhythmic Behaviors. *Experientia* **1992**, *48*, 456–463.

495. Wendelaar-Bonga, S. E. Ultrastructure and Histochemistry of Neurosecretory Cells and Neurohaemal Areas in the Pond Snail *Lymnaea stagnalis* (L.). *Z. Zellforsch.* **1970**, *108*, 190–224.

496. Wendelaar-Bonga, S. E. Formation, Storage, and Release of Neurosecretory Material Studied by Quantitative Electron Microscopy in the Freshwater Snail *Lymnaea stagnalis* (L.). *Z. Zellforsch.* **1971a**, *113*, 490–517.

497. Wendelaar-Bonga, S. E. Osmotically Induced Changes in the Activity of Neurosecretory Cells Located in the Pleural Ganglia of the Fresh Water Snail *Lymaea stagnalis* (L.), Studied by Quantitative Electron Microscopy. *Neth. J. Zool.* **1971b**, *21*, 127–158.

498. Wendelaar-Bonga, S. E. 1972. Neuroendocrine Involvement in Osmoregulation in a Freshwater Mollusc, *Lymnuea stagnalis*. *Gen. Comp. Endocrinol.* **1972**, *Suppl. 3*, 308–316.

499. Wendelaar-Bonga, S. E.; Boer, H. H. Ultrastructure of the Reno-pericardial System in the Pond Snail *Lymnaea stagnalis* (L.). *Z. Zellforsch. Mikrosk. Anat.* **1969**, *94*(4), 513–529.

500. Wentzell, M. M.; Martinez-Rubio, C.; Miller, M. W.; Murphy, A. D. Comparative Neurobiology of Feeding in the Opisthobranch Sea Slug, Aplysia, and the Pulmonate Snail, Helisoma: Evolutionary Considerations. *Brain Behav. Evol.* **2009**, *74*, 219–230.

501. Werner, G. D.; Gemmell, P.; Grosser, S.; Hamer, R.; Shimeld, S. M. Analysis of a Deep Transcriptome from the Mantle Tissue of *Patella vulgata* Linnaeus (Mollusca: Gastropoda: Patellidae) Reveals Candidate Biomineralising Genes. *Mar. Biotechnol.* **2013**, *15*, 230–243.

502. White, B. H.; Kaczmarek, L. K. Identification of a Vesicular Pool of Calcium Channels in the Bag Cell Neurons of *Aplysia californica*. *J. Neurosci.* **1997**, *17*, 1582–1595.

503. Wickham, L.; Desgroseillers, L. A Bradykinin-like Neuropeptide Precursor Gene is Expressed in Neuron L5 of *Aplysia californica*. *DNA Cell Biol.* 199, 10(4), 249–258.

504. Widjenes, J.; Runham, N. W. Studies on the Control of Growth in *Agriolimax reticulatus* (Mollusca, Pulmonata). *Gen. Comp. Endocrinol.* **1977**, *31*, 154–156.

505. Wijdenes, J.; van Elk, R.; Joosse, J. Effects of Two Gonadotropic Hormones on Polysaccharide Synthesis in the Albumen Gland of *Lymnaea stagnalis*, Studied with the Organ Culture Technique. *Gen. Comp. Endocrinol.* **1983**, *51*(2), 263–271.

506. Wiens, B. L.; Brownell, P. H. Neurotransmitter Regulation of the Heart in the Nudibranch *Archidoris montereyensis*. *J. Neurophysiol.* **1995**, *74*, 1639–1651.

507. Wilbur, K. M. *Shell Formation and Regeneration*. In *Physiology of Mollusca*; Wilbur, K. M., Yonge, C. M., Eds.; Academic Press, Inc.: New York, 1964, Chapter 8.

508. Wilbur, K. M.; Saleuddin, A. S. M. Shell Formation. In *The Mollusca. Vol. 4. Physiology Part 1*. Saleuddin, A. S. M., Wilbur, K. M., Eds. Academic Press, Inc.: New York, 1983.

509. Willoughby, D.; Yeoman, M. S.; Benjamin, P. R. Cyclic AMP is Involved in Cardioregulation by Multiple Neuropeptides Encoded on the FMRFamide Gene. *J. Exp. Biol.* **1999a**, *202*, 2595–2607.

510. Willoughby, D.; Yeoman, M. S.; Benjamin, P. R. Inositol-1,4,5-trisphosphate and Inositol-1,3,4,5-tetrakisphosphate are Second Messenger Targets for Cardioactive Neuropeptides Encoded on the FMRFamide Gene. *J. Exp. Biol.* **1999b**, *202*, 2581–2593.

511. Wilson, G. F.; Magoski, N. S.; Kaczmarek, L. K. Modulation of a Calcium-sensitive Nonspecific Cation Channel by Closely Associated Protein Kinase and Phosphatase Activities. *Proc. Natl. Acad. Sci. U.S.A.* **1998**, *95*, 10938–10943.

512. Wood, C. M.; Kelly, S. P.; Zhou, B.; Fletcher, M.; O'Donnell, M.; Eletti, B.; Pärt, P. Cultured Gill Epithelia as Models for the Freshwater Fish Gill. *Biochim. Biophys. Acta.* **2002**, *1566*(1–2), 72–83.

513. Yasuda-Kamatani, Y.; Nakamura, M.; Minakata, H.; Nomoto, K.; Sakiyama, F. A Novel cDNA Sequence Encoding the Precursor of the D-Amino Acid-containing Neuropeptide Fulicin and Multiple Alpha-amidated Neuropeptides from *Achatina fulica*. *J. Neurochem.* **1995**, *64*, 2248–2255.

514. Yasuda-Kamatani, Y.; Kobayashi, M.; Yasuda, A.; Fujita, T.; Minakata, H.; Nomoto, K.; Nakamura, M.; Sakiyama, F. A Novel D-Amino Acid-containing Peptide, Fulyal, Coexists With Fulicin Gene-related Peptides in *Achatina* Atria. *Peptides* **1997**, *18*, 347–354.

515. Yokotani, S.; Matsushima, A.; Nose, T. Bioactive Conformation of a D-Trp-containing Cardioexcitatory Tripeptide Islated from the Sea Hare Aplysia. In *Shimohigashi Y Peptide Science*. Peptide Research Foundation: Osaka, 2004.

516. Young, K. G.; Chang, J. P.; Goldberg, J. I. Gonadotropin-releasing Hormone Neuronal System of the Freshwater Snails *Helisoma trivolvis* and *Lymnaea stagnalis*: Possible Involvement in Reproduction. *J. Comp. Neurol.* **1999**, *404*, 427–437.

517. Zhang, S. M.; Nian, H.; Wang, B.; Loker, E. S.; Adema, C. M. Schistosomin from the Snail *Biomphalaria glabrata*: Expression Studies Suggest No Involvement in Trematode-mediated Castration. *Mol. Biochem. Parasitol.* **2009**, *165*, 79–86.

518. Zhang, L.; Tello, J. A.; Zhang, W.; Tsai, P. S. Molecular Cloning, Expression Pattern, and Immunocytochemical Localization of a Gonadotropin-releasing Hormone-like Molecule in the Gastropod Mollusk, *Aplysia californica*. *Gen. Comp. Endocrinol.* **2008a**, *156*, 201–209.

519. Zhang, L.; Wayne, N. L.; Sherwood, N. M. Postigo, H. R.; Tsai, P. S. Biological and Immunological Characterization of Multiple GnRH in an Opisthobranch Mollusk, *Aplysia californica*. *Gen. Comp. Endocrinol.* **2000**, *118*, 77–89.

520. Zhang, Y.; Helm, J. S.; Senatore, A.; Spafford, J. D.; Kaczmarek, L. K.; Jonas, E. A. PKC-induced Intracellular Trafficking of Ca(V)2 Precedes Its Rapid Recruitment to the Plasma Membrane. *J. Neurosci.* **2008b**, *28*, 2601–2612.

521. Zhang, Y.; Magoski, N. S.; Kaczmarek, L. K. Prolonged activation of Ca^{2+}-activated K+ Current Contributes to the Long-lasting Refractory Period of Aplysia Bag Cell Neurons. *J. Neurosci.* **2002**, *22*, 10134–10141.

522. Zhao, X.; Wang, Q.; Jiao, Y.; Huang, R.; Deng, Y.; Wang, H.; Du, X. Identification of Genes Potentially Related to Biomineralization and Immunity by Transcriptome Analysis of Pearl Sac in Pearl Oyster *Pinctada martensii*. *Mar. Biotechnol.* **2012a**, *14*, 730–739.

523. Zhao, X.; Yu, H.; Kong, L.; Li, Q. Transcriptomic Responses to Salinity Stress in the Pacific Oyster *Crassostrea gigas*. *PLoS ONE* **2012b**, *7*, e46244. DOI:10.1371/journal. pone.0046244.

CHAPTER 11

Ascidian Neuropeptides and Peptide Hormones

HONOO SATAKE

Suntory Foundation for Life Sciences, Bioorganic Research Institute,
8-1-1 Seikadai, Seika-Cho, Soraku-gun, Kyoto 619-0284, Japan,
E-mail: satake@sunbor.or.jp

11.1 INTRODUCTION

Ascidians (sea squirts), invertebrate deuterostome marine animals, belong to the same phylum to vertebrates, chordata, and subphylum tunicate or urochodata [1]. Their critical phylogenetic position as the sister group of vertebrates has made them attractive and useful in researches on the evolution of the nervous, neuroendocrine, and endocrine system in chordates [2, 3]. To date, various neuropeptides and peptide hormones of ascidians, particularly the cosmopolitan species, *Ciona intestinalis*, have been characterized by conventional purification, database-searching followed by molecular cloning, and peptidomoic approaches [4–9]. These findings will lead to the investigation not only of peptidergic regulatory mechanisms of *C. intestinalis* but also of the evolutionary processes of neuropeptides, peptide hormones, and their receptors. In this chapter, we provide essential and recent findings on primary structures and several biological functions of neuropeptides, peptide hormones, and their receptors in of ascidians.

11.2 OVERVIEWS OF ASCIDIAN NEUROPEPTIDES AND PEPTIDE HORMONES

Ascidian peptides and/or their genes have thus far been characterized by conventional purification, molecular cloning, genome database-searching, and peptidomics, since the discovery of the first *C. intestinalis* neuropeptide in 1990. Recently, two approaches have been employed. The first one is homology search on a genome/EST database of an organism of interest. This procedure is useful for the detection of peptides containing long sequences and/or conserving consensus motifs. For example, genes of *Ciona* GnRHs, calcitonin (CT), insulin-like peptides (INS-Ls), and corticotrophin-releasing factor (CRF) were identified by homology search on the *C. intestinalis* genome/EST database [11–14]. In contrast, such database search is frequently useless for identification of small peptides or their genes, given that many neuropeptides and peptide hormones contain short sequences, and that their precursors have poor sequence homology, even though homologous peptide hormones and neuropeptides share complete consensus motifs. Additionally, novel peptides or peptide homologs with partial consensus motifs cannot be detected by any form of homology search. The second one is mass spectrometry (MS)-based peptidomic approach [7]. This strategy enables the direct characterization of peptide sequences present in the tissues of interest. The most important step in peptidomic analysis to discriminate neuropeptides and peptide hormones with protein fragments. In general, neuropeptide or peptide hormone precursors harbor a hydrophobic signal peptide sequence at the N-terminus, and the mature peptide sequences there are flanked by endo-proteolytic mono- or dibasic sites (Lys-Lys/Lys-Arg/Arg-Arg/Arg). These criteria lead to the discrimination of neuropeptides and peptide hormones from non-significant peptide fragments. Currently, a genome/EST database or next-generation sequencer-based transcriptomes enable us to detect precursor sequences of the peptide candidates by referencing the obtained sequences followed by cloning of full-length cDNAs. Our previous Peptidomic study [7] identified 33 neuropeptides in the *C. intestinalis* central nervous systems (neural complex) (Table 11.1). In conclusion, a combination of MS-based peptidomic analysis and database-referencing of the resultant sequence is the most reliable and efficient methods for characterization of neuropeptides and peptide hormones.

TABLE 11.1 Major Peptides Identified by Peptidomes of the *C. intestinalis* Cerebral Ganglion

Gene	Peptide	Peptide Sequence
ci-gnrh-1	t-GnRH-3	pQHWSKGYSPGa
	t-GnRH-5	pQHWSYEFMPGa
	t-GnRH-6	pQHWSYEYMPGa
ci-gnrh-x	Ci-GnRH-X	pQHWSNWWIPGAPGYNGa

TABLE 11.1 *(Continued)*

Gene	Peptide	Peptide Sequence
ci-ntlp-A	Ci-NTLP-1	pQLHVPSIL
	Ci-NTLP-2	MMLGPGIL
	Ci-NTLP-3	GMMGPSII
	Ci-NTLP-4	FGMIPSII
ci-ntlp-B	Ci-NTLP-5	NKLLYPSVI
	Ci-NTLP-6	SRHPKLYFPGIV
ci-galp	Ci-GALP	PFRGQGGWTLNSVGYNAGLGALRKLFE
ci-lf	Ci-LF-1	FQSLF
	Ci-LF-2	YPGFQGLF
	Ci-LF-3	HNPHLPDLF
	Ci-LF-4	YNSMGLF
	Ci-LF-5	SPGMLGLF
	Ci-LF-6	SDARLQGLF
	Ci-LF-7	YPNFQGLF
	Ci-LF-8	GNLHSLF
ci-yfv/l	Ci-YFV-1	ELVVRDPYFV
	Ci-YFV-2	NNQESYFV
	Ci-YFV-3	DDEPRSYFV
	Ci-YFL-1	DAARPNYYFL
ci-vp	Ci-VP	CFFRDCSNMDWYR
ci-tk	Ci-TK-I	HVRHFYGLMa
	Ci-TK-II	SIGDQPSIFNERASFTGLMa

pyro-Glutamate and C-terminal amide are shown by "pQ" and "a."

The *Ciona* peptides are largely classified into three categories: (i) prototypes and homologs of vertebrate peptides such as cholecystokinin (CCK), GnRHs, TK, CT, INS-Ls, CRF, galanin/galanin-like peptide (GALP), (ii) peptides partially homologous with vertebrate peptides including vasopressin (VP), GnRH-related peptide, NTLPs, (iii) novel peptides such as LF-family peptides and YFL/V peptides. Of particular significance is that *C. intestinalis* has been shown to possess the prototypes and homologs of vertebrate peptides, including CCK, TK, CT, and CRF, galanin/GALP. These findings suggest that many peptide prototypes originated from common ancestor chordates of ascidians and vertebrates, not from ancient vertebrates.

This view is compatible with the fact that ascidians occupy a phylogenetic position as the sister group of vertebrates in evolutionary lineages of animals. It should be noteworthy that the *Ciona* homologs of vertebrate neuropeptides and/or peptide hormones from various glands including the hypothalamus have been identified, but no homologs of vertebrate pituitary hormones have ever been reported [7–9]. This is consistent with the fact that ascidians are not endowed with an organ corresponding to a pituitary [1]. Collectively, these findings suggest that *C. intestinalis* shows applicability to the investigation of biological functions and their evolutionary aspects of neuropeptides and non-pituitary peptide hormones of vertebrates. These findings shed light on the advantages of *C. intestinalis* in studies on both the functional conservation and diversification of neuropeptides and hormones throughout chordates. In the following sections, we summarize major *C. intestinalis* peptides and their receptors.

11.2.1   CIONIN

Cionin (Table 11.2) is an octapeptide that was first isolated from ascidians in 1990 [15]. The primary sequence of cionin contains the C-terminal four amino acids Trp-Met-Asp-Phe-NH$_2$ and two sulfated Tyr residues at position 6 and 7 from the C-terminus, which are typical of the primary sequence of CCK/gastrin peptides, and thus, this peptide is believed to belongs to the CCK/gastrin family peptides that are brain/gut peptides in vertebrates and are involved in various physiological events including digestion and feeding behavior [16, 17]. In addition, an orthologous peptide, sulfakinin [18–20]. Cionin and sulfakinins share a sulfated tyrosine at the corresponding position, whereas the CCK/ gastrin C-terminal consensus is conserved only in cionin (Table 11.2). Such sequence conservation suggests that the prototype of the CCK/gastrin family might have emerged in common ancestors of invertebrates and vertebrates followed by diversification of their sequences in differential lineages: protostome lineages and chordate lineages.

TABLE 11.2 Sequences of Cholecystokinin/ Gastrin Family Peptides

Peptide	Sequence
cionin	NYYGWMDFa
CCK	DYMGWMDFa
gastrin	pQGPWLEEEEEAYGWMDFa
Dsk1 (fruitfly)	FDDYGHMRFa
NLP12 (nematode)	DGYRPLQFa

Sulfated tyrosines are indicated by a double line. The CCK/ gastrin C-terminal consensus is underlined.

CCK and gastrin exert their activities through specific interaction with two cognate receptors, CCK receptor-A (CCKR-A or CCK1R) and CCKR-B (CCK2R) [16, 17]. CCKR-A is a CCK-specific receptor, and sulfation of CCK is required for binding to CCKR-A [16, 17]. In contrast, CCKR-B interacts with both CCK and gastrin, and sulfation of these peptides is not required for binding [16, 17]. In *C. intestinalis*, two cognate cionin receptors, CioR1 and CioR2, which are homologous to vertebrate CCKR-A, have thus far been shown to activate intracellular calcium mobilization in cultured cells to the same extent [21, 22]. Additionally, mono-sulfated cionin derivatives exhibited much less potent activity at CioRs than authentic cionin and, non-sulfated cionin abolished any activity [22]. These findings showed that both of two sulfated tyrosines are prerequisites for biological activities of cionin. CioRs were found to be expressed in the neural complex, digestive organs, oral/atrial siphons, and ovary, suggesting the multiple functions of cionin [22]. However, the biological roles of cionin and differences in biological functions between CioR1 and -2 awaits further studies.

11.2.2 GNRHS

GnRH (previously designated luteinizing-hormone-releasing hormone = LH-RH), plays a central role in reproductive development and function, which is released via the hypothalamic-hypophysial portal system to regulate the synthesis and release of pituitary gonadotropins that in turn trigger the steroidogenesis and stimulate gonadal maturation in vertebrates [23, 24]. Furthermore, GnRHs are involved in diverse neuroendocrine, paracrine, autocrine, and neurotransmitter/neuromodulatory functions in the central and peripheral nervous systems, and various peripheral tissues [23, 24]. To date, GnRHs have been identified in non-insect protostomes, echinoderms, urochordates as well as vertebrates (Table 11.3). Vertebrate GnRHs, except lamprey GnRH-I and –III (Table 11.3), are composed of 10 amino acids with the consensus sequences of pyro-Glu^1-His/Tyr^2-Trp^3-Ser^4, Gly^6, and Pro^9-Gly^{10}-amide [23–27].

TABLE 11.3   Sequences of GnRHs

Animal	Species	Peptide	Sequence
Ascidians	*Chelyosoma productum*	t-GnRH-1	pQHWSYGLRPGa
		t-GnRH-2	pQHWSLCHAPGa
	Ciona intestinalis	t-GnRH-3	pQHWSYEFMPGa
		t-GnRH-4	pQHWSNQLTPGa
		t-GnRH-5	pQHWSYEYMPGa
		t-GnRH-6	pQHWSKGYSPGa
		t-GnRH-7	pQHWSYALSPGa
		t-GnRH-8	pQHWSLALSPGa
	Ciona savignyi	t-GnRH-9	pQHWSNKLAPGa
	Ciona intestinalis	Ci-GnRH-X	pQHWSNWWIPGAPGYNGa
	Halocynthia roretzi	t-GnRH-10	pQHWSYGFSPGa
		t-GnRH-11	pQHWSYGFLPGa
Human	*Homo sapiens*	GnRH1	pQHWSYGLRPGa
		GnRH2	pQHWSHGWYPGa
Guinia pig	*Cavia porcellus*	GnRH1	pQHWSYGVRPGa
Trout	*Oncorhynchus mykiss*	GnRH3	pQHWSYGWLPGa
Lamprey	*Petromyzon marinus*	l-GnRH-I	pQHYSLEWKPGa
		l-GnRH-II	pQHWSHGWFPGa
		l-GnRH-III	pQHWSHDWKPGa
Amphioxus	*Branchiostoma floridae*	Amph.GnRHv	pQEHWQYGHWYa
		Amph.GnRH	pQILCARAFTYTHTWa
Deuterostome invertebrates (Sea urchin and starfish)			
Sea urchin	*Strongylocentrotus purpuratus*	Sp-GnRHP	PQVHHRFSGWRPGa
Starfish	*Asterias rubens*	Ar-GnRH	pQIHYKNPGWGPGa
Protostomes (Molluscs and annelids)			
Octopus	*Octopus vulgaris*	Oct-GnRH	pQNYHFSNGWHPGa

TABLE 11.3 *(Continued)*

Animal	Species	Peptide	Sequence
Cuttlefish	*Sepia officinalis*	Oct-GnRH	pQNYHFSNGWHPGa
Swordtip squid	*Loligo edulis*	Oct-GnRH	pQNYHFSNGWHPGa
Oyster	*Crassostrea gigas*	Cg-GnRH	pQNYHFSNGWQPa
Yesso scallop	*Patinopecten yessoensis*	Py-GnRH	pQNFHYSNGWQPa
Sea hare	Aplysia californica	Ap-GnRH	pQNYHFSNGWYAa
Owl limpet	Lottia gigantean	Lg-GnRH	pQHYHFSNGWKSa
Marine worm	Capitella teleta	Ca-GnRH	pQAYHFSHGWFPa
Leech	Helobdella robusta	Hr-GnRH	pQSIHFSRSWQPa

The N-terminal pyroglutamic acid and C-terminal amide are shown by "pQ" and "a," respectively. Chordate GnRH consensus motifs are underlined.

Two types of GnRH, GnRH1 and -2, were characterized in vertebrates. In teleost, an additional subtype, GnRH3, was found [23, 24]. In lamprey, one GnRH1 orthlog (l-GnRH-II) and two paralogs (l-GnRH-I and –III) were identified [28]. Phylogenetic molecular tree analysis indicated that these lamprey GnRHs were generated in the lamprey-specific evolutionary lineage [28]. Protostome and echinoderm GnRHs are composed of 11 to 12 amino acids, and partially conserve vertebrate GnRH characteristic (Table 11.3). These GnRHs possess two amino-acid insertions after position 1 (Table 11.3). Moreover, the C-terminal Pro-Gly-amide is replaced with Pro/Ser/Ala-amide in non-cephalopod (octopus and squid) protostome GnRHs, whereas this consensus motif is found in echinoderm GnRHs (Table 11.3). In contrast, protostome, and echinoderm GnRHs completely conserve Gly^8, corresponding to Gly^6 in vertebrate GnRHs (Table 11.4). All GnRHs are encoded as a single copy in the precursor, whose organization is conserved in vertebrates and protostomes [25–28].

To date, 11 ascidian GnRHs have been identified in *Chelyosoma productum* (t-GnRH1, 2), [29], *C. intestinalis* (t-GnRH-3 to -8) and *Ciona savignyi* (t-GnRH-5 to -9) [10], and *Halocynthia roretzi* (t-GnRH-10, 11) [30]. All ascidian GnRHs completely conserve the consensus sequences of pyro-Glu-His-Trp-Ser and Pro-Gly-amide (Table 11.3). A striking feature is that, unlike any other species GnRH genes, two *Ciona* GnRH genes, *Ci-gnrh-1* and *-2*, encode three different GnRH peptide sequences [10]: t-GnRH-3, -5, and -6 are encoded in *Ci-gnrh-1*, whereas t-GnRH-4, -7, and -8 are encoded in *Ci-gnrh-2*. All GnRH sequences are flanked by mono- or dibasic endoproteolytic sites at their N- and C-termini [10]. These triplet

GnRH sequence organizations were also observed in *Cs-gnrh-1* and *2* of *C. savignyi*, an ascidian species closely related to *C. intestinalis* [10]. In *Halocynthia roretzi*, the *H. roretzi* GnRH gene encodes t-GnRH-10 and -11 [30]. These findings have established the basis that multiple copies of a GnRH sequence are present in one precursor in ascidians, indicating conservation and species-specific diversification of GnRHs in urochordates.

TABLE 11.4 Tachykinin Family Peptides

Species	Peptide	Sequence
Chordate tachykinins		
Ascidian		
(*Ciona intestinalis*)	Ci-TK-I	HVRHFYGLMa
	Ci-TK-II	SIGDQPSIFNERASFTGLMa
Mammals	Substance P	RPKPQQFFGLMa
	Neurokinin A	HKTDSFVGLMa
	Neurokinin B	DMHDFVGLMa
Rat and mouse	Hemokinin-1	SRTRQFYGLMa
Human	Endokinin A/B	GKASQFFGLMa
	Endokinin C	KKAYQLEHTFQGLLa
	Endokinin D	VGAYQLEHTFQGLLa
zebrafish	Neurokinin F	YNDIDYDSFVGLMa
Protostome Tachykinin-Related Peptides (TKRPs)		
Locust (*Locusta migratoria*)	Lom-TK-I	GPSGFYGVRa
Fruitfly	DTK-1	APTSSFIGMRa
(*Drosophila melanogaster*)	Natalisin-1	EKLFDGYQFGEDMSKENDPFIPPRa
Echiuroid worm (*Urechis unicinctus*)	Uru-TK-I	LRQSQFVGARa

C-terminal amide is shown by "a." The Chordate TK consensus motif and its related sequences are underlined by plain and dotted lines, respectively. The protostome TKRP consensus motif and its related sequence are indicated by wavy lines.

A novel GnRH-like peptide, Ci-GnRH-X [31], was also characterized by the neural complexes of *C. intestinalis*. This *Ciona* neuropeptide harbors the GnRH consensus sequences, including the N-terminal pGlu-His-Trp-Ser and a C-terminal amidated Gly. However, Ci-GnRH-X is composed of 16 amino acid residues, and lacks the common Pro at position 2 from the C-termini of the GnRH family peptides (Table 11.3). Moreover, the *t-gnrh-X* gene, like non-ascidian GnRH genes, encodes a single sequence of Ci-GnRH-X [31].

Most GnRH receptors (GnRHRs) induce elevation of intracellular calcium ion [23–29]. GnRHRs display some species-specific distribution, although at least one GnRHR was identified in all vertebrates [23–29]. Type-I GnRHRs show high affinity to both GnRH-I and II, whereas the type-II receptor is specific to GnRH-II. In protostomes, an octopus GnRHR was the only receptor that was shown to be responsive to the cognate ligand and to trigger mobilization of intracellular cellular calcium ions [25–29, 32]. Recently, Ar-GnRH specifically activated intracellular Ca^{2+} mobilization of a cognate receptor, ArGnRHR in the starfish, *Asterias rubens* [33].

To date, four GnRH receptors, namely, Ci-GnRHR-1, -2, -3, and -4, have been identified in *C. intestinalis* [34, 35]. Ci-GnRHR-1 share 70%, 38%, and 36% sequence homology to Ci-GnRHR-2, -3, and -4, respectively, and all Ci-GnRHRs display approx. 30% sequence identity to human Type I GnRHR [34, 35]. The phylogenetic molecular tree demonstrated that Ci-GnRHR-2, -3, and -4 are *Ciona*-specific paralogs of Ci-GnRHR-1 [34, 35]. Of particular interest is the unique ligand-selective production of second messengers mediated by Ci-GnRHRs. The elevation of intracellular calcium, which is typical of other GnRHRs [23–29, 32, 33, 36], was only observed in the administration of t-GnRH-6 to Ci-GnRHR-1 [35]. Instead, Ci-GnRHR-2 produce cAMP in responsive to t-GnRH-7, 8, 6 in this order of potency, while t-GnRH-3 and -5 specifically activate cAMP production at Ci-GnRHR-3 to a similar extent. Ci-GnRHR-4 exhibited neither elevation of intracellular calcium nor cAMP production [35]. However, Ci-GnRHR4 was shown to play a very unique functional role in *C. intestinalis* GnRH signaling (Figure 11.1). Ci-GnRHR4 forms a heterodimer with Ci-GnRHR-1 and then potentiates the elevation of intracellular calcium via both calcium-dependent and –independent protein kinase C subtypes, and ERK phosphorylation in a fashion specific to the t-GnRH6 and Ci-GnRHR1 pair [37]. Ci-GnRHR-4 was also found to heterodimerize with Ci-GnRHR-2 [38]. The Ci-GnRHR-2/-4 heterodimer decreased cAMP production by 50% in a non-ligand selective manner by shifting of activation from Gs protein to Gi protein by Ci-GnRHR-2 in response to t-GnRH7, and -8, compared with the Ci-GnRHR-2 monomer/homodimer [38]. These findings substantiated that Ci-GnRHR-4 serves as a protomer of GPCR heterodimers rather than a ligand-binding receptor [39]. In addition, molecular phylogenetic analysis demonstrated that Ci-GnRHRs are included in vertebrate GnRHR clades but form an independent cluster in chordate GnRHRs, suggesting that these receptors have evolved within the *Ciona* species [26, 27, 35, 39]. Collectively, these findings indicate ascidian-specific molecular and functional diversity of ascidian GnRH signaling systems.

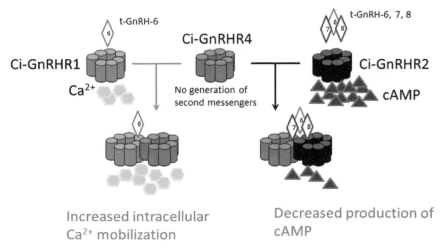

FIGURE 11.1 Functional consequences of *Ciona* GnRH receptor heterodimerization. Ci-GnRHR4 heterodimerizes with Ci-GnRHR1, and Ca^{2+} mobilization is 10-fold more potently upregulated than Ci-GnRHR1 alone in a t-GnRH6-specific fashion. Ci-GnRHR4 also heterodimerizes with Ci-GnRHR2, and the R2-R4 heterodimer decreases cAMP production in response to t-GnRH6, 7, and 8.

Ci-GnRH-X was also found to possess unique functions on Ci-GnRHRs. This *Ciona*-specific GnRH-related peptide moderately (10%–50%) inhibited the elevation of the intracellular calcium and cAMP production by t-GnRH-6 at Ci-GnRHR-1, and cAMP production by t-GnRH-3and-5 via Ci-GnRHR-3 [31]. In contrast, no inhibitory effect of Ci-GnRH-X at Ci-GnRHR-2 was detected [31]. These findings provide evidence that t-GnRHs and Ci-GnRHRs have not redundant but specific biological roles.

No ascidian GnRHs have been shown to exhibit any reproductive activities. However, t-GnRH-3 and -5 increased the gemate spawning activity and water flow [10]. Moreover, tGnRH-3 and -5 suppressed the growth of adult organs by arresting cell cycle progression and the promotion of tail absorption in the *Ciona* larval development [40]. Combined with the fact that *C. intestinalis* possesses no pituitary and gonadotropins, these findings suggest that t-GnRHs are responsible for the regulation of various biological events that are distinct from gonadotropin release. Furthermore, Ci-GnRHRs were detected in test cells of vitellogenic oocytes as well as in the cerebral ganglion [35, 37–39]. Altogether, it is presumed that *C. intestinalis* GnRHs play various reproduction-related roles as multifunctional neuropeptides. Such neuropeptidergic GnRH-directed regulation of the ovary is highly likely to be a functional ancestor of vertebrate GnRHergic endocrine and neuroendocrine systems.

11.2.3 TACHYKININ

TKs are vertebrate multi-functional brain/gut peptides regulating smooth muscle contraction, vasodilation, nociception, inflammation, neurodegeneration, and neuroprotection in a neuropeptidergic or endocrine fashion [41–43]. All TKs are featured by the C-terminal consensus -Phe-X-Gly-Leu-Met-NH$_2$ [41–43]. The mammalian TK family consists of four major peptides: Substance P (SP), Neurokinin A (NKA), Neurokinin B (NKB), and Hemokinin-1/Endokinins (HK-1/EKs), as summarized in Table 11.4. TKs are encoded by three genes, namely *TAC1* (or *PPTA*), *TAC3* (or PPTB), or *TAC4* (PPTC) gene [41–44]. *TAC1* generates four splicing variants which produce SP alone or SP and NKA [41–44]. Similarly, several EK isoforms are produced from the *TAC4* gene. *TAC3* gene yields only NKB [41–44]. Additionally, teleost NKB genes were found to harbor additional NK peptides, NKF [45].

Three subtypes of TK receptors, namely NK1, NK2, and NK3, have been identified in vertebrates [41–45]. NK1, 2, and -3 were shown to induce both elevations of intracellular calcium and production of cAMP. NK1, 2, and -3 possess moderate ligand-selectivity: SP>NKA>NKB for NK1, NKA>NKB>SP for NK2, and NKB>NKA>SP for NK3, respectively [41–45]. HK-1 and the 10-amino acid C-terminal common sequence of EKA and EKB (EKA/B) exhibited potent binding activity on all NK1-3 with the highest selectivity to NK1 [41–44], whereas EKC and EKD are devoid of any activity on NK1–3 [41–44]. The teleost-specific NKB-related peptide, NKF, was shown to activate NK3 [45].

TK-related peptides (TKRPs) have been identified in the central nervous system of various protostomes [41, 46–48]. However, they contain the analogous Phe-Xaa1-(Gly/Ala/Pro)-Xaa2-Arg-NH$_2$ consensus (Table 11.4), and TKRP precursors encode multiple TKRP sequences [41, 46–48], which are totally distinct from those of vertebrate TKs. These findings suggest that TKRP and TK genes might have evolved in distinct evolutionary lineages. TKRP receptors were characterized by several protostomes [41, 46–48]. These receptors display high similarity to vertebrate TK receptors in the sequence and exon/intron organization [41, 46–48]. Consequently, TK receptors and TKRP receptors share the common original GPCR gene. Moreover, TKRP receptors were shown to stimulate the increase in intracellular calcium or generation of cAMP in response to cognate ligands [41, 46–48].

Nevertheless, TKRP receptors, unlike NK1–3, exhibit no significant ligand selectivity to their endogenous TKRPs [41, 46–48].

Two authentic TK peptides, Ci-TK-I and -II, were identified in the neural complex of *C. intestinalis* using mass spectrometric analysis [7, 46, 48, 49]. Unlike protostome TKRPs, the vertebrate TK consensus motif is completely conserved in of both the *Ciona* TKs (Table 11.4). The Ci-TK precursor encodes both Ci-TK-I and Ci-TK-II, and shows structural organization similar to γ-TAC1 [41, 46, 48, 49]. These findings revealed that the TK family is conserved as neuropeptides throughout Olfactores (vertebrates and urochordates), not only in vertebrates, and indicate that the Ci-TK gene is a direct prototype of vertebrate TKs. However, Ci-TK-I, and -II sequences are encoded in the same exon, indicating that no alternative splicing of the Ci-TK transcript occurs. Consequently, alternative production of TK peptides might have been: established during the evolution of vertebrates [49].

The cognate Ci-TK receptor, Ci-TK-R, was also identified in *C. intestinalis*. Ci-TK-R displayed high amino acid sequence similarity (30%–43%) to mammalian TK receptors [9, 41, 46, 48, 49]. The molecular phylogenetic tree indicated that Ci-TK-R belongs to the vertebrate TK receptor clade, not to the protostome TKRPR clade [9, 41, 46, 48, 49]. Ci-TK-I evoked a typical intracellular calcium mobilization at Ci-TK-R [49]. Moreover, SP and NKA also exhibited comparable responses to Ci-TK-I [49]. These data revealed that Ci-TK-R lacked the ligand selectivity typical of NK1–3 [41, 46, 48, 49].

As depicted in Figure 11.2, Ci-TK-I was shown to specifically enhance the development of vitellogenic follicles via up-regulation of enzymatic activities of cathepsin D, chymotrypsin, and carboxypeptidase B1 [8, 9, 50, 51]. This unique biological function is consistent with the intraovarian specific expression of the Ci-TK-R in test cells (partially corresponding to granulosa cells in vertebrate follicles) of vitellogenic follicles [8, 9, 50, 51]. This is the only report on the biological role of TKs in the ovary in an animal, and thus, of interest is whether such TKergic ovarian follicular developmental system is conserved throughout chordates.

11.2.4 VASOPRESSIN (VP)/OXYTOCIN (OT)

Oxytocin (OT) and VP are composed of nine amino acids, and share Asn^5, Pro^7, Gly^9, C-terminal amidation, and a circular structure formed by a disulfide bridge between conserved Cys^1 and Cys^6 (Table 11.5). They have so far been characterized by a great variety of animal species from protostomes to mammals [52–58]. OT is responsible for reproductive behaviors including uterine contraction, milk ejection, and male reproductive tract stimulation [59–61], and VP participates in osmoregulation, control of blood pressure and anti-diuretic effect [59, 60]. OT and VP are also shown to be involved in highly advanced central functions and disorders, including learning, social behavior, anxiety, and autism [58, 59, 62].

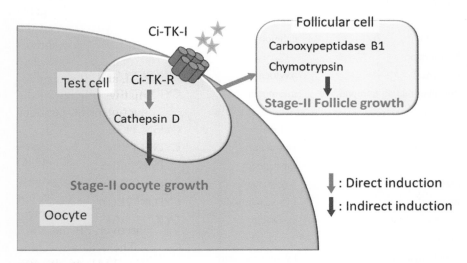

FIGURE 11.2 Regulation of the growth stage-II follicles by Ci-TK. Ci-TK stimulates Ci-TK-R exclusively expressed in test cells of vitellogenic (stage II) follicles, and the up-regulates directly transcription and the resultant enzymatic activities of *Ciona* cathepsin D and indirectly those of carboxypeptidase B1 and chymotrypsin in follicular cells, eventually leading to the growth of stage-II oocytes and follicular cells.

OT and VP are discriminated by an amino acid present at position 8; VP family peptides contain a basic amino acid (arginine or lysine), whereas a neutral amino acid (leucine, isoleucine, valine, or threonine) is located at this position of OT family peptides (Table 11.5). Likewise, the non-mammalian vertebrate VP homolog, vasotocin, has an Arg at position 8, and an Ile is found in the non-mammalian vertebrate OT homologs, mesotocin, and isotocin (Table 11.5). One VP family peptide and one OT family peptide are present in most jawed vertebrates [52, 53], although only one type of the OT/VP superfamily peptide has ever been found in cyclostomes and most invertebrates [52, 53]. Consequently, the OT family and VP family are likely to have occurred via gene duplication of the common ancestral gene during the evolution from jawless fish to jawed fish.

TABLE 11.5   Oxytocin/ Vasopressin Family Peptides

Peptide	Species	Sequence
Deuterostome invertebrate		
Ascidian	Ci-VP	CFFRDCSNMDWYR
(*C. intestinalis*) ascidian	SOP	CYISDCPNSRFWSTa
(*Styela plicata*)		CYIINCPRGa
amphioxus	[Ile⁴]-vasotocin	CFISNCPKGa
sea urchin	echinotocin	
Vertebrate		
mammal	oxytocin	CYIQNCPLGa
non-mammalian tetrapod, etc.	mesotocin	CYIQNCPIGa
lungfish		
Osteichthyes	isotocin	CYISNCPIGa
(bony fish)	valitocin	CYISNCPVGa
dogfish	aspargtocin	CYINNCPLGa
mammal	Arg-vasopressin	CYFQNCPRGa
mammal	Lys-vasopressin	CYFQNCPKGa
other vertebrates	vasotocin	CYIQNCPRGa
Protostome		
Leech, Sea Snail, Sea hare	Lys-conopressin	CFIRNCPKGa
Sea Snail	Arg-conopressin	CIIRNCPRGa
Earthworm	annetocin	CFVRNCPTGa
Octopus	cephalotocin	CYFRNCPIGa
Octopus	octopressin	CFWTSCPIGa
Locust, Red flour beetle	inotocin	CLITNCPRGa
Daphnia	Crustacean OT/ VP-like	CFITNCPPGa
Nematode	Nematocin	CFLNSCPYRRYa

'a' denotes C-terminal amide.

OT/VP superfamily peptides have also been identified in invertebrates [52–58]. Interestingly, protostome, and non-ascidian deuterostome inverte-brate OT/VP superfamily peptides are all composed of nine amino acids, and share the OT/VP consensus amino acids (Table 11.5).

Both OT and VP precursors are organized with major three regions: a signal peptide, an OT or VP sequence flanked by a putative glycine C-terminal amidation signal and the following dibasic endoproteolytic site, and a neurophysin featured by 14 highly conserved cysteines [52–58]. Seven disulfide bridges between each of the 14 cysteines are essential for the adoption of correct tertiary structure to interact with OT/VP [52–58]. This structural organization is also conserved in the OT/VP precursors except for *Ciona* homologs.

Two ascidian OT/VP superfamily peptides were identified in different species. Ci-VP is the first deuterostome invertebrate OT/VP family peptide from *C. intestinalis* [55], and SOP was characterized by another ascidian, *Styela plicata* [64]. The most outstanding characteristics of these ascidian OT/VP superfamily peptides are an elongated sequence of the C-termini, compared with other OT/VP superfamily peptides (Table 11.5). Ci-VP and SOP consist of 13 and 14 amino acids, respectively, whereas typical OT/VP superfamily peptides are comprised of 9 amino acids (Table 11.5). In particular, Ci-VP is the only OT/VP superfamily peptide that bears the non-amidated C-terminus (Table 11.5). In contrast, the N-terminal circular region of these ascidian peptides displays high sequence homology to other OT/VP superfamily peptides (Table 11.5). These findings suggest that Ci-VP and SOP might have diverged in the respective evolutionary lineages. The Ci-VP precursor, like other species homologs, also encoded a neurophysin-like domain, but the *Ciona* neurophysin possesses only 10 cysteines. However, 10 cysteines in the *C. intestinalis* neurophysin, including two cysteine doublets, are conserved almost identically to those of other neurophysin domains [55]. Additionally, the 13-residue and C-terminally non-amidated Ci-VP peptide and the 10-cysteine neurophysin domain were detected in the genome database of the closely related species, *Ciona savignyi*. In contrast, the 14-cysteine neurophysin domain is completely conserved in the SOP gene [64]. These findings indicate the intraphyletic molecular diversity of neurophysin domains as well as the hormone sequences within ascidian species.

The OT/VP superfamily peptides manifest their activities through their receptors, which belong to the Class A GPCR family [53, 59, 62]. The OT/VP superfamily peptide receptors display high sequence similarity with one another, indicating that they originated common ancestors [53]. To date, three VP receptors (V1aR, V1bR, and V2R) and one OT receptor (OTR) have been identified in mammals. V1aR, V1bR, and OTR have been shown to trigger an increase in the intracellular calcium ions [53, 59, 62], whereas V2R induces the production of cAMP [53, 59, 62].

C. intestinalis was shown to possess a single Ci-VP-receptor, Ci-VP-R, which also displayed high amino acid sequence similarity (35%–56%) to those of vertebrate and protostome OT/VP superfamily peptide receptors

[55]. Molecular phylogenetic analysis confirmed that Ci-VP-R belongs to the OT/VP superfamily peptide receptor family. Furthermore, Ci-VP-R specifically evoked an intracellular calcium elevation in response to Ci-VP [55].

No biological function of Ci-VP has been ever observed. However, the expression of SOP mRNA in hypotonic seawater was 2-fold greater than those in isotonic and hypertonic seawater [64]. Furthermore, SOP evoked contractions with increased tonus in the siphon of the ascidian [64]. These results suggested the functional correlation of SOP with osmoregulation.

11.2.5 NEUROTENSIN-LIKE PEPTIDES (NTLPS)

Neurotensin (NT) is a vertebrate brain/gut peptide involved in dopamine transmission, pituitary hormone secretion, hypothermia, and analgesia [65]. NT is encoded in a single precursor with a structurally related peptide, neuromedin N. These family peptides are featured by a Pro-Tyr-Ile-Ile C-terminal consensus sequence (Table 11.6). The NT family has so far been characterized only in mammals and birds [65]. NTR1, 2, and 3 have been identified in mammals. NTR1 and -2 are Class A GPCRs, which trigger an elevation of intracellular calcium ion [65]. NTR3 (or sortilin) is a non-GPCR membrane protein, and bound to NT. However, the resultant signaling remains unclear.

TABLE 11.6 Sequences of Neurotensin, Neurotensin-Related Peptides, and *Ciona* Neurotensin-Like Peptides

Peptide	Sequence
Neurotensin (rat)	pQLYENKPRRPYIL
Neurotensin (chicken)	pQLHVNKARRPYIL
Neuromedin N (rat)	KIPYIL
LANT 6(chicken)	KNPYIL
Ci-NTLP-1	pQLHVPSIL
Ci-NTLP-2	GMMGPSII
Ci-NTLP-3	MMLGPGIL
Ci-NTLP-4	FGMIPSII
Ci-NTLP-5	NKLLYPSVI
Ci-NTLP-6	SRHPKLYFPGIV

In *C. intestinalis*, six NT-like peptides, Ci-NTLP-1 to -6, have been identified by a peptidomic approach [7]. Ci-NTLPs bear the Pro-Ser/Gly-Ile/

Val-Ile-Leu, which is reminiscent of the NT C-terminal consensus (Table 11.6). However, The N-terminal sequences of Ci-NTLPs 2–4 are distinct from those of vertebrate NTs and neuromedin N (Table 11.6). Thus, whether or not Ci-NTLPs are authentic homologs of vertebrate NTs and neuromedin N remains to be concluded. Ci-NTLPs are encoded by two genes; Ci-NTLPs were encoded in two precursors; *ci-ntlp-A* encodes Ci-NTLP-1 to -4, while *ci-ntlp-B* encodes Ci-NTLP-5 and -6 [7]. *ci-ntlp-A* is expressed exclusively in neurons of the cerebral ganglion, whereas the abundant expression of *ci-ntlp-B* is detected in the ovary[7]. Consequently, it is presumed that Ci-NTLP-1 to 4 serves as neuropeptides, while Ci-NTLP-5 and -6 are ovarian hormones. This is the first characterization of NT-related peptides in invertebrates. Interestingly, Ci-NTLP-6 was found to suppress the Ci-TK-directed upregulation of cathepsin D, carboxypeptidase B1, and chymotrypsin [7], indicating that Ci-NTLP6 is a negative regulator of the growth of vitellogenic follicles.

11.2.6   CALCITONIN (CT)

CT is a 32-amino acid peptide, and is synthesized in various tissues including the C cells of the thyroid gland in mammals and the ultimobranchial gland in non-mammalian vertebrates expect cyclostomes [66, 67]. CTs are responsible for calcium metabolism via suppression of osteoclasts activity in bones and teleost scales. CTs are featured by a C-terminally amidated proline and N-terminal circular structure formed by a disulfide bridge between Cys^1 and Cys^7. In vertebrates, the CT family includes CT, CT gene-related peptide (CGRP), Amylin (AMY), Adrenomedullin (AM), and CT receptor-stimulating peptide (CRSP) [66–71]. CGRPs are generated from the CT gene via alternative splicing (α-CGRP) and another CGRP gene (β-CGRP) in the central and peripheral neuron, and acts not only as a vasodilator but also as a neuromodulator [66, 67]. Other family peptides are encoded by different genes.

Two Class B GPCRs, CT receptor (CTR) and CTR-like receptor (CRLR) have so far been shown to be the receptors for CT/CGRP family peptides [66, 67]. Moreover, three receptor activity-modifying proteins (RAMPs), single-transmembrane spanning proteins, have been shown to form a heterodimer with CTR or CRLR, and then to modulate the ligand-receptor specificity [66, 67].

The *Ciona* CT (Ci-CT) gene was cloned from the adult *Ciona* neural complex [72]. As summarized in Table 11.7, the deduced amino acid sequence of Ci-CT shows that the N-terminal circular region formed by a disulfide bond between Cys[1] and Cys[7] and the C-terminally amidated Pro are almost completely conserved, indicating that Ci-CT possesses the essential sequence characteristics of vertebrate CTs [72]. In contrast, the *Ci-CT* gene showed the four-exon/three-intron structure, whereas the most vertebrate CT genes are composed of six exons and five introns [72]. CT or CGRP peptide is generated from the CT gene via alternative splicing [66, 67], whereas the Ci-CT gene encodes a Ci-CT peptide sequence alone, and no splicing variant was detected [72]. No candidate for AM, AMY, CRSP, and β-CGRP genes was detected in the *Ciona* genome [72], suggesting that Ci-CT is the sole *Ciona* peptide of the CT/CGRP family peptides.

TABLE 11.7   The CT/CGRP Superfamily Peptides

Ascidian CT/CGRP Family Peptide	
Ci-CT	CDGVSTCWLHELGNSVHATAGGKQNVGFGPa
Amphioxus CT/CGRP family peptides	
Bf-CTFP1	DCSTLTCFNQKLAHELAMDNQRTDTANPYSPa
Bf-CTFP2	GKIACKTAWCMNNRLSHNLSSLDNPTDTGVGAPa
Bf-CTFP3	KCESGTCVQMHLADRLRLGLGHNMFTNTGPESPa
Vertebrate CT/CGRP family peptides	
Human CT	CGNLSTCMLGTYTQDFNKFHTFPQTAIGVGAPa
Human CGRP	ACDTATCVTHRLAGLLSRSGGVVKNNFVPTNVGSKAFa
Human Adrenomedullin	YRQSMNNFQGLRSFGCRFGTCTVQKLAHQIYQFTDKD-KDNVAPRSKISPQGYa
Human Amylin	KCNTATCATQRLANFLVHSSNNFGAILSSTNVGSNTYa
Pig CRSP	SCNTATCMTHRLVGLLSRSGSMVRSNLLPTKMGFKVFGa

Consensus motif is underlined. The *Ciona* peptide (Ci-CT) is indicated in boldface.

Ci-CT mRNA is localized to the endostyle and the neural gland, the latter of which is a non-neuronal ovoid body spongy texture lying immediately ventral to the cerebral ganglion, suggesting that Ci-CT serves as an endocrine/paracrine factor, not as a neuropeptide, in the neural gland and endostyle of ascidians [72]. Such tissue-distribution of Ci-CT implied that an original CT/CGRP family peptide might have played various physiological roles of current vertebrate CT/CGRP family peptides in common ancestral chordates, and current tissue-specific gene expressions and physiological roles of CT/CGRP family peptides diverged from those of a Ci-CT-like ancestor in concert with multiplication of the family gene members via gene duplications and advances of tissue organizations during the evolution of protochordate to vertebrates. Unfortunately, direct evidence for the interaction of Ci-CT with the endogenous receptor candidate, Ci-CT-R, has yet to be obtained [72]. On the other hand, three CGRP-like peptides, Bf-CTFP1 to -3 (Table 11.7) were identified and localized to nervous tissues in another invertebrate chordate, an amphioxus *Branchiostoma floridae* [73, 74]. Furthermore, unlike *C. intestinalis*, one BF-CTFP receptor and three RAMPs were identified in *B. floridae*, and the RAMP-dependent ligand selectivity was elucidated [73]. These findings confronted us further puzzling evolutionary processes of the CT family in chordates. Elucidation of the biological roles and molecular and functional evolution of invertebrate chordate CT/CGRP family peptides awaits further investigation.

11.2.7 GALANIN-LIKE PEPTIDES

Galanin and galanin-like peptide (GALP) are vertebrate brain/gut peptides involved in reproduction and feeding in vertebrates [75]. The N-terminal Gly-Trp-Thr-Leu-Asn-Ser-Ala-Gly-Tyr-Leu-Leu-Gly-Pro sequence is conserved among vertebrates including mammals, quail, and goldfish (Table 11.8), and plays an essential role in binding to the cognate receptor. GALP shares the identical consensus sequence in the N-terminal region (Table 11.8). On the other hand, GALP has a longer sequence than galanin and, GALP is N-terminally elongated by a Pro-Ala-His-Arg-Gly-Arg-Gly sequence upstream of the consensus sequence (Table 11.8).

TABLE 11.8 Galanin and Galanin-Like Peptides (GALP)

Peptide	Sequence
Ci-GALP	-PFRGQGGWTLNSVGYNAGLGALRKLFE
Galanin (quail)	-------GWTLNSAGYLLGPHAVDNHRSFNDKHGFTa
Galanin (goldfish)	-------GWTLNSAGYLLGPHAIDSHRSLGDKRGVAa
Galanin (human)	-------GWTLNSAGYLLGPHAVGNHRSFSDKNGLTSa
GALP (human)	PAHRGRGGWTLNSAGYLLGPVLHLPQMGDQDGKRETA-LEILDLWKAIDGLPYSHPPQPS

A single copy of galanin and GALP is encoded by a separate gene [75]. Galanin and GALP share two Class A GPCRs, GLR1 and 2, while galanin- or GALP-specific receptors have not ever been identified [75].

In *C. intestinalis*, a galanin/GALP-related peptide, Ci-GALP was identified [7]. This peptide sequence bears the galanin/GALP consensus-like sequence, Gly-Trp-Thr-Leu-Asn-Ser-Val-Gly-Tyr-Asn-Ala-Gly-Leu, whereas it has a

C-terminally shortened sequence compared with vertebrate galanin and GALP (Table 11.8). Furthermore, the *Ciona* peptide possesses a Pro-Phe-Arg-Gly-Gln-Gly sequence at the N-terminus, which is similar to the Pro-Ala-His-Arg-Gly-Arg-Gly sequence in GALP (Table 11.8). These homologous sequences suggest that Ci-GALP is a 'hybrid peptide' of galanin and GALP, and vertebrate galanin and GALP diverged from a Ci-GALP-like ancestor during the evolution. The cognate receptor for Ci-GALP has yet to be characterized.

11.2.8 NOVEL PEPTIDES

Peptidomic analysis of the *C. intestinalis* neural complex have detected not only homologs of vertebrate neuropeptides and peptide hormones but also two novel peptide families [7]. The first one is LF family. The family includes eight peptides, Ci-LF1 to -8 (Table 11.9). These peptides share the Leu-Phe sequence at their C-termini. The Ci-LF precursor gene encodes all Ci-LF sequences flanked by a typical dibasic endoproteolytic sites at their both termini [7]. The second one is YFV/L family. The family includes four peptides, Ci-YFV1 to -3 and Ci-YFL1 (Table 11.10). These peptides share the Tyr-Phe-Val or Tyr-Phe-Leu sequence at their C-termini. Moreover, all of these peptides are encoded by a single precursor gene, in which their sequences are flanked by a typical dibasic endoproteolytic sites at their both termini [7]. No peptides containing the C-terminal Leu-Phe sequence and Tyr-Phe-Val/Leu have ever been characterized from any other animal species, providing evidence that Ci-LFs and Ci-YFV/Ls are novel peptides. Moreover, the Ci-LF gene and Ci-YFV/L gene were shown to be expressed exclusively in the neural complex, indicating that these novel peptides serve as neuropeptides. Elucidation of their cognate receptors and biological roles awaits further study.

11.3 CONCLUSION

As reviewed in this chapter, *C. intestinalis* has been found to conserve more homologs and prototypes of mammalian neuropeptides and peptide hormones, including CCKs, GnRHs, TKs, CT, VP, and galanin than protostome model organisms. Moreover, *C. intestinalis* has been shown to possess species-specific peptides, including Ci-LFs and Ci-YFV/Ls. These findings shed light on the usefulness of *C. intestinalis* for research not only on peptidergic regulatory mechanisms underlying various biological events but also on conservation and diversification of the endocrine, neuroendocrine, and nervous system in chordates. Ci-TKs and Ci-NTLP6 have been shown to control the growth of vitellogeninc follicles via regulation of proteases. Consequently, of particular interest is that the TKergic follicle growth is conserved in vertebrates or other invertebrates. Except for a few homologous peptides, many cognate receptors for *C. intestinalis* peptides have yet to be identified. Identification of receptors for peptides will lead to the elucidation of the signaling mechanisms and localization, which will eventually pave the way for investigating the net peptidergic regulatory systems in ascidians and the evolutionary processes of the endocrine, neuroendocrine, and nervous system.

TABLE 11.9   Sequences of Ci-LFs

Peptide	Peptide Sequence
Ci-LF-1	FQSLF
Ci-LF-2	YPGFQGLF
Ci-LF-3	HNPHLPDLF
Ci-LF-4	YNSMGLF
Ci-LF-5	SPGMLGLF
Ci-LF-6	SDARLQGLF
Ci-LF-7	YPNFQGLF
Ci-LF-8	GNLHSLF

TABLE 11.10   Sequences of Ci-YFV/Ls

Peptide	Peptide Sequence
Ci-YFV-1	ELVVRDPYFV
Ci-YFV-2	NNQESYFV
Ci-YFV-3	DDEPRSYFV
Ci-YFL-1	DAARPNYYFL

ACKNOWLEDGMENTS

This study is financially (in part) supported by the Japan Society for the Promotion of Science to HS (16K07430).

CONFLICTS OF INTEREST

The author declares no conflict of interest.

KEYWORDS

- **corticotrophin-releasing factor**
- **CT gene-related peptide**
- **CT receptor-stimulating peptide**
- **galanin and galanin-like peptide**
- **Neurotensin-Like Peptides**
- **receptor activity-modifying proteins**
- **tachykinin-related peptides**

REFERENCES

1. Burighel, P., & Cloney, R. A., (1997). Urochordata: Ascidiacea, In: Harrison, W. F., & Ruppert, E. E., (eds.), *Microscopic Anatomy of Invertebrates* (Vol. 15, pp. 221–347). Wiley-Liss, New York.
2. Satoh, N., & Levine, M., (2005). Surfing with the tunicates into the post-genome era. *Genes Dev., 19*, 2407–2411.
3. Satoh, N., (2009). An advanced filter-feeder hypothesis for urochordate evolution. *Zoolog. Sci., 26*, 97–111.
4. Satake, H., & Kawada, T., (2006). Neuropeptides, hormones, and theirreceptors in ascidians: Emerging model animals, in: Satake, H., (ed.), *Invertebrate Neuropeptides and Hormones: Basic Knowledge and Recent Advances* (pp. 253–276). Transworld Research Network, Kerala, India.
5. Sherwood, N. M., Tello, J. A., & Roch, G. J., (2006). Neuroendocrinology of protochordates: Insights from *Ciona* genomics. *Comp. Biochem. Physiol. A Mol. Integr. Physiol., 144*, 254–271.
6. Kawada, T., Sekiguchi, T., Sakai, T., Aoyama, M., & Satake, H., (2010). Neuropeptides, hormone peptides, and their receptors in *Ciona intestinalis*: An update. *Zoolog. Sci., 27*, 134–153.
7. Kawada, T., Ogasawara, M., Sekiguchi, T., Aoyama, M., Hotta, K., Oka, K., & Satake, H., (2011). Peptidomic analysis of the central nervous system of the protochordate, *Ciona intestinalis*: Homologs and prototypes of vertebrate peptides and novel peptides. *Endocrinology, 152*, 2416–2427.
8. Satake, H., Sekiguchi, T., Sakai, T., Aoyama, M., & Kawada, T., (2013). Endocrinology and neuroendocrinology of protochordates: Evolutionary views and potentials as new model organisms. *Recent Res. Devel. Endocrinol., 5*, 1–19.
9. Matsubara, S., Kawada, T., Sakai, T., Aoyama, M., Osugi, T., Shiraishi, A., & Satake, H., (2016). The significance of *Ciona intestinalis* as a stem organism in integrative studies of functional evolution of the chordate endocrine, neuroendocrine, and nervoussystems. *Gen. Comp. Endocrinol., 227*, 101–108.

10. Adams, B. A., Tello, J. A., Erchegyi, J., Warby, C., Hong, D. J., Akinsanya, K. O., Mackie, G. O., Vale, W., Rivier, J. E., & Sherwood, N. M., (2003). Six novel gonadotropin-releasing hormones are encoded as triplets on each of two genes in the protochordate, *Ciona intestinalis*. *Endocrinology, 144*, 1907–1919.

11. Olinski, R. P., Lundin, L. G., & Hallbook, F., (2006). Conserved synteny between the *Ciona* genome and human paralogons identifies large duplication events in the molecular evolution of the insulin-relaxin gene family. *Mol. Biol. Evol., 23*, 10–22.

12. Olinski, R. P., Dahlberg, C., Thorndyke, M., & Hallbook, F., (2006). Three insulin-relaxin-like genes in *Ciona intestinalis*. *Peptides, 27*, 2535–2546.

13. Sekiguchi, T., Suzuki, N., Fujiwara, N., Aoyama, M., Kawada, T., Sugase, K., Murata, Y., Sasayama, Y., Ogasawara, M., & Satake, H., (2009). Calcitonin in a protochordate, *Ciona intestinalis*-the prototype of the vertebrate calcitonin/calcitonin gene-related peptide superfamily. *FEBS, J., 276*, 4437–4447.

14. Lovejoy, D. A., & Barsyte-Lovejoy, D., (2010). Characterization of a corticotropin-releasingfactor (CRF)/diuretic hormone-like peptide from tunicates: Insight into the origins of the vertebrate CRF family. *Gen. Comp. Endocrinol., 165*, 330–336.

15. Johnsen, A. H., & Rehfeld, J. F., (1990). Cionin: A disulfotyrosyl hybrid of cholecystokinin and gastrin from the neural ganglion of the protochordate *Ciona intestinalis*. *J. Biol. Chem., 265*, 3054–3058.

16. Johnsen, A. H., (1998). Phylogeny of the cholecystokinin/gastrin family. *Front. Neuroendocrinol., 19*, 73–99.

17. Noble, F., Wank, S. A., Crawley, J. N., Bradwejn, J., Seroogy, K. B., Hamon, M., & Roques, B. P., (1999). International union of pharmacology. X. X., I., Structure, distribution, and functions of cholecystokinin receptors. *Pharmacol. Rev., 51*, 745–781.

18. Nichols, R., (2003). Signaling pathways and physiological functions of *Drosophila melanogaster* FMRFamide-related peptides. *Annu. Rev. Entomol., 48*, 485–503.

19. Dickinson, P. S., Stevens, J. S., Rus, S., Brennan, H. R., Goiney, C. C., Smith, C. M., Li, L., Towle, D. W., & Christie, A. E., (2007). Identification and cardiotropic actions of sulfakinin peptides in the American lobster *Homarus americanus*. *J. Exp. Biol., 210*, 2278–2289.

20. Janssen, T., Meelkop, E., Lindemans, M., Verstraelen, K., Husson, S. J., Temmerman, L., Nachman, R. J., & Schoofs, L., (2008). Discovery of a cholecystokinin-gastrin-like signaling system in nematodes. *Endocrinology., 149*, 2826–2839.

21. Nilsson, I. B., Svensson, S. P., & Monstein, H. J., (2003). Molecular cloning of a putative *Ciona intestinalis* cionin receptor, a new member of the CCK/gastrin receptor family. *Gene, 323*, 79–88.

22. Sekiguchi, T., Ogasawara, M., & Satake, H., (2012). Molecular and functional characterization of cionin receptors in the ascidian, *Ciona intestinalis*: The evolutionary origin of the vertebrate cholecystokinin/gastrin family. *J. Endocrinol., 213*, 99–106.

23. Millar, R. P., Lu, Z. L., Pawson, A. J., Flanagan, C. A., Morgan, K., & Maudsley, S. R., (2004). Gonadotropin-releasing hormone receptors. *Endocr. Rev., 25*, 235–275.

24. Millar, R. P., (2005). GnRHs and GnRH receptors. *Anim. Reprod. Sci., 88*, 5–28.

25. Kah, O., Lethimonier, C., Somoza, G., Guilgur, L. G., Vaillant, C., & Lareyre, J. J., (2007). GnRH and GnRH receptors in metazoa: A historical, comparative, and evolutive perspective. *Gen. Comp. Endocrinol., 153*, 346–364.

26. Kawada, T., Aoyama, M., Sakai, T., & Satake, H., (2013). Structure, function, and evolutionary aspects of invertebrate GnRHs and their receptors. In: Scott-Sills, E., (ed.),

Gonadotropin-Releasing Hormone (GnRH): Production, Structure and Functions (pp. 1–16). Nova Science Publisher Inc., New York, U.S.A.

27. Sakai, T., Shiraishi, A, Kawada, T, Matsubara, S., Aoyama, M., & Satake, H., (2017). Invertebrate gonadotropin-releasing hormone-related peptides and their receptors: An update. *Front. Endocrinol (Lausanne).*, *8*, 217.

28. Kavanaugh, S. I., Nozaki, M., & Sower, S. A., (2008). Origins of gonadotropin-releasing hormone (GnRH) in vertebrates: Identification of a novel GnRH in a basal vertebrate, the sea lamprey. *Endocrinology, 149*, 3860–3869.

29. Powell, J. F., Reska-Skinner, S. M., Prakash, M. O., Fischer, W. H., Park, M., Rivier, J. E., Craig, A. G., Mackie, G. O., & Sherwood, N. M., (1996). Two new forms of gonadotropin-releasing hormone in a protochordate and the evolutionary implications. *Proc. Natl. Acad. Sci. USA, 93*, 10461–10464.

30. Hasunuma, I., & Terakado, K., (2013). Two novel gonadotropin-releasing hormones (GnRHs) from the urochordate ascidian, *Halocynthia roretzi*: Implications for the origin of vertebrate GnRH isoforms. *Zoolog. Sci., 30*, 311–318.

31. Kawada, T., Aoyama, M., Okada, I., Sakai, T., Sekiguchi, T., Ogasawara, M., & Satake, H., (2009). A novel inhibitory gonadotropin-releasing hormone-related neuropeptide in the ascidian, *Cionaintestinalis. Peptides., 30*, 2200–2205.

32. Kanda, A., Takahashi, T., Satake, H., & Minakata, H., (2006). Molecular and functional characterization of a novel gonadotropin-releasing-hormone receptor isolated from the common octopus (*Octopus vulgaris*). *Biochem. J., 395*, 125–135.

33. Tian, S., Zandawala, M., Beets, I., Baytemur, E., Slade, S. E., Scrivens, J. H., & Elphick, M. R., (2016). Urbilaterian origin of paralogous GnRH and corazonin neuropeptide signaling pathways. *Sci. Rep., 6*, 28788.

34. Kusakabe, T., Mishima, S., Shimada, I., Kitajima, Y., & Tsuda, M., (2003). Structure, expression, and cluster organization of genes encoding gonadotropin-releasing hormone receptors found in the neural complex of the ascidian *Ciona intestinalis. Gene, 322*, 77–84.

35. Tello, J. A., Rivier, J. E., & Sherwood, N. M., (2005). Tunicate gonadotropin-releasing hormone (GnRH) peptides selectively activate *Ciona intestinalis* GnRH receptors and the green monkey type II GnRH receptor. *Endocrinology, 146*, 4061–4073.

36. Tello, J. A., & Sherwood, N. M., (2009). Amphioxus: Beginning of vertebrate and end of invertebrate type GnRH receptor lineage. *Endocrinology, 150*, 2847–2856.

37. Sakai, T., Aoyama, M., Kusakabe, T., Tsuda, M., & Satake, H., (2010). Functional diversity of signaling pathways through G protein-coupled receptor heterodimerization with a species-specific orphan receptor subtype. *Mol. Biol. Evol., 27*, 1097–1106.

38. Sakai, T., Aoyama, M., Kawada, T., Kusakabe, T., Tsuda, M., & Satake, H., (2012). Evidence for differential regulation of GnRH signaling via heterodimerization among GnRH receptor paralogs in the protochordate, *Ciona intestinalis. Endocrinology, 153*, 1841–1849.

39. Satake, H., Matsubara, S., Aoyama, M., Kawada, T., & Sakai, T., (2013). GPCR Heterodimerization in the reproductive system: Functional regulation and implication for biodiversity. *Front. Endocrinol.* (Lausanne.), *4*, 100.

40. Kamiya, C., Ohta, N., Ogura, Y., Yoshida, K., Horie, T., Kusakabe, T. G., Satake, H., & Sasakura, Y., (2014). Nonreproductive role of gonadotropin-releasing hormone in the control of ascidian metamorphosis. *Dev. Dyn., 243*, 1524–1535.

41. Satake, H., & Kawada, T., (2006). Overview of the primary structure, tissue-distribution, and functions of tachykinins and their receptors. *Curr. Drug Targets., 7*, 963–974.

42. Steinhoff, M. S., Von Mentzer, B., Geppetti, P., Pothoulakis, C., & Bunnett, N. W., (2014). Tachykinins and their receptors: Contributions to physiological control and the mechanisms of disease. *Physiol. Rev., 94*, 265–301.

43. García-Ortega, J., Pinto, F. M., Fernandez-Sanchez, M., Prados, N., Cejudo-Roman, A., Almeida, T. A., Hernandez, M., Romero, M., Tena-Sempere, M., & Candenas, L., (2014). Expression of neurokinin B/NK3 receptor and kisspeptin/KISS1 receptor in human granulosa cells. *Hum. Reprod., 29*, 2736–2746.

44. Page, N. M., (2006). Characterization of the gene structures, precursor processing and pharmacology of the endokinin peptides. *Vascul. Pharmacol., 45*, 200–208.

45. Biran. J., Palevitch, O., Ben-Dor, S., & Levavi-Sivan, B., (2012). Neurokinin Bs and neurokinin B receptors in zebrafish-potential role in controlling fish reproduction. *Proc. Natl. Acad. Sci. US, A., 109*, 10269–10274.

46. Satake, H., Kawada, T., Nomoto, K., & Minakata, H., (2003). Insight into tachykinin-related peptides, their receptors, and invertebrate tachykinins: A review. *Zoolog. Sci., 20*, 533–549.

47. Jiang, H., Lkhagva, A., Daubnerová, I., Chae, H. S., Šimo, L., Jung, S. H., Yoon, Y. K., Lee, N. R., Seong, J. Y., Žitňan, D., Park, Y., & Kim, Y. J., (2013). Natalisin, a tachykinin-like signaling system, regulates sexual activity and fecundity in insects. *Proc. Natl. Acad. Sci. USA., 110*, 3526–3534.

48. Satake, H., Aoyama, M., Sekiguchi, T., & Kawada, T., (2013). Insight into molecular and functional diversity of tachykinins and their receptors. *Protein Pept. Lett., 20*, 615–627.

49. Satake, H., Ogasawara, M., Kawada, T., Masuda, K., Aoyama, M., Minakata, H., Chiba, T., Metoki, H., Satou, Y., & Satoh, N., (2004). Tachykinin and tachykinin receptor of an ascidian, *Ciona intestinalis*: Evolutionary origin of the vertebrate tachykinin family. *J. Biol. Chem., 279*, 53798–53805.

50. Aoyama, M., Kawada, T., Fujie, M., Hotta, K., Sakai, T., Sekiguchi, T., Oka, K., Satoh, N., & Satake, H., (2008). A novel biological role of tachykinins as an up-regulator of oocyte growth: Identification of an evolutionary origin of tachykininergic functions in the ovary of the ascidian, *Ciona intestinalis. Endocrinology., 149*, 4346–4356.

51. Aoyama, M., Kawada, T., & Satake, H., (2012). Localization and enzymatic activity profiles of the proteases responsible for tachykinin-directed oocyte growth in the protochordate, *Ciona intestinalis. Peptides, 34*, 186–192.

52. Hoyle, C. H. V., (1998). Neuropeptide families: Evolutionary perspectives. *Regul. Pept., 73*, 1–33.

53. Kawada, T., Sekiguchi, T., Sugase, K., & Satake, H., (2009). Evolutionaryaspects of molecular forms and biological functions of oxytocinfamily peptides. In: Jastrow, H., & Feuerbach, D., (eds.), *Handbook of Oxytocin Research: Synthesis, Storage and Release, Actions and Drug Forms* (pp. 59–85). Nova Science Publishers, Hauppauge, New York.

54. Satake, H., Takuwa, K., Minakata, H., & Matsushima, O., (1999). Evidence for conservation of the vasopressin/oxytocin superfamily in Annelida. *J. Biol. Chem., 274*, 5605–5611.

55. Kawada, T., Sekiguchi, T., Itoh, Y., Ogasawara, M., & Satake, H., (2008). Characterization of a novel vasopressin/oxytocin superfamily peptide and its receptor from an ascidian, *Ciona intestinalis. Peptides, 29*, 1672–1678.

56. Elphick, M. R., & Rowe, M. L., (2009). NGFFFamide and echinotocin: Structurally unrelatedmyoactive neuropeptides derived from neurophysin-containing precursors in seaurchins. *J. Exp. Biol., 212*, 1067–1077.

57. Stafflinger, E., Hansen, K. K., Hauser, F., Schneider, M., Cazzamali, G., Williamson, M., & Grimmelikhuijzen, C. J., (2008). Cloning and identification of an oxytocin/vasopressin-likereceptor and its ligand from insects. *Proc. Natl. Acad. Sci. USA., 105,* 3262–3267.

58. Beets, I., Janssen, T., Meelkop, E., Temmerman, L., Suetens, N., Rademakers, S., Jansen, G., & Schoofs, L., (2012). Vasopressin/oxytocin-related signaling regulates gustatory associative learning in, *C. elegans. Science, 338,* 543–545.

59. Gimpl, G., & Fahrenholz, F., (2001). The oxytocin receptor system: Structure, function, and regulation. *Physiol. Rev., 81,* 629–683.

60. Fujino, Y., Nagahama, T., Oumi, T., Ukena, K., Morishita, F., Furukawa, Y., et al., (1999). Possible functions of oxytocin/vasopressin-superfamily peptides in annelids with special reference to reproduction and osmoregulation. *J. Exp. Zool., 284,* 401–406.

61. Kawada, T., Kanda, A., Minakata, H., Matsushima, O., & Satake, H., (2004). Identification of a novel receptor for an invertebrate oxytocin/vasopressin superfamily peptide: Molecular and functional evolution of the oxytocin/vasopressin superfamily. *Biochem. J., 382,* 231–237.

62. Frank, E., & Landgraf, R., (2008). The vasopressin system--from antidiuresis to psychopathology. *Eur. J. Pharmacol., 583,* 226–242.

63. Aikins, M. J., Schooley, D. A., Begum, K., Detheux, M., Beeman, R. W., & Park, Y., (2008). Vasopressin-like peptide and its receptor function in an indirect diuretic signaling pathway in the red flour beetle. *Insect Biochem. Mol. Biol., 38,* 740–748.

64. Ukena, K., Iwakoshi-Ukena, E., & Hikosaka, A., (2008). Unique form and osmoregulatory function of a neurohypophysial hormone in a urochordate. *Endocrinology., 149,* 5254–5261.

65. Evers, B. M., (2006). Neurotensin and growth of normal and neoplastic tissues. *Peptides, 27,* 2424–2433.

66. Hull, K. L., Fathimani, K., Sharma, P., & Harvey, S., (1998). Calcitropic peptides: Neural perspectives. *Comp. Biochem. Physiol. C Pharmacol. Toxicol. Endocrinol., 119,* 389–410.

67. Conner, A. C., Simms, J., Hay, D. L., Mahmoud, K., Howitt, S. G., Wheatley, M., & Poyner, D. R., (2004). Heterodimers and family-B GPCRs: RAMPs, CGRP and adrenomedullin. *Biochem. Soc. Trans., 32,* 843–846.

68. Katafuchi, T., Yasue, H., Osaki, T., & Minamino, N., (2009). Calcitonin receptor-stimulating peptide: Its evolutionary and functional relationship with calcitonin/calcitonin gene-related peptide based on gene structure. *Peptides, 30,* 1753–1762.

69. Katafuchi, T., Kikumoto, K., Hamano, K., Kangawa, K., Matsuo, H., & Minamino, N., (2003). Calcitoninreceptor-stimulating peptide, a new member of the calcitonin gene-related peptide family. Its isolation from porcine brain, structure, tissue distribution, and biological activity. *J. Biol. Chem., 278,* 12046–12054.

70. Kubo, A., Minamino, N., Isumi, Y., Katafuchi, T., Kangawa, K., Dohi, K., & Matsuo, H., (1998). Production of adrenomedullin in macrophage cell line and peritoneal macrophage. *J. Biol. Chem., 273,* 16730–16738.

71. Lafont, A. G., Fitzpatrick, T., Cliff Rankin, J., Dufour, S., & Fouchereau-Peron, M., (2006). Possible role of calcitonin gene-related peptide in osmoregulation via the endocrine control of the gill in a teleost, the eel, *Anguilla anguilla. Peptides, 27,* 812–819.

72. Sekiguchi, T., Suzuki, N., Fujiwara, N., Aoyama, M., Kawada, T., Sugase, K., Murata, Y., Sasayama, Y., Ogasawara, M., & Satake, H., (2009). Calcitonin in a protochordate,

Ciona intestinalis--the prototype of the vertebrate calcitonin/calcitonin gene-related peptide superfamily. *FEBS, J., 276*, 4437–4447.

73. Sekiguchi, T., Kuwasako, K., Ogasawara, M., Takahashi, H., Matsubara, S., Osugi, T., Muramatsu, I., Sasayama, Y., Suzuki, N., & Satak, E. H., (2016). Evidence for conservation of the calcitonin superfamily and activity-regulating mechanisms in the basal chordate *Branchiostoma floridae*: Insights into the molecular and functional evolution in chordates. *J. Biol. Chem., 291*, 2345–2356.

74. Sekiguchi, T., Shiraishi, A., Satake, H., Kuwasako, K., Takahashi, H., Sato, M., Urata, M., Wada, S., Endo, M., Ikari, T., Hattori, A., Srivastav, A. K., & Suzuki, N., (2017). Calcitonin-typical suppression of osteoclastic activity by amphioxus calcitonin superfamily peptides and insights into the evolutionary conservation and diversity of their structures. *Gen. Comp. Endocrinol., 246*, 294–300.

75. Lang, R., Gundlach, A. L., Holmes, F. E., Hobson, S. A., Wynick, D., Hokfelt, T., & Kofler, B., (2015). Physiology, signaling, and pharmacology of galanin peptides and receptors: Three decades of emerging diversity. *Pharmacol. Rev., 67*, 118–175.

Note:

During the editorial process, cognate receptors for Ci-GALP, Ci-NTLP2, Ci-LF-1, 2, 5, 6, 8, and Ci-YFV-1 and 3 were elucidated (Shiraishi et al. Proc Natl Acad Sci U S A. 9116, 7847-7856, 2019). Moreover, Ci-VP was shown to play crucial roles in oocyte maturation and ovulation (Matsubara et al., eLife 8, pii: e49062, 2019).

Index

For Product Safety Concerns and Information please contact our EU
representative GPSR@taylorandfrancis.com Taylor & Francis Verlag GmbH,
Kaufingerstraße 24, 80331 München, Germany

Printed and bound by CPI Group (UK) Ltd, Croydon, CR0 4YY

01/05/2025
01858567-0001